HANDBOOK OF SMART ANTENNAS FOR RFID SYSTEMS

HANDBOOK OF SMART ANTENNAS FOR RFID SYSTEMS

Edited by

NEMAI CHANDRA KARMAKAR

WILEY

A JOHN WILEY & SONS, INC., PUBLICATION

Published by John Wiley & Sons, Inc., Hoboken, New Jersey.
Published simultaneously in Canada.

Library of Congress Cataloging-in-Publication Data:

Karmakar, Nemai Chandra, 1963–
Handbook of smart antennas for RFID systems / Nemai Chandra Karmakar.
p. cm.
Includes bibliographical references.
ISBN 978-0-470-38764-1 (cloth)
1. Radio frequency identification systems–Design and construction–Handbooks, manuals, etc. 2. Adaptive antennas–Design and construction–Handbooks, manuals, etc. 3. Phased array antennas–Design and construction–Handbooks, manuals, etc. I. Title.
TK6570.I34K37 2010
681′.2–dc22

2010008433

Printed in Singapore

10 9 8 7 6 5 4 3 2 1

To my eldest brother, Mr. Hirendra Nath Karmakar, who constructively influenced my childhood and supported me tirelessly in all stages of my life

CONTENTS

FOREWORD

Radio-frequency identification (RFID) is one of the fastest growing wireless technologies in recent decades. The market volume of the RFID-related hardware and software exceeded $5 billion in 2009 and is expected to have an exponential growth of $25 billion within a decade. Contrary to other wireless mobile terrestrial and satellite communications that have only a few dedicated sectors of applications, RFID enjoys an infinite number of applications of tracking items, resources movement, supply chain management and logistics, and even monitoring the settlement of an implanted organ in a human body. The derivatives of developing RFID for goods and services accelerated after the largest retail chain, Wal-Mart of USA, made it mandatory to tag each item they purchase from their vendors. The objective is to track the goods and services from their origin to the end of sale when boxes are crushed after the goods are sold. The process needs a huge amount of data gathering and processing. However, the benefit is enormous because the data may provide not only the health of the goods and their inventory control and logistics, but also the customers' buying patterns that can leverage the sales of items in a timely manner. Other organizations such as the US Department of Defense, K-Mart, and Myer in Australia followed Wal-Mart's practice. The outbreak of mad cow disease motivated Australia to implement a mandatory national livestock information management system. This is another step forward for mandatory RFID applications on a massive scale.

Every technology that grows very fast will put forward technological and management challenges. RFID is no exception. With the increased volume of RFID development and its emerging applications, there is a need for solving the issues of efficient reading and retrieval of data from the read tags. Another goal is to remove the chip from the tag in order to lower the cost of the tag and compete with the optical barcodes that have been dominating the market for about the last four decades. If the tag can be made chipless, it has the potential to replace trillions of barcodes printed each year. IDTechEx, a respected RFID market research and forecast company based in the United Kingdom, has predicted that 60% of the total tag market will be dominated by chipless tags. However, without a chip the tag becomes dumb and its data processing capacity will be limited. To mitigate this problem, the reader needs to be smart enough to read and process the data from the dumb tag. For issues such as improving throughput and system capacity, as well as mitigating collisions of proximity tags, smart antennas will play significant roles in RFID technology. Prudent research on the smart antennas for RFID reflects that researchers have been trying to implement smart antennas in the readers in all possible ways to improve the performance of the reader. The advent of smart antennas with their capabilities to provide spatial, temporal, and polarization diversities and improved signal to interference and signal to noise ratio (SINR) will significantly advance RFID technology. Implementation of smart antennas in RFID is still in the research phase. Only recently, Omron Corporation based in Japan has announced the development of smart antennas in Omron's readers. However, the system has not yet become a mainstream commercial solution. Therefore, the *Handbook of Smart Antennas for RFID Systems* is a timely publication. The book covers a broad spectrum of topics: the historical perspective and comprehensive review of modern development of RFID; RFID reader architecture where the smart antennas will be implemented; the physical layer development of smart antennas for RFID systems; directional of arrival and localization of RFID tags using smart antennas; multi-antenna RFID tags for system capacity improvement, MIMO antennas for RFID; and, finally, anti-collision protocols using smart antennas. This book, which includes comprehensive coverage on smart antennas applied to emerging RFID technology, will be a fantastic resource for the research community.

TAKASHI OHIRA (IEEE FELLOW)

Toyohashi University of Technology (TUT)
Tempaku, Toyohashi, Japan

PREFACE

Radio-frequency identification (RFID) is a contactless, usually short distance, wireless data transmission and reception technique for identification of items, asset tracking, surveillance, access control, electronic ticketing, car immobilizers, toll collection, and many other emerging applications. With the recent advent and accelerated development of RFID technologies, and strong patronization by giant retail chains and their suppliers such as Wal-Mart, Kmart, and the US Department of Defense, the application areas have also been increasing from simple identification and security to the retail markets, military, original part manufacturing, medicine, animal tagging, and space applications. The reliable prediction of IDTechEX was for an RFID market value of \$5.56 billion in 2009. This prediction relates to the total sales of RFID tags, readers, and related software. The applications of RFID are also increasing with the developments of new technologies. Again referring to the IDTechEX prediction, more than 60% development will encompass low-cost, fully printable chipless RFID tags. The current bottleneck for implementation of RFID system in a new business and its return on investment is the cost of tags.

Our industry partner, FE Technologies Pty Ltd., based in Geelong, Victoria, Australia, has been marketing their Smart Library® RFID system in Australia and overseas. In February 2009, FE Technologies demonstrated their automated library database management system to a group of librarians from Monash University. Smart Library®, which comprises an automatic checkout kiosk, a smart trolley, and a magic wand for inventory checking and misplaced items, is a fantastic solution for the library. Monash University's library possesses more than 3 million books to cater to about 10,000 staff and 50,000 students in Australia and overseas campuses in Malaysia and South Africa. With a book tag costing 50 cents a piece, Monash University immediately needs to invest about \$2 million to implement their RFID system.

While the existing optical barcodes for books cost less than 10 cents per unit and the existing library database management system based on the optical barcode works very well within the existing infrastructure and operational culture, a question always remains about the return on investment of more than $2 million to implement the RFID system for the library database system for Monash University's libraries. This is a big question mark and an uphill battle to persuade management to finance RFID for the library. This is only one example. The huge potential of RFID in many other applications is hindered by the high price of the chipped tags. The viable solution is the low-cost printable chipless RFID tag that will cost less than 10 cents and can compete with the optical barcode. The chipless RFID tags developed by the editor's research group at Monash University are simple and passive printable microwave electronic circuits, which can be printed with inkjet printer or other printing methods with conductive inks. Some conductive inks are invisible. How fantastic will that be if an RFID tag can be made invisible, but work very well with a compatible RFID reader? This technology will open up a full new spectrum of applications starting with Australian polymer banknotes, library books, apparel, shoes, and tagging of low-cost and perishable items such as apples, bananas, and so on. Now imagine the market volume if low-cost tags can be delivered and reliably read. To make the tag chipless and simple in operation, the bulk of the operation will be bestowed on the reader electronics. Certainly, the reader should be built more powerfully than the conventional chipped tag readers to process the returned echoes of the tags and encode the unique identification and location of the tag. The smart antennas in the reader will play a major role in improving the reading of the tags. Parallel to the RFID development and deployment, we also have been observing the explosive growth of wireless mobile communications and wireless ad hoc networks for portable electronic communication devices such as notebook PCs, plum tops, PDAs, and even mobile phones. In every aspect we can implement RFID system. With the increase in the subscribers' demand and the invent of value added services alongside the conventional voice communications, the questions of capacity improvement, the quality of services, and the throughput always agitate technologists. The smart antenna came into play once the technologists realized that the multiplexing schemes such as time division multiplexing (TDMA), code division multiplexing (CDMA), frequency division multiplexing (FDMA), and other advanced modulation schemes were not adequate to meet the requirements. Technologists looked into the electromagnetic signals and antennas to enhance the capacity within the existing available bandwidth, throughput, and quality of services. *Necessity is the mother of invention*! And the new invention that has significant footprints in the existing mobile communications is the *smart antenna*!! So beautiful!! The problem could not be solved alone with the advanced signal processing algorithms and modulation schemes; however, the problem has been significantly reduced by dealing with electromagnetic propagation with the smart antennas. As happened in the mobile communication industries about a couple of decades ago, which has now reached maturity and physical implementation, *smart antennas* paved the way to *dreams-come-true technologies* for mobile subscribers. A recent book edited by my former PhD student Dr. Chen Sun entitled *Handbook on Advances in Smart Antenna Technologies for Wireless Networks* by IGI in 2008 has presented the most recent development of smart antennas for wireless communications. An invited chapter in the book on RFID Smart Antennas has motivated the editor to publish the current book.

RFID is an emerging technology that has been going through various development phases in terms of technological developments and businesses (applications), the potential as well as the challenges are huge. As for the example of the implementation of RFID in Monash University's Library above, the bottleneck is the cost of the tag and its mass deployment. The answer to the problem lies in the development of new materials and printing technologies that can appropriately address the problem and produce a sustainable solution in terms of economy and technological advancements. When the tags become dumb, the reader should be smart. The smartness will come from the smart signal capturing capabilities from the dumb tags and the post-processing of the returned echoes, which are the signals from the uniquely identifiable tags. Again the answer lies in the implementation of the smart antennas in the reader and, if feasible, in the tag. Durgin and Griffin (2007)* proved that multiple antennas in the tag can significantly improve the throughput of the tag. Searching the open literature on the topic of smart antennas specifically dedicated to RFID applications was a frustrating experience. Only one article was found in a scholarly conference in the IEEExplore database. The rest came in the form of patents. The information obtained from a patent could not be as good as writing a book chapter. The information provided in the patents was not presented in technical detail but was, instead, written in plain English. The editor has undertaken the daunting task of editing a book wholly dedicated to *Smart Antennas for RFID*. The initial responses from the contributors were not at all promising. In the first phase of the invitation, only three contributions from Spain, Singapore, and the United States were received. Then in the later phase of personal contacts and repeated invitations, a few more contributions were obtained from Taiwan, Australia, and Japan. The low responses from potential authors and researchers indicate the very specialized area of the topic.

The smart antennas for RFID have exploited all possible features of smart antennas, as was done for the wireless telecommunications and networks. The work presented in this book focuses on the following main categories: Fundamentals of RFID and smart antennas, RFID reader architecture, smart antenna physical layer development, RFID position location using electronically steerable parasitic array radiator (ESPAR), RFID multiple-input multiple-output (MIMO) antenna systems, multi-antenna RFID tags, anti-collision and throughput improvement, and, finally, ultra-wideband (UWB) RFID direction of arrival (DOA) estimation. Besides the contributions from outside, the members of the editor's research group at Monash University have contributed significantly to the physical layer development of RFID reader architectures for chipped and chipless RFID tag systems, RFID smart antennas, and the anti-collision algorithm. The research group was supported by the Australian Research Council Discovery Project Grant DP665523: *Chipless RFID for Barcode Replacement* in the Department of Electrical and Computer Systems

*G. D. Durgin and J. D. Griffin: Reduced fading for RFID tags with multiple antennas, IEEE Antenna and Propagation Society International Symposium Digest, July 2007, Honololu, USA

Engineering, Monash University from 2006 to 2009. The dedication of the postgraduate students Dr. Sushim Mukul Roy, Dr. Stevan Preradovic, and Mr. Isaac Balbin under the supervision of the editor has made the chipless RFID tag system a viable commercial product for the Australian polymer banknote (ARC Linkage Project LP0989652: Printable Multi-bit Radio Frequency Identification for Banknotes) and library database management systems (ARC Linkage Project LP0991435: Backscatter based RFID system capable of reading multiple chipless tags for regional and suburban libraries) and possibly the diagnostic RFID tags for partial discharge from faulty power apparatus (ARC LP0989355: Smart Information Management of Partial Discharge in Switchyards using Smart Antennas). The editor has been supervising five RFID-related Australian Research Council Discovery and Linkage Projects that are worth more than AUD 2 million.

The dramatic growth of the RFID industry has created a huge market opportunity. Patronage by Wal-Mart alone has prompted more than two thousand suppliers to implement RFID systems for their products and services. The motto is to track the goods, items, and services from their manufacturing point until the boxes are crushed once the goods are sold. How fantastic the idea is! The RFID system providers are searching all possible technologies that can be implemented in the existing RFID system (GEN2 becomes a worldwide standard) that can be made inexpensively, can be implemented to provide high accuracy in multiple tags reading with minimum errors and extremely low false alarm rate, location finding of tags for inventory control and asset tracking. Employing smart antennas in the reader and, if possible, in tags presents an elegant way to improve the performance of the RFID system.

By deploying smart antennas in the reader architecture and network, there may be outstanding improvement in throughput, high-speed reading, and position detection of tagged items. These facilities can be obtained with an efficient beamforming scheme and diversity techniques. Positioning of tagged items has many applications in industry, thanks to the direction finding ability of the smart antennas.

Smart antennas can also be used in handheld RFID readers, making the reading more efficient and long range. The beamforming and interference suppression abilities of smart antennas enable the reader to increase throughput. In a networked RFID environment where each reader represents a node and where the smart antenna is in a node with packet routing protocols, the direction finding and suppression of interference abilities from the neighboring nodes may provide the optimum reading capability of multiple tags hence efficacy of the reader.

A MIMO wireless communication channel can be built by installing antenna arrays that provide uncorrelated signal outputs at both readers and tags. The MIMO system provides a large number of channels with antenna elements in both transmit and receive chains. The MIMO system enhances the channel capacity, and hence the throughput, of the RFID reader. Even multiple antennas are proposed in the RFID tags by pushing the operating frequency at the 5.8-GHz frequency band to incorporate multiple antennas in a credit card size tag (Durgin and Griffin, Chapter 18). The benefit is the high-speed tag reading and significant throughput improvement. MIMO also enhances the anti-collision capability and capturing effect of the tag when the reader reads multiple tags in close proximity.

To take advantage of the smart antennas' abilities to improve the effectiveness of the RFID system, researchers in both academia and industry recently have envisaged all possible ways of designing smart antennas, modulation and diversity techniques. One very good example is Lia et al.'s patent (Lia et al. 2005)* for redundant networked multimedia RFID systems incorporating both wireless local area network and Ethernet connections. The smart antenna for the RFID reader has a wide variety of capabilities such as frequency hopping, timeslotting, antenna positioning, beam scanning, subset antenna switching, and polarization diversity to exploit the maximum signal readability from multiple tags.

This book aims to provide the reader with comprehensive information about recent developments of smart antennas for RFID systems both in the physical layer development and the software algorithms and protocols. To serve this goal, the book features 24 chapters authored by leading experts in both academia and industry. They offer in-depth descriptions of terminologies and concepts relevant to RFID systems and smart antennas related to RFID. The chapters of the handbook are organized into seven distinct topics. The first two chapters present a comprehensive overview of RFID fundamentals. A smart antenna overview and recent developments of smart antennas specifically applied to RFID system are presented next. These chapters form the foundation for the subsequent chapters in the book. Usually researchers ignore the physical layer development of smart antennas, with the perception that smart antennas require the algorithms to calculate the weight vectors and maximize the signal-to-noise ratio. However, a smart physical layer implementation of a smart antenna can make the antenna more efficient and can save significant cost and implementation of the intelligence. One good example is the electronically steerable parasitic array radiator (ESPAR) antenna, which needs only one RF port and one A/D converter. If the process can be followed in the RF and microwave layers, the baseband processing can be simplified, the speed of the processing can be enhanced, and the processing cost can be minimized. The types of practical smart antennas are presented: a planar fixed-beam high-gain antenna with delay line beamforming, a smart radial power divider-based switched beam smart antenna for handheld RFID readers, a phased-array antenna with 3D scanning capabilities, an optically controlled phased-array antenna, and finally an adaptive-array antenna. All developments were done as the smart RFID reader antennas. In the next section, RFID DoA estimation and position location using two types of smart antennas—ESPAR and conventional smart antennas—are presented. Position location of tags is vital in asset tracking, security, and surveillance. Therefore, these chapters will offer efficient and elegant solutions to the tracking problems of tagged items. Next, multiple antennas for RFID tags are presented. A chipless multi-antenna tag with large bit size in phase-encoded mode is reported first. The fading channel statistics and multiple antenna RFID tag are presented next. These tags are for the throughput and system capacity improvement. RFID MIMO antennas are reported for optimum power allocation under independent Rayleigh fading, low cost, and compact RF-MIMO transceivers for RFID readers and blind

*Y. Lia et al. Radio Frequency Identification (RFID) system, United States Patent Application 20060261938.

channel estimation in MIMO using multi-carrier reception. The low-cost solution for RF-MIMO and the optimum power-handling MIMO system are useful for handheld readers. In the final section, three chapters report anti-collision algorithms: slotted ALOHA, frame-slotted ALOHA using MIMO antennas, and capture effect analysis using Agilent's Advance Design System (ADS) simulation. The capture effect takes care of the power budget issues where the tag with a higher power level is read first and then the tag with a lower power level is read. This is contrary to conventional anti-collision algorithms where data are discarded when collision between tags are detected in the reader.

In this book, utmost care has been paid to keep the sequential flow of information related to the RFID system-based smart antennas. I hope that the book will serve as a good reference for smart antennas for RFID and will pave the way for further motivation and research in the field.

NEMAI CHANDRA KARMAKAR

Monash University
August 2010

ACKNOWLEDGMENTS

I would like to thank Professor Kai Chang, Professor of Texas A&M University and the Editor in Chief of Wiley Interscience Microwave and Optical Engineering Book Series, for his invitation to write a book on smart antennas for RFID. Appreciation also goes to the reviewers who reviewed the book proposal and chapters. Dr. Chen Sun's support by delivering information to write a proposal and a book chapter was instrumental for the book. His invitation to write a book chapter on smart antennas for RFID for his edited book inspired me to go this far to edit this special book dedicated to RFID systems. Generous support from the authors and their timely responses for submission of chapters are highly acknowledged. Special thanks to those authors who submitted their chapters on time, but had to wait for a long time until the completion of the manuscript. Special thanks to my current and former students Isaac Balbin, Stevan Preradovic, Abdur Rahim, Parisa Zakavi, Maneesha Kumbukage, Qi Jing Teoh, Parisa Zakavi, and Sushim Roy for their generous support and chapter contributions. I would like to thank my colleague Professor Jeffrey Fu for his inspiration and contributions to the book. Authors reviewed the chapters of the book. I acknowledge their support. I must acknowledge Ms. Lucy Hitz, Editorial Assistant, Christy Michael, and Rishi Chawla Production Managers and George Telecki, Editor of Wiley-Blackwell for their continuous support and patience throughout the editing and writing process of the manuscript. Special thanks to Professor Arokiaswami Alphones for his special contribution on optically controlled phased-array antennas. It was a surprise when we discussed the book in our return flight from the European Microwave Conference 2009 in Rome. He instantly agreed to contribute a chapter. This special chapter has unique significance in the book. I would also like to offer special thank you to my former student and current research assistant, Parisa Zakavi, for the nice illustration on the front cover of the book. Special thanks also go to Hamza Msheik for his constant assistance for collating the electronic copies of chapters of the book. Yang Yang collated all index terms and communicated all authors for final checking of the manuscript by the lead authors. Special thank you goes to Yang for his nice cooperation and hard work in this endeavor.

Finally, the research funding support from Australian Research Council's Discovery Project Grants and Linkage Project Grants and Monash University's internal research grants are highly acknowledged.

During editing of the book, my family experienced my absence and gave me the moral support. Thanks to my wife Shipra and daughters Antara and Ananya. Special thanks to Shipra and Antara for taking my portrait and selection of the photo for the back cover of the book.

Nemai Chandra Karmakar

Monash University
August 2010

CONTRIBUTORS

KAZI M. AHMED, Telecommunications Program, School of Engineering and Technology, Asian Institute of Technology, Pathumthani, Thailand.

AROKIASWAMI ALPHONES, School of Electrical and Electronic Engineering, Nanyang Technological University, Singapore.

ISAAC BALBIN, Department of Electrical and Computer Systems Engineering, Monash University, Clayton, Victoria, Australia.

JIN CHENG, Nanyang Technological University, School of Electrical and Electronic Engineering, Singapore.

GREGORY D. DURGIN, School of Electrical and Computer Engineering, Georgia Institute of Technology, Atlanta, Georgia.

RALF EICKHOFF, Chair for Circuit Design and Network Theory, Technische Universitaet Dresden, Dresden, Germany.

VICTOR ELVIRA, Department of Communication Engineering, University of Cantabria, Santander, Spain.

MENG HWA ER, School of Electrical and Electronic Engineering, Nanyang Technological University, Singapore.

JEFFREY S. FU, Department of Electronic Engineering, Chang Gung University, Taoyuan, Taiwan, Republic of China.

JOSHUA DAVID GRIFFIN, Disney Research, Pittsburgh, Pennsylvania.

HIROSHI HARADA, National Institute of Information and Communications Technology (NICT), Japan.

Jesús Ibáñez, Department of Communication Engineering, University of Cantabria, Santander, Spain.

Behnam Jamali, School of Electrical and Electronic Engineering, The University of Adelaide, Adelaide, Australia.

Nemai Chandra Karmakar, Department of Electrical and Computer Systems Engineering, Monash University, Clayton, Victoria, Australia.

Maneesha Kumbukage, Department of Electrical and Computer Systems Engineering, Monash University, Clayton, Victoria, Australia.

Taekyoung Kwon, Seoul National University, Seoul, Korea.

Vasileios Lakafosis, Department of Electrical and Computer Engineering, Georgia Institute of Technology, Atlanta, Georgia, USA.

Jeongkeun Lee, Hewlett-Packard Laboratories, Palo Alto, California, USA.

Joni Polili Lie, School of Electrical and Electronic Engineering, Nanyang Technological University, Singapore.

Chin-Heng Lim, Temasek Laboratories, Nanyang Technological University, Singapore.

Weixian Liu, School of Electrical and Electronic Engineering, Nanyang Technological University, Singapore.

Uwe Mayer, Chair for Circuit Design and Network Theory, Technische Universitaet Dresden, Dresden, Germany.

Boon Poh Ng, School of Electrical and Electronic Engineering, Nanyang Technological University, Singapore.

Jesús Pérez, Department of Communication Engineering, University of Cantabria, Santander, Spain.

Stevan Preradovic, Department of Electrical and Computer Systems Engineering, Monash University, Clayton, Victoria, Australia.

Abdur Rahim, Department of Electrical and Computer Systems Engineering, Monash University, Clayton, Victoria, Australia.

Sushim Mukul Roy, Department of Electrical and Computer Systems Engineering, Monash University, Clayton, Victoria, Australia.

Chong Meng Samson See, School of Electrical and Electronic Engineering, Nanyang Tecnological University, Singapore.

Ignacio Santamaría, Department of Communication Engineering, University of Cantabria, Santander, Spain.

Chen Sun, National Institute of Information and Communications Technology (NICT), Japan.

MANOS M. TENTZERIS, Department of Electrical and Computer Engineering, Georgia Institute of Technology, Atlanta, Georgia, USA.

QI JING TEOH, Department of Electrical and Computer Systems Engineering, Monash University, Clayton, Victoria, Australia.

PHAM QUANG THAI, School of Electrical and Electronic Engineering, Nanyang Technological University, Singapore.

MATTHEW TRINKLE, Department of Electrical and Electronic Engineering, The University of Adelaide, Adelaide, Australia.

JAVIER VÍA, Department of Communication Engineering, University of Cantabria, Santander, Spain.

RUSHI VYAS, Department of Electrical and Computer Engineering, Georgia Institute of Technology, Atlanta, Georgia, USA.

WENJIANG WANG, School of Electrical and Electronic Engineering, Nanyang Technological University, Singapore.

PARISA ZAKAVI, Department of Electrical and Computer Systems Engineering, Monash University, Clayton, Victoria, Australia.

PART I

INTRODUCTION TO RFID

CHAPTER 1

THE EVOLUTION OF RFID

BEHNAM JAMALI
School of Electrical and Electronic Engineering, University of Adelaide, Adelaide, Australia

1.1 INTRODUCTION

Radio-frequency identification (RFID) is a relatively new technology. Some believe that its concept might have originated in military plane identification during World War II and that it really started to be intensively developed for tracking and access applications during the 1980s. These wireless systems allow for noncontact and non-line-of-sight reading of data from electronic labels by the means of electromagnetic signals, and consequently they are attractive for numerous tracking and tagging scenarios. For example, they are effective in hostile environments such as manufacture halls, where bar code labels could not survive. Furthermore, RFID tags can be read in challenging circumstances when there is no physical contact or direct line of sight. RFID has established itself in a wide range of markets, including livestock identification and automated vehicle identification systems, because of its ability to track moving objects. RFID technology is becoming a primary player in automated data collection, identification, and analysis systems worldwide.

RFID, its application, its standardization, and its innovation are constantly changing. It is a new and complex technology that is not well known and well understood by the general public, or even by many practitioners. Many areas of RFID operation need development to achieve a longer reading range, larger memory capacity, faster signal processing, and more secure data transmission.

1.2 ELECTROMAGNETIC TIMELINE

In this section we will provide an anecdotal history of the most important electromagnetic personalities in chronological order. A short biography of each scientist is also provided along with their main contribution to this field

Handbook of Smart Antennas for RFID Systems, Edited by Nemai Chandra Karmakar

Charles-Augustin de Coulomb (1736–1806) was a military civil engineer, retired from the French army because of ill health after years in the West Indies. During his retirement years he became interested in electricity and discovered that the torsion characteristics of long fibers made them ideal for the sensitive measurement of magnetic and electric forces. He was familiar with Newton's inverse-square law, and in the period 1785–1791 he succeeded in showing that electrostatic forces obey the same rule.

$$\mathbf{F}_{e12} = \frac{Q_1 Q_2}{4\pi \varepsilon r^2}\mathbf{u}_{r12} \tag{1.1}$$

Luigi Galvani (1737–1798) was an Italian physician who, in the 1770s, began to investigate the nature and effects of what he conceived to be electricity in animal tissue and of muscular stimulation by electrical means. He discovered that contact of two different metals with the muscle of a frog resulted in an electric current.

Alessandro Giuseppe Antonio Volta (1745–1827) was a professor at the University of Pisa. He was a close friend of Galvani. After he heard about Galvani's discovery, Volta began experimenting in 1794 with metals alone and found that animal tissue was not needed to produce a current. His invention and demonstration of the electric battery in 1800 provided the first continuous electric power source.

Hans Christian Oersted (1777–1851) was born in a village without a school. He was educated by the villagers and went on to become a professor at the University of Copenhagen. In 1820 he was performing a classroom demonstration of the heating effect of electric currents when he observed the deflection of a nearby compass. He had discovered a connection between electricity and magnetism.

Andre-Marie Ampere (1775–1836) learned about Oersted's discovery in 1820 that a magnetic needle can be deflected by a nearby current conducting wire. He then prepared within a week the first of several papers on the theory of this phenomenon, formulating the law of electromagnetism, known as Ampere's Law, which describes mathematically the magnetic force between two current-conducting elements.

$$\oint_C \mathbf{B} \cdot d\mathbf{l} = \mu_0 I \tag{1.2}$$

Jean-Baptiste Biot (1774–1862), along with Felix Savart, formulated the Biot–Savart law of magnetic fields:

$$d\mathbf{B} = \frac{\mu_0}{4\pi} \cdot \frac{I dr \times \mathbf{u}_r}{r^2} \tag{1.3}$$

Karl Friedrich Gauss (1777–1855) ranks as one of the greatest mathematicians of all time. Beginning in 1830, Gauss worked closely with Weber. Gauss lived to an advanced age; and having systematically studied the financial markets and invested accordingly, he died a very wealthy man. Gauss' law of electrostatics states that the

TABLE 1.1. Maxwell's Equation

Gauss' law of electrostatics	$\nabla \cdot E = \frac{\rho}{\varepsilon}$
Gauss' law of magnetism	$\nabla \cdot B = 0$
Faraday's law of induction	$\nabla \times E = -\frac{dB}{dt}$
Ampere's law with Maxwell's displacement current	$\nabla \times B = \mu_0 J + \mu_0 \varepsilon_0 \frac{dE}{dt}$

total electric flux through a closed surface is proportional to the total electric charge enclosed within the surface:

$$\oint_s \varepsilon_0 \mathbf{E} \cdot d\mathbf{s} = \int_v \rho \, dv \tag{1.4}$$

Michael Faraday (1791–1867) was born in a village near London. Faraday became the greatest experimentalist in electricity and magnetism of the nineteenth century. He produced an apparatus that was the first electric motor, and in 1831 he succeeded in showing that a magnet could induce electricity. Faraday's law of induction describes an important basic law of electromagnetism:

$$\mathbf{E} = -\frac{d\phi}{dt} \tag{1.5}$$

James Clerk Maxwell (1831–1879) is ranked with Newton and Einstein for the fundamental nature of his many contributions to physics. Most importantly, he originated the concept of electromagnetic radiation, and his field equations (1873) led to Einstein's special theory of relativity. In classical electromagnetism, Maxwell's equations are a set of four partial differential equations that describe the properties of the electric and magnetic fields and relate them to their sources, charge density, and current density. Maxwell used the equations listed in Table 1.1 to show that light is an electromagnetic wave.

Heinrich Rudolf Hertz (1847–1894), a German physicist, was the first to broadcast and receive radio signals. He applied Maxwell's theories to the production and reception of radio waves. In 1884, He rederived the Maxwell's equations by a new method, casting them in modern form as shown in Table 1.1. He produced electromagnetic waves in the laboratory and measured their wavelength and velocity. He showed that the nature of their reflection and refraction was the same as those of light, confirming that light waves are electromagnetic radiation obeying Maxwell's equations.

Guglielmo Marconi (1874–1937), an Italian physicist, is the inventor of radio. He was granted a patent for a successful system of radio telegraphy in 1896. In 1909 he received the Nobel Prize in Physics. Marconi's great triumph was in 1901, when he successfully received radio signals transmitted across the Atlantic Ocean. This sensational achievement was the start of the vast development of radio communication and broadcasting the way we know it today.

1.3 RADAR

The use of electromagnetic waves to identify the range, altitude, direction, or speed of both moving and fixed objects was first contemplated in the early 1900s. The term RADAR was coined in 1941 as an acronym for radio detection and ranging [1].

A radar system consists of a transmitter that emits either radio waves that are reflected by the target and detected by a receiver, typically in the same location as the transmitter. Although the signal returned is usually very weak, the signal can be amplified. This enables radar to detect objects at ranges where other emissions, such as sound or visible light, would be too weak to detect. Radar's potential in determining the speed and position of an object was quickly understood by the military, leading to its significant development during World War II era.

1.4 GENESIS OF RFID

Many authors date the origin of RFID to the 1940s during World War II. The Germans discovered that if their pilots rolled their planes as they were approaching their base, they could establish a secret handshake. The roiling of the planes modulated the reflected radar signal. The British, on the other hand, developed the first active identify friend or foe (IFF) system [2]. They placed a transmitter on each British plane that upon detecting a radar signal would broadcast back a signal that would identify the aircraft as friendly. An RFID system basically works on the same principles. A base station (RFID reader) sends a signal to a transponder (tag) that either reflects back the received signal (passive RFID) or broadcasts a signal back to the reader (active RFID).

A major milestone toward modern RFID was the work by Harry Stockman in his 1948 paper, entitled "Communication by Means of Reflected Power." Stockman stated in his paper that ". . . considerable research and development work has to be done before the remaining basic problems in reflected-power communication are solved and before the field of useful applications is explored [3]." The discovery of semiconductor transistors in the 1950s enabled Stockman's vision of reflected power-coded communication to become a reality.

The main era of exploration of RFID technology began in the 1950s by work done by F. L. Vernan [4] and D. B. Harris [5]. The first patent on RFID technology was granted to Mario Cardullo in 1973 [6]. Cardullo's invention was the first true ancestor of modern RFID: "A passive radio transponder with memory." Cardullo's RFID tag was designed to be used as a toll device, and there were a number of potential users, including the New York Port Authority.

In early 1960 many companies began commercializing Electronic Article Surveillance (EAS) or anti-theft systems that are based on a very simple RFID concept. These RFID tags that are still in use today have 1 bit of digital information. The bit of an unpaid (unscanned) item is originally set to "on"; and when a patron pays for that item, the bit is set to off by the cashier. A switched-off tag will not trigger the alarm system when the item leaves the store (i.e., passes through the interrogation zone, located at the exit gate.)

Those earlier RFID developments have paved the way for today's booming deployment of RFIDs in industrial and commercial applications. The varieties of application-specific requirements have led to operation of RFID tags mostly in three main frequency bands today: the industrial (low-frequency), scientific (high-frequency), and medical (ISM—ultra-high-frequency) bands.

1.5 OPERATING FREQUENCIES

There are three main varieties of RFID tags in use today. They all operate at Industrial, Scientific, and Medical (ISM) band.

1.5.1 Low Frequency

In the early days of RFID, low-frequency (LF) tags were the most common. The LF tags operate at 125 kHz and 134.2 kHz. Because of the electromagnetic properties at LF frequencies, those tags can be read while attached to objects containing water, animal tissues, metal, wood, and liquids. They are only suitable for proximity applications, because they can be interrogated from a very short range of only a few centimeters.

They have the lowest data transfer rate among all the RFID frequencies and usually store a small amount of data. The LF tags have no or limited anti-collision capabilities, therefore, reading multiple tags simultaneously is almost impossible. The LF tag antennas are usually made of a copper coil with hundreds of turns wound around a ferrite core.

Because of these properties of LF tags, they are well-suited for specific applications such as access control, asset tracking, animal identification, automotive control, vehicle immobilizer, health-care, and various point-of-sale applications. In particular, LF tags have been intensively used for animal tracking since the early 1980s. Nowadays, the automotive industry is the largest user of LF tags. For example, in an automobile vehicle immobilizer system, an LF tag is embedded inside the ignition key. When that key is used to start the car, an RFID interrogator placed around the key slot reads the tag ID. The car can be started only if the correct ID can be read from the key.

1.5.2 High Frequency

The high-frequency (HF) tags operate at 13.56 MHz. Their operating principles are similar to LF tags they use near-field inductive coupling as source of power to communicate with the interrogator. HF tags have a better read range than LF tags and can be read from up to half a meter away. They have a better data transfer rate and larger memory size (up to 4 kbyte) compared to LF tags. The HF tags may have anti-collision capability that facilitates reading of multiple tags simultaneously in the interrogation zone. However, since the read range of many HF tags and interrogators is small, anti-collision features are usually not implemented to reduce the complexity and consequently its cost.

HF tag antennas are usually made of several turns of conductive materials such as copper, aluminum, or silver as a flat spiral. Therefore, HF tags are usually very thin and almost two-dimensional (as thick as paper). They can be made in different sizes, some only a couple of centimeters in diameter. Simple antenna design translates in a low-cost fabrication. HF tags can be easily read while attached to objects containing water, tissues, metal, wood, and liquids. Their performance, however, is affected by metal objects in the close vicinity.

Higher data transfer rate of HF RFIDs, along with their limited read range (which provides privacy against eavesdroppers), makes these systems an ideal choice for applications such as credit cards, smart cards, library book tags, airline baggage tags, and asset tracking. Due to those properties, HF tags are currently the most widely used RFID tags around the world.

1.5.3 Ultra-High Frequency

In the ultra-high-frequency (UHF) band, 433 MHz and 860–960 MHz, are used for UHF RFID applications. The 433-MHz frequency is used for active tags, while the 860 to 960-MHz range is used mostly for passive tags. In contrast to LF and HF tags, UHF tags and interrogators use far-field coupling or backscatter coupling to communicate with one another. Therefore UHF tags have a read range of up to 20 m under good conditions. All the protocols in the UHF range have some type of anti-collision capability, allowing multiple tags to be read simultaneously.

The UHF tag antennas are mostly based on dipole antennas and made of copper, aluminum, or silver deposited on the substrate. The length of a resonant half-wave dipole antenna at 900 MHz is approximately 15 cm. However, the overall antenna size can be reduced through proper design techniques such as folding, fractals, double layering, and so on. The UHF antennas are easy to manufacture and, as LF tags, can be made thin using planar design, and they are almost two-dimensional.

While UHF tags offer more memory size and better read range, their performance is severely degraded when attached to objects containing water, biological tissues, and metals. The proximity of those materials will degrade the efficiency of the tag through absorption and detuning. The range will also be affected by propagation effects in unfavorable environments. UHF tags cannot be read if water or any conductive material is placed between the interrogator antenna and the tags.

1.5.4 Ultra-Wideband

According to the Federal Communications Commission (FCC) regulations, any radio technology that communicates over a bandwidth exceeding the lesser of 500 MHz or 20% of the arithmetic center frequency is classified as ultra-wideband (UWB). Due to the extremely low emission levels currently allowed by regulatory agencies, UWB systems tend to be short-range. The benefit of a UWB system is a high data rate that can enable wireless communication between devices at a very high speed. UWB is also used in real-time location systems.

UWB operates by emitting a series of extremely short pulses (billionths of a second or shorter) across a band of frequencies simultaneously. The FCC has cleared 3.1 GHz to 10.6 GHz to be used by UWB devices. UWB devices are less prone to interfere with existing devices (usually less than 1 GHz) because of their operation frequency range. All the UWB RFID tags currently in the market are active.

1.6 CHIPLESS RFID

RFID tags that do not contain a silicon chip are called chipless RFID tags. The primary potential benefit of chipless tags is the possibility of printing them directly on products and packaging with a cost of a fraction of a cent. This would replace 10 trillion barcodes printed yearly with something far more versatile and reliable. As of 2008, there are few chipless technologies potentially available, including acousto-magnetic, swept RF inductor capacitor arrays, and electromagnetic RF sputtered film. Others have been proposed in the form of diode arrays, surface acoustic wave (SAW) devices, and nonlinear chemicals particles that emit high frequencies when radiated with radio waves. However, only acousto-magnetic tags for theft protection and SAW tags for road toll collection have been successfully implemented and are in common use today. The reasons for the relative lack of success of the conceptually attractive chipless technologies can be understood by considering some technical limitations that still need to be overcome.

Acousto-magnetic (AM) tags are manufactured by Sensormatic [7]. They have a relatively long detection range and remain usable at high moving speed. The operation of the system is based on a reader transmitting pulses of radio-frequency signals at about 58 kHz, which energizes a tag in the surveillance zone. When the pulse ends, the tag responds by emitting a single frequency signal like a tuning fork. As the transmitter is off between pulses, the tag signal is detected by the reader. To reduce false detection, the reader checks the tag signal frequency, its time-synchronization relative to the exciting pulse, the signal power level, and the repetition frequency. If all predefined conditions are met, the system will assess this as identification and raise an alarm.

CrossID Communication Material [8] offers a chipless RFID made of a collection of chemical pigments with varying degrees of magnetism that resonate when exposed to electromagnetic waves from a reader. Each chemical emits its own distinct radio frequency that is picked up by the reader. Up to 70 different chemicals are available and can be mixed in different ways to provide a coding of the tag. All the notes emitted by a specific mix of different chemicals are then interpreted as a binary number. CrossID has not specified the operating frequency of their system.

Inkode Technology Group [9], a Vienna-based company, has a patent for a chipless RFID technology that involves embedding tiny metal fibers in paper, plastic, and other materials that radio-frequency waves can penetrate. The fibers reflect radio waves back to the reader, forming a unique resonant signature. These can be converted into a unique serial number. The Inkode readers operate at either 24–25 GHz or 60–66 GHz.

Although chipless RFID tags enjoy the benefit of being simple, small, and inexpensive to manufacture, they also suffer from weaknesses; one is that the resonant signatures of two or more tags in the same field can interfere with one another, preventing the reader from converting the signatures into the right serial number. Also, such tags are read-only; once manufactured, their id number cannot be changed, so readers have to be tied into an IT infrastructure that can associate the unique serial number with information in a database.

1.7 RECENT DEVELOPMENT

The last two decades were a significant time in the development of RFID. IBM engineers developed and patented an UHF system in the early 1990s [10]. The patent addresses the basic RFID system: chip, tag, and reader, and how they interface with each other. Other companies followed suit and very soon UHF RFID became popular. The world's first highway electronic toll collection was installed in Oklahoma in 1991, allowing vehicles to pass toll collection point at highway speed and without being impeded by a toll plaza or barriers. The first fully automated library system was installed in Singapore Library by ISD in 1994 [11], enabling patrons to check in or out items on the fly.

Until recently, the Universal Product Code (UPC), also known as the barcode, has been the primary means of identifying products. Barcodes were designed to provide an open standard for product labeling. It has helped businesses reduce costs, increase efficiency, and drive innovation for the benefit of consumers, manufacturers, and retailers. There are, however, several shortcomings of barcodes that can be solved through their replacement by RFID tags, one being the need for a line of sight from the scanner to the barcode. Additionally, the barcode, because of its limited capacity to hold information, can only track product categories. This means, for example, that the barcode cannot distinguish between a can of soda and another of the same brand and make, whereas with RFID, using EPC's 96-bit numbering, it is possible to give unique identification number to every single product.

Realizing the benefits of RFID, several major companies, including the Uniform Code Council, EAN International, Procter & Gamble, and Gillette, put up funding to establish the Auto-ID Center at the Massachusetts Institute of Technology (MIT) in 2000. There, Professor David Brock and Professor Sanjay Sarma had envisioned the possibility of putting low-cost RFID tags on all products made to track them through the supply chain. This can be achieved by establishing an association between the serial number on the tag and a database that would be accessible over the Internet.

One of the first academics who was asked to join the project was Professor Peter Cole from the University of Adelaide (Australia) along with Professor Duncan McFarlane from the University of Cambridge, UK [12].

The establishment of the Auto-ID Center essentially changed the way people thought about RFID in the supply chain. Previously, tags were a mobile database that carried information about the product or container they were on as they traveled. The Auto-ID Center turned RFID into an Internet of Things by linking objects to the Internet through the tag. For businesses, this was an important change: This opened the possibility for a manufacturer to let a business partner know in real-time when a shipment was leaving the dock at a manufacturing facility or warehouse, and subsequently the retailer could automatically let the manufacturer know when the goods arrived.

By 2003 the Auto-ID Center had already gained the support of more than 100 large end-user companies, as well as from the U.S. Department of Defense and many key RFID vendors. It had three laboratories at MIT (USA), Adelaide (Australia) [13], and Cambridge (UK). It developed two air interface protocols (Class 1 and Class 0) [14], the Electronic Product Code (EPC) [15] numbering scheme, and a network architecture for looking up data associated to RFID tags on the Internet.

The technology was licensed to the Uniform Code Council in 2003, and the Uniform Code Council created EPCglobal, as a joint venture with EAN International, to commercialize EPC technology. The Auto-ID Center was transformed into two entities, Auto-ID Labs and GS1 in October 2003. During the transition, its research responsibilities were passed on to Auto-ID Labs. Currently, there are seven research laboratories that are part of Auto-ID Labs in Australia, the United Kingdom, the United State, Switzerland, Japan, China, and Korea [16].

Some of the biggest retailers in the world (Albertsons, Metro, Target, Tesco, WalMart) and the U.S. Department of Defense are planning to use EPC technology to track goods in their supply chain. Other industries such as the pharmaceutical and tire manufacturers are also moving to adopt the technology. EPCglobal ratified a second-generation standard in early 2005, paving the way for broad adoption of RFID.

1.8 SUMMARY

In this chapter we look at the progressive evolution of RFID since its origin. We have discussed how advancement of electronics micro-circuit has made the RFID a reality. However, the concept of RFID is more than just electronic circuits; RFID is a technology that spans across many disciplines, such as electromagnetic, circuit theory, antenna theory, software engineering, mechanical engineering, material engineering, and business. The full potential of RFID will not be realized unless we see more advancement in all of the above fields.

There still are a lot of myths about RFID circulating the media. Many of them are due to a misunderstanding of the technology or pure media hypes. As the development and innovation in the field of radio communication and electrical engineering continues, the pace of developments in RFID will continue to accelerate. As such, the future looks very promising for this technology.

REFERENCES

1. C. Hülsmeyer, Radar World, http://www.radarworld.org/huelsmeyer.html, Retrieved May, 2007.
2. D. Barrett, All you ever wanted to know about British air defence radar in The Radar Pages http://www.radarpages.co.uk/index.htm, Retrieved December 2007.
3. H. Stockman, Communication by Means of Reflected Power in *Proceedings of the IRE*, Vol. 36, pp. 1196–1204, Oct 1948.
4. Jr. F. Vernon, Application of the microwave homodyne in *Transactions of the IRE Professional Group on Antennas and Propogation*, Vol. 4, pp. 110–116, Dec 1952.
5. D. B. Harris, Radio transmission systems with modulatable passive responder, USA patent office 2927321, 1960.
6. M. Cardullo, U.S. Patent: Transponder Apparatus and System. Patent number: 3,713,148, 1973.
7. Sensormatic, http://www.sensormatic.com.
8. CrossID Identification Technology, http://crossid.innovya.com/.
9. Morton Greene. Radio frequency automatic identification system, April 1999, U.S. Patent No. 5,891,240.
10. IBM, Radio frequency circuit and memory in thin flexible U.S. Patent: package, Patent number: 5528222.
11. A. Shameen, Singapore Seeks Leading RFID Role, http://www.rfidjournal.com/article/articleview/1024/1/1/.
12. Cambridge Auto-ID Lab http://www.autoidlabs.org.uk/
13. Adelaide Auto-ID Lab http://autoidlab.eleceng.adelaide.edu.au
14. Class 1 Generation 2 UHF Air Interface Protocol Standard. http://www.epcglobalinc.org/standards/.
15. EPCIS - EPC Information Services Standard. http://www.epcglobalinc.org.
16. Auto-ID Lab website, http://www.autoidlabs.org/.

CHAPTER 2

INTRODUCTION TO RFID SYSTEMS

SUSHIM MUKUL ROY
Department of Electrical and Computer Systems Engineering, Monash University, Clayton, Victoria, Australia

NEMAI CHANDRA KARMAKAR
Department of Electrical and Computer Systems Engineering, Monash University, Clayton, Victoria, Australia

2.1 INTRODUCTION

Radio-Frequency Identification (RFID) is a wireless data capturing technique from a tagged item. An RFID system comprises an interrogator (reader) and a tag or transponder. A middleware is a buffer stage that encodes the data captured from the tag in meaningful identification codes. RFID tags or radio transponders are high-frequency electronic circuits that broadcast the position or attributes of items to which they are attached. This allows these items to be remotely detected, identified, and tracked. In a broader perspective, RFID fall into the specialized category of Automatic Identification (Auto ID) that uses an electromagnetic signal to communicate between the reader and the transponder.

In the era of "silent commerce," most of the business processes are run by various forms of Auto ID technologies. Auto ID collects data related to objects and feeds that data into the database management system without much human intervention. The process of identification is preprogrammed and runs like clockwork with high level of efficiency and reduced cost. This advantage of automatic identification makes the Auto ID technology so attractive to different business processes in recent decades.

Auto ID technology is a big superset of different technologies such as Magnetic Ink Character Recognition (MICR), Voice Recognition, Biometrics, Barcodes, and RFID [1, 2]. Although Auto IDs are supposed to work without any human intervention, still technologies like Biometrics, MICR, and optical barcodes require considerable human intervention for their operation. Thus these technologies are mostly confined in security-related usage such as customs, immigration, and so on. Among the various

Handbook of Smart Antennas for RFID Systems, Edited by Nemai Chandra Karmakar

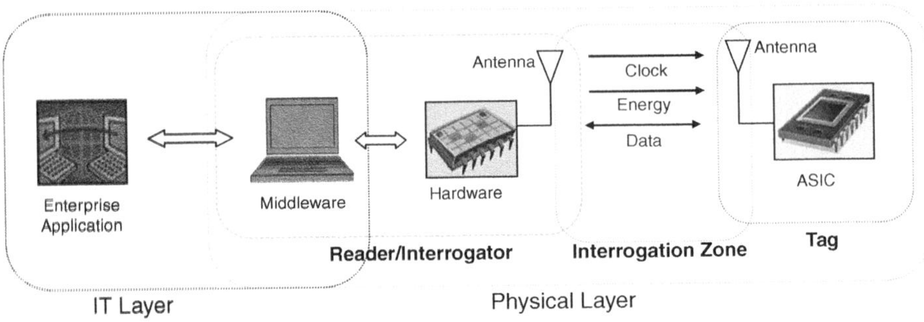

FIGURE 2.1. Overview of a generic RFID system.

forms of Auto ID, optical barcodes have been dominating the Auto ID market and are widely used in almost everything and everywhere in the present-day world. The main reason for the omnipotent application of barcodes is their cost, which is almost negligible. However, barcodes are limited in memory storage capabilities; and due to their "line-of-sight" operation, an operator must be present to read a barcode. Due to these limitations, RFID is coming into the Auto ID market with huge potential. Though the RFID tags remove the barrier of the "line-of-sight" reading and thus remove the human intervention in the reading process, RFID tags require a silicon chip to store the data. This makes the tags expensive to be implemented in the Auto ID market. Hence the momentum arises for the need of low-cost RFID and what makes RFID technology of such big demand around the world.

In any process such as manufacturing, supply chain, logistics, quality control, inventory management, hospital management, and so on the best possible service can be ensured if each and every item can be tracked. Different forms of Auto ID are developed to ensure this. Some numbers are used to identify an item uniquely; and throughout the life cycle of the item, that number should be traceable. As the item passes through different manufacturing and logistics processes, there should be provision for more and more information to be incorporated and retrieved. In this regard, RFID have been playing pivotal roles in the Auto ID market. According to the reliable prediction of IDTechEx, the 2009 RFID market volume will be $5.56 billion, up from $5.25 billion in 2008. This includes the tags, readers, and related software sales. The growth of RFID market is exponential with the patronization of government departments and giant retail chains. The growth forecast with quick return of investment (ROI) is in the sectors like apparel tracking, manufacturing, asset tracking, book tagging, and so on [RFID Market Forecasts 2009–2019, http://www.printedelectronicsworld.com/articles/rfid_market_forecasts_2009_2019_00001377.asp].

RFID technology as shown in Figure 2.1 deploys automatic identification and tracking facilities with the help of electromagnetic waves and an integrated circuit (IC) chip. The basic RFID system [1] consists of three components: (i) a small and mobile "tag" unit (or transponder) that is attached to items of interest, as well as (ii) a reader (or transceiver) whose location is generally fixed and which contains

(iii) an attached antenna (Figure 2.1). The system operates as follows: Signals are broadcast by the reader via its attached antenna. The tag receives these signals and responds either by reading or writing the data or by replying with another signal containing some data, such as an identity code or a measurement value. The tag may also rebroadcast the original signal received from the reader, sometimes with a predetermined time delay.

In the IC, the unique identification of the object to be tracked is stored. In that IC only, there is provision for writing more information related to that product as it passes through different manufacturing, warehousing, and transportation processes. This writing, subsequent writings, retrieval, and subsequent retrievals of data take place through electromagnetic waves. The major difference among the different Auto ID technologies is in how identification is stored and retrieved and how less frequent is the human intervention. This is where RFID technology has triumphed over all other existing Auto ID technologies in terms of ease and areas of application and subsequently became a major topic of research in these current years.

2.2 BRIEF HISTORY

Although RFID has been of great interest of research in recent years, it started as early as World War II, where airplanes used to be identified as "friend or foe" using this technology. However, the optical barcode, the nearest rival of RFID, came to commercial usage in the 1960s or 1970s. Due to their inexpensive implementation and benefits over contemporary technologies, it became a huge success and is still prevalent in almost everything we see around us. However, from the late 1970s, due to increase in complexity and volume of business, requirement of a new technology came to the forefront and hence started the journey of RFID.

In 1948, Harry Stockman [3] first showed communication by means of reflected power, and in 1950 the first patent was lodged for passive transponders. Until 1979, researches related to RFID remained within different laboratory research. For the first time, it had commercial usage in animal tracking in the United States. This was followed by motor vehicle toll collection in Norway in 1987, followed by RFID tracking of US rail cars in 1994. With the establishment of an Auto ID center at the Massachusetts Institute of Technology in 1999, research in RFID technology received a huge boost up.

Absence of other related technologies like database management restricted the proper development of this technology in the initial phase. Another big deterrent factor was the absence of any global standards. Everything that used RFID was localized in nature and used proprietary technology. So there was almost no interoperability among the different players. This problem was solved when EPC Global systems came to being in 2003 and they started standardizing RFID from all possible directions. This was the biggest boost up that this technology received, followed by the mandates and demands of the giants like Wal-Mart, US Department of Defense, Gillette, and so on.

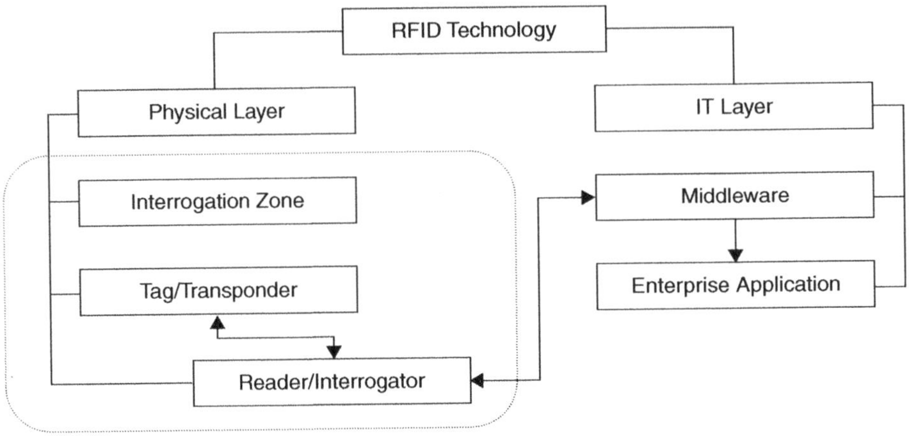

FIGURE 2.2. Generic division of any RFID system into layers.

2.3 OVERVIEW OF RFID TECHNOLOGY

Like any other multidisciplinary system, an RFID system provides a complete solution and can be deployed independently or in compatibility with other existing systems—for example, optical barcodes. Nowadays, RFID tags are printed alongside the optical barcodes so that automatic identification and tracking can be augmented with the existing infrastructures of the optical barcode identification and tracking. The basic focus of an RFID system is to make operations accurate, and user-friendly for the business process and the people associated with it; this includes an increase of efficiency in monitoring with reduced human intervention, better quality control, fault analysis, and lessening of shrinkage of products (loss of items) and, in the process, an increase the profit of the business. As shown in Figure 2.2, the RFID system is divided into two layers: physical layer and IT layer. As can be seen in the figure, the physical layer comprises tag, reader/interrogator, and interrogation zone (IZ). Following are the detailed discussion of each component of the physical layer if the RFID system.

Tag. Tags are similar to the optical barcodes, which are attached to the item/case and which store the unique identification of the item/case. Tags are called "transmitter responders (transponders)" too. The tags primarily consist of two components: the antenna and the IC chip. In some cases, depending on the business process involved, they have environmental sensors for measurement values such as temperature, humidity, and so on. The tag antenna communicates with the reader/interrogator by means of electromagnetic waves. Also in semiactive and passive tags, antennas scavenge power from the interrogator to operate the on-board IC chip of the tag. The IC stores the unique identification of the item/case in the form of some numbers. Also, depending on the business process involved, they have provision for subsequent read and write of data and their retrieval. If there is any environmental sensor included in the tag, they communicate directly with the IC chip.

Reader/Interrogator. The reader of the RFID system is compared to the scanner used for optical barcodes. They come in different forms such as handheld, mobile, or stationary. Readers are made up of primarily two components: the antenna and the interrogator circuitry. The antenna is used for communication with the tag using electromagnetic waves. For semiactive and passive tags, the reader antenna is used to supply power to the tags for the operation of their IC. The interrogator circuitry is a conduit or intermediary between the reader antenna and the IT layer. Interrogator circuitry performs the task of sending data through the reader antenna and also receiving data and then sending it to the back end for processing. Interrogator circuitry also performs the task of coordination between different reader antennas for the efficient and successful reading of tags. The detailed of the reader architecture is presented in a later chapter.

Interrogation Zone (IZ). The interrogation zone consists of the area in which the reader reads/writes data from/to a tag.

- This is the three-dimensional physical space consisting of everything in the vicinity of the tag and the reader where the electromagnetic (EM) waves travel between them.
- IZ is included in the physical layer because successful communication between the tag and the reader is greatly dependent on the interferences from other electromagnetic sources, reflection of the waves, presence of other stationary/mobile objects in the IZ, and so on.

IT Layer

This consists of ***Middleware*** and ***Enterprise Applications***.

Middleware. This is the intermediate between the interrogator and the enterprise layer.

- Middleware sends and collects data directly from the interrogator, performs a business-related process regarding the data, stores the data, and, as per the requirement, sends data to the enterprise applications.
- Middleware also comprises the software used to monitor, configure, and manage the hardware of the interrogator.

Enterprise Application. Data gathered from middleware are used in here, and relevant business processes such as the creation of an invoice are carried out using those data in the required formats.

Depending on the business process involved, any RFID process is designed where selection of tags, readers, middleware, and so on, are very important and application-specific. Criteria such as read range, electromagnetic power involved, frequency, protocol, form factors (shape and size of the tags), and so on, have to be carefully selected, depending on the application.

Like any other business process operating globally, intercompatibility of the different components in any single system and between different systems is very important. Hence the standards in the RFID system came into being. Interoperability, safety, quality, and a specific set of guidelines to follow globally are the basic idea behind any standards. Standards in the RFID are there in both international level and national level, where each and every aspect of the system and its working are well-defined. The key players in making the standards in international level are bodies such as ISO, EPC Global, and so on.

After the standards are made, regulatory authorities come to play where they make it mandatory for the RFID operations to oblige their regulations if they have to operate in their territory. Regulatory bodies are national, such as FCC (USA), ERO (Europe), Australian Communication Authority (Australia), and so on. Regulations are like a subset of standards.

Over and above that, some industrial mandates are there for the big companies like Wal-Mart, US DoD, and so on, where their vendors and suppliers have to oblige them to carry on business with them. Mandates are proprietary in nature and can be considered to be a subset of regulations.

2.4 APPLICATION AREAS OF RFID

Probably only dreams can limit the areas of application of RFID. Nowadays, RFID tags are applied in almost every business process and are projected to be applied everywhere that exists in real world—every single item. Manufacturing, logistics, supply chain, and tracking are the broad areas where RFID is used currently in the largest volume. Health care, pharmaceutical, document management, sports (timing), livestock, baggage handling, finance (cashless), and access control are only some of them [1, 2].

Manufacturing and supply chain being the largest areas of application of RFID technology, two examples would make it clearer how the system is used in the business process.

Manufacturers send their products to the shipping dock for transportation. At the manufacturer's end, the products are tagged (item level and case level) with their unique identification numbers. At the dock, the items are read and information such as time, place, date, and so on, are stored against the unique identifications in the database system of both the manufacturer and the supply-chain people. The manufacturer sends an advance ship notice to the supply-chain people containing the item numbers of the products to be tagged (in this case it is normally the case level identifications—pallets), and these data are used by the supply-chain company to prevent loss or theft in the transportation process. When these items are received at the retailer end, a similar process is carried out, both at case level and item level. In the whole process, reading (also writing) of data into the tags take place at definite intervals. Any discrepancy is promptly reported, and it becomes easier to track the lost items or to decide the person for whom loss or damage of any item took place [2].

For manufacturing process, especially for the big business, the raw materials or accessories coming from the vendors or suppliers are tagged in the very beginning as per the mandate of the manufacturer. Then the items are tracked (by read or read/write) as it goes across the different processes of the business. Different environment sensors such as temperature, humidity, time, and so on, also assist in this process. The database of the manufacturer keeps track of every movement and process the items undergo, and therefore any discrepancy is promptly recorded. In this way, quality is assured. Also along the lifetime of the product/products, if any fault is discovered later on, the tags associated with them help in understanding which process they underwent and which process they bypassed and thus improve quality for the later products [2].

2.5 BENEFITS OF RFID

Before any new system or technology comes to be in vogue or is proposed, the benefit of using that technology is the first thing that is looked upon. That is why return of investment (ROI) is a jargon in RFID world. In looking for the benefits, the biggest emphasis is given on comparison with its nearest rival in the market. Special emphasis is given on certain points such as cost cutting, better fault findings, quality control, and so on. Eventually, it is the overall profit that is counted before any technology is taken up in businesses. Following the same rule, some of the benefits of RFID are presented with special emphasis to its nearest rival the optical barcode. The disadvantages of RFID would be given too to have a fair idea of the technology and its applicability in the market.

The salient advantages of the RFID technology are delineated as follows:

- In RFID technology, no line-of-sight operation is required. The tags can have any orientation. This reduces human intervention to a large extent.
- Identification using RFID is possible at item, case, and pallet levels, although the items may be inside cases that are stacked up to form a pallet.
- Identification can be done simultaneously along with read/write capability at each station of monitoring; thus any loss/damage of items can be traced to their location.
- It has more data storage capacity along with worldwide standardization.
- The tags are rugged in nature and hence operable in a wide range of environmental conditions as well as business processes.
- It utilizes real-time locating of items possible without going through the process of opening the pallets, cases, and so on. These data can be stored either in central database and/or in the tag itself and can be used downstream in the business process.

Although RFID has a lot of benefits, it has some shortcomings too, which are being worked on currently throughout the world. The shortcomings are:

- RFID technology, although it has global standards, is still an immature technology compared to barcodes. So, still there is room for discrepancies when the tags have to go through different regions of the world, in terms of standards.
- Cost of the tag is a major factor. Cost of printing and attachment of tag and also maintenance of databases makes it nonprofitable for inexpensive everyday items in the market/supermarket, which is supposed to be the biggest market for this technology.
- Presently, RFID technology is a two-step technology where a tag is first created and then it is attached. This incorporates some delay and extra cost compared to barcodes, which are printed directly on the item/case/pallet casings.

However, one should keep an eye on the overall revenue generated in any business, after proper weighing of the pros and cons, RFID technology holds a bigger promise and hence there is a worldwide rush for the development of this technology.

2.6 THE RF IN RFID TECHNOLOGY

Before going into anything in RFID technology, the first thing that comes into mind is the radio-frequency (RF) part in the RFID technology. RF is of prime importance due to two reasons [1,2]:

- RF is the means of communication between the tag and the reader.
- The nature of the Interrogation zone (IZ) is primarily determined and affected by the RF waves that are present or operating in the IZ.

Out of the vast electromagnetic spectrum starting from audio waves and ending in the gamma rays, only parts of it are used in RFID technology, depending on the nature of use, standards, regulations, and mandates. Electromagnetic waves can be described in terms of a stream of photons that travel at the velocity of light in vacuum or air. As the photons travel, they radiate energy that is called electromagnetic (EM) radiation. The difference in different electromagnetic radiation is the amount of energy the photons carry and how they dissipate that energy, depending on the distance and the media through which they are traveling. The EM spectrum is divided into different segments, depending on the wavelength and frequency of the EM waves. Any EM wave is composed of mutually interchanging electric and magnetic fields that are perpendicular to each other as well as the direction of propagation. When the EM waves propagate, they tend to radiate their energy in three dimensions in space; hence with increase of distance, the energy carried by any wave decreases in magnitude.

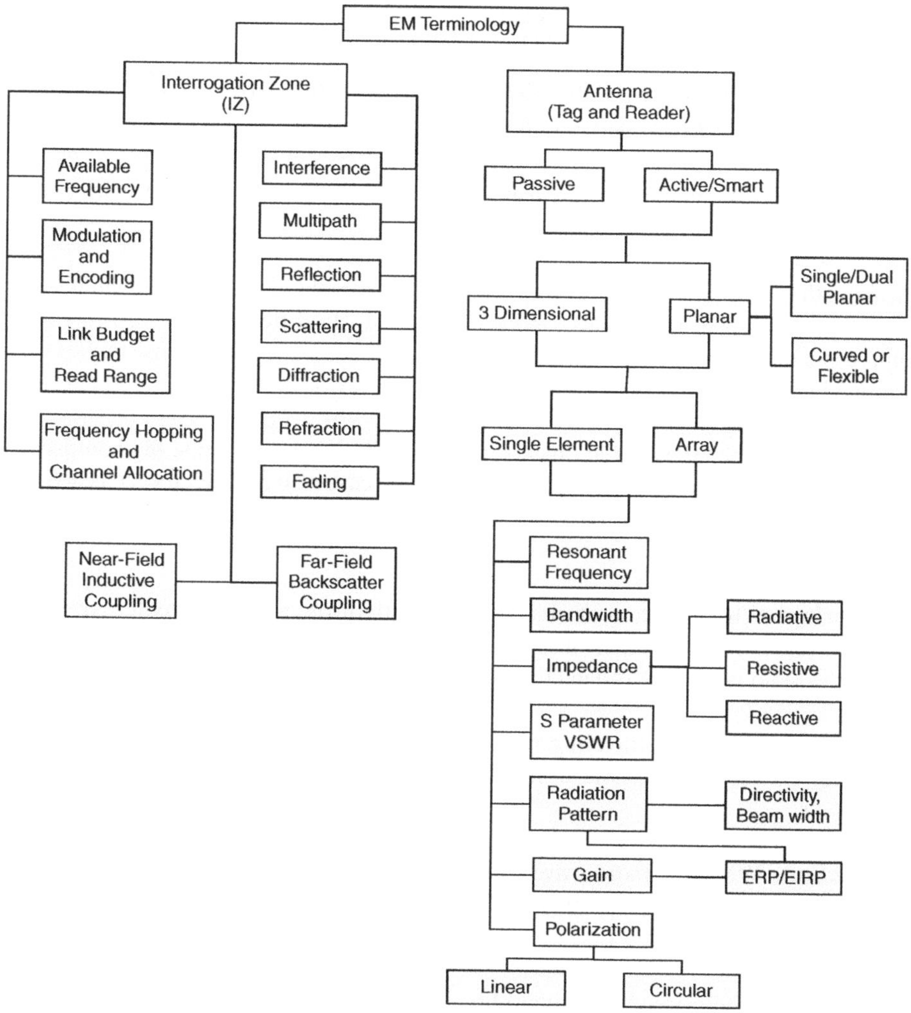

FIGURE 2.3. EM terminology associated with RFID systems.

2.7 EM TERMINOLOGY IN RFID

When designing any RFID system, one has to be familiar with the following EM wave and RF-related terms [1, 2]. They are broadly classified in Figure 2.3.

Available Frequencies. For RFID technology, by virtue of standards, regulations, and mandates, only some parts of the EM spectrum are available for use. Designing of any RF system has to comply with the following frequencies:

- Low frequency (LF): 125–134 kHz
- High frequency (HF): 13.56 MHz
- Ultra-high frequency (UHF): 433 MHz and 860–960 MHz
- Microwave: 2.4 GHz and 5.8 GHz

Interference. When two or more waves having different wave characteristics combine to give an entirely new/different wave.

Multipath. When a propagating wave takes different paths and then after phenomena such as interference, reflections, and so on, add up in different phases, it creates unprecedented areas in space where the waves are strong and weak compared to the original incident one.

Reflection. When the area of objects in the path of the wave is quite big compared to the wavelength, the waves get reflected as in the case of light. Metals act as a very good reflector, whereas nonmetals (dielectrics) let the wave pass after some amount of loss of energy.

Scattering. When the area of objects in the path of the wave are small compared to the wavelength and number of those objects are quite large, scattering takes place where the radio waves get reflected in all sort of directions.

Diffraction. When the traveling radio waves encounter any sharp object in their path, they tend to bend/deviate from their original path. This phenomenon is called *diffraction*.

Refraction. When the radio waves travel from one medium to another medium of different density (dielectrics), their velocity changes along with direction. This is called *refraction*.

Fading. When the strength of the radio signal varies in time, it is called "fading." This phenomenon is completely random in nature and dependent on all the above-mentioned phenomena and hence it is very difficult to predict. Fading is not taken into account when designing an RFID system.

Modulation and Encoding. For communication in RFID technology, the frequencies mentioned above are the carrier frequencies. Data in binary form are encoded into the EM waves at the mentioned frequencies, by means of modulation (change). This modulated wave is then transmitted, received, and then decoded for the data it is carrying. Modulation can be of different kinds, such as amplitude, frequency, phase, and so on. Encoding schemes are like NRZ, Manchester, and so on.

Near-Field/Inductive Coupling. Coupling is basically the transfer of EM energy from one medium (reader/tag) to the other medium (tag/reader). Near field is the three-dimensional space surrounding an antenna where the plane wave (the phase front of the wave is assumed a plane) has not yet fully developed and separated from the antenna. The distribution of the near field is fairly omnidirectional, and the power attenuates at the sixth power of the distance from the antenna ($\propto\frac{1}{d^6}$). This is basically a transfer of energy through shared magnetic field and hence operation is limited to only LF and HF frequencies and only RFID tags. The wavelength has to be much bigger than the antenna and the interrogation zone (IZ), and a change of current flow in one device induces current flow in the other device in a "push–pull" manner. The antenna used in this technology is actually not a transducer in its true sense, but can be rightly referred to as a transformer. That's the reason why the antennas operating in the near-field region are always "inductive coil." Change in the coupling magnetic field with movement of the interrogator also affects the coupling that can be referred to as unintentional "cross-talk." This type of RFID technology has its application in animal tagging, item tagging, library database management system, smart shelves in the supermarket, and so on.

Far-Field/Backscatter Coupling. The area in the three-dimensional space beyond the near field is called "far field," and communication between tag and reader takes place by backscatter coupling by EM radiation. In the far field, EM energy is continuously transmitted "away" from the antenna in a radial manner and the power drops, obeying the inverse square law of distance from the antenna ($\propto\frac{1}{d^2}$). The EM energy transmits from the reader's transmitting antenna and encounters the tag's antenna, where it is reflected or absorbed depending on the tag antenna's radar cross section (RCS) or reflection cross section. If the tag is terminated with a matched load, almost no energy is reflected back, whereas if the tag has an open/short-circuit termination, most of the energy is transmitted back. The tag IC, depending on the data to be transmitted to the reader, switches between a load and an open/short circuit and thus controls the reflected EM wave. This technique of changing RCS of the tag's antenna is called the "antenna load modulation." This reflected EM wave, which is much smaller in magnitude compared to the incident wave, is detected at the reader antenna by means of a directional coupler/circulator and then amplified, decoded to extract the data sent by the tag. This kind of communication is prevalent in the UHF and microwave frequency ranges of passive RFID tags.

Link Budget and Read Range. Link budget calculation in wireless communications specifies the power budgeting for the transmitter and receiver, the antenna gain, and effective isotropic radiated power (EIRP) of the reader antenna to obtain a certain link distance (reading range). The link budget helps to calculate the required antenna gain and related specification to obtain a robust and viable communication in the specific conditions of wireless communications. The communication from the reader to the tag is called the "downlink," and the communication from the tag to the reader is called the "uplink." Therefore, the link budget is a calculation used to specify the read range in any RFID system. It encounters all the losses and gains taking part in the communication and thereby, depending on the sensitivity of the components, determines the distance at which reliable communication can take place between the tag and the reader in any RFID system. Factors like noise floor, cable losses, free space losses, and so on, are taken into account in calculating the link budget.

Frequency Hopping and Channel Allocation. All the RFID systems operating globally have to follow the standards, regulations, and mandates, so if the scenario is such that all the readers and tag operating in any particular IZ start communication simultaneously at one particular frequency, then due to interference, multipath, reflection, and so on, communication would come to a standstill very soon. As a result, the available frequencies are again subdivided into narrower spectra called channels and there are regulations that specify how many channels should be there in any particular frequency range and how long any particular reader would be allowed to stay in any channel. To reduce interference, the spread spectrum method is used for frequency hopping following the present regulations. One very common RFID technology "Zigbee" uses direct sequence spread spectrum (DSSS) for this purpose.

2.8 ANTENNA CHARACTERISTICS

In an RFID system, electromagnetic waves are used as the medium of communication. Antennas are the spatial filters that couple guided electromagnetic energy to free space electromagnetic energy (vice versa) to enable communication in an RFID system. Any conducting structure can be termed to be an antenna, but the efficiency with which the structure can transform the energy is the key determining factor on how well communication can be established in the system.

Passive antennas (i.e., without active components like amplifiers) are mostly used in RFID systems and are reciprocal in their behavior. This means that the antenna behaves in a similar fashion irrespective of its transmission and reception modes. According to Faraday's law, when an antenna is placed in a time-varying electromagnetic field, it induces electrical potential at its terminals and thus acts like a receiver. Like any component, when designing any antenna, certain characteristics such as antenna radiation patterns, bandwidth, gain, and polarization are the key ones that have to be kept in mind when designing it. They are discussed as follows.

Resonant Frequency. Any antenna transmits or receives EM waves in a most efficient way at one or a few frequencies, depending on the design. Those frequency/frequencies are called its "resonant frequency."

Bandwidth. The range of frequency surrounding the resonant frequency where the efficiency of the antenna in transmitting or receiving EM waves is close to 90% (−10 dB) is supposed to be the bandwidth of the antenna. Under different requirements, the above-mentioned efficiency may vary, thus varying the bandwidth of the antenna.

Impedance. Any antenna has three different resistances: radiative, resistive, and reactive. Together, these are referred to as the impedance of the antenna. Power absorbed by the radiative resistance is transmitted as EM energy or vice versa. Radiative resistance is proportional to the antenna length; that is, a bigger antenna is a better radiator. The power absorbed by the resistive resistance is dissipated as heat. The reactive energy is the unwanted energy that does not do any useful work and is composed of inductance and capacitance. It inhibits the transfer of energy and thus acts like a barrier. At resonant frequency, the inductance and capacitance mutually cancel each other to almost zero, and hence the antenna is the most efficient transducer at its resonant frequency.

Scattering Parameter/VSWR. This is an estimate that shows how efficiently an antenna is transmitting (or receiving). The S_{11} parameter (in decibels) measures the ratio of the reflected wave to the incident wave. The voltage standing wave ratio (VSWR) is another form of expressing the matching condition expressed by S_{11} parameter. S_{11} (in decibels) is called the return loss (RL). The industry standard of acceptable RL of 10 dB or VSWR 2:1 over the frequency band means that 90% of the signal is transmitted and only 10% is reflected back.

Gain. An isotropic antenna is an imaginary antenna that radiates in an omnidirectional fashion (spherical) with equal intensity in all directions. Because an isotropic antenna is an imaginary concept, the nearest real existing type is a dipole antenna that radiates uniformly in one plane (donut shape radiation pattern). Gain of any antenna is the ratio of the power transmitted in any given direction with respect to the power transmitted in that direction by an isotropic antenna (unit dBi) or a dipole antenna (unit dBd). It should be kept in mind that any passive antenna might have higher gain in any given direction, but this has to be compensated by lower gain in another direction to comply with the rule of conservation of energy.

Radiation Pattern. For any real antenna, the radiation is never spherical or omnidirectional in nature. It is more powerful in certain directions in space, whereas it is less powerful in the other directions. The graphical three-dimensional view of the power pattern of any antenna is called its *radiation pattern*. The radiation pattern consists of (a) a main lobe where the intensity is maximum and (b) some side lobes that are pointed at different directions and don't contribute to the antenna performance. Radiation pattern is very important in determining the interrogation zone for any RFID operation where it shows the area of optimal operation (main lobe) and sources of interferences (side lobes).

Directivity, Beamwidth. When the radiation pattern of any antenna is plotted in the direction of its main lobe, the angle (three-dimensional) at which the power level falls to half of maximum power is called the *beamwidth* (3-dB beamwidth). This determines the directivity of the antenna. In RFID operations, the reader antennas are required to have the maximum gain and directivity, whereas the tag antennas are supposed to achieve (as much as possible) omnidirectional nature in their radiation pattern. It should be kept in mind that with more directional antenna, more side lobes come into being which result in interference among the RFID readers working in close proximity.

ERP/EIRP. Power transmitted by any antenna is given in terms of effective radiated power (ERP) and EIRP. ERP is the product of the power supplied at the antenna (via cables) and its gain with respect to a dipole antenna. EIRP is the product of the power supplied to the antenna and its gain with respect to an isotropic radiator. The unit of ERP/EIRP is watts of dBm (30 dBm = 1 W). The terms ERP and EIRP are very important because all the antenna-related regulations in any RFID system strictly mention either of these two.

The ERP and EIRP refer to the reader/interrogator antennas. Now although ERP or EIRP are the product of the incident power and the gain, it is always advisable to have an antenna of higher gain for the two following reasons:

1. It would reduce power supply, which is directly related to cost of power and life of the components.
2. When the reader/interrogator receives the backscattered signal, higher gain would ensure better signal quality and, hence, a more accurate system.

Polarization. The EM wave radiated from any antenna has electric field and magnetic field perpendicular to each other. They increase and decrease in their amplitudes in a sinusoidal manner as the EM wave propagates. Based on the orientation of the electric (E) field, the polarization of any antenna is defined which may be either linear or circular.

In linear polarized antennas, the E-field is either horizontal or vertical with respect to the ground. In circularly polarized antennas, the antenna is designed in such a way that both vertically and horizontally polarized E-fields add up with 90° phase delay between them. This results in the creation of a rotating E-field with time which may rotate either in clockwise or anticlockwise manner in the direction of the propagation of the EM wave. Thus a few situations arise:

- If two linear polarized antennas are there with similar polarization, maximum transfer of energy takes place between them (co-polarized antennas).
- If the polarization of one of the antennas is changed (rotating the antenna by 90°), then power is no longer transferred between them (cross-polarized antennas).
- If one of the above-mentioned antennas is replaced by a circularly polarized antenna, keeping everything else the same, then transfer of half (−3 dB) the power takes place.
- If both the antennas are made circular-polarized (both right/left-hand), then they behave like the linear (co-polarized) antennas.
- If one of the above ones is right-hand circular-polarized and the other is left-hand circular-polarized, then no power is transmitted among them anymore (cross-polarized antennas).

These five situations are very important while designing the antenna for any RFID system. For linearly polarized antennas, the beam is narrow and hence directivity and penetration of the beam is higher. For circularly polarized antennas, the beam is wide, making the penetration lesser. However, with a wide beam, the area of coverage (reading area) increases. Thus, depending on the requirement—for example, reading tags inside the cases (high penetration) or reading tags in a toll collection point on highway (wider reading area)—suitable antenna should be designed.

Circularly polarized antennas are good in the sense that tags can be read from in any direction. But in a business process where tags location and orientations are predetermined (e.g., on the case), linearly polarized antennas are good for better penetration (item level monitoring without opening the cases). Another very big advantage of the circularly polarized antennas is the fact that they can operate side by side in very close proximity without any interference if their polarizations are different. Readers with opposite polarizations placed alternately are very helpful in dense reader environments where lot of readers have to be accommodated to accurately and quickly read any tag that might be passing through their IZ.

Antenna Types. Antennas used in RFID technology are normally flat planar-type, typically microstrip antennas. However, depending on the nature of usage, they may be horn, tunnel, loop, dual planar, and so on. The antennas can be single elements or the array of elements, depending on the purposes and the natures of deployments. The antennas can be passive or active with provisions of beam squinting too. Several antennas placed at different angles and distances can be controlled by one interrogator that would be synchronizing their duty cycle and thus carry on effective reading of tags.

Antenna Accessory (Cables and Connectors). In RFID readers, the cables and connectors make the bridge between the antenna/antennas and the interrogator. At times the cables are connected permanently to the antenna and the interrogator, and at times there are provisions of detaching any of these. When options of detaching are present, the cables are terminated with specialized terminations called "connectors." The cables and connectors are mostly coaxial in nature and must have impedance that matches the components it connects for the purpose of maximum transfer of power. Cables with shorter cross-sectional areas and longer lengths contribute to more losses due to radiation leakage. Cables should not be bent more than their specified bending ratios to prevent any damage within. Typically, cables of RG series such as RG-174, RG-58, and so on, and cables of LMR series such as LMR-195, LMR-240, and so on, are used. Connectors such as SMA are used mostly for high frequency operation.

2.9 TAG

In any RFID installation, the readers and interrogators are one-time investment only, whereas the tags are the consumables that are required in thousands, millions, or billions, depending on the business process. Tags are placed on the items to be monitored. Hence, for proper running of any RFID installation, detailed knowledge of the tag and its performance is essential.

At the component level, the tag is made up of an integrated circuit (IC) and an antenna.

Integrated Chip (IC). This is a semiconductor circuitry normally designed by a chip manufacturer by silicon lithography. Tag manufacturers normally buy them. The big players of RFID ICs are Texas Instruments and Philips. The IC consists of logic circuitry that controls the unique identification number and all the other information carried by the tag. The IC is roughly divided into the following parts:

- A part of the IC is dedicated for controlling power. This power may come from a battery (semipassive/active) or radiated energy from the reader/interrogator (passive).
- Modulation/demodulation of the signals, encoding/decoding of the digital bits, and implementation of the communication protocol take place in the IC.
- The memory is divided into blocks called "banks," which may be read only or read write enabled depending on the usage. The unique identification number, error checking codes, public and private passwords, and so on, are stored in the IC memory.

Tag Antenna. This is the largest part of the tag which is directly connected to the IC. Communication and also flow of power between the reader and the tag occur through the antenna. Different designs of antennas are available in the market. They are manufactured by screen-printing, foil-stamping, or copper-etching methods; screen-printing is the fastest and most common one. One important thing to be kept in mind is that the interconnection between the antenna and the IC of the tag is the weakest link in any tag.

Substrate/Laminate. This is the common platform on which the tag antenna, IC, and any other necessary associated circuitry are placed. Substrates should be thin, flexible, and able to withstand the different harsh environmental conditions. Substrate material should provide a smooth printing surface for the antenna, dissipate any static charge buildup, and provide a mechanically stable platform for the entire life of the tag. Commonly used materials for RFID laminates include PVC, PET, polyesters, FR-4, and even paper.

Tag Packaging. With the manufacture of the tag components, the most important thing that follows next is packaging for proper functioning of the tag and protection of the tag components. Regarding proper packaging, the following terms are important:

- *Strap.* Because the pads of the IC are too small, two pads are given by the manufacturer to facilitate their attachment to the antenna.

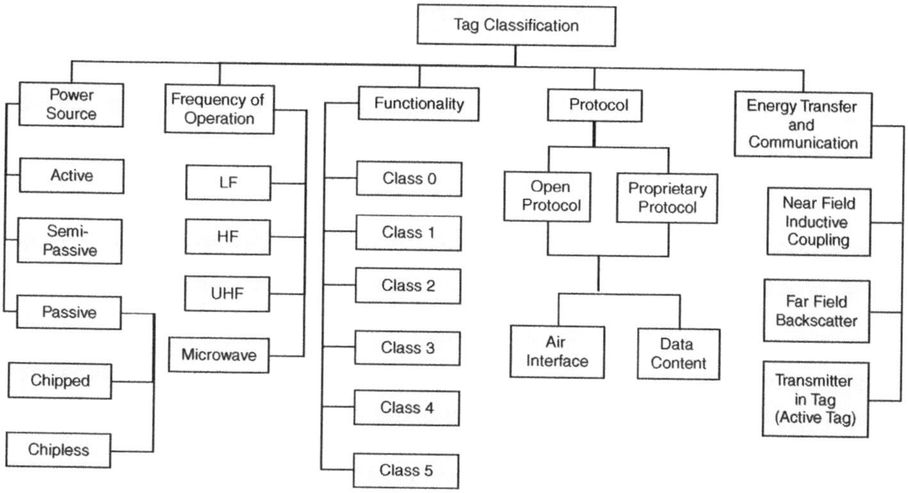

FIGURE 2.4. Classification of RFID tags.

- *Inlay.* The strap, when added to the antenna and with some extra spacing of substrate to accommodate any other associated circuitry, is called inlay. Electrostatic discharges developed during the manufacture of the inlays tend to destroy the tag.
- *Smart Label.* The smart label consists of the inlay, which is inserted inside a paper label, and (b) other necessary readable information printed outside, like barcode, EPC logo, and so on.
- *Encapsulated tag.* For working in harsh environments or where the items are in a closed loop—that is, reused in a business process (e.g., trolleys, containers, etc.)—the inlays are encapsulated in hard RF translucent outer covers such as polypropylene, polyacetate, and so on, to increase the lifetime of the tag.

Tag Types. Depending on the basis of classification, tags can be subdivided into different categories as shown in Figure 2.4

2.9.1 Power Source

Depending on the power source, tags can be classified into **passive**, **semi-passive**, and **active** tags. Detailed discussion of each tag type is presented below.

Passive Tags. These tags have no on-board power source for operation of the IC. Power transmitted from the reader is used for both powering up the IC and also communicating back with the reader. Due to its high requirement of power, the read range of the tags are usually lower ranging for up to 2 ft for inductively coupled and up to 20 ft for backscattered tags. When the passive tag is not communicating with the tag, it is in "dead" condition. Hence it doesn't contribute to any radio noise and can't support any on-board environmental sensors such as temperature and humidity. These tags have theoretically infinite lifetime. Passive tags are the least expensive tags and are normally used in places where the tags are not reusable—that is, consumable items. The lifetime of the tag ends with the lifetime of the item. Passive tags normally have limited data storage capability

Semi-Passive Tags. These are also called battery assisted tags (BATs). They have on-board power supply to provide power to the IC to keep it alive, but it doesn't carry any transmitter. Thus it communicates with the reader by backscatter coupling. Longer read ranges up to 100 ft are possible compared to the passive tags. With the IC being kept alive (i.e., supplied with power all the time), environmental sensors can be accommodated for use in temperature-controlled business processes such as frozen food, blood for transfusion, and so on. Due to the absence of any active transmitter, these tags don't contribute to any radio noise. These tags have more memory capacity than the passive tags. Semipassive tags are more expensive, bigger in size, and heavier, depending on the size of the battery. One major disadvantage of a semi-passive tag is the fact that the life of the tag is determined by the life of the battery. Also, under some harsh environmental conditions, the battery may fail to function, thus rendering the tag virtually dead in that environment.

Active Tags. These tags have an on-board battery and a transmitter. The battery supplies power to both the IC and the transmitter. Due to the presence of transmitter, it doesn't have to rely on an interrogator to transmit its data by backscatter coupling. In fact it can act as an interrogator itself. Longest read range of up to kilometers is available, depending on the battery and the transmitter. On-board environmental sensors and data processing capability are present in the ICs. The memory capacity in active tags is the highest. In fact, these tags can locally accept data from other tags/sensors, process them, and then broadcast them (e.g., nodes in Zigbee protocol for communication). These types of tags can be used in Real-Time Location Systems (RLTS). Active tags can have a sleep mode when it consumes the least power in the idle stage. When it is awakened by an awakening signal, it wakes up, performs what it is being asked to do, and again goes back to sleep. Thus the battery life and the tag's life can be elongated. Also, they can be programmed to wake up at specific intervals and send their identities and sensor data, thus acting like "beacons." Active tags have one major advantage over semi-passive tags. Here the battery life can be monitored. As a result, if the battery life is coming near its end or the environment condition is becoming harsher, it can send a signal to the interrogator and thus make it alert before it actually becomes inoperative. At times these tags use two different frequencies for transmission and reception of data (downlink and uplink). The active tags are, however, the most expensive type of tags and thus are limited in their usage by cost factor. Due to the presence of an on-board transmitter, these tags contribute largely to radio noise.

2.9.2 Frequency of Operation

Depending on the available frequencies, tags around the world operate at the four following frequency ranges [1, 2, 4].

Low-Frequency (LF) Tags. Tags operating between 125 kHz and 134 kHz fall under this category. These tags are mostly used for animal tracking, for patients in hospitals, for a vehicle immobilizer, and so on. This is the most mature technology among all the RFID tags. Generally, these tags are passive, have short read ranges ($\sim$ few inches), use near-field inductive coupling, use copper wire coil antennas (several hundreds of turns), have short read ranges, and possess minimum data handling capacity. They have very limited or no anti-collision capability. However, they can be very easily read when attached to objects containing water, metal, tissue, and so on.

High-Frequency (HF) Tags. RFID tags operating at 13.56 MHz with similar power level globally fall under this category. These tags are passive tags using near-field inductive coupling with a bit longer read range of about 3 ft, and they have larger data handling capacity with greater data transfer rate compared to the LF tags. This technology is less mature than the LF tags, and it has provision for anti-collision; however, because the read range is low, this feature is rarely used. Antennas here are aluminum, copper, or silver coils (three to seven turns only) and two-dimensional in appearance, thus reducing their cost compared to the LF tags. These tags can be used effectively when attached to objects containing water, tissue, wood, and so on. When they have metals in vicinity, their behavior changes, however. The interrogators are comparatively less complex and the interrogation zone is a uniform one, thus making these tags very useful in applications such as smart shelves in supermarkets, baggage handling, library database management systems, and so on. These tags are in maximum usage round the world currently.

Ultra-High-Frequency (UHF) Tags. Tags using 433 MHz (active tags) or 860–960 MHz (passive and semi-passive tags) with center frequency of around 915 MHz fall under this category. It is important to note that in any particular IZ, use of both frequencies is never advisable due to harmonic interference (433 MHz $\times$ 2 $=$ 866 MHz). The passive and semi-passive tags at this frequency use backscatter coupling to communicate with the reader, while the active ones use their own transmitter. For the passive and semi-passive ones, the read range is quite high, ranging to about 20 ft. Tags operating at this frequency have well-defined anti-collision protocols to enable the reading of multiple tags simultaneously. The antennas are generally planar dipole antennas and made by copper, silver, or aluminum deposition on the substrate. Thus the tags are two-dimensional in shape. These tags do not perform well in the vicinity of water and metals. In particular, water absorbs the UHF frequencies. The UHF antennas are generally directional, thereby defining a well-defined interrogation zone (IZ). However, phenomena such as multipath, interference, fading, and so on, affect the IZ, making it nonuniform in nature. This technology is less mature than the HF tags; however, with the introduction of the EPC Global, Gen 2 protocols, a huge boost has been given to tags operating at this frequency. Because the technology is comparatively immature, the frequency of operation varies in different countries. To overcome that, the communication takes place by breaking up the entire available frequency range into different channels and then hopping across those channels. In Australia the available frequency band is 918–928 MHz, with maximum power transmitted being 4 W EIRP and number of channels being 16. These figures vary with different countries of the world.

Microwave Tags. The tags operating at frequencies of 2.4 GHz and 5.8 GHz ISM band fall under this category. Passive, semi-passive, and active tags are available in this frequency range. The tags operating at this frequency are the smallest in size. Read range for the passive tags are around 15 ft, semi-passive ones are around 100 ft, and the active ones are around 350 ft. Because the antennas are more directional in this frequency range, the IZ is much better defined. These tags are easier to work with metallic objects due to shorter wavelength. Anti-collision protocols are existent; and due to wider available frequency ranges, more channels can be used, thus making way for more effective reading. Other microwave devices such as WLAN, cordless phones, microwave ovens, and so on, provide RF noise at this frequency. Because regulations, standards, and so on, are available minimally at this frequency, this spectrum has a lot of scope and freedom. Highway toll collection, fleet identification, Real-Time Location System, and so on, have application at this frequency.

2.9.3 Functionality (EPC Global Classes)

Depending on the tag functionality, EPC Global, a standardization body has classified RFID tags into six different classes. The main criteria for differentiation are read/write capability, source of power, capacity of memory, and capability of communication [2, 5, 66]. The classification is as follows:

Class 0. Passive tags with Write Once Read Many (WORM) IC chips written at factory during manufacture.

Class 1. Passive tags with WORM chips, but this writing can be done either at the factory or at site of operation for the first time.

Class 2. Passive tags, read/write capability with memory available for user, and possibility of encryption of data.

Class 3. Semi-passive tags with on-board environmental sensors, read/write capability, and memory space availability for user.

Class 4. Active tags with on-board sensors, read/write capability, user memory, and provision for peer communication with other similar active tags and interrogator.

Class 5. This class defines the reader/interrogator, which can communicate and power the tags belonging up to the aforesaid classes.

2.9.4 Protocols

Protocols explicitly define the grammar of the language used in communication between the reader and the tag/tags. Although protocols are not confined within the tags only, the reader and tag have to have the same protocols to communicate with each other. Protocols can be first divided into two broad categories: open and proprietary.

- *Open Protocols.* These are developed by standardization bodies such as ISO 18000-6(A/B) and are available in equal terms globally for those who want to use them.
- *Proprietary Protocols.* These are developed by particular manufacturers for their own business purpose—for example, Texas Instrument's TI Tag-IT, Intermec IntelliTag, and so on.

Readers can be fitted with single protocol or multiple protocols, depending on the nature of tags it encounters. However, reading with multiple protocols makes the reading process very slow. Again, in any of the aforesaid protocol types there are two kinds of subdivisions: air interface and data content protocols.

- *Air Interface Protocols.* These decide how the tag and data communicate using EM waves and include frequency of operation, emission levels, bit rate, anti-collision algorithms, modulation, encoding, and so on.
- *Data Content Protocols.* These include the division, definition, and layout of the memory in the IC, mandatory information that should be included there with their specific locations, and so on.

The EPC Global Gen 2 protocols are the maximum worked-upon ones globally; and like all other protocols, they ultimately aim is to bring about uniformity between different RFID systems when operating at different locations around the world.

2.10 CHIPPED AND CHIPLESS TAGS

The biggest inhibiting factor in deploying an RFID system in everyday items is the cost of the tags. This is the sole reason why barcodes still reign supreme in everyday consumer items, despite the fact that an RFID system has a lot of benefits over barcodes. The main cost of RFID tags comes from the IC chip embedded in the tag. For everyday consumer items where the identification of the items is enough for the business process, the solution to this problem can be RFID tags without any chips, called "chipless RFID tags." The market of the chipless RFID tags is supposed to be the largest one in the coming years (IDTechEx). However, the technology is still in the nascent stage and holds a great promise and scopes. Later in this chapter, chipless RFID tags later will be discussed in more detail.

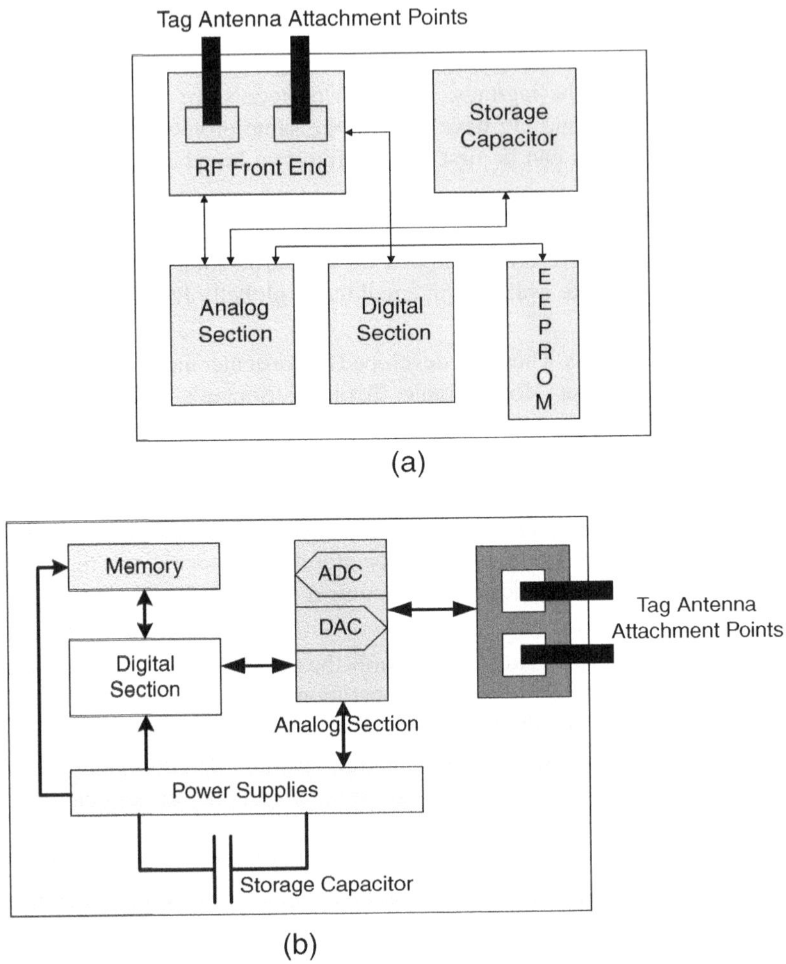

FIGURE 2.5. (**a** and **b**) Block diagrams of a generic passive RFID tag [6].

Architecture of a Passive Tag. As shown in Figures 2.5a and 2.5b, an RFID tag consists of an application-specific IC (ASIC) and an antenna. Figure 2.5a is reported in reference 6, whereas Figure 2.5b is a generalized approach of the various reported literature. The architecture is discussed in the following section.

Tag ASIC. The ASIC works in three phases [7–9]:

- *Decoding Phase.* Decodes the modulated carrier wave of the reader.
- *Processing Phase.* Acts as per its circuitry to modulate the downlink, thereby sending its unique identity.
- *Reflecting Phase.* Communicates with the reader in this phase.

During the first and last phase, power is largely consumed and this is supplied by the on-chip "charging pool." To reduce burden of charge pool, the phases are separated by a small time gap when the charge pool recharges again.

The ASIC comprises the following units [6, 10, 11]:

- *RF Front-End.* Tuning capacitor, voltage rectifier, voltage limiter, and clock recovery.
- *Analog Front-End.* Voltage reference/power management unit, regulator, power on reset.
- *Digital Part.* Modulator and demodulator. This part deals with the several anti-collision algorithms and tags multifunctionality part. It carries the memory of the ASIC.

RF Front-End and Analog Part. A tag interface should be there for power matching of the Tag antenna with the ASIC. Full-wave rectifier using n^+/p^- or p^+/n^- is used to convert the incident RF signal from reader to direct current usable by the ASIC. The oscillator is the source of system clock as well as the two FSK reverse link frequencies. The power management unit (PMU) provides regulated voltage and current to other circuit blocks of the chip. Efficient PMU in TIRIS (Texas Instrument Passive Tag—134.2 kHz) makes it outperform many active counterparts because especially in hazardous environments (like coal mines), the output power of the reader plays a crucial role in safety issues [12].

Digital Part. Modulation of the uplink (reader-tag) and demodulation of the downlink (tag-reader) is done digitally. The digital part consumes the maximum power of the ASIC. It also does cyclic redundancy checks (CRC) to enhance air interface link reliability. Minimization of switching energy and operation in a subthreshold region is used to reduce overlap current, thus reducing energy waste. For memory, electrically erasable programmable read-only memory (EEPROM) or Static Random Access Memory (SRAM) are used. However, chips can be operated at low voltages of 1.5 V (analog part) and 1.2 V (digital part) [10]. Use of a thin-film transistor circuit (TFTC) or polymer bases and silver or gold ink reduces this cost considerably [13–15].

2.10.1 Tag Antenna

In low-frequency systems, normally copper wire is wounded in huge number of turns and in UHF frequencies, single turn or dipole or patch antenna is preferred. However, some properties of the tag antenna are being presented which must be given special care [1, 16, 17]:

- Be small enough to be attached to the required object.
- Have omnidirectional or hemispherical coverage to ensure non-line-of-sight operation of the tag.
- Must provide maximum possible signal to the ASIC.

TABLE 2.1. Comparative Study of Different Tag Antenna Configurations [16]

Antenna	Pattern Type	Free-Space Bandwidth (%)	Impedance (Ohms)	Polarization
Dipole	Omnidirectional	10–15	50–80	Linear
Folded dipole	Omnidirectional	15–20	100–300	Linear
Printed dipole	Directional	10–15	50–100	Linear
Printed patch	Directional	2–3	30–100	Linear/circular
Log spiral	Directional	100	50–100	Circular

- Have a polarization such as to match the enquiry signal regardless of the physical orientation of the tagged object.
- Be robust and very cheap.

The major considerations in choosing an antenna are:

- The type of antenna
- Impedance of the antenna
- RF performance when applied on the object
- RF performance when other substances are around it.

In Table 2.1, some possible antennas are listed with their properties [16].

2.10.2 Recent Research and Development in Tag Antennas

Some recent developments in the tag antenna with a special view to (a) reduction of the size and cost, (b) improvement of polarization purity, (c) impedance matching, and (d) performance degradation due to bending, and so on, are presented here. In resonant-type wire antennas, length of wire primarily determines frequency of operation as well as efficiency of radiation. In nonfractal antennas, reduction of length results in reduction of radiation efficiency. Fractal antennas, however, can operate with high efficiency, although their length is reduced significantly. A planar slot ring antenna with microstrip feedline has been designed. It has been observed that by the second iteration only, the resonant frequency could be reduced to 40% of the original 2.45-GHz resonant frequency. This antenna has a bidirectional pattern, making it more suitable for RFID applications [18].

In RFID systems, bandwidth is of less importance in comparison to cost, profile, and size. In UHF frequencies, printed dipoles are used but have large size for resonant frequencies below 1 GHz. To meet the demands of both low resonant frequency and size, a meander line antenna (MLA) [19] has been designed. Gain is up to 0.45 dBi, reading range is 0.5 m, and bandwidth is a few megahertz. In another case, MLA has been developed using text, thereby serving both the purpose of antenna and hi-tech advertisement. Here, lowering of the resonant frequency has been achieved at the cost of gain [20].

A rectangular patch antenna operating at its dominant mode is considered. Considering electric field to vanish around the middle of the patch, the patch has been shorted along the middle line by a metal wall without significantly changing the resonant frequency of the antenna. Thus a shorted patch antenna is obtained. Next the shorted patch together with the ground plane is folded, thus arriving at a folded shorted patch antenna, whose physical length is half but the electrical length is almost unchanged. The gain of the antenna is 0 dBi [21]. For operation in 24-GHz frequency range, a microstrip-fed slot antenna has been designed. The slot and the feed line are perpendicular, and slot width plays an important role in impedance behavior of the antenna. The antenna has 10% impedance bandwidth and 2.3-dBi gain [22].

When the cost of the tag has to be reduced, conductive polymeric thick film pastes offer a valuable alternative. The fillers are silver flakes that get connected during the curing process and the paste is applied by screen printing. The conductivities of the components in polymer-based antenna coils are lower than copper or aluminum. As a result, higher ohmic losses, lower Q factor, and, hence, lower operating range are observed. However, this problem can be solved by lamination of the coils [23]. The ASIC in the tag has varying impedance that can go as high as 1200 ohms. So the antenna has to be developed suitably to match that impedance for maximum transfer of power to the ASIC. A folded dipole structure has been designed whose folded end is open. Great freedom is there to achieve the required impedance by adjusting the open folded end suitably [24].

Metals in the vicinity strongly lower the radiation efficiency of an antenna. Also, in the UHF range, metals reflect the electromagnetic waves fully, thus increasing the directivity of the antennas. This reflected wave is at 180° phase shift with the incident one, and thus they cancel each other. A wire planar inverted-F antenna (PIFA) for the RFID tag has been designed which can be mounted on metallic objects. A small decrease in return loss and radiation efficiency and high improvement in directivity is obtained [25].

With metals in vicinity of the antenna, another antenna also performs remarkably well. A patch antenna with an electromagnetic bandgap (EBG) ground plane and a two-layer substrate is designed. A two-layer substrate is used to improve the insulation and also to add metal to the antenna structure so that mounting it on a metal won't hamper its radiation performance significantly. Return loss and directivity improve significantly with slight decrease in bandwidth. The reading rate of the tag is decreased compared to the case for the normal ground plane, but the reading rate is enhanced in the vicinity of metal. Also, the resonant frequency of the antenna decreases, which implies that reduction of the antenna size is possible to get the same resonant frequency [26].

Tags in higher frequencies tend to use dipoles as their antennas. The tags are placed on the objects, which are of various shapes ranging from spherical to conical, and so on. Also, paper-mounted tags are affixed at the corners of objects. This leads to bending and distortion of the tag antenna and thereby results in degradation

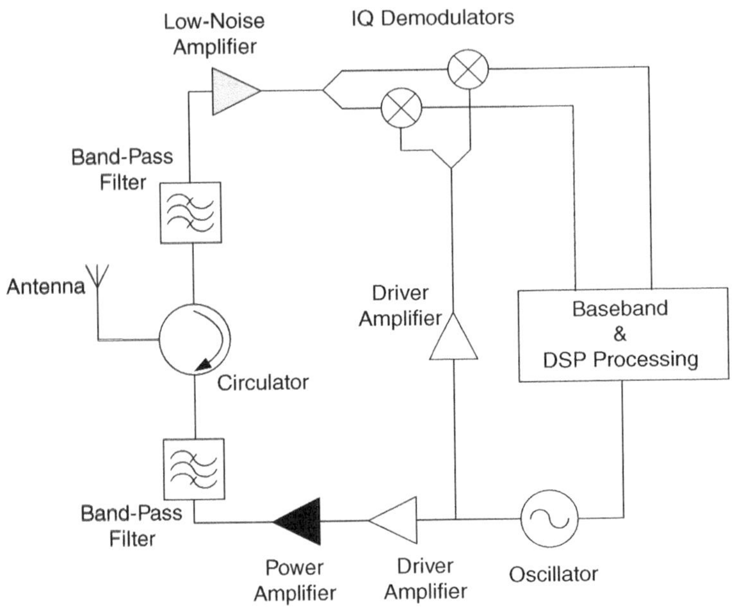

FIGURE 2.6. Circuit diagram of a generic RFID reader following EPC Global Gen 2 standards [28].

of radiation performance. A folded dipole has been taken as a tag antenna, and the radiation parameters of the antenna are studied. It has been observed that the operating range is reduced to 40% when the dipole is bent 90° at the feed, whereas bending at 3/8 of the dipole's total length from the center has no significant effect on the operating range [27].

2.11 READER/INTERROGATOR

The most important part of an RFID system after the tag is the reader/interrogator along with the interrogation zone. They must be configured appropriately, depending on the business process for effective reading operation. Like the tags, they also have to comply with the existing standards, regulations, and mandates (see Figure 2.6) [1, 2, 8–30]. The functions of the RFID reader are as follows:

- Readers read and write data to and from the tags.
- They also provide power remotely for operation of the IC in the tags (passive).
- As a result, bidirectional communication is established between the tag and the interrogator, which involves signal processing such as modulation, encoding, analog-to-digital conversion, and so on.

- Interrogators are at times connected with peripheral devices such as light stack, horns, and so on, to provide feedback on the condition of the interrogation zone and associated sensors, and so on.
- Interrogators are also connected to back-end devices to provide necessary business information as an outcome of their interrogation process. Either they process the information themselves and render the filtered data (smart interrogators) or they just simply report the readings (dumb interrogators).
- The interrogators must have graphical user interface through which various processes such as power settings, antenna settings, synchronization of the antennas, firmware upgrade capabilities, and so on, can be controlled.
- The reader antenna is a part of the interrogator on the whole and is connected to the interrogator by means of cables. The functionalities of the reader antenna are controlled by the hardware and software of the interrogator.

Interrogator Components. The interrogator components can be divided into three broad categories: (i) transmitter, (ii) receiver, and (iii) processor. The detailed descriptions of the three parts are given as follows.

Transmitter. The transmitter is the RF section of the reader. The main components of the transmitter are:

- *RF transmitter.* This transmits the carrier waves.
- *Oscillator.* This produces the alternating current at the required carrier frequency, which is later transmitted by the transmitter.
- *Power amplifier.* This is required to amplify the signal produced by the oscillator.
- *Modulator.* This is required for modulating the carrier signal to carry the useful data.

Receiver. The receiver is the second RF section of the reader. The main components of the receiver are:

- *Amplifier.* Preferably a low-noise amplifier (LNA) to amplify the weak reflected signal coming from the tag.
- *Demodulator.* Required to demodulate the signal coming from the tag to extract the useful data.
- *Circulator or Directional Coupler.* This is required to differentiate and canalize the weak received signal and the strong transmitted signal. Interrogators that use a single antenna use a circulator operating in very rapid succession to carry out this operation.

The antenna in the interrogator is used to transmit and receive signal using EM waves. The antenna's radiation pattern, polarization, and gain play a very important role in proper functioning of the interrogator and properly defining the IZ.

Processor. This part is controlled by a processor that is essentially some digital signal processing chip inside the interrogator. This part controls the communication with the middleware or back-end, runs the operating system of the interrogator, and controls the memory of the interrogator.

Besides the three main parts, the reader has an external interface. The interrogator has some external interfaces that are communication ports such as RS232, USB, Bluetooth, Wireless network interfaces, local area network (LAN) ports, and so on. Proper selection and configuration of the external interface part is very important for mobile readers such as handheld, vehicle-mounted, and so on.

2.11.1 Reader Antennas

Reader antennas in general are similar to tag antennas. But here gain has to be larger, and size is no big issue. A larger reader antenna does not necessarily mean a longer read range. In fact the optimal antenna radius for a given read range is $R = \sqrt{2}x$, where R is the reader antenna radius and x is the interrogation distance [1]. The same principle applies for the inductance of the reader coil antenna. Increasing the number of turns does not increase read range infinitely. Optimization is also required in this case.

The read range of the reader is affected by the following:

- Operating frequency and performance of antenna coils
- The Q-factor of the antenna and the tuning circuit of the antenna
- Antenna orientation
- The excitation current of the antenna
- The sensitivity of receiver
- Coding and decoding algorithm
- The number of data bits and detecting algorithm
- The operating environment (electrical noise, etc.) of the RFID system

Some reported developments in research related to the reader antenna are presented below:

A dual-polarized broadband aperture-coupled microstrip patch antenna has been designed whose operating frequency is 5.9 GHz. The antenna has two orthogonal polarizations and maintains a high degree of isolation between transmitter and receiver ports. It has a 7-dBi gain and 70° half-power beam width [31]. A very low profile spiral curl antenna is designed using an EBG ground plane. Good circular polarization can be achieved and antenna is a thin one. The antenna has certain drawbacks such as reduced gain, axial ratio bandwidth, and input match bandwidth [32]. A dual linearly polarized, compact, aperture coupled microstrip patch antenna array has been designed at 2.45 GHz. The 2 × 2-array uses a dog-bone-shaped symmetric coupling aperture. The array yields 14.6-dBi gain and 36° half-power beam width [33]. A five-element rectangular patch antenna array has been used for 2.45-GHz operation as an intelligent beam scanning array for RFID readers [34].

Types of Interrogators. Depending on their size, mobility, and fixture, interrogators are basically of the three following types. The nature of the antenna varies vastly, depending on the interrogator types [2].

Stationary/Fixed Interrogators. These are generally kept at different points of the manufacturing process: dock doors, the point of exit, and so on. The whole structure is fixed and hence there is a lot of room for providing antenna arrays or a number of synchronized antennas kept at different angles. Here the reading operation is generally quick and independent of the orientation of the tags. Two different antennas are generally used for transmitting and receiving data to and from the tags.

Mobile Handheld Interrogators. These are handheld instruments that have the smallest size of antenna. Read rate is slower and reading has to depend considerably on the wrist movement of the interrogating person. Normally, linear antennas connected to circulators are used in the readers. These interrogators are connected to the back-end computer system through external interfaces.

Vehicle-Mounted Interrogators. These are generally fixed to vehicles such as forklifts, cargo trucks, and so on, to facilitate automated movements of goods in the business process. They can have large or small shapes, depending on the vehicle. As the vehicles have to undergo considerable jerks, the interrogators have to be rugged and given protective covering. These interrogators are connected to the back-end computers by wireless network or LAN.

Interrogator Interference. In the IZ, the interrogator often comes across various types of RF-related interference that has to be properly resolved [1, 2]. In a *single-interrogator environment*, interferences from multipath reflection plays the dominant role, and proper RF shielding with ground chain and anechoic foam, and so on, can mitigate this problem. In a *multiple-interrogator environment* where only shielding cannot stop the interference, assigning of specific channels, time slices for operation, proper synchronization of the duty cycle of the antennas, and so on, can solve the problem. In a *dense reader environment* where 50 or more readers (USA standard) are operating in the same area, the interrogators have to be capable of operating with proper spectral allocation, time slicing, hardware timing, software synchronizing, and "listen before talk (LBT)" mode of communication. A dedicated antenna should be kept whose job is to listen to the interrogators and thereby permit any interrogator to start its interrogation procedure only when others are not interrogating. Also, power levels of the adjacent antennas have to be tuned to minimize overlap of the interrogation area of each antenna.

2.12 STANDARDS, REGULATIONS, AND MANDATES

Like any other mature technology, to offer operability between different manufacturers and geographical areas and to also provide quality, it was necessary for standards, regulations, and mandates to come into existence in RFID technology. These standards started being formulated only about the last three years, but with their coming in vogue the technology got a huge boost globally [1, 2].

- *Standards.* These are created by different organizations to facilitate operability among different components manufactured by different manufacturers. They are applicable to hardware, software, and their usage and thus define a definite quality of the products. Standardization organizations may be international such as ISO, IEC, ITU, and so on, or national such as ANSI (USA) and JISC (Japan), and so on, or may be industrial such as EPC Global, AIAG, and so on. In RFID technology, EPC Global Gen 2 Class 1 RFID operation conforms to the ISO 18000–6C standard and is the global standard for UHF operation.
- *Regulations.* Regulatory bodies make the regulations. They are national in nature, and standards have to comply with the regulations to operate in that particular geographical area. Regulatory bodies are ETSI, ERO (Europe), FCC (USA), and so on.
- *Mandates.* These are industry-specific where the different suppliers have to comply with these to carry on businesses with the concerned industry—for example, mandates by Wal-Mart and US DoD, and so on.

2.13 ANTI-COLLISION

One important part of the standards, regulations, or mandates is the anti-collision features available in RFID technology. When operating with multiple tags in the vicinity of a single reader, multiple readers in the vicinity of a single tag, and multiple readers in the vicinity of multiple tags, the efficient reading of tags is impossible without a suitable anti-collision algorithm [1, 2, 35, 36]. The algorithms facilitate the proper reading of the tags without the phenomena such as no-read, ghost read, multiple read, false read, silent-tag, and so on. Classification of different anti-collision algorithms present in the RFID technology is shown in Figure 2.7.

Anti-collision algorithms can be broadly divided into "deterministic" and "probabilistic."

- In a **deterministic algorithm** such as binary tree search, each bit of the tags is queried individually, and identification of all the tags takes place as a result of that. This is a slow reading process.
- In a **probabilistic algorithm** such as ALOHA, the tags transmit their identities at random intervals of time and thereby read as a whole. The process is quicker but leads to more collision and keeps a bit of room for ambiguities.

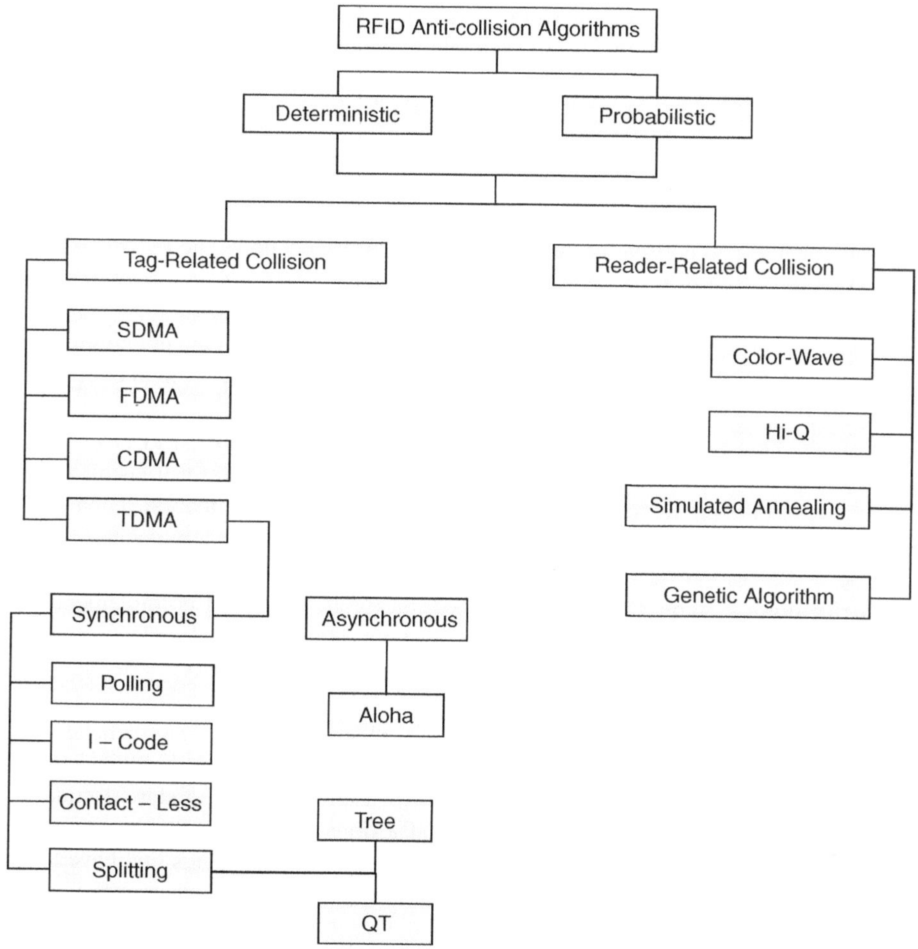

FIGURE 2.7. Classification of different anti-collision algorithms present in RFID systems.

The other methods of classifying anti-collision protocols are tag-driven and reader-driven.

Tag-Driven

These protocols govern the reading of multiple tags in the vicinity of a single reader. Combined with reader-driven protocols, they carry on the running of the entire RFID system.

- *SDMA*. Space Division Multiple Access, where the tags are identified in space by steering the beam electronically (smart antennas) or manually (handheld readers).

- *FDMA*. Frequency Division Multiple Access, where several transmission channels at different frequencies are available to the tags in the IZ.
- *CDMA*. Code Division Multiple Access or Spread Spectrum method, where the modulation technique uses pseudorandom codes spread over the entire available frequency spectrum. This technique is, however, used rarely due to added complexity of the system and limited data-handling capacity of the tags.
- *TDMA*. Time Division Multiple Access where the channels for communication are available at definite periodic timeslots. TDMA occupies the bulk of RFID systems currently.

In TDMA systems, when the protocols are driven by the tags, it is **asynchronous** in nature like Aloha. When the protocols are driven by the reader, they are **synchronous**.

- *Aloha*. This is a tag-driven protocol where the "Tags Talk First (TTF)." This system is probabilistic in nature too, where the tags randomly transmit their data until the reader acknowledges the receipt of their identification. Then they are muted or slowed down. This is thus a slow and inflexible approach.

In **synchronous** systems the "Readers Talk First (RTF)" protocol is implemented. Further classifications of the synchronous systems are: Polling, I-code protocol, Contactless protocol, and Splitting. The protocols are described below.

- *Polling*. This is a deterministic approach where the reader has to have the database of all the possible tag identifiers. This database may be dynamic in nature too. The reader takes up any identification number and starts to query all the tags with only one binary digit at a time for any specific position of the identification number. Responses of the tags are noted. Tags that have that specific digit in that specific position await further response, whereas the ones that have not are muted for the time being. This is also a slow process.
- *I-code*. This is a probabilistic approach based on slotted Aloha, which is used for identification of passive tags. Tags transmit their information on slots that are chosen randomly by the reader. The process is wholly a stochastic one where the frame size and number of slots are determined by the reader after every reading response cycle.
- *Splitting*. This is basically a deterministic type of protocol where the tags are grouped under different nodes, depending on the presence of "0" and "1" in any particular position of their identification numbers. Starting from the left or right, the reader continues the search for "0" or "1" in any particular position of the tag's identification number. In case of the presence of multiple tags with the same binary digit at any particular position, the ones with "0" (zeroes) are treated as a node at a specific level and are made mute/ignored for the time being and the ones with "1" (ones) are further queried upon for the next position. The query process goes until the end of the tree and then it retraces back using recursive methods until all the tags in the interrogation zone are traced.

- *Tree.* This protocol is essentially a subclass of the splitting protocol. Here the tags in the IZ are grouped into a number of subsets, depending on "0" or "1" in any particular position. This is a continuous approach where the number of tags in any subset decreases in number until it reaches "1." It is read, and then retracing takes place on "first in last out" basis.
- *Query Tree (QT).* This is another subclass of the splitting protocol where the current response of the tag only depends on the current query of the reader. The reader starts the query with a particular prefix and waits for a response. If single or multiple tags respond for a given prefix, the reader sends the next query with a longer prefix until all the tags in the IZ have been uniquely identified. In case there is no response for any given prefix, the entire search excludes any query that might follow that prefix and thus saves a lot of time. When a tag is identified in any given read cycle, its quiet bit is set to "1" to prevent it from responding in that given read cycle. At the next read cycle, the first response of the reader to all the tags is Answer to Reset (ATR); that is, quiet bits are reset to "0."

In EPC Global Gen 2 RFID technology [1], the QT protocol is the standard one. One improvement done in this standardization is to remove the quiet bit and instead replace it with a flag identifier that instead says whether it has been identified or not. This change is done to improve the read timing and remove any read errors that might arise from the failure of the tags to wake up before the query process starts.

- *Contactless.* This protocol is based on splitting algorithms where one bit of the identification code is identified in each query step. As a result, after the entire query process, all the tags in the IZ are identified uniquely.

Reader-Driven

These protocols define the query process of the readers when multiple readers operate in close proximity, and their interrogations zones overlap. These protocols are designed to minimize reader interference and multiple read of a single tag by several readers in its proximity.

- *Color-Wave.* In this protocol, colors symbolize the available timeslots for the readers to carry on their queries. The readers are synchronized and randomly choose timeslots to carry on their queries. If the timeslots of two readers collide, both of them are stopped from querying and they have to send new queue request. A dedicated antenna is kept to listen to the readers for that reason (Listen before Talk).
- *Hierarchical Query (HiQ).* This is a probabilistic approach. The readers are kept in hierarchy and synchronized. Before starting its query process, any reader should "ping" the neighboring readers to know of their reading status. The information received by "ping" is passed up the hierarchy for synchronization purpose.

- *Simulated Annealing (SA).* This is a kind of local search where the algorithm starts with a initial solution and thereafter makes random changes to channel assignment, depending on the response from the neighboring readers.
- **Genetic Algorithm (GA).** This is a kind of blind local search. This algorithm takes up a number of solutions initially and, based on their performance, carries on computation until a solution is found where no interference is present.

2.14 RFID MARKET TREND

So far the comprehensive technical aspect of the RFID technology has been presented. Starting from the brief historical perspective of FFF and Stockman's data transmission by means of wireless communication to the detailed technical specifications, tag and reader architectures, standards, and, finally, the anti-collision protocols to improve multiple tag reading in a complex environment have been covered in the preceding sections. No doubt that the RFID technology has already created a significant impact in the market. The reliable prediction of IDTechEx for 2009's RFID market volume was of $5.56 billion [37]. This includes the sales of RFID tags, readers, and related software and services. In this section the underpinning market trend to support such a large market volume is presented.

As stated earlier, currently the HF tags occupy the biggest RFID market with big acceleration in the UHF RFID tags after its global standardization. However, none of the RFID tags are currently able to compete with barcodes in terms of price, even if they are produced in billions [2]. As a result, the new proposed RFID technology is "chipless RFID" [38–42]. In chipless RFID, the most expensive part of the RFID tag—the IC chip—is removed. Definitely, as a result, the functionality of the tag is reduced to quite some extent, but what is left is more than enough for most of the business processes concerning everyday consumer items. The bulk prices of the tags are projected to be a fraction of a cent.

Additionally, chipless RFID tags can be printed directly on the casing and thereby reduce extra overhead and valuable manufacturing time and expense. In 2006, sale of chipless RFID was around 0.4% of the total RFID sales, which accumulated to 100 million tags [38]. Item level tagging of trillions of consumer goods ranging from everyday items to letters, documents, postal stamps, banknotes and so on, is projected to have a meteoric shoot by 2017 and is projected to be 62% of the total RFID market [38, 40, 42]. Figure 2.8 shows the current market share of the different kinds of RFID systems. As can be seen from the figure, because the RFID systems range from the conventional access control or car immobilizers to the item level identification in supermarkets, the application of the RFID technology could have been so huge if the technology providers could provide tags with less than 0.1 cents. This will create revolution in the RFID market, replacing trillions of barcodes printed each year [38]. Only a chipless RFID system can provide this cost-saving revolutionary technology.

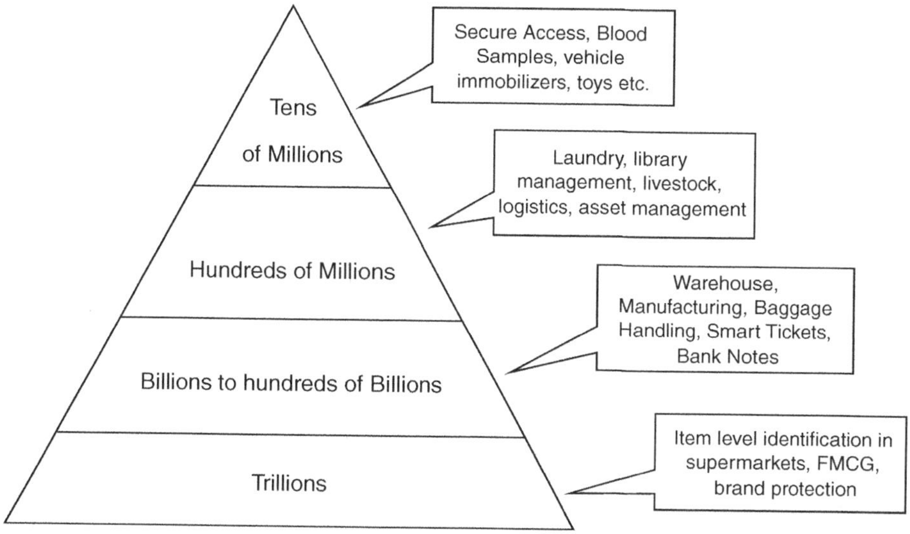

FIGURE 2.8. Market segments of RFID systems.

2.15 CHIPLESS RFID SYSTEMS

Because chipless RFID is a very nascent technology, most of the research or development is either confined within the laboratories or is proprietary in nature. This technology requires a real smart reader that handles all the relevant business operation just on obtaining the specific identification of the tag. Because this technology is currently confined to a small read range, no literature is available for any anti-collision algorithm that might be used in this technology [38–47]. A classification of the present advancements in this technology is given in Figure 2.9. However, the smart antennas with agile beams and nulls steering capabilities, adaptive directionality, and capability to enhance signal-to-interference ratio have significance for the successful reading of chipless RFID tags. The aforementioned smartness of the RFID reader for the chipless tags may come from the smart antennas.

Development in chipless RFID technology is divided into the two following generations:

- First Generation—single service providers without any standards.
  - Very little memory handling capacity and used for anti-counterfeit, anti-tamper, in-house tracking, automated error prevention, and so on.
  - Acoustomagnetic, electromagnetic, and LC arrays are some of the technology involved in these tags.
- Second Generation—multiple service providers and conformity to global standards.

- Higher data handling capacity and targeted for barcode replacement.
- Surface acoustic wave (SAW), TFTC (both polymer- and silicon film-based), secret VTT/Panipol printed pyridine labels are some of the technologies involved in these tags currently.

Over 35 big giants such as Canon, Xerox, Philips, IBM, and so on, are involved currently on the development of printed organic electronics on plastic film. However, with electromagnetic technologies being a mature technology, a lot more companies are directly involved in innovation and development of EM chipless RFID tags. Companies such as Flying Null, Link Sure, Zebra Technologies, Holotag, Fuji, and so on, are the big players for this technology. The main aim of these companies is to produce second-generation chipless RFID tags using EM techniques.

Both the first- and second-generation chipless RFID tags work on the principle of remote magnetic (including EM technology), transistorless, or transistor circuits (Figure 2.9). Chipless tags working on the Barkhausen principle [48, 49] or the magnetostriction principle [50, 51] rely solely on the magnetic properties of the material of the tag. However, chipless tags working on EM technology [43–45] rely on EM waves for the transmission or reception of the tag identity. Chipless RFID tags using EM technology have variations such as microstrip/printed resonator circuits [39], chemical resonator circuits [45], SAW [43] circuits, and arrays of printed radars [45]. Some chipless RFID tags use thin-film transistors, which are organic or polymer in nature [39, 42]. However, diode-based [52] or coil capacitor (LC) based [53] tags do not use any transistors inside the tags.

Some of the techniques involved in chipless RFID are as follows:

Barkhausen Effect. In these tags [48, 49, 54], specially processed cylindrical ferromagnetic wires of very small diameter (~0.2 mm) are used as the data carriers. The data-carrying wires are arranged at different distances with respect to a reference line. The reader has a reference wire. Introduction of a magnetic field (from the reader) and a subsequent sharp reversal of the magnetic field results in a very short duration (~10 μs) induced voltage (Barkhausen effect) in the reference wire of the reader.

The different data-carrying elements induce different voltages in the reference wire due to their different distance and orientation. The induced voltage of the individual wire elements is the bit representation of the tag's identity. The tags are single bit and work in very close proximity (2.5-cm range).

Magnetostrictive. In this chipless RFID technology, the magenetostriction property of ferromagnetic materials is utilized [50, 51]. Scientific Generics Ltd. of Cambridge, UK [14, 54] has developed a commercial chipless RFID tag utilizing this technology, also known as the Programmable Magnetic Resonance (PMR) tag. The tag contains both data (identity)-carrying hard magnetic material and soft magnetostrictive material in very close proximity. The data on the hard magnetic material is written by the contact method. Small strips of magnetic foil are used for multiple data content.

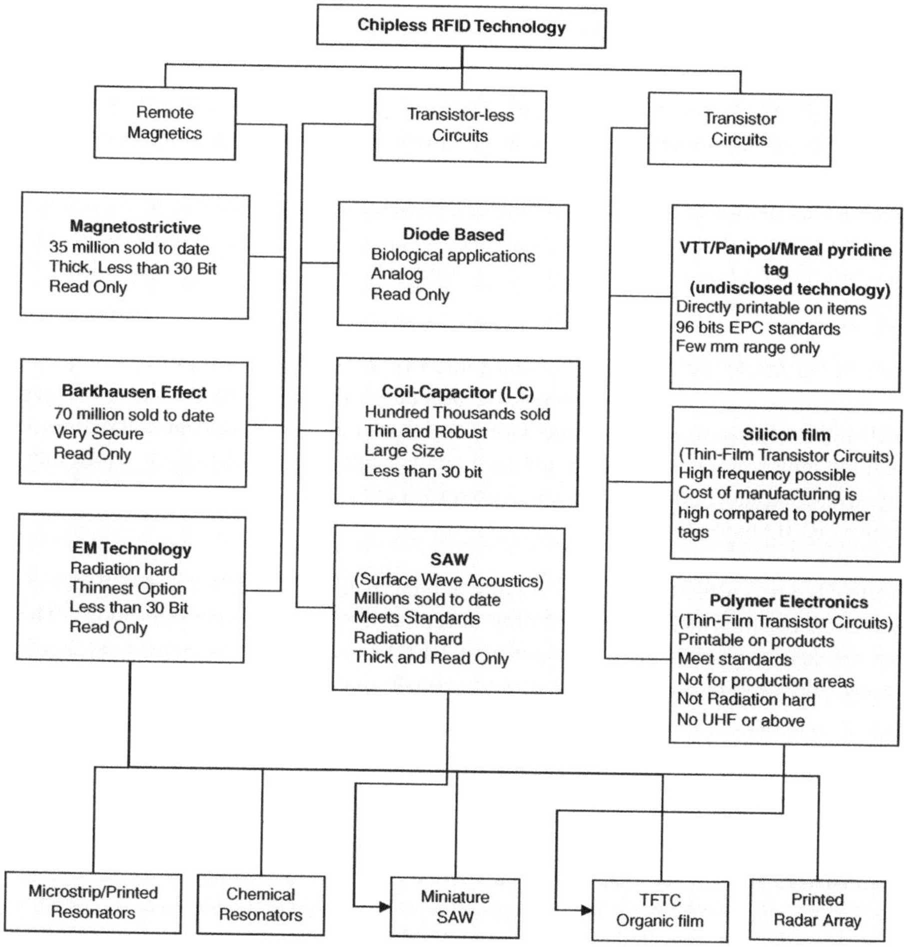

FIGURE 2.9. Classification of different kinds of chipless RFID systems and technology.

The tag may be interrogated using a ramped magnetic field or an ac field or a combination of the two. In response to the interrogation signal, the soft magnetostrictive material generates mechanical vibration (oscillation). Based on the data content of the hard magnetic material, certain harmonics of the vibration are disabled. The disabled or enabled vibrations are detected by the reader as the identity of the tag. The resonators in this tag need space to vibrate freely. Hence, these tags cannot be printed directly on the item to be tagged [14].

Microstrip/Printed Resonators. In this technology [46], the different resonant structures are printed using conductive ink on polymer surfaces. The resonant structures reflect the EM waves at definite predetermined frequencies. The presence or absence of the definite structures determines the frequency signature of the RFID tag [55–57].

Chemical EM Resonant Structures. A company named CrossID of Israel [44] has developed a chipless RFID technology in which certain chemical fibers are used to carry the frequency signature of the tags. The fibers are tiny chemical particles with varying degrees of magnetism. When impinged with broadband EM waves, the fibers resonate at definite predetermined frequencies. Seventy-bit encoding of the tag is reported to be possible using this technology. However, the read range is limited to within a few centimeters.

Printed Radar Arrays. A US company named Tapemark [45] has reported a chipless RFID tag that uses only arrays of passive antennas. The antennas are made of very small fibers (5 μm in diameter and 1 mm in length) and have a proprietary design. Each antenna resonates at a different frequency ranging from 24 GHz to more than 60 GHz. Companies like Inksure and Vubiq [39] are also reported to have developed chipless RFID using this technology.

Diode-Based. A New Zealand company named Sandtracker [52] developed a chipless RFID tag in which the silicon chip is replaced with quartz crystal diodes. The tags are less expensive than conventional RFID tags and have a long read range (18 m), and 4000 tags can be read simultaneously in the vicinity of a single reader.

TFTC. Thin-film transistor circuits with conventional printed antennas are currently being developed by companies such as Phillips and Motorola [39, 42] under the name of "Plastic Tags." However, the plastic/organic semiconductors have an inherent problem of lower electron mobility that inhibits the incorporation of a large number of transistors and slower read rates [13, 14].

Miniature SAW Tags. This is a hybrid technology in which the piezoelectric effect is combined with printed EM technology for chipless RFID tags. This technology offers enhanced memory and longer read range [43].

Flying-Null Inc. (www.flying-null.com) [54, 58] has developed a chipless RFID system in which the tag uses a strip of very-high-permeability magnetic material. The data are encoded in the high-permeability material by reducing the magnetic permeability at selected regions of the material (null points). The selected reduction of the permeability is achieved by the placement of a data-carrying hard magnetic strip in close proximity to the high-permeability material. In the interrogation process, a sufficiently high-intensity alternating (~2 kHz) magnetic field is used to saturate the high-permeability material. However, at the null points, saturation does not take place due to the two opposing fields.

An additional low-amplitude and low-frequency (1/100th) alternating magnetic field is used for scanning the data (scanning field). This scanning field drives the null region into and out of saturation, making the null regions radiate at harmonics of the scanning frequency. The arrivals of the harmonics are precisely recorded in time leading to the estimation of the spatial distribution of the nulls. The relative spatial distribution of the nulls carries the identity of the tag. The estimation is accurate up to 50-μm distance between two consecutive nulls. The technology is deemed to be a magnetic equivalent of the optical barcode [58].

Holotag Inc. (www.holotag.co.uk) has developed a chipless RFID system based on thin-film magnetic materials. The tags consist of multiple thin layers of metal alloy arranged at small incremental angles (~3–4°) [54] and capable of magnetization in a very directional manner. The data are encoded by the presence or absence of directional magnetization in the individual layer. During the reading process, the reader rotates a magnetic interrogation field and detects the magnetic interaction in any given direction.

Mukherjee [59, 60] has reported a novel chipless RFID system where a chirped (short duration) RF signal of wide bandwidth is impinged on a chipless RFID tag. The backscattered signal from the tag is further mixed with the transmitted chirped RF signal to generate an intermediate frequency (IF) signal. Next, the IF signal is further processed with a narrow detection bandwidth to identify the given tag. The tag is identified by uniquely matching both phase and frequency signatures.

Jalaly and Robertson [61] have reported a replacement for the barcode using capacitively tuned multi-resonator circuits. They utilized a cascade of open-circuited printed dipole resonators. Each dipole resonator has a narrow gap at the center. The outer dimension of all the dipole resonators was kept the same. However, the width of the central gap was varied. The gap contributed to the capacitance, whereas the thin metal arms contributed to the inductance. Thus, a series of resonances of different resonant frequencies were obtained. The resonators were bombarded with high-power RF signals of wide bandwidth. The backscattered signal from the resonators was captured by a high-gain microstrip reader antenna array. The captured signal was analyzed in the frequency domain. The presence or absence of any particular resonator signified 0 or 1. However, smeared detection of almost overlapping and not very distinct resonances calls for further improvement of the design.

Vemagiri et al. [62] reported a novel technique for chipless RFID tag where transmission-line-based delays are incorporated in the tag. The tag identification is done in real time, where the signal through the tag is superimposed with the signal through the delay line, thereby creating a unique identity.

Zheng et al. [63, 64] developed a fully printed RFID tag on flexible substrate using distributed circuit elements and discontinuities. Their principle can be compared with a high-resolution time-domain reflectometer where the ID is coded in the form of discontinuities. Ultra-wideband (UWB) signal was used for interrogation, and tag detection was carried out in the time domain.

From the preceding literature review on chipless RFID tags, it is clear that significant emphasis should be placed on the development of very-low-cost chipless RFID tags. The main technologies currently involved are magnetic materials, thin-film polymer transistors, and RF SAW and EM tags involving chemical and printed microstrip technology. Microstrip resonant dipoles, delay lines, and discontinuities are generally used in printed EM chipless tags. However, these technologies are evolving and need refinement to reach maturity. Each existing technology has its limitations

TABLE 2.2. Comparative Study of RFID Technology with Other Present Wireless Communication Technologies [65]

	SRD		WLAN		WWAN	
Parameters	RFID	Bluetooth	802.11b	802.11a	GSM/GPRS	IS2000
Multiple access	FHSS/TDMA	FHSS/TDMA	FDM/CSMA	FDM/CSMA	FDMA/TDMA	CDMA
Frequency band	125 kHz to 2.5 GHz	2402–2484 MHz	2402–2484 MHz	5.2–5.8 GHz	800–2000 MHz	1885–2200 MHz
Data rate	1–200 kbit/s	1 Mbps	1–11 Mbps	1–54 Mbit/s	14–115 kbit/s	100–2000 Mbit/s
RF modulation	ASK/FSK/PSK	FSK	DQPSK	OFDM	GMSK	QPSK BPSK
Transmission power	1 mW to 4 W	1–100 mW	100 mW	100 mW	2 W	600 mW
Typical range	20′ (passive) 400′ (active)	30′–100′	400′	200′	20 miles	12 miles

and advantages. Therefore, there is a demand for a new approach to chipless RFID tag development that can truly replace optical barcodes by offering features that will address the limitations of optical barcodes such as the need for human intervention and line-of-sight of the reader.

2.16 CONCLUSION

After making a detailed survey of the RFID technology, a conclusion can be arrived that the technology holds ample promise for the future and has lots of room for its development. Although an overview of the whole technology has been presented, emphasis has been given on the physical layer where EM waves are concerned. A comparative study between RFID and other existing communication technologies are presented in Table 2.2 [65] to have an overall idea of the standing of RFID technology compared to the other wireless communication technologies around us.

REFERENCES

1. K. Finkenzeller, *RFID Handbook*, 2nd edition, John Wiley & Sons, Hoboken, NJ, 2003.
2. M. Myer, M. Brown, S. Patadia, and S. Dua, *Passport for Comptia RFID+ Certification*, McGraw-Hill, New York, 2007.
3. H. Stockman, Communication by means of reflected power, *Proc. IRE*, pp. 1196–1204, October 1948.
4. B. Scher, Understanding RFID Frequencies, internet whitepaper, Dynasys Technologies Inc., www.dyna-sys.com.
5. C. Diorio, Class 1 Gen 2 UHF RFID, proprietary presentation copy of Impinj, Inc., Seattle, USA, internet archive.
6. K. V. S. Rao, An overview of backscattered radio frequency identification system (RFID), in *IEEE Asia Pacific Microwave Conference*, Vol. 3, December 1999, pp. 746–749.
7. L. Junhua, Y. Kun, Z. Chun, and W. Zhihua, A transponder IC for wireless auto identification system, in *5th International Conference on ASIC,* Vol. 2, October 2003, pp. 1114–1116.
8. F. M. Yasin, M. K. Khaw, and M. B. I. Reaz, Radio frequency identification: Evolution of transponder circuit design, *Microwave J.*, Vol. 49, No. 6, pp. 56–66, 2006.
9. J. Curty, M. Declercq, C. Dehollain, and N. Joehl, *Design and Optimization of Passive UHF RFID Systems*, Springer, Switzerland, 2007.
10. P. Villard, C. Bour, E. Dallard, D. Lattard, J. De Pontcharra, G. Robert, and S. Roux, A low-voltage mixed-mode CMOS/SOI integrated circuit for 13.56 MHz RFID applications, in *IEEE International SOI Conference*, October 2002, pp. 163–164.
11. R. Glidden, C. Bockorick, S. Cooper, C. Diorio, D. Dressler, V. Gutnik, C. Hagen, D. Hara, T. Hass, T. Humes, J. Hyde, R. Oliver, O. Onen, A. Pesavento, K. Sundstrom, and M. Thomas, Design of ultra-low-cost UHF RFID tags for supply chain applications, *IEEE Commun. Mag.*, Vol. 42, No. 8, pp. 140–151, August 2004.
12. D. J. Hind, Radio frequency identification and tracking systems in hazardous areas, in *Fifth International Conference on Electrical Safety in Hazardous Environments*, April 1994, pp. 215–227.

13. P. F. Baude, D. A. Ender, T. W. Kelley, M. A. Haase, D. V. Muyres, and S. D. Theiss, Organic semiconductor RFID transponders, in *IEEE International Electron Devices Meeting*, December 2003, pp. 811–814.
14. D. Redinger, S. Molesa, Y. Shong, R. Farschi, and V. Subramanian, An ink-jet-deposited passive component process for RFID, in *IEEE Trans. Electron Devices*, Vol. 51, Issue 12, pp. 1978–1983, December 2004.
15. M. Usami, An ultra-small RFID chip: /spl mu/-chip, in *(Advanced System Integrated Circuits) Proceedings of IEEE Asia-Pacific Conference*, August 2004, pp. 2–5.
16. N. Raz, V. Bradshaw, and M. Hague, Applications of RFID technology, in *IEE Colloquium on RFID Technology (Ref. No. 1999/123)*, 25 October 1999.
17. J. R. Tuttle, Traditional and emerging technologies and applications in the radio frequency identification (RFID) industry, in *IEEE Radio Frequency Integrated Circuits (RFIC) Symposium*, June 1997, pp. 5–8.
18. S. K. Padhi, G. F. Swiegers, and M. E. Bialkowski, A miniaturized slot ring antenna for RFID applications, in *15th International Conference on Microwaves, Radar and Wireless Communications*, Vol. 1, May 2004, pp. 318–321.
19. G. Marrocco, A. Fonte, and F. Bardati, Evolutionary design of miniaturized meander-line antennas for RFID applications, in *IEEE Antennas and Propagation Society International Symposium*, Vol. 2, June 2002, pp. 362–365.
20. M. Keskilammi and M. Kivikoski, Using text as a meander line for RFID transponder antennas, *Antennas Wireless Propag. Lett.*, Vol. 3, pp. 372–374, 2004.
21. R. L. Li, G., DeJean, M. M. Tentzeris, and J. Laskar, Integrable miniaturized folded antennas for RFID applications, in *IEEE Antennas and Propagation Society Symposium*, Vol. 2, June 2004, pp. 1431–1434.
22. S. K. Padhi, N. C. Karmakar, C. L. Law, S. Aditya, Z. Shen, and P. Hui, Microstrip-fed slot antenna for millimetre-wave RFID system, in *IEEE Asia-Pacific Microwave Conference*, December 2000, pp. 1396–1399.
23. S. Cichos, J. Haberland, and H. Reichl, Performance analysis of polymer based antenna-coils for RFID, in 2nd *IEEE Conference on Polymers and Adhesives in Microelectronics and Photonics*, June 2002, pp. 120–124.
24. Q. Xianming and Y. Ning, A folded dipole antenna for RFID, in *IEEE Antennas and Propagation Society Symposium*, Vol. 1, June 2004, pp. 97–100.
25. L. Ukkonen, D. Engels, L. Sydanheimo, and Kivikoski, M., Planar wire-type inverted-F RFID tag antenna mountable on metallic objects, in *IEEE Antennas and Propagation Society Symposium*, Vol. 1, 20–25 June 2004, pp. 101–104.
26. L. Ukkonen, L. Sydanheimo, and M. Kivikoski, Patch antenna with EBG ground plane and two-layer substrate for passive RFID of metallic objects, in *IEEE Antennas and Propagation Society Symposium*, Vol. 1, 20–25 June 2004, pp. 82–87, 93–96.
27. J. Siden, P. Jonsson, T. Olsson, and G. Wang, Performance degradation of RFID system due to the distortion in RFID tag antenna, in *11th International Conference on, Microwave and Telecommunication Technology*, 2001, pp. 371–373.
28. EPC Global UHF Architecture of RFID Reader, Internet archive of RFID Solutions Online at http://images.vertmarkets.com/crlive/files/Images/B693AB14-8A17-4F2C-9694-022CB52744491/MA-Com%20Architecture.jpg.
29. T. Cameron, UHF RFID industry growth powered by RF technology, *Microwave Journal, Wireless Technologies 2005*: Special report, internet archive.

30. N. Roy, A. Trivedi, and J. Wong, Designing an FPGA-based RFID reader, *Xcell Journal*, 2006, proprietary of Xilinx Inc., internet archive.

31. S. K. Padhi, N. C. Karmakar, C. L. Law, and S. Aditya, A dual polarized aperture coupled microstrip patch antenna with high isolation for RFID applications, in *IEEE Antennas and Propagation Society International Symposium*, Vol. 2, July 2001, pp. 2–5.

32. G. Marrocco, Gain-optimized self-resonant meander line antennas for RFID applications, *Antennas and Wireless Propag. Lett.*, Vol. 2, pp. 302–305, 2003.

33. S. K. Padhi, N. C. Karmakar, and C. L. Law, Dual polarized reader antenna array for RFID application, in *IEEE Antennas and Propagation Society International Symposium*, Vol. 4, June 2003, pp. 265–268.

34. P. Salonen, M. Keskilammi, L. Sydanheimo, and M. Kivikoski, An intelligent 2.45 GHz beam-scanning array for modern RFID reader, in *IEEE International Conference on Phased Array Systems and Technology*, May 2000, pp. 407–410.

35. D. Shih, P. Sun, D. Chen, and S. Huang, Taxonomy and Survey of RFID anti-collision protocols, *Comput. Commun.*, Vol. 29, pp. 2150–2166, 2006.

36. J. Waldrop, D. W. Engels, and S. Sharma, Colorwave: A MAC for RFID reader networks, in *IEEE Wireless Communications and Networking Conference*, 2003, pp. 1701–1704.

37. RFIDNews IDtechEx forecasts 2009 growth for RFID market, www.rfidnews.org/2009/04/23/idtechex-forecasts-2009-growth-for-rfid-market, accessed 23 April 2009.

38. R. Das, Chipless RFID—The End Game, proprietary report of IDTechEX, 2006, internet archive, www.idtechex.com/products/en/articles/00000435.asp.

39. R. Das, Printed and Chipless RFID—Addressing the Biggest RFID Opportunity, proprietary report of IDTechEx, USA archived at www.idtechex.com/pRFID.

40. Chipless RFID, proprietary report of TagSense Inc., archived at www.tagsense.com/ingles/tec/chipless-rfid.html.

41. K. Heires and A. Kambil, Tracking RFID's Next Wave to Gain Strategic Advantage, proprietary research report of Deloitte, USA, ISBN 1-892383-25-X, Internet archive, www.deloitte.com/research.

42. R. Das, Chip versus Chipless for RFID Applications, proprietary report of IDTechEx, internet archive, www.soc-eusai2005.net/documents/presentations/pres_84.pdf.

43. C. S. Hartmann, A global SAW ID tag with large data capacity, in *IEEE Ultrasonics Symposium*, Vol. 1, October 2002, pp. 65–69.

44. M. Glickstein, Firewall protection for paper documents, internet whitepaper, www.Rfidjournal.com/article/articleprint/790/-1/1/.

45. R. Omonhndro, RFID fibers for secure applications, internet whitepaper, www.rfidjournal.com/article/articleprint/845/-1/1/.

46. RFID in Banknotes, internet whitepaper, IDTechEx Web Journal, Issue 19, August 2002, www.idtechex.com/journal/upload/sla19.html.

47. M. Philipose, J. Smith, B. Jiang, A. Mamishev, S. Roy, and K. S. Rajan, Battery-free wireless identification and sensing, *Pervasive Computing published by IEEE CS and IEEE Com Soc*, January March 2005, pp. 37–45.

48. Transponder News—EAS systems, Internet article archived at http://www.transpondernews.com/easbasic.html, accessed 10 October 2008.

49. M. J. Brady, T. A. Cofino, R. J. Gambino, P. A. Moskowitz, A. G. Schrott, and R. J. von Gutfeld, US Patent 7084770, Combination radio frequency identification transponder (RFID tag) and magnetic electronic article surveillance (EAS) tag, Intermec IP Corp.

50. L. Reindle, R. Steindl, C. Hausleitner, A. Pohl, and G. Scholl, Wireless passive radio sensors, in *Proceedings, SENSOR*, Miami, FL, 2001, pp. 331–336.

51. G. Scheerschmidt, K. J. Kirk, and G. McRobbie, Investigation of magnetostrictive microdevices, *IEEE Trans. Magetics*, Vol. 43, No. 6, pp. 2722–2724, June 2007.

52. J. Collins, New tags use diodes not chips, *RFID J.*, December 2004, Online archive: http://www.rfidjournal.com/article/view/1263/1/1, accessed 10 October 2008.

53. S. N. Carney, G. L. Lauro, E. L. Krenz, and S. Ghaem, US Patent 5446447, RF tagging system including RF tags with variable frequency resonant circuits.

54. RFID Compendium, Association for Automatic Identification and Mobility (AIM), Denmark, Internet archive at www.aimdenmark.dk/RFID_compendium.pdf, accessed 10 October 2008.

55. I. Balbin and N. C. Karmakar, Phase-encoded chipless RFID transponder for large-scale low-cost applications, *IEEE Microwave and Wireless Compon. Lett.*, accepted 28 March 2009.

56. S. Preradovic, I. Balbin, N. C. Karmakar, and G. Swiegers, Multiresonator based chipless RFID system for low cost item tracking, *IEEE Trans. Microwave Theory Tech.*, accepted February 18, 2009.

57. S. Preradovic, I. Balbin, N. C. Karmakar, and G. Swiegers, Chipless frequency signature based RFID transponders, in *38th European Microwave Conference*, Amsterdam, Netherlands, 28–31 October 2008, pp. 1723–1726.

58. M. Crossfield, Have null, will fly [data tagging], *IEE Rev.*, Vol. 47, No. 1, pp. 31–34, January 2001.

59. S. Mukherjee, Chipless Radio Frequency Identification (RFID) device, in *Proceedings, 1st Annual RFID Eurasia*, Istanbul, Turkey, September 2007, pp. 1–4.

60. S. Mukherjee, Chipless radio frequency identification by remote measurement of complex impedance, in *Proceedings, 10th European Conference on Wireless Technology*, October 2007, Munich, Germany, pp. 249–252.

61. I. Jalaly and I. D. Robertson, Capacitively-tuned split microstrip resonators for RFID barcodes, in *European Microwave Conference*, Vol. 2, Rome, October 2005, CD ROM.

62. J. Vemagiri, A. Chamarti, M. Agarwal, and K. Varahramyan, Transmission line delay-based radio frequency identification (RFID) tag, *Microwave Optical Technol. Lett.*, Vol. 49, No. 8, pp. 1900–1904, August 2007.

63. L. R. Zheng, M. B. Nejad, S. Rodriguez, L. Zhang, and H. Tenhunen, System-on-flexible-substrates: Electronics for future smart-intelligent world, in *Conference on High Density Microsystem Design and Packaging and Component Failure Analysis (HDPC)*, Paris, January 2006, pp. 29–36.

64. L. Zhang, S. Rodriguez, H. Tenhunen, and L. R. Zheng, An innovative fully printable RFID technology based on high speed time domain reflections, *Proceedings of HDPC*, Paris, January 2006, CD ROM.

65. R. Bridgelall, Enabling mobile commerce through pervasive communications with ubiquitous RF tags, in *IEEE Wireless Communications and Networking*, Vol. 3, March 2003, pp. 2041–2046.

66. EPC Global Homepage: www.epcglobalinc.org/home, accessed 10 October 2008.

CHAPTER 3

RECENT PARADIGM SHIFT IN RFID AND SMART ANTENNAS

NEMAI CHANDRA KARMAKAR
Department of Electrical and Computer Systems Engineering, Monash University, Clayton, Victoria, Australia

3.1 INTRODUCTION

In the preceding chapter, a comprehensive review of the RFID technology was presented. The definition of the enabling automatic identification technology (RFID), its tag and reader architectures, different generations of RFID tags and readers, global standards and mandates, and, finally, the emerging application areas of the tag have been explained. In the very last section of the preceding chapter, the significance of the chipless tags compared to the expensive chipped tags has been highlighted. It is imperative that the tag must be chipless to compete with the trillions of optical barcodes printed per year. Figure 3.1 shows the projection of the RFID tag price versus the total usage of tags over 16 years starting from 2004 as reported by IDTechEx [1]. As can be seen, the RFID tag will replace the trillions of barcodes if they can be made for less than a cent. This will only be possible if the tags can be made possibly by direct printing on product or package like optical barcodes. According to the prediction, this will not happen soon and it may take more than a decade to reach to that level from its current status of more than 10 cents a tag (2009). Technical challenges in designing printed electronics at RF frequencies are the main hindrance of producing a tag for less than a cent.

The conclusion is that the tag must cost less than a cent to coexist or eradicate the barcode completely. As mentioned earlier, barcodes have known limitations—they cannot be read non-line-of-sight (NLoS)—behind an obstacle and when they are stained with soil and grease or are printed on some transparent substrates and in sunlight and dark. Each barcode needs personal care to be read, which makes the inventory control and warehouse logistics a nightmare. In addition to these limitations, optical barcodes are printed for a class of items, not for unique identification of

Handbook of Smart Antennas for RFID Systems, Edited by Nemai Chandra Karmakar

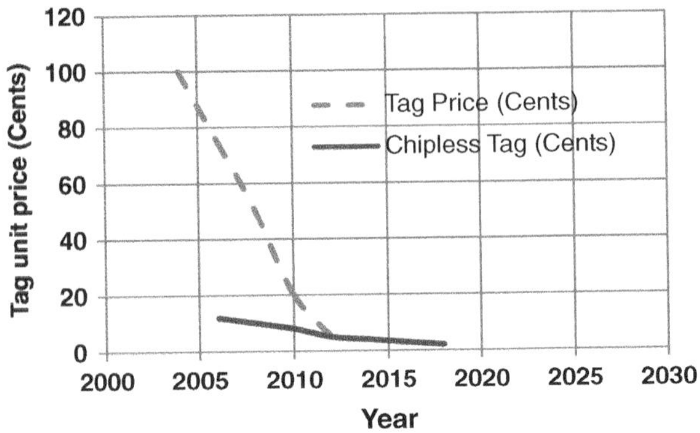

FIGURE 3.1. Prediction of enabling RFID to replace barcode. (Adapted from IDTechEx, 2009).

individual items. Contrary to the above limitations of the optical barcodes, RFID tags provide much more flexibility in terms of (a) NLoS reading capability and (b) reading behind an obstacle. Soiled tags can be read automatically, tags can be printed on many laminates, all-weather condition reading is feasible, and finally no or minimum human intervention is required in the reading process. Therefore, warehouse managements and logistics are much easier and made automated with the RFID technology. However, RFID tags still suffer from tough competition from the barcode only due to the cost. Though big RFID tag manufacturers such IBM Corp, Texas Instruments-RFID, Motorola, and Hitachi have been claiming that the tags can be made for less than 10 cents if purchased in billions, still there is a need for the development of a fully printable chipless RFID tag to compete with the barcode. The current low-cost RFID tags are made of silicon transistors in either an organic or nonorganic form. In order to create the transistor, it requires a substrate that gives the transistor with sufficient electron mobility for fast switching and modulation, and conductivity to "print" the transistor. Silicon has mobility in the order of 1000 (cm^2/V)—as opposed to that of organic polymers, which is 4 or less. Nonorganic (nano-silicon) transistors such as Kovio have mobility ranges from 100 to 500. Our Industry partner FE Technologies Pty. Ltd., based in Geelong, Victoria, Australia [2], has identified the following low-cost tags after attending Printed Electronics and Photovoltaics Europe 2009 Conference in Dresden, Germany in April 2009. The current development has offered some promises to lower the cost of the tag and make the tag more competitive for mass deployment in billions. However, the ideal solution of the low-cost tagging is the chipless fully printable tags that can be printed with inkjet and other forms of printing techniques. The biggest challenges of realizing the chipless tags are that they are microwave passive circuits that require very high precision in track widths. It is required to reduce the width and length* of the circuit

*Length is related to wavelength hence the frequency. A 60-GHz frequency band would be a good solution for multibit printable tags.

in order to maximize the information storage capacity within a credit card and/or a postage-stamp-size area. Current circuit printing techniques offer the following accuracy: offset printing 127 μm, inkjet printing minimum 20 μm, flexo printing 5 μm, and rotogravure printing 1–2 μm. It will also be possible to achieve a similar result with a paper substrate, which is much more inexpensive (1/7th) in the price compared to that of the printed electronics. However, the electromagnetic properties of paper changes with the moisture contents in air. The downside of print other than inkjet is that a method of etching the resonant structures is to be developed in order to make them unique. There are a number of suppliers that print the chipless tag roll to roll. These include GSI, Acreo, Prelonic Technologies, Inktec, and Soligie. Table 3.1 shows the current status of the low-cost tags.

3.2 PATHWAY TO LESS-THAN-A-CENT CHIPLESS TAG

According to the above discussion on the implementation of the low-cost optical barcode replaceable chipless tag, massive technical challenges must to be solved to achieve the target of a less-than-a-cent tag. The main reason is that the tag will be "dumb" without any intelligent circuitry and power source. To read the "dumb" tag, the design of the reader should carry the bulk of the intelligence. Figure 3.2 shows the pathway to achieve the goal. The reader will possess a highly sensitive receiver and advanced protocol to read multiple tags when they will be in close proximity. Advanced-level anti-collision protocol and smart antenna algorithm will alleviate the problem of tag reading, system capacity, and throughput improvement. The application-specific system will need specific designs for the RFID tag, the RFID reader, and the middleware. The book is about these designs and the outcomes of the research and studies recently churned out to mitigate many technical challenges relevant to the RFID system implementation. In the following sections, some of the outcomes are discussed. First, the generic reader architecture is presented. The antennas of the tag and readers are presented next with elaborate classifications of the antennas for the readers. The fundamental definition and classifications of the smart antennas are presented in the next section, followed by the application-specific smart antennas developed for various RFID systems. Finally, the conclusion closes the chapter.

3.3 RFID READER ARCHITECTURE

As mentioned earlier, the chipless RFID is a fully passive device. According to the fundamentals of the radar technology, the passive dumb tags are targets that need to be illuminated with EM signals by the reader's transmitting antenna so that the tag returns its distinct echoes. These echoes are received by the reader's receiving antenna either monostatically* with an antenna connected to a duplex switch or

*Monostatic radars use one antenna for transmission and reception. The radar needs a duplexer switch to switch between the two modes or a coupler to isolate the transmit and the receive paths.

TABLE 3.1. Low-Cost RFID Tag Current Status as of April 2009

Organization	Thinfilm www.thinfilm.se	PolyIC www.polyic.com	Kovio www.Kovio.com	Silicon[a]	Vubiq and Inksure: Radar ID: 60 GHz[b]
Memory Size:	15 bits	4 bits	128 bits	256-bit UHF known technology	64 bits
Make-up:	Organic printed with silver based ink, no antenna	Organic printed with silver-based ink PE, with printed chip strap attached	Nonorganic (stainless steel) with attached memory printed with silicon	Organic substrate with strap attached memory chip	Current organic substrate printed with silver ink
Source of Substrate/ Ink:	Outside suppliers include printing through Inktec	Outside suppliers, but print internally	Develop own ink and outside suppliers, but inkjet single layer (20 μm) print internally	Not applicable	Not known yet
Reader:	Proprietary near field (NF) reader	Proprietary reader, moving to HF	Claiming standard HF reader	HF/UHF	Proprietary 2.4–5 GHz
Release Date:	April 2009	2010–2012, likely due to need to develop substrate	2009	Available in market	TBD
Price:	Not known yet	Not known yet	5–10 cents, long-term 3 cents	5.8 cents in millions	TBD
Target Market:	Gaming	Gaming, brand protection, also adding value	Gaming, also wanting to work on transport	Versatile	Library

[a]NXP (the biggest silicon supplier) have just released a 1-cent chip. The current makeup of the price of making millions of the silicon-based chip is 1 cent for strap attachment, 1c for the antenna. This is the likely lowest base cost that will be achieved. With NXP releasing the 1-cent chip, this will enable the silicon-based suppliers to get their cost under 5 cents keeping Kovio out of the transport market.

[b]Other potential printed RFID technology options include one based on a magnetic compound with the makeup of the compound yielding different readings, a two-step process where you print where you want the conductive material to go and then lay a metal coating in this position.

Source: Courtesy of Robert Reeds, CEO, FE Technologies Pty Ltd., Geelong, Victoria, private communications, April 2009.

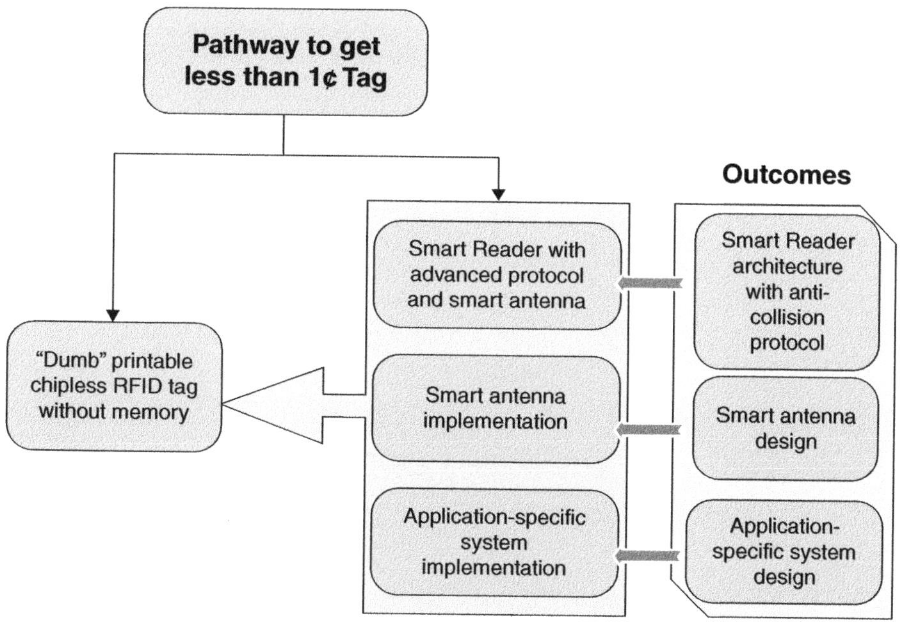

FIGURE 3.2. Document map for pathway to less-than-a-cent chipless tag.

bistatically* with a separate transmitting antenna and a separate receiving antenna. The echo from the tag is processed in the reader to a meaningful identification code. Since the passive tag has no amplifier to amplify the signal that it transmits, antennas with moderate to high gain can play a very significant role in improving the link budget and the ability to capture the optimum radar cross section (RCS) of the tag. RCS is a complex signal containing both amplitude and phase. Therefore, the returned echo may have very unpredictable characteristics due to the environmental factors (clutters). However, the agile beam high-gain directional antenna—the smart antenna—can mitigate multipath interference from the returned echoes of the tag, improve the signal-to-interference ratio, localize the tag reading, and offer many other extra features that a conventional reader with a fixed, beam antenna cannot offer. Figure 3.3 shows the generic architecture of a RFID reader [3]. The detailed working principle, design, and implementation of the reader architecture will be presented in a later chapter. An RFID reader comprises five parts. These main five functional blocks are:

1. Antenna
2. Radio-frequency (RF) transceiving electronics
3. RF interface
4. Digital control section
5. Host application and middleware

*Bistatic radars use separate transmitting and receiving antennas.

Radio-

to the reader. A part of the energy of the interrogation signal is used to energize the microcontroller and switches in a passive tag. The reader's receiving antenna receives the reply signal from the tag and amplifies the signal with an LNA. The LNA increases the received signal's amplitude before the signal is decoded in the demodulator. Another LNA can be used to amplify the signal after demodulation of the signal. Different demodulation techniques are used to decode the received data from the transponder. Most RFID systems operate using binary phase shift keying (BPSK) [4] and amplitude shift keying (ASK) [4].

3.4 RFID READER/TAG ANTENNAS

The efficacy of the RFID reader is highly dependent on the performance of the reader antenna. An efficient antenna controls the link margin and the interference for a wireless system. This is very vital for an RFID system where the transmitted signal is returned from the tag with a fourth-order reduction in magnitude of the reading distance (R^{-4}). Therefore, a slight improvement in antenna gain and directivity plays a significant role in improving reading range, accuracy, and localization of the tag. A comprehensive literature review by the author and a colleague [3] has suggested two types of RFID reader antennas:

1. Fixed-beam arrays
2. Scanned arrays

The fixed-beam antenna for the RFID reader is commonplace, and a plethora of antenna vendors are selling RFID antennas to cater the RFID market demand. The fixed-beam antenna has a unique and fixed-beam radiation pattern. Mostly panel antennas, which are microstrip patch antenna arrays, are used as the reader antennas. Most RFID readers are equipped with omnidirectional or wide-beamwidth antennas in order to cover as much area as possible as their interrogation zones. Several fixed-beam antennas are also used and can be commonly found in Alien Technology [5] and Omron [6] readers. The fixed-beam antennas are easy to install and do not need any switching electronics and associated logic control to steer their beams. However, they pick up multipath signals, interferences and are incapable of localizing the tag's position when receiving transponders' backscattered signals. This situation may lead to reading errors during interrogations. The smart antenna for the RFID reader can mitigate these problems offering directional high-gain beam toward the desired tags and nulls toward the interferers. Ingram [7] suggests that RFID tags are subject to multipath interference—an inherent problem of electromagnetic signals. Multipath interference makes the reading unreadable even if it is within the reading range of the reader. The smart antenna can solve the interference problem by electronically controling the main beam emitted from the reader's antenna toward the desired tags and nulls toward the interfering signals. This smart antenna technology incorporated in RFID readers reduces reflections from surroundings and thus minimizes degradation of the system performance due to multipath and other undesired effects.

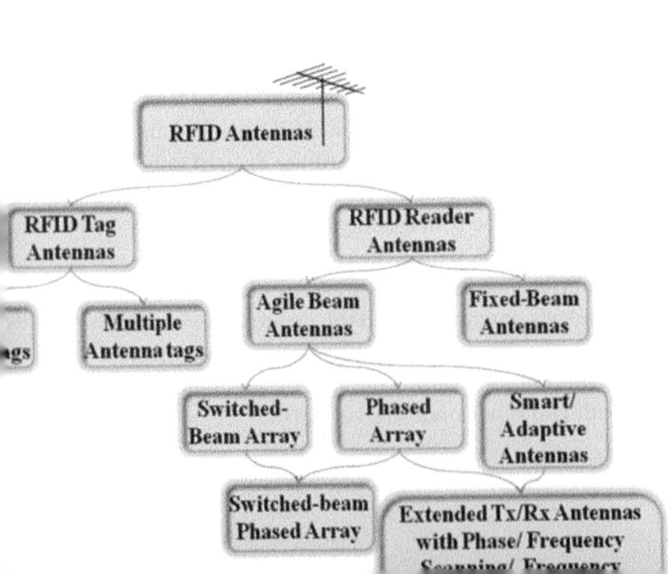
RFID Antennas
RFID Tag Antennas
RFID Reader Antennas
Multiple Antenna tags
Agile Beam Antennas
Fixed-Beam Antennas
Switched-Beam Array
Phased Array
Smart/ Adaptive Antennas
Switched-beam Phased Array
Extended Tx/Rx Antennas with Phase/ Frequency

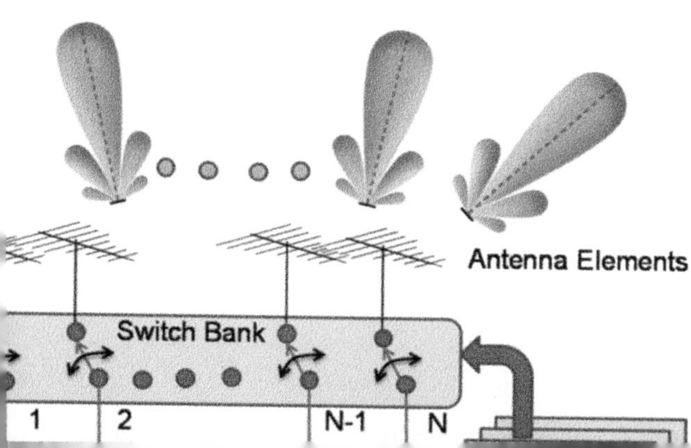

Antenna Elements
Switch Bank
1
2
N-1
N

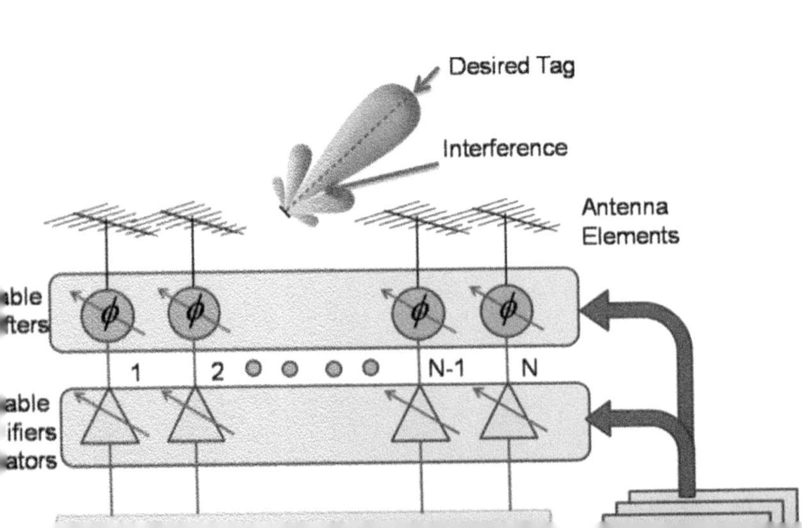

Desired Tag
Interference
Antenna
Elements
able
fters
1
2
N-1
N
able
ifiers
ators

of the phased-array antenna. As can be seen, the phased-array antennas have a set of elements which are connected to the phase shifter matrices and variable attenuator (amplifier) as opposed to the switches for the switched-beam antenna. The phase shifters are pin diode digital phase shifters. Various types are: (i) loaded line phase shifters, (ii) switched delay line phase shifters, and (iii) hybrid coupled phase shifters. These phase shifters are digital in nature because the pin diodes are used in tandem to get quantized phase bits as low as $\pm 11.25^\circ$. These digital phase shifters have the quantization errors due to its discrete nature of phase shifts. Moreover, these phase shifters are narrowband. Each phase shifter is controlled by the bias currents provided by the switching electronics and the program loaded in the CPU. The beamforming algorithm is controlled by the maximum power received from the signal, and the nulls can be steered toward the interference if the predefined directions of the interferers are known from the prior knowledge.

Switched-Beam Phased-Array Antenna

Combining the phase shifters and switches in the switched-beam antenna, a switched-beam Phased-Array antenna can be designed. A switched-beam Phased-Array antenna* composed of eight antenna elements in a conical and circular grid generates 24 beams instead of only eight beams [8]. Thanks to the phase shifters, they can control the beams in more fine angular intervals from a subarray of two elements than does the conventional switched-beam antenna. When the phase difference of the subarray elements is zero, the beam is fully on boresight direction. Introducing relative phase differences between the elements two extra beams—left and right beams—are generated.

Adaptive (Smart) Antenna

The third and most comprehensive type of the RFID reader antennas are adaptive (smart) antennas. According to Frank Gross [11], the smart/adaptive antenna refers to any antenna array that can adjust or adapt its beam pattern toward the desired signal and nulls toward interfering signals with the means of sophisticated signal processing algorithm. Adaptive arrays have the capabilities to steer the beam at any direction of interest and simultaneously steering the pattern nulls toward the interfering signal. If we assume the beam pattern of the adaptive antenna to be an inflated balloon on a child's hand, then the algorithm performs the function of the child's hands squeezing the balloon to any form factor. This means the antenna can form multiple adaptive main lobes as well as multiple adaptive nulls. This flexibility of the beam pattern by the adaptive antenna array is obtained by the complex weight vectors generated from the adaptive beamformer. As stated by Gross [11], "The smart antenna concept is opposed to the fixed-beam 'dumb antenna,' which does not attempt its radiation pattern to be adjusted to the ever-changing electromagnetic environment."

*The author and a colleague designed such a switched beam phase array (N. C. Karmakar and M. E. Bialkowski, A compact switched-beam array antenna for mobile satellite communication, *Microwave Optical Technol. Letts.*, Vol. 21, No. 3, pp. 186–191, May 5, 1999.)

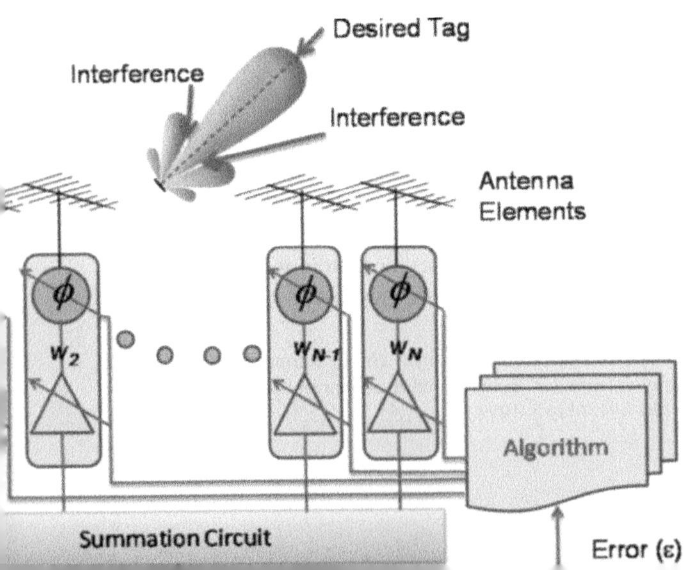
Desired Tag
Interference
Interference
Antenna
Elements
w_2
w_{N-1}
w_N
Algorithm
Summation Circuit
Error (ε)

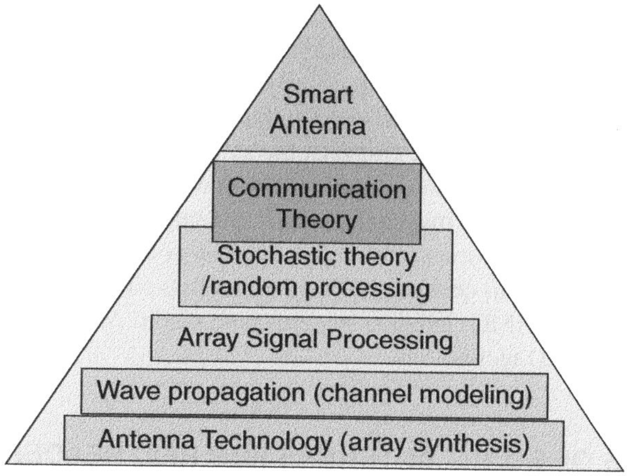

FIGURE 3.8. Supporting disciples to realize smart antenna.

structure and the weight vectors for individual antenna elements in an array are known and the radiation patterns are predicted based on the design parameters and weight vectors. The cost function represents the difference between the desired signal and the array output. The adaptive antenna is an adaptive technique of minimizing the cost function until the array output (pattern) matches the desired pattern. In other words, the goal of the signal is set at the weight vector of the antenna array and is adjusted until the goal is obtained. Thus adaptive antenna is fully a mathematical process resembling the iterative process of Newton-Raphson [12] or the Runga–Kutta method [12]. In this iterative process the initial value of a variable is assumed, and the value of the variable is changed with the algorithm until the error between the desired value and the computed values is minimized close to zero. This is the saddle point of the optimization process, and the computation is stopped once the solution is reached.

The physical realization of the smart antenna requires multidisciplinary expertise and resources. As shown in Figure 3.8, the smart antenna needs knowledge bases starting from antenna technology such as an efficient antenna element design, array synthesis and analysis, wave propagation (channel modeling), array signal processing, stochastic and random processing and communication theories such as modulation and coding, and so on. As for examples, each antenna element is connected to an analog-to-digital converter (ADC), and the algorithm is processed in a high-speed signal processor. The antenna element needs to be well-matched and located in a defined array configuration such as linear, rectangular, and circular grids. Therefore, an integrated realization of the smart antenna needs experts from all these disciplines. The cost of the antenna increases with the number of antenna elements as the number of ADC and process power increase, thanks to the burgeoning high-speed ADC and digital signal processors [11]. Now a days, ADCs with up to 1000 Gigasamples/s (GSa/s) are available in the commercial market, which makes direct digitization of most radio-frequency signals possible. Recently, high-speed signal processing is commonplace with the field programmable gate arrays (FPGAs). Most universities teach FPGA programming to their undergraduate electronics and computer engineering students. FPGA has a speed of up to 256 BMACS (billion multiply accumulates per second) [11]. Therefore, with the exponential growth of the digital technology (Moore's law) and readily available chipsets, the smart antenna integration continues to flourish with the implementation of a powerful new algorithm. This advancement in both hardware and software makes the smart antenna implementation possible in mobile wireless communications, mobile ad hoc network, mobile wireless sensor nodes, wireless metropolitan network (WMAN), satellite communications, multiple-input multiple-output (MIMO) antenna systems, and so on. And this will come soon for the burgeoning RFID technology.

3.6 BENEFITS OF SMART ANTENNAS FOR RFID READERS

Antennas are interfaces between the free space waves and the guided waves. An antenna is a transducer that converts guided electromagnetic waves from transmission lines to a free space unbounded wave in its transmission mode and converts free space waves to guided waves in its reception mode. As stated by Balanis [13] in a review on antennas, antennas are the eyes of the electronics. As we—*Homo sapiens*—communicate with the outer world with our two eyes, the radio-frequency (RF) electronics communicate wirelessly with the outer world with antennas. More precisely, smart antennas are like our ears. Our ears have an adaptive capability to listen to the desired sounds by filtering out the noises and interferences. This technique is more vivid in many animals that can twist and change forms of their ear lobes to listen to the desired sound signals. Smart antennas for RFID systems are scanned-beam antenna systems, which point beams to selective transponders within their main-lobe radiation zones, thus reducing reading errors and collisions among tags. This technique exploits *spatial diversity,* and in many cases polarization *diversity* of the tag and the reader antennas. The directed beam also reduces the effects of multipath fading [7]. Due to these salient features of the smart antennas, smart antennas can provide higher system capacity, along with reduced power consumption by directing narrow beams toward the users of interests while nulling other users not of interest. These characteristics of the smart antennas allow high signal-to-interference ratio, lower power requirement for system operation, and greater reuse of frequency bandwidth by offering the space division multiple access (SDMA) diversity. These features of the smart antenna offer significant system capacity improvement by localizing the beams in a narrow area of interest. Multipath fading and co-channel interference can be mitigated simultaneously with the smart antenna by pointing the beams in the desired signal and nulling the unwanted signals; hence, system capacity and speed (high data rate) can be improved. With the angle of arrival detection capabilities of the smart antenna, the localization of tags can be very accurately determined. This helps us to trace lost items and misplaced items and to track assets. With the RFID reader installed with multiple-input multiple-output (MIMO) antennas, the reader can exploit the uncorrelated/independent signals from individual antenna elements and improves throughput and anti-collision of multiple tag reading. In summary, smart antennas offer many benefits as follows [11]:

- Improved system capacities
- Higher permissible signal bandwidths
- Space division multiple access (SDMA)
- Higher signal-to-interference ratios
- Increased frequency reuse
- Sidelobe canceling or null steering
- Multipath mitigation
- Constant modulus restoration of phase-modulated signals
- Blind adaption
- Improved angle-of-arrival estimation and direction finding
- Instantaneous tracking of moving sources
- Reduced speckle in radar imaging
- Clutter suppression
- Increased degrees of freedom
- Improved array resolution
- MIMO capability in both communications and radar

In many RFID systems the above features of the smart antennas are exploited for various applications [14].

3.7 APPLICATION-SPECIFIC SMART ANTENNAS FOR RFID READERS

Although smart antennas for telecommunications have been in existence for some time, the smart antennas for RFID applications found their niche areas of application only recently. Therefore, the deployment of smart antennas in the RFID reader is bourgeoning and has not come as the mainstream technology for RFID. The mainstream deployment of the smart antenna in the RFID system is very slow, and only a few reported works are available in the open literature. The low acceptance of the smart antenna in RFID is due to the following reasons: (i) Smart antennas are complex systems encompassing a large group multidisciplinary expertise; (ii) RFID is a low-cost technology compared with the mobile communications. In mobile communication the system environment and scale of deployment is very different from those for the RFID system; (iii) RFID technology is application-specific—the application and the deployment of RFID for the specific application may vary significantly from other applications. Therefore, the antenna design and deployment may require individual cares; (iv) The operating frequency of the RFID system spans from extremely low frequency (ELF) of a few hundred kilohertz up to millimeter-wave frequencies of

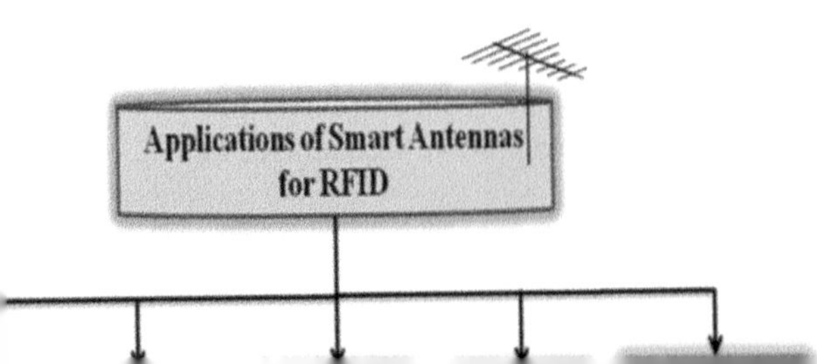

Applications of Smart Antennas for RFID

pupils, monitoring delegates past the sponsor's booths in conferences, tracking of individual animals in herds, tracking trolleys in parking lots near supermarkets and airports, monitoring residents in old edged homes, watching children in playgrounds in childcare centers, monitoring children in theme parks and shopping centers, locating cars in vast parking lots, monitoring old artifacts in museum and galleries so that they are not misplaced or stolen, reporting the movement of parcels for distribution in a courier system, tracking of luggage and finding the missing luggage in airports, monitoring the progress of commuters in the underground transit systems, tracking the golf balls in a golf course, tracking of goods in retail systems, and checkout of goods from a supermarket. The list can be endless; hence it is safe to state that the emergence of RFID technology is thriving.

The most fundamental ways of position location of tagged items are to use multiple readers in different points of entries. When a tagged item passes near one RFID scanner, the location of the object to somewhere within the read range of the scanner is determined. A more precise method of positioning is the hyperbolic triangulation. In this method an object is scanned by multiple antennas with measurable time delays in signal response to an emitted RF signal, and the comparison of the time delays of multiple scanners are used in determining the position using hyperbolic triangulation. The third approach is to move the RFID readers that identify the presence of the RFID tags in their proximity. In this case the position of the moving reader at the instance of time is known. The fourth approach is the miniature GPS-based device that can read the location of the tagged object and transmits an RF signal giving its GPS determined location. This GPS-based device is more expensive than the devices based on other methods stated above. The fifth method of position location of tagged objects is to use an item with reader which moves with time and reads fixed tags at different referenced locations. For example, a forklift with an RFID reader passes through tags embedded in the floor of a warehouse. When the forklift moves along the path, the reader reads the tag and transmits the position of the vehicle in real time. This method can also be used for tracking of emergency response personnel (such as nurses and doctors in hospitals) inside a building. The current commercial method of locating the emergency personnel in hospitals is PanGo [16]. PanGo Locator® [16] is the real-time position locator of asset and personnel in an organization. It is based on active tags, IEEE 802.11 b/g radio, and enterprise software. The system can be hooked up to the Internet for global reach between the different outlets of the same organization. Another approach to locating the tagged items is to install a reader on a transport vehicle, a set of reference tags are positioned in known locations; the tagged items to be tracked are read by the reader along with the reference tags to give the position of the tag. Ultra-wideband (UWB) tags used for position location up to 300 ft are reported [17]. Finally, a directional antenna is used to improve the range of the RFID locators. An agile-beam directional antenna scans selected areas to identify objects based on their RFID codes. For these readers, highly advanced signal processing electronics are used to recover weak signals from the tags at a larger distance. Thus the smart-antenna-enabled RFID reader can locate the tagged objects from a much larger distance than does the conventional RFID locator. For example, the US military has deployed [18] such smart-antenna-based readers to increase the read range of the passive tags to larger distances to detect military tanks, vehicles, containers, and so on, in battlefields as well as in warehouses.

Trolley Scan Pty Ltd. [19] has developed an advanced locating system comprised of a sophisticated ranging beacon capable of fixing locations of many passive tags in a reading zone. The reader has three microstrip patch antenna array modules that map the location of all the tags in its interrogation zone. The reader is capable of scanning many tags in 1D, 2D, and 3D spaces and provides an angle of arrival (AoA) of the tags with accuracy better than 1°. A chapter is dedicated on AoA using electronically steerable parasitic array radiator (ESPAR) in this book. The reader possesses a fast processing module that updates the positions of the tags at 1-s intervals. The reader interfaces with other computer networks with an RS232 serial communication interface or a local area network (LAN). Of the three antenna array modules, one antenna array measures the range of the tags, the second antenna array module transmits RF power at 860–960 MHz using a 10-kHz bandwidth spectrum to energize the passive tags, and the third antenna module is connected with the receiver. For 1D scanning, two antennas are used; for 2D scanning, three antennas are used; and for 3D scanning, four antennas are used. Two reader systems can operate within 4 m of each other and consolidate a huge amount of data to process the information about identity, range, position, and movement of the tags. Location accuracy up to 0.5 m, pointing (bearing) accuracy of 1°, and range up to 100 m for active tags are recorded with the reader. The reader is also able to track moving objects by measuring the range many times per second for each transponder. The reader also can plot the path of moving object within its interrogation zone.

Authentication and tracking of gaming chips in casinos may save tens of millions of dollars of the casino owners. In April 2009, a floor worker detected a fake gaming chip in Melbourne's Crown Casino [20]. This detection forced the authority of Crown Casino to check the authenticity of their $13.7 million worth of gaming chips after the discovery of counterfeit tokens. The casino has so far identified 36 fake $1000 chips out of 13,700 in total. Crown Casino instantly changed the color of chips (probably the RFID-enabled chips with state-of-the-art detection capabilities) used on its gaming tables to further limit the financial damage. This recent news may trigger other casinos in Australia and overseas to replace their expensive gaming chips with RFID-enabled chips so that counterfeiters cannot copy the chip. And with the smart-antenna-engineered RFID reader the location of the genuine chips can be traced in real time.

Smart antennas can play a major role to stop replacing the original chips with fake chips and monitor the movement of the chip on the table when the game is on. As shown in Figure 3.9, a sequentially fed switched beam antenna array [21] is proposed to find the location of gaming chips with embedded RFID tags for gambling applications. The antenna array is composed of multiple planar antenna elements that are located under the gaming table. The antenna generates many subsets of beams in different directions. The RFID-tagged gaming chips are located on a Blackjack, poker, or other gaming table. A subset of antenna elements is sequentially turned on and off to form sectorized directional beams to illuminate the RFID-tagged gaming chips,

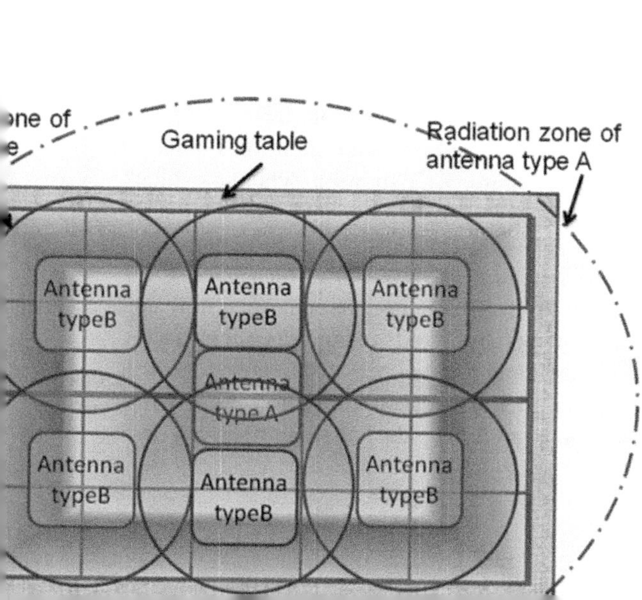

one of
e
Gaming table
Radiation zone of
antenna type A
Antenna
typeB
Antenna
typeB
Antenna
typeB
Antenna
type A
Antenna
typeB
Antenna
typeB
Antenna
typeB

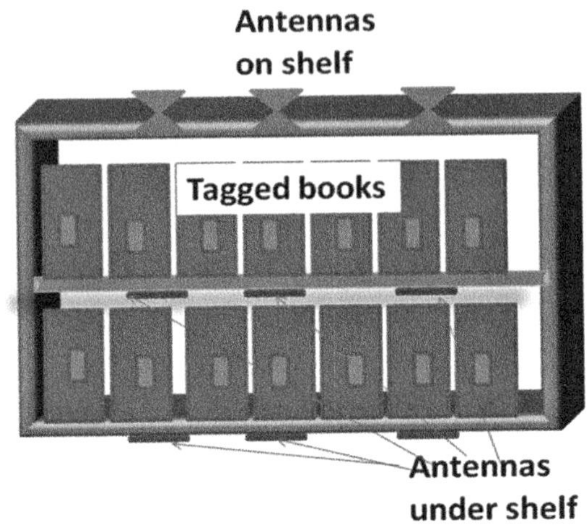

FIGURE 3.11. Library smart shelf.

Smart Shelves

Smart shelves augment the inventory checking and replenishment of items in a retail store and finding misplaced items to restock in appropriate positions. Smart shelves are also very useful in library database management for misplaced books, after-hour return chutes, stock take in, least used books, depository checking and control, and so on. Figure 3.11 shows a proposed smart shelf with multiple antennas and readers for library database management system. With the RFID-enabled smart database management system, librarians can devote more time to serve their patrons with advising reading lists and many other useful advices rather than mundane tasks. Smart shelves are instrumented shelves installed with two or more readers in suitable locations or with a system of antennas connected to a single reader. While the system with dumb antennas attached to many readers was the original development, the smart-antenna-augmented single reader provides low-cost intelligent information management for items. The development of Applied Wireless Identification Group, Hollister, California [24] reported a smart shelf with a phased-array antenna system. The antenna is in planar form and is coupled with "fast look-ahead decay sensing system" [25]. The antenna can be hidden under paper or other materials.

Lee [22] proposes a switched-beam coil antenna array configuration for RFID readers for smart shelves for retail shops. Figure 3.12 shows a schematic block diagram of the smart array reader. The reader has many coil antennas that are placed under the platform of the racks as shown in Figure 3.12. The items to be identified will be placed on the platform of the coil antenna. The switching electronics of the smart shelf reader will sequentially switch on one of the coil antennas at a time and will go through frequent recalibrations for matching and tuning depending on the load changes on the coil antenna. This load changing happens when shoppers pick up items from the shelves and then replace them after inspection, as well as when shopkeepers replenish items once the goods are sold. The intelligent sensing unit

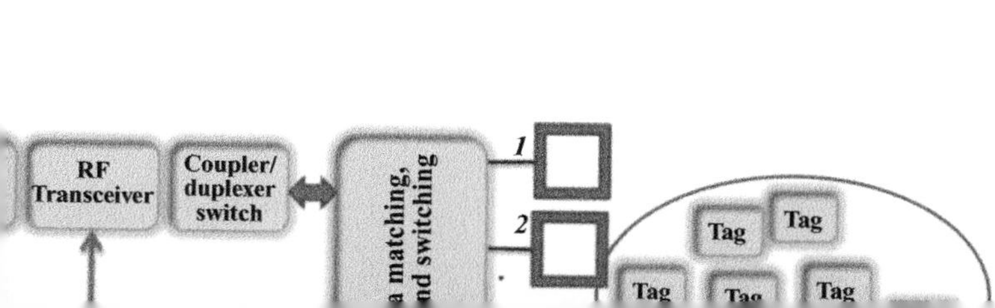
RF
Transceiver
Coupler/
duplexer
switch
a matching,
nd switching
1
2
Tag
Tag
Tag
Tag
Tag

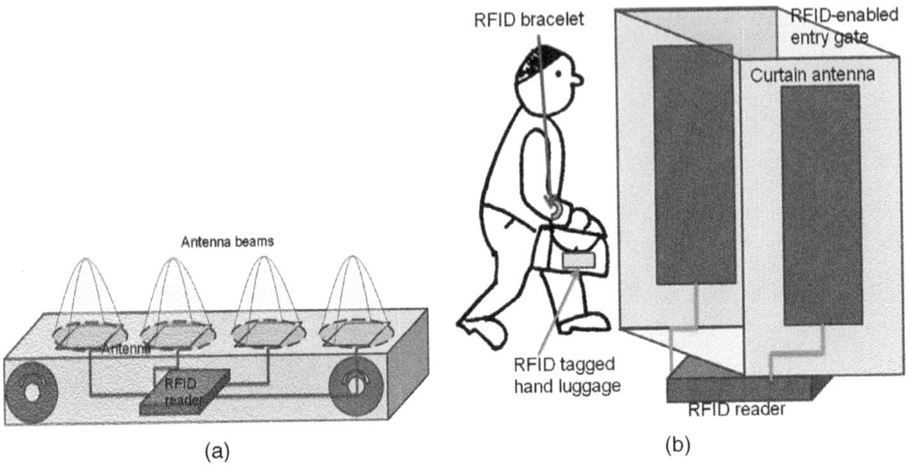

FIGURE 3.13. (**a**) RFID-enabled conveyor belt for luggage tracking and (**b**) entry gate with switched-beam curtain antennas.

tracing and tracking capabilities, and higher reliability and accuracy, RFID will impact the largest market in airlines, airports, and customs. Again, smart-antenna-based RFID readers will enhance the efficacy of the system. Chung and Liu [29] propose a few types of loop antenna arrays for detection of objects. The loop antennas are placed at different angles in different planes in either a 2D or a 3D rectangular volumetric space so that the diversity of the individual loops can be exploited to maximize the reading of RFID tags. Figure 3.13 shows such arrangements of the smart antennas for an RFID reader for the luggage conveyor belts (Figure 3.13a) and switched-beam curtain-antenna-enabled entry gate (Figure 3.13b).

These antennas are suitable for baggage and personnel tracking at checkpoints and ports of entry, inventory tracking in a warehouse scenario, factory or warehouse inventory control, security identification, and access control. Chung and Lui [29] also propose a curtain antenna with five loop antenna elements and associated matching and switching circuits similar to the configuration shown in Figure 3.11. In an extended version of the antenna configuration, an elongated antenna element with back-to-back loop arrangement connected to appropriate filters is also proposed to maximize the reading range. The antennas can operate over wide frequency bands of 125 kHz, 13.56 MHz, 915 MHz, or 2.45 GHz, depending on the tuning capability of the loop antenna and its configurations (physical dimensions). A reading distance of 1 m with two antenna arrays with a power of 25 W is claimed. In another embodiment of the switched loop antenna for RFID readers for metal containers, the loop antenna is made of cables with three undulations and the loop portions all in series and coupled to the tuner circuit. This mechanism well defines the detection region approximating the volume defined by the base and walls of the container. The coupling means of the antenna include a tuning circuit, a filter, and a switch for selectively connecting a particular loop antenna to the external processor.

Logistics, Security, Customs

Logistics means the efficient movement of goods and information. RFID has created a tremendous impact on the logistics industry with enhanced security features, real-time vehicle location and tracking, shipment tracking and tracing of the progression of shipment, designing optimum routing of the vehicles, and electronic toll collection [30]. RFID also provides increased security and inspection procedure with multimedia contents authentication for customs at the port of entry and borders of a country. Combining RFID and multimedia helps to facilitate the process of associated useful information with RFID identification codes. The associated multimedia contents care digital photographs, video clips, digitized voice recording animated graphs, and computer files [25]. Lai et al. [31] propose a redundant networked multimedia RFID system that uses wireless local area network (LAN) and Ethernet connections simultaneously. The system is capable of providing output videos, graphics, or image data so that after reading RFID tags, one or more pictures preloaded into the system databases can be downloaded instantly for visual verification by logistics handlers, security guards, or customs officers in real time. The reader is equipped with an active smart antenna for transmission and reception. In the transmitting antenna a power amplifier is incorporated for cable loss compensation and to boost the transmitting power. In the receiver chain a low-noise amplifier (LNA) is added to the antenna to improve the received signal's quality and power. The smart antenna for the reader has wider varieties of intelligence such as frequency hopping, timeslotting, antenna positioning and beam scanning, subset antenna switching, and polarization diversity to exploit the maximum signal readability from multiple tags. A master synchronization controller controls all functionalities mentioned above for the antenna arrays. The antenna system can operate in multiple transmit and single receive mode or single transmit and multiple receive mode. The various transmitting antennas are configured in such a way that either spatial diversity collimation or temporal diversity collimation is achieved via sequential switching of the subsets of the antenna array. These spatial and temporal diversities will increase the reading probability of moving tags. As an extension of the above method, the author proposes multiple transmit multiple receive, in which the active transmit antennas operate in different frequency channels or hopping frequency channels. In this arrangement, the multiple transmitting antennas are not timeslotted in operation, but all are actually working at the same time. This will enable true simultaneous multidetection bistatic reader operation.

3.9 CONCLUSION

RFID is an enabling technology that is transforming the identification, asset tracking, security surveillance, medicine, and many other industries at an accelerated pace. However, the main bottleneck for the implementation of the RFID technology in mass market to relinquish trillions of barcodes each year is the cost. According to an IDTechEx research and survey [1], only the printable technology can provide the envisioned chipless RFID tags, which are projected to be available in 2020. This means that we will have to wait about a decade to materialize the dream of replacing the optical barcodes with the powerful RFID technology. The motivation arises from the fact that when the dumb chipless tags will not have any on-board power source and intelligent processing circuits, the smart antennas can enhance performance significantly by mitigating multipath, concentrating more energy toward the desired tag and nulls toward the undesired tag. The significantly enhanced and advanced processing power of the smart antenna not only improves the chipless tag reading, but also improves the signal-to-interference ratio, system capacity, and throughput improvement. This chapter has also presented a comprehensive review of smart antennas for modern RFID readers. This comprehensive review has covered new ground as follows:

Analysis, synthesis, and classification of various smart antennas for RFID readers developed recently, mainly by various commercial organizations. This classification is based on the unique features of the application-specific RFID readers. They are: authentication and location finding of chipped tags in gaming application, gas bottle tracing, smart shelves, baggage handling, tracking, logistic control, security, and customs. Asset tracking and management in both retail chains and warehouse scenarios can be performed in smart shelves. This will give the database system with information of real-time health of the inventory, misplaced items, warning of replenishments of items, and so on.

Finally, tremendous strides have been evolving around the RFID technology. These developments will make the RFID technology more ubiquitous, accurate, high speed, and user friendly. Smart antennas for RFID readers have shed light on the solution of existing problems for various applications such as collision avoidance, location determination of tags, and reading capacity improvement of RFID readers. Almost every type of smart antenna has been developed to fulfil the demands of RFID applications. Now it is time for antenna designers to reap the benefit of this new enabling technology with innovative and active participation in this field of research.

REFERENCES

1. RFID Forecasts, Players and Opportunities 2009–2019, Executive Summary and Collusions, IDTechEx, 2009.
2. R. Reeds, CEO, FE Technologies Pty Ltd., private communications, April 2009.
3. S. Preradovic and N. C. Karmakar, Modern RFID readers, *Microwave J.*, September 13, 2007 (on-line URL: http://www.mwjournal.com/Journal/Issues.asp?Id=68).
4. K. Finkenzeller, *RFID Handbook: Fundamentals and Applications in Contactless Smart Cards and Identification*, 2nd edition, John Wiley & Sons, Hoboken, NJ, 2003.
5. Alien Technology Corporation, BAP ALR-2850 reader data sheet, 2005. Retrieved from http://www.alientechnology.com/products/rfid_readers.php in February 2006.
6. T. Nakamura and J. Seddon, Omron develops world's first antenna technology that boosts UHF RFID tag read performance, Omron Corporation press releases, accessed from http://www.omron.com/news/n_270306.htmlC in June 2007.
7. M. A. Ingram, Smart reflection antenna system and method, US Patent No. US 6,509,836, B1, January 21, 2003.

8. J. D Griffin and G. D. Durgin, Reduced fading for RFID tags with multiple antennas, in *IEEE Antenna Propagation Society International Symposium Digest* Honolulu, HI, July 2007.
9. I. Balbin and N. C. Karmakar, Phase-encoded chipless RFID transponder for large-scale low-cost applications, accepted for publication in *IEEE Microwave Wireless Compon. Lett.* (26/03/09).
10. Smart antennas—Network up to 32 RFID readers to a single PLC I/O, board/port. http://www.rfidinc.com/smart-antennas-video.html accessed.
11. F. Gross, *Smart Antennas for Wireless Communications with MATLAB*, McGraw-Hill, New York, 2005.
12. D. Nelson (editor), *The Penguin Dictionary of Mathematics*, 3rd edition, Penguin Books, London, 2003.
13. C. A. Balanis, Antenna theory: A review, *Proce. IEEE*, Vol. 80, No. 1, pp. 7–23, January 1992.
14. N. C. Karmakar, Smart antennas for automatic radio frequency identification readers, Chapter 21 in *Handbook on Advancements in Smart Antenna Technologies for Wireless Networks*, Idea Group Inc., Chicago, 20xx, pp. 449–473.
15. J. Tuovinen and T. Vaha-Heikkila, MMID-millimetre-wave identification using RF-MEMS and nanocomponents, in *2006 European Microwave Conference Workshop Notes WS14 RF MEMS and RF Microsystem Application and Development in Europe*, Manchester, UK, 2006.
16. PanGo Active RFID Tag® Wi-Fi Asset Tracking Tag for Enterprise Asset Visibility, www.pango.org.
17. J. Lindsay, W. Reade, M. Vicksta, and B. Kressner, RFID Locating Systems for Linking Valued Objects with Multimedia Files, issue 32, sep. 2003, pp. 15–16. Source: http://www.jefflindsay.com/rfid1.shtml.
18. RFID sensors: From battlefield intelligence to consumer protection, *RFID J.*, Aug. 12, 2002, available online at http://www.rfidjournal.com/article/view/182. in Jeff Lindsay, Walter Reade, Maryellen Vicksta, and Bernie Kressner, RFID Locating Systems for Linking Valued Objects with Multimedia Files, http://www.jefflindsay.com/rfid2.shtml, November 25, 2003.
19. http://www.rfid-radar.com access (3/11/2005).
20. Fake gaming chips scam at Crown Casino, April 24, 2009, *The Age*, http://www.theage.com.au/national/fake-gaming-chips-scam-at-crown-casino-20090424-ah9r.html.
21. K. Hercht and L. Storch, Sequenced antenna array for determining where gaming chips with embedded RFID tags are located on a Blackjack, poker or other gaming tables and for myriad other RFID application, US Patent No. US 2007/0035399 A1, February 15, 2007.
22. D. V. Lee, RFID reader with multiple antenna selection and automated antenna matching, US Patent No. 6,903,656 B1, June 7 2005.
23. J. Kowalski, D. Serra, and B. Charrat, RFID-UHF integrated circuit, US Patent No. US2005/0186904 A1, August 25, 2005.
24. http://www.awid.com/ accessed April 2009.
25. J. Lindsay, W. Reade, M. Vicksta, and B. Kres, RFID locating systems for linking valued objects with multimedia files, http://www.jefflindsay.com/rfid2.shtml, accessed May 13, 2009.
26. S. Shepard, *RFID Radio Frequency Identification*, McGraw-Hill Professional, New York, 2005.

27. Delta has success in RFID baggage tag test, http://www.computerworld.com/action/article.do?command=viewArticleBasic&articleId=88390.

28. Las Vegas airport to implement RFID baggage-tag system, http://www.computerworld.com/industrytopics/travel/story/0,10801,86909,00.html.

29. K. K. T. Chung and S. Liu, Antenna arrangement for RFID smart tags, US Patent No. US 6,703,935 B1, March 9, 2004.

30. http://www.softlogistics.com/Transportation/Transportation.html, accessed April 2009.

31. K. Y. A. Lai, O. Y. T. Wang, T. K. P. Wan, H. F. E. Wong, N. M. Tsang, P. M. J. Ma, P. M. J. Ko, and C. C. D. Cheung, Radio frequency identification (RFID) system, European Patent Application EP 1 724 707 A2, 22 July 2005.

PART II

RFID READER SYSTEMS

CHAPTER 4

RFID READERS—REVIEW AND DESIGN

STEVAN PRERADOVIC and NEMAI CHANDRA KARMAKAR
Department of Electrical and Computer Systems Engineering, Clayton, Victoria, Australia

4.1 INTRODUCTION

Radio-frequency identification (RFID) is a wireless data capturing technique that utilizes radio-frequency (RF) waves for automatic identification of objects. RFID relies on RF waves for data transmission between the data-carrying device, called the RFID tag, and the interrogator [1, 2].

A typical RFID system is shown in Figure 4.1. An RFID system consists of three major components: a **reader** or **interrogator**, which sends the interrogation signals to an RFID tag, which is to be identified; an RFID **tag** or **transponder**, which contains the identification code; and **middleware software**, which maintains the interface and the software protocol to encode and decode the identification data from the reader into a mainframe or personal computer. The RFID reader can read tags only within the reader's interrogation zone. The reader is most commonly connected to a host computer that performs additional signal processing and has a display of the tag's identity. The host computer can also be connected via Internet for global connectivity/networking.

RFID was first proposed by Stockman [1] in his landmark paper "Communication by Means of Reflected Power" in 1948. Stockman advocates that by alternating the load of the tag antenna it is possible to vary the amount of reflected power (also called "antenna load modulation") and therefore perform modulation. This new form of wireless technology is now known as RFID. For many years, researchers and engineers have been working on developing low-cost RFID systems. In the following section the applications of RFID technology are presented.

Handbook of Smart Antennas for RFID Systems, Edited by Nemai Chandra Karmakar

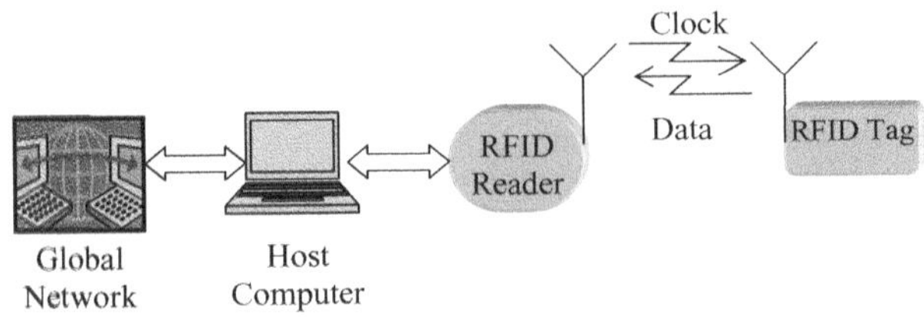

FIGURE 4.1. Block diagram of a typical RFID system.

4.2 RFID TECHNOLOGY AND APPLICATIONS

RFID is a multidisciplinary technology that encompasses a variety of disciplines. These include RF and microwave engineering, RF and digital integrated circuits, and a big chunk of load bestowed on the antenna technology followed by software and computer engineering for encoding and decoding of analog signals into meaningful codes for identifications and real-time tracking. Due to the flexibility and numerous advantages of RFID systems compared to barcodes and other identification systems, RFID is now becoming a major player in mass market. Patronization for RFID technology by major retail chains like Wal-Mart, K-Mart, US Department of Defense, and similar consortia in Europe and Asia has accelerated the progress of RFID technology significantly in the new millennium. As a result, significant advancement on RFID technology has been gained within a short period of time. The RFID market has surpassed the few-billion-dollar mark recently, and this growth is exponential with its diversified applications in all sectors like medicine and health care, agriculture, livestock, logistics, retail chains, and so on. Today, RFID is being researched and investigated by both industry and academic scientists and engineers around the world. The Massachusetts Institute of Technology (MIT) founded the AUTO-ID center to standardize RFID, thus enabling faster introduction of RFID into the mainstream [3–5] of retail chains, identification, and asset management.

Automatic identification (Auto-ID) of items and livestock is required by many areas of government, commerce, science, industry, and farming. For example, Australia has its comprehensive National Livestock Information System (NLIS) [6]. Under the umbrella of NLIS, every animal and pet in Australia must be e-tagged. Recent outbreak of mad cow disease and ban of imported beef from some countries have forced Australia to adopt NLIS. The recent enforcement by private and government sectors has pushed RFID technology in the forefront of development and implementation in almost all sectors of businesses and service industries. Therefore, it can be concluded that the applications of RFID involve the tracking of livestock, personal identification, area security, road tolling, supply chain management, and logistics. Large corporations such as Wal-Mart have taken significant steps to automate their supply chain management and logistics using Auto-ID [7, 8]. As a result, thousands of manufacturers, suppliers, and logistic support organizations of Wal-Mart are adopting the RFID technology for their goods, products, and services. For any new technology, return of investment (IOR) is very crucial for any organization. Being a wireless system,

RFID has been going through harsh scrutiny for reliability and security. With the advent of new anti-collision and security protocols, along with efficient antenna and RF/microwave systems, these problems are being delineated and solved.

4.3 LIMITATIONS OF BARCODES AND EMERGENCE OF RFID AS AN ENABLING TECHNOLOGY

Barcode labels have been used to track items and stocks for sometime after their inception in the early 1970s. Though barcodes are printed in marks and spaces and very inexpensive to implement, they impose undeniable obstacles in terms of their short-range readability and unautomated tracking. These limitations are costing large corporations millions of dollars per annum [9]. The growing tendency today is to replace the barcodes with RFID transponders that have unique ID codes in order to identify items. Hence, the obstacles of reading range and automation would be solved using RFID. The only reason why RFID tags haven't replaced the barcode is due to the price of the tag, which is still much higher when compared to the price of the barcode. Hence, a huge amount of investments and investigations focusing on lowering the price of the transponder have been put in motion, and the price of the RFID tag is getting lower and lower every year. The development of chipless transponders without silicon-integrated circuits (ICs) has lowered the cost of the tags even more, but at the expense of requiring smarter and agile RFID readers.

4.4 RFID READER SYSTEM ARCHITECTURE

Today's RFID readers are composed of (i) smart antenna systems, (ii) dedicated digital signal processing units, and (iii) embedded systems alongside with (iv) middleware and (v) networking features. These features allow easy integration of RFID readers in data networks complying with standardized data transfer protocols. With versatile applications, along with range and speed of data communications, the operational frequency and coding techniques used by RFID readers are very widely spread—from low-frequency (LF) systems to microwave systems. This creates a problem in having a ubiquitous RFID system and acceptance of global standards. In UHF Gen-2 tags a unified approach has been formulated to remove the bottleneck and make the tag universal like the barcodes. To this end, RFID readers play a significant role in reading tags with various formats and standards.

Figure 4.2 shows the data flow of the RFID system architecture. RFID readers are devices that perform the interrogation of RFID tags. In an RFID system, the RFID reader detects the tag by using signal processing demodulation techniques to extract data from the tag's signal. A (passive) RFID tag cannot generate a signal without the reader sending an interrogation signal to the tag. Therefore, the reader and tags are in a master–slave relationship in which the reader acts as a master while the tags act as slaves. Nevertheless, RFID readers themselves are also in a slave position. A software application, also called middleware, processes data from the RFID

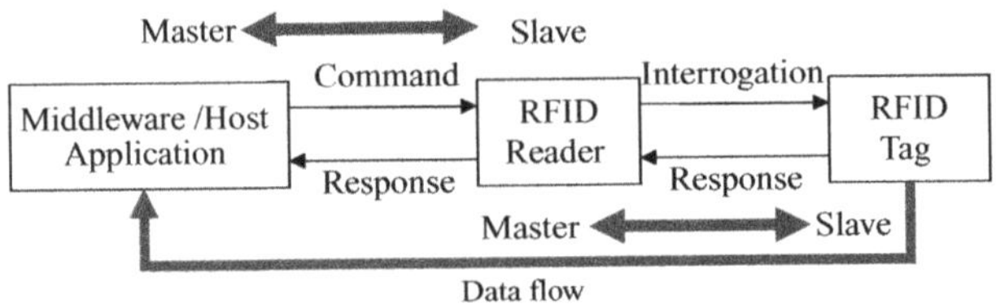

FIGURE 4.2. Master–slave principle between the application software and reader and between the reader and transponders.

reader, acts as the master unit, and sends commands to the reader [10] as shown in Figure 4.2.

RFID systems can be classified as passive and active tag systems [11]. Powered by batteries, active tags have on-board signal amplification and modulation facilities. Passive tags are batteryless and rely on the reader's radiated energy to do some minor tasks of modulation and transmission of data. Therefore, active tags do not rely on the reader as much as passive tags do. As a result, active tags enjoy much superior performances in interrogation and data transmission due to their independence of power supply on a reader. In a passive tag system, a reader needs to have a powerful radiation field so that the passive RFID transponder can effectively cull sufficient power to perform some minor signal processing functions.

An RFID reader consists three main parts shown in Figure 4.3. These main three components are:

- Control section
- High-frequency (HF) interface
- Antenna

At the user end, the reader is connected to the host application such as enterprise software.

Figure 4.4 shows the block diagram of the reader control section. The control section of the RFID reader performs digital signal processing and procedures over

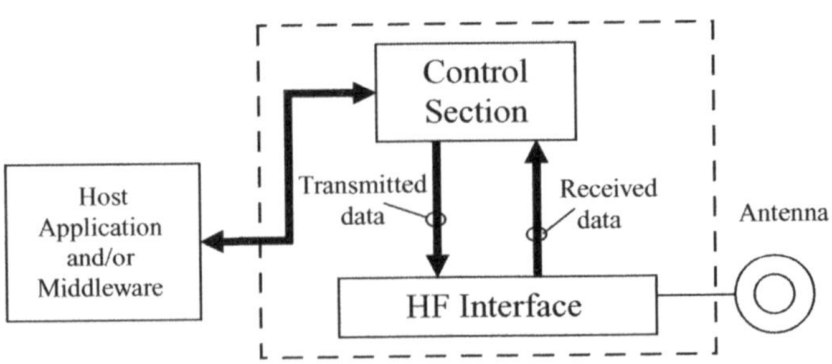

FIGURE 4.3. Block diagram of a typical RFID reader.

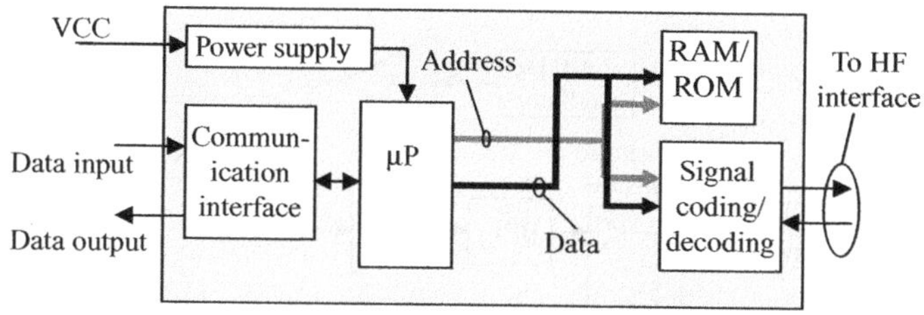

FIGURE 4.4. Block diagram of the RFID reader control unit.

the received data from the RFID transponder. Also, the control section enables the reader to communicate with the transponders wirelessly by performing modulation, anti-collision procedures and decoding the received data from the transponders. These data are usually used to interrogate tags (read) or to reprogram the tag (write). This section usually consists, of a microprocessor, a memory block, a few analog-to-digital converters, and a communication block for the software application.

Figure 4.5 shows the HF interface module of a RFID reader. The high-frequency interface of the reader is used for RF signal transmission and reception. HF interfaces consist of two separate signal paths to correspond with the two directional data flows from and to the transponder.

The local oscillator generates the RF carrier signal, a modulator modulates the signal, the modulated signal is amplified by the power amplifier, and the amplified signal is transmitted through the antenna. A directional coupler separates the system's transmitted signal and the received weak backscattered signal from the tag. A directional coupler consists of two continuously coupled homogeneous wires; and if all ports are matched, the power of the incoming and outgoing signal is divided in the coupler [12]. The received backscattered signal is weak, and the low-noise amplifiers increase the signal's amplitude before and after the signal is decoded in the demodulator. The HF interface is one of the most complex sections of the reader. Most HF interfaces are protected from EM interference using metal cages. Different demodulation

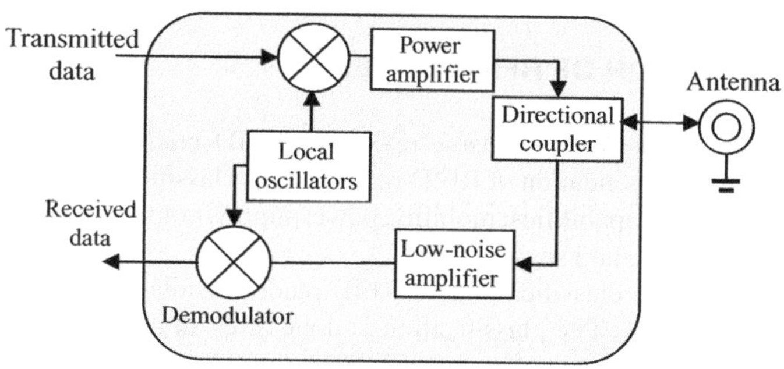

FIGURE 4.5. Block diagram of the HF interface of a RFID reader operating in the 2.4-GHz ISM band.

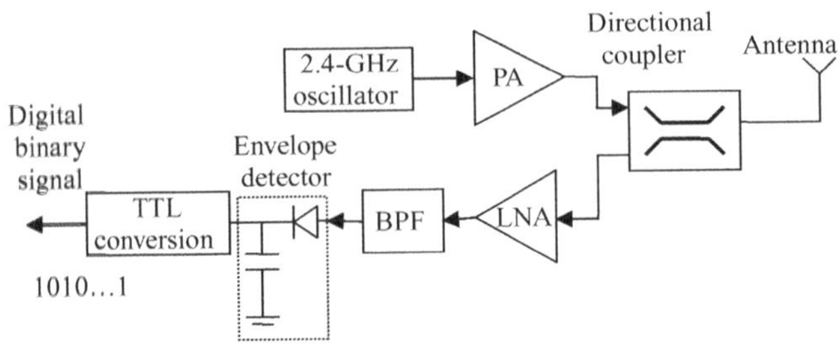

FIGURE 4.6. Block diagram of an RFID reader operating at 2.45 GHz with ASK demodulation technique.

techniques are used when decoding the data received from the transponder. Most RFID systems operate using binary phase shift keying (BPSK) [13] and amplitude shift keying (ASK) [14]. A simple HF interface for a continuous-wave (CW) RFID reader operating in the 2.45-GHz industrial, scientific, and medical (ISM) band using ASK demodulation is shown in Figure 4.6. The reader transmits a CW signal at 2.4-GHz ISM band. Therefore, the transmit section need not to be connected with the control section as shown in Figure 4.5. The CW signal is amplified with a PA, and the amplified continuous-wave signal is transmitted via the antenna to the tag.

The radiation intensity of the reader antenna determines the interrogation range and zone. Depending on the RFID system's applications, the RFID reader can be designed in different ways where the antenna's resonating frequency, gain, directivity, and radiation pattern can vary. Antennas are spatial filters. Adaptive antennas are a promising technique for implementing this spatial diversity into RFID readers. Reference 15 reports an adaptive antenna for the RFID reader. This antenna is a five-element rectangular patch antenna array with an intelligent beamforming network at 2.45 GHz. A number of different reader antennas have been developed throughout the years based on microstrip patch antennas [16–18].

Following is a detailed discussion on various RFID reader systems available in the open literature.

4.5 CLASSIFICATION OF RFID READERS

This chapter presents a comprehensive review of RFID readers found in today's market with a novel classification of RFID readers. This classification is based on the RFID readers' reading capabilities, mobility, power supply, communication interface, data encoding protocols, and so on.

Figure 4.7 shows the classifications of RFID readers available in open literature and commercial markets. The classification is done after an analysis and synthesis of a comprehensive literature review on RFID readers. The classification is based on the power supply, communication interface, mobility, tag interrogation, frequency response, and the supporting protocols of the reader.

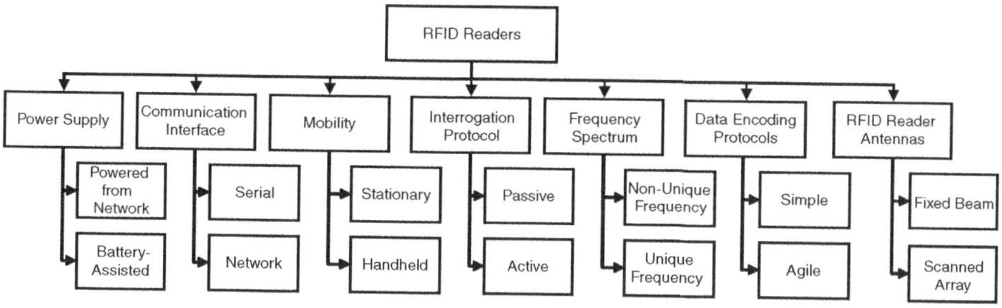

FIGURE 4.7. Classification of RFID readers available in the market.

(a) Power Supply. The classification of the RFID reader based on their power supply brings forth two types of readers:

- Readers supplied from the power network
- Battery-powered (BP) readers

Readers supplied by the **power network** (Figure 4.8) generally use a power cord connected to an appropriate external electrical outlet. Most readers using this type of power supply are fixed stationary readers, and their operating power supply ranges from 5 V to 12 V [19], but there are examples of readers that operate at voltage levels as high as 24 V [20].

Battery-powered (BP) readers are light in weight and portable. The battery is mainly used to power up the motherboard of the reader. Most BP readers are handheld, but there are also stationary readers that are battery-assisted. BP readers use 5- to 12-V batteries for their power supply [21].

FIGURE 4.8. BP Alien Technology reader ALR-2850. (Courtesy of Alien Technology Corporation, 2006. www.alientechnology.com.)

FIGURE 4.9. UHF stationary Mercury4 network reader from ThingMagic. (Courtesy of ThingMagic. www.thingmagic.com.)

(b) Communication Interface. RFID readers can be classified on the basis of the interface that a reader provides for communication. Based on this communication interface, readers can be classified as

- Serial
- Network

Serial readers use a serial communication link to communicate with their host computers or software applications. The reader is physically connected to a host computer using the RS-232 [22], RS-485, IC2, or USB serial connection [23–25]. The disadvantage of serial readers is that they are limited by the number of serial ports at the host side. It might be necessary to have a large number of host computers to connect all of the serial readers. Another problem is data transfer rate that is lower than network data transmission rate.

Network readers (Figure 4.9) are connected to the host computer via a wired or wireless network. These types of readers behave like a standard network device. Today's RFID readers support multiple network protocols such as Ethernet, TCP/IP, UDP/IP, HTTP, LAN, WLAN, and others [26, 27]. This allows easier tracking and maintenance, along with a better data rate, and it results in a smaller number of hosts for the installation of a large number of readers in comparison with serial readers.

(c) Mobility. The next classification of RFID readers can be made on their mobility. Hence, we distinguish two types of readers:

- Stationary
- Handheld

Stationary RFID readers are also known as fixed readers. This term comes from the reader's ability to be mounted on walls, portals, doors, or other objects where they can perform effective transponder readings and are not meant to be moved or carried. Fixed RFID readers are mainly used for wireless data capture in supply chain management, asset tracking, and product control [28]. Today's fixed RFID readers are used for personnel identification and authentication for restricted access areas installed and mounted on portals and doors [29, 30] as well. Most stationary readers support multiple protocols for transponder reader communication and can operate in standalone and in networking mode. A new trend involving the design of stationary readers places multiple antenna connections for connecting more than one antenna to the reader, allowing the user to achieve greater and diverse radiation patterns of the reader's interrogation zone. They use power supply ranging from 12-V DC up to 24-V DC, weigh 1.5 kg up to 5 kg, and can achieve reading ranges up to 300 m [31].

Handheld RFID readers are mobile readers that can be carried and operated by users as handheld units. Handheld readers have built-in antennas and usually do not have connectors for additional antennas. They are battery-powered and are lightweight (from 82–700 g). They have shorter reading ranges than fixed readers (up to 100 m) [32]. Handheld readers are used in tracking livestock (farm animals such as pigs, sheep, goats, and cows), locating items in stores and in stock, and so on [33]. They communicate with the host computer using wireless communication protocols and contain memory blocks to save data. After the user finishes data capturing, it enables data transfer from the reader to a database via wired communication. Most handheld readers have the ability to call out to a specific transponder, and even they can locate a transponder in regard to the location of the handheld RFID reader. Handheld RFID readers are also integrated with barcode scanners so that users can perform both tag and barcode identification simultaneously. A handheld reader by SAVI Technology [33] is shown in Figure 4.10.

(d) Interrogation Protocol. Another classification of RFID readers can be made upon the reader's interrogation protocols in terms of being

- Passive
- Active

Passive readers are limited to only "listening" and do not perform additional tag interrogations. A passive reader's block diagram is shown in Figure 4.6. When interrogating the tags, the reader sends CW signal as a power source for data processing for the RFID transponder. The transponder provides data transmission, which is a unique ID code; hence no message or command is needed from the RFID reader. Protocols or data transmission techniques between the tag and reader that are characterized in this manner are called transponder-driven protocols due to the passive role of the reader in the interrogation process. Transponder-driven protocols for communication with the reader are implemented in tags in the form of ALOHA-based protocols [34].

Active readers are true interrogators that interrogate and listen to tags depending on the number of tags and communication protocols embedded in the system for data transmission between the tags and the readers. Active readers are more capable and "agile" than passive readers. Other than just providing a power source and medium for

The transceiver is responsible for sending the reader's signal to the surrounding environment and receiving the response back from the tags. The transmitter sends RF power (effective isotropic radiated power—EIRP) and the reader's command via its antenna to the tag which is situated in the reading zone. The receiver receives the analog signals from the tag via the reader's antenna. It then sends the signals to the reader microprocessor, where it is converted to an equivalent digital signal. The microprocessor thus decodes the received analog signal and performs data processing. It also encodes and modulates the reader's carrier signal when it wants to send out a message to one particular tag or toward all of the tags in the interrogation zone. In addition, the microprocessor could contain custom logic for doing low-level filtering and additional data processing [35]. The memory block is used for storing data such as the reader's configuration parameters and a list of tag reads. Hence, if the connection between the reader and host computer goes down, not all read tag data will be lost. The communication interface provides the communication instructions for the reader and allows it to interact with external entities and modules. A reader could have a serial as well as a network interface communicating with the external modules such as the host computer. These types of readers have been analyzed in detail in the previous section.

(e) Frequency Spectrum. Besides using their carrier frequencies to perform data transmission and reception, RFID reader's also use other frequency bands. Most readers send out commands to tags on a certain frequency band and receive the transponder's response on a different frequency band (usually a second harmonic or a frequency division of the original reader's interrogation frequency). Therefore we can classify readers based on the transponder frequency responses that they listen to as

- Unique frequency-response-based readers
- Non-unique frequency-response-based readers

Unique frequency response based readers operate at a unique (or short bandwidth <80 MHz) frequency range and use this frequency for both data transmission and reception. The vast majority of RFID readers that can be found on the market today are unique frequency response based readers.

Non-unique frequency response-based readers operate using one frequency for sending a command or simply provide a carrier signal at a certain frequency and listen for an integer multiple of its carrier frequency, generally in the form of a second harmonic, or a frequency-divided signal as the tag's response [36]. Two RF frequencies used for communication by the reader to the RFID system allow fast and reliable full-duplex communication, but this system needs a more complex RF front end for both the reader and tag module [37]. Figure 4.12 shows a multifrequency RFID system.

(f) Data-Encoding Protocols. RFID readers can use multiple frequencies for data transmission and reception, but they can also use multiple protocols to communicate

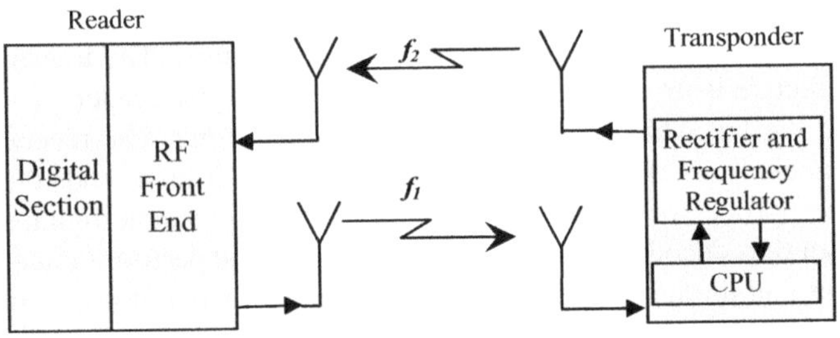

FIGURE 4.12. Block diagram of a RFID system operating at multiple frequencies.

with transponders. Therefore we can classify RFID readers based on their ability to communicate with transponders using diverse protocols.

We can distinguish between two types of RFID readers based upon their ability to communicate with transponders with regard to data-encoding protocols:

- Simple RFID readers
- Agile RFID readers

Simple RFID readers use a unique protocol for communication and data transmission between tags in the reader's interrogation zone [38, 39]. When a tag that supports the reader's interrogation protocol is set in the interrogation area of the reader, the tag is automatically recognized and detected. When a tag that operates using a different protocol is put into the interrogation area, no data transmission will occur, because of the unfamiliarity of the reader's interrogation protocol to the tag.

Agile RFID readers can operate and perform interrogations and data transmission with tags using multiple protocols. The most commonly used protocols for data transmission between tags and readers include EPC Gen1 [40], EPC Gen2 [41], ISO 18000 [42], and TIRIS Bus Protocol. The majority of RFID readers that can be found in the market are designed for multiple protocols and multi-tag readings [43].

(g) RFID Reader Antenna. The efficiency of the RFID reader in terms of transponder interrogation and detection is highly dependant on the reader antenna. We can distinguish between two types of RFID reader based on their antennas:

- Fixed-beam RFID reader
- Scanned-array RFID reader

Fixed-beam antennas are characterized by a unique and fixed-beam radiation pattern [17]. Most RFID readers are equipped with omnidirectional or wide-beamwidth antennas in order to cover as much area as possible. Several fixed-beam antennas are used as well and can be commonly found in alien technology readers. The advantage of using such antennas is that they are easy to install and do not need any logic to control their radiation patterns. The disadvantage of these antennas is that they pick up multipath signals along with the transponders' signal, which can result in errors during interrogation.

Scanned-array RFID readers use smart antenna systems in order to reduce the number of transponders within their main lobe radiation zone, thus reducing reading errors and collisions among tags. This technique exploits *spatial diversity* among tags' locations. The direct beam also reduces the effects of multipathing [44]. This is because phased-array antennas can have a controlled radiation pattern and are highly directional, so it is possible to introduce spatial diversity into transponders' interrogations. The effects of multipathing can be reduced by the use of directional antennas. This new approach to RFID antenna technology is being incorporated by a few RFID manufacturers such as Omron Corporation, Japan [45], RFID Inc. [46], and RFSAW, USA [39]. In March 2006, Omron Corporation developed a new electronic control antenna technology in their UHF band RFID reader systems [45]. RFID tags are subject to multipath interference–an inherent problem of electromagnetic signals. Multipath interferences make an RFID tag unreadable even if it is within the reading range of the reader. To solve this problem, a new type of antenna technology is used that can electronically control the main beam emitted from the reader's antenna. This smart antenna technology incorporated in RFID readers has reduced reflections from surroundings and thus minimized the degradation of the system performance due to multipath and other undesired effects. RFID Inc. [46] has introduced a 32-element compact package smart antenna with associated switches at 125 kHz. This new product advances RFID applications in factory automation, process controls, and original equipment manufacturer (OEM) markets.

So far we have presented various types of modern RFID readers available in the market. Next we are going to present the next generation of RFID readers which are based on universal RFID reader chip sets.

4.6 UNIVERSAL READER DESIGN

Barcodes have been enjoying a robust global standard and have free movements across boundaries. RFID, being an enabling and maturing technology, lacks this flexibility. Manufacturers in different countries have been following their own standards and procedures, but Gen 2 RFID tag systems have set standards so tags can be read across boundaries. An interesting trend has been observed in RFID reader manufacturers in developing universal RFID readers and modules [47, 48]. These readers are generally designed as independent circuit boards with defined output pins and a communication interface and can be equipped with different antennas, depending on the users needs. The main aim of this type of reader is to enable maximum design flexibility and diversity to the user and to allow easy-to-implement reader modules to speed up the design and setup of modules of an RFID system. Most reader modules operate on 5- to 9-V DC power supply and use RS-232/422/485 communication interface, while some have WLAN and Bluetooth protocols embedded in them, allow full-duplex (FDX) or half-duplex (HDX) communication, have a short read range (up to a few meters), and weigh from 5 g up to 100 g. Their applications are aimed mainly toward the design

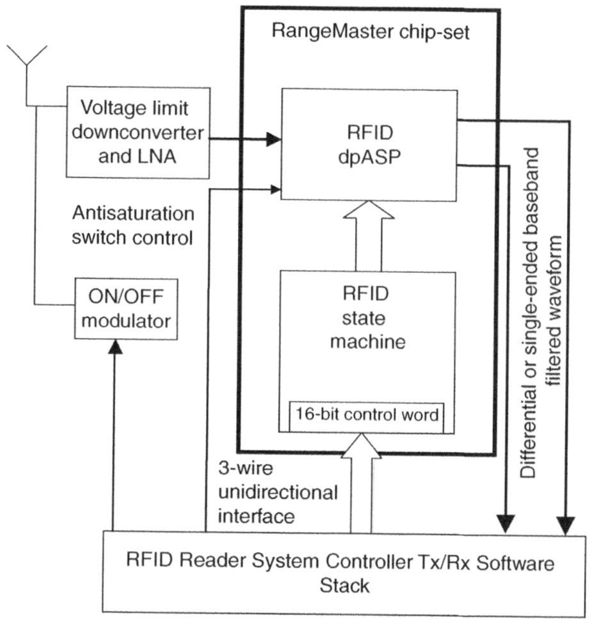

FIGURE 4.14. System-level overview of the RangeMaster-embedded RFID reader [51].

price comparable with the low cost of the optical barcode (below 1 cent). The vast majority of RFID transponders are usually composed of an antenna and integrated circuit (IC) [52]. The cost of the entire RFID system is dependent on the cost of the tag which is dependent on the cost of its IC. Hence, efforts have been put in developing chipless RFID tags with no ICs, which mean that the main cost of the tag is being removed. The authors have reported a novel multiresonator-based chipless RFID tag/system, which is presented in reference 53. The signal flow diagram of the novel chipless RFID system based on multiresonators is presented in Figure 4.15.

The chipless tag encodes data in the frequency spectrum, thus encoding the spectrum with its unique spectral signature. The spectral signature is obtained by the RFID reader by interrogating the tag by a multifrequency signal. The tag encodes its spectral signature into the interrogation signal spectrum using a multiresonating circuit that is a multistop band filter. The multiresonator is a set of cascaded spiral resonators designed to resonate at particular frequencies and create stop bands. The stop band resonances introduce magnitude attenuation and phase jumps to the transmitted interrogation signal at their resonant frequencies that are detected as abrupt amplitude attenuations and phase jumps by the RFID reader. In order to provide isolation between the transmitting and receiving signal, the reader and tag antennas are cross-polarized. As a result, cross-talk between the transmitting and receiving antennas is minimized at the cost of introducing restrictions in tag positioning and orientation.

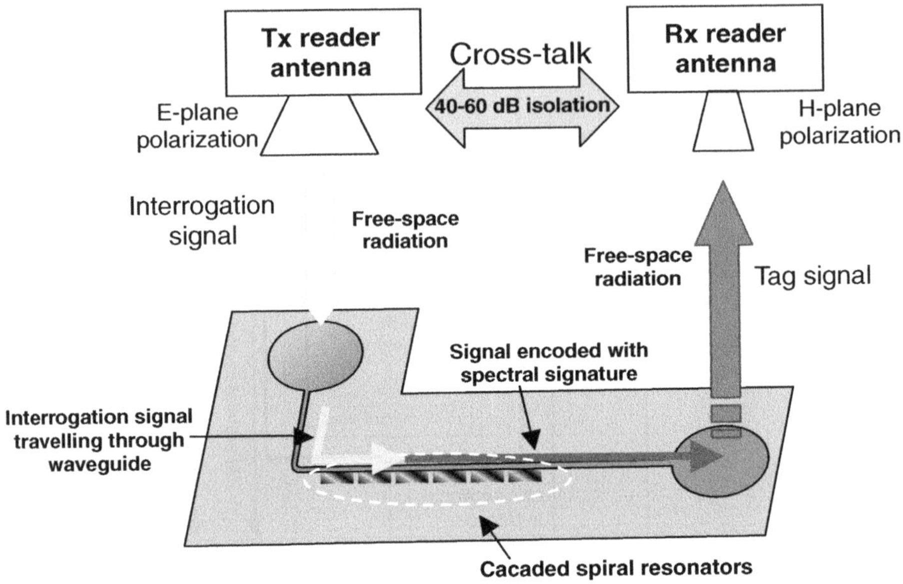

FIGURE 4.15. Chipless RFID system signal flow diagram.

Although the use of conventional off-the-shelf RFID readers would be preferable, the new chipless RFID tags demanded a completely new development of the reader from scratch. Three main differences between the developed chipless RFID tag reader and conventional off-the-shelf RFID reader are as follows: (i) Conventional RFID readers operate mostly at HF (13.56 MHz), UHF (915 MHz), and microwave (2.45 GHz) bands, while the chipless tag reader operates outside these bands; (ii) conventional readers use amplitude shift keying (ASK) and binary phase shift keying (BPSK) time-domain-based demodulation techniques, while the presented reader decodes the tag by sweeping the microwave frequency spectrum and acquiring the tag's spectral signature; and finally (iii) the proposed reader can process the tag data even after the tag has been read and has left the interrogation zone, while conventional readers require the tag to be in the reader's interrogation zone due to handshaking algorithms between tag and reader [53].

The initial step in the design procedures of the chipless RFID tag reader was to formulate the specifications for a particular application. The conveyor belt application for tagging Australian polymer banknotes with the fully printable chipless RFID tags has been presented in [54]. The main governing parameters for formulating the specifications are bandwidth, design frequency, and reading range. Specifications for the chipless tag RFID reader transceiver are given in Table 4.1. The RFID system operates between 2 and 2.5 GHz and is a proof-of-concept 6-bit chipless RFID system intended for implementation on low-cost plastic/paper items such as banknotes, passports, and envelopes.

TABLE 4.1. Specifications for the Chipless Tag RFID Reader Transceiver

Electrical Specifications	
Frequency of operation	2–2.5 GHz
Transmitting power	15 dBm
Interrogation signal type	Frequency sweep (CW)
Detection type	Amplitude and phase
Tx/Rx isolation	60 dB
Receiver sensitivity	−35 dBm
Maximum power consumption	2 W
Commercial	
Cost	Less than AU $200 (guide only)

The chipless tag RFID reader is designed as a reader prototype operating between 1.9 and 2.5 GHz. The two main parts of the RFID reader are the digital section and the RF section. A block diagram of the RFID reader is illustrated in Figure 4.16.

The central processing unit (CPU) sends data to the RF section using a digital-to-analog (DAC) converter. The analog data is the tuning voltage for the voltage-controlled oscillator (VCO) which generates the RF signal for interrogating the chipless tag. The received signal from the tag is amplified and filtered before being sent to the RF-to-baseband (RTB) circuit, where it is converted to a DC analog value and sent to the digital section. The analog signal is then converted to a digital signal using the analog-to-digital converter (ADC) and sent to the CPU for tag ID decoding. The tag ID is displayed on a seven-segment LED display section and/or sent to the host computer.

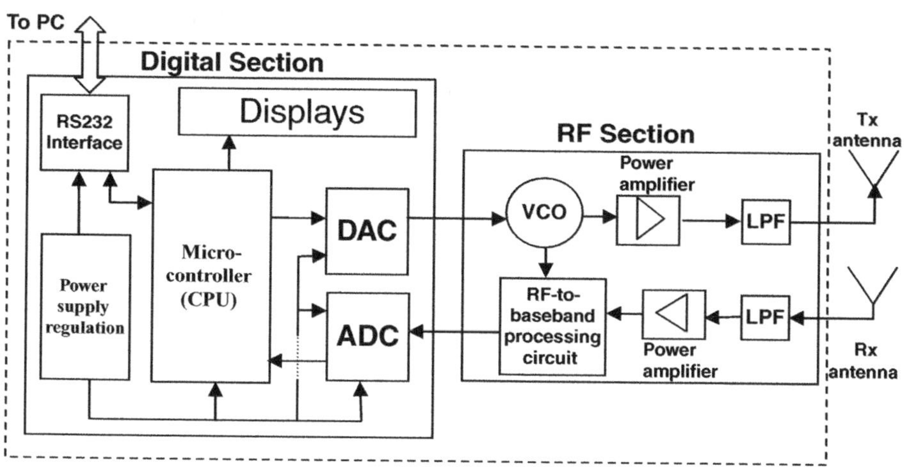

FIGURE 4.16. Block diagram of chipless tag RFID reader.

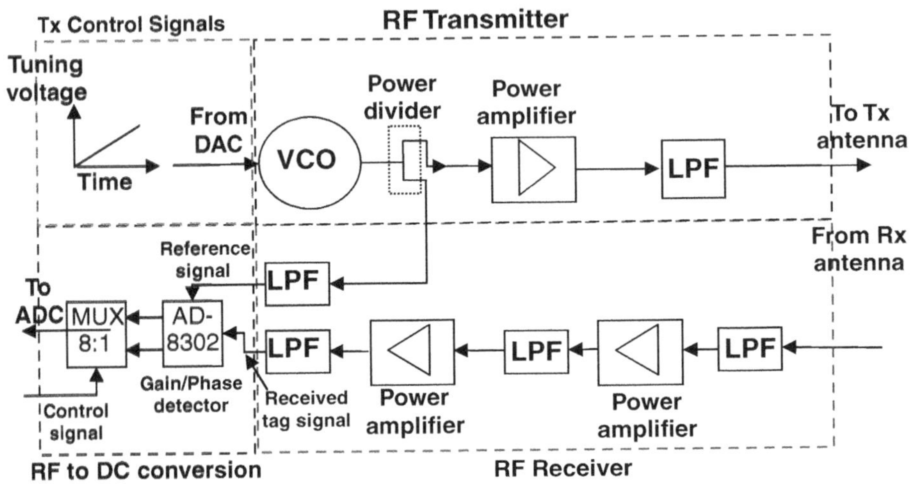

FIGURE 4.17. Block diagram of chipless tag RFID reader transceiver.

The digital section is designed around an 8-bit Atmel AT89C52 microcontroller that performs the major signal processing and data decoding algorithms. The RF section consists of two RF paths: transmitter and receiver. The transmitting circuit is comprised of a VCO that generates the RF interrogation signal and is used as a reference signal for the receiver. The interrogation signal is thus amplified and filtered and transmitted by a wideband directive reader antenna. The signal is received by a cross-polarized receiver antenna and processed (filtered and amplified) before being sent to the RTB circuit. The RTB circuit utilizes a gain/phase detector AD8302 as the RTB. The AD8302 compares the received RF signal from the tag with the reference signal from the VCO and yields DC equivalent values of magnitude and phase difference between the two RF signals. The two DC values are multiplexed and sent to the digital section of the reader for further processing (digitizing) and decoding.

The RF transceiver (section) is designed to operate between 2 and 2.5 GHz and detect both the amplitude and phase variation of the tag's spectral signature. In order to accomplish this, the receiver circuit utilizes an AD8302 gain/phase detector circuit that generates a DC voltage output corresponding to the amplitude and phase difference between the reference RF signal and the received tag signal. The block diagram and photograph of the RF transceiver are shown in Figures 4.17 and 4.18, respectively. The RF components used for the design of the chipless tag RFID reader RF section (transceiver) are shown in Table 4.2.

The digital section is designed around an 8-bit Atmel AT89C52 microcontroller that performs the major signal processing and data decoding algorithms. The architecture of the digital section is presented in Figure 4.19. A microcomputer system has been used. The whole concept is based on using a CPU (AT89C52) and the

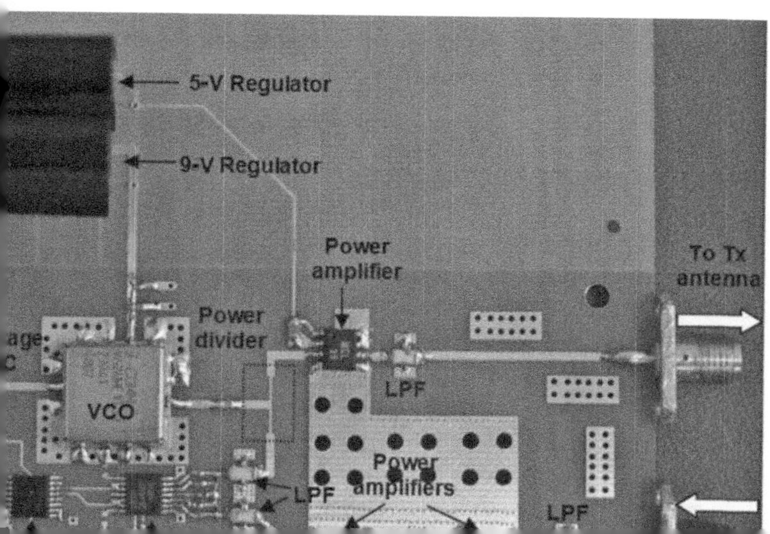

5-V Regulator
9-V Regulator
Power amplifier
To Tx antenna
Power divider
VCO
LPF
LPF
Power amplifiers
LPF

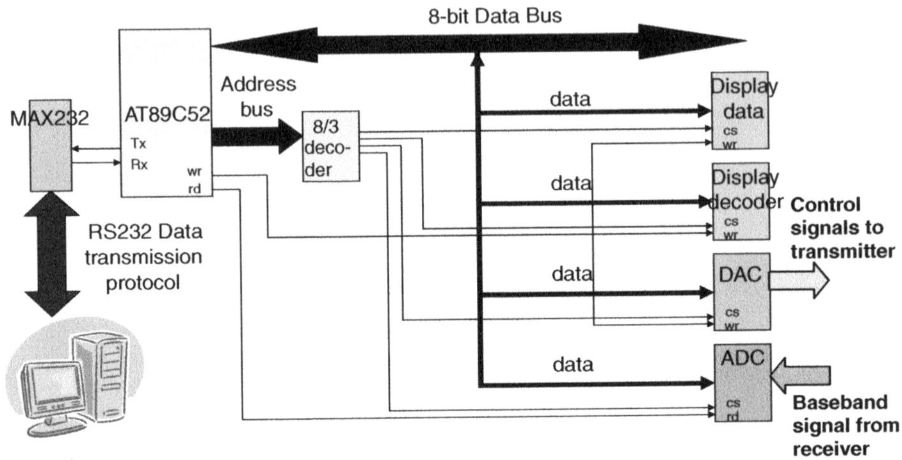

FIGURE 4.19. Block diagram of chipless RFID reader digital control section.

chip select (CS) pin which activates the specific IC. After activating the peripheral unit, data are put on the data bus and the write signal is clocked, hence loading the IC with data from the CPU through the data bus. A photograph of the digital section is shown in Figure 4.20. The components used for the short-range RFID reader are:

- AT89S52 8-bit CMOS microcontroller—Atmel
- AD7533JN DAC—Analog Devices
- AD817AN High-Speed Amplifier—Analog Devices
- AD7819 ADC—Analog Devices
- MAX232 RS232 Driver—Maxim-Dallas
- 6 Seven-Segment Common Anode Displays
- Buffers and DFF Latches—Texas Instruments

The following sections present the results of the RFID reader performance trials and field trials.

4.8 CHIPLESS TAG RFID READER PERFORMANCE

The measurement results of the RFID reader transceiver designed for the chipless tag RFID reader are presented in this section. The transceiver was tested in a wired scenario. The wireless field trials of the integrated tag-reader system will be presented in the next section. The wired tests were performed in order to validate the operation of the RF transceiver circuits only.

erface
Tag ID displays
DAC
Control signals
to transmitter
CPU

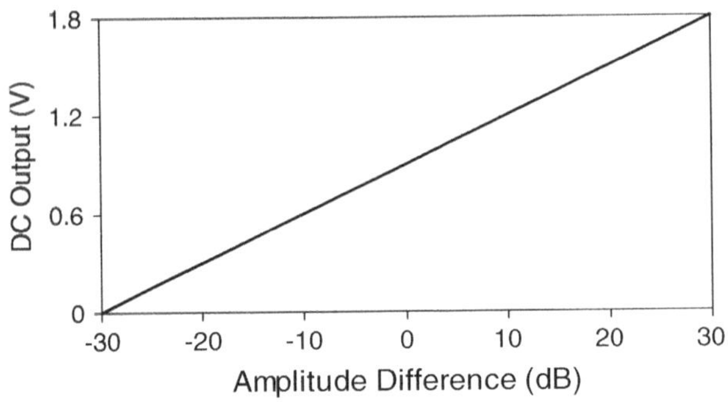

FIGURE 4.22. Measured AD8302 DC output based on amplitude difference.

To verify that the developed RF transceiver sections works correctly, an Agilent performance network analyzer (PNA) E8361A was used. The two sets of data were then compared. Upon obtaining satisfactory agreement between the two results, the above procedure was implemented to digitize the analog data from the transceiver circuit.

The following characteristic parameters of the transceiver were investigated and their effects on the performance of tag detection recorded:

- Transmitter output power
- Transmitting signal higher order harmonic distortion
- Isolation between transmitter and receiver
- Phase noise
- Receiver sensitivity

The 5-bit multiresonator is interrogated by varying the frequency of the CW signal from 2 to 2.5 GHz. The gain/phase detector requires two RF inputs in order to detect the phase and amplitude difference between the two RF signals. One RF input signal is the received signal passing through the multiresonator, while the second RF input is directly from the transmitter VCO and acts as a reference signal. The amplitude and phase difference of the two RF signals are displayed as different DC voltage outputs. The DC voltage relationships with the amplitude and phase differences between the two RF signals are shown in Figures 4.22 and 4.23. It is clear that the AD8302 has a 60-dB dynamic range for amplitude difference detection. The phase difference detection is performed in such a way that a difference of $\pm 180^\circ$ gives a DC output of 0 V, while in-phase signals give a voltage output of 1.8 V by the AD8302.

The measured transmitter output power is shown in Figure 4.24. The RF output power was measured using Agilent's E8257D Spectrum Analyzer. The output power of the receiver is around 14 dBm due to the power divider implemented after the VCO in the transmitter to provide a coherent reference signal for the receiver amplitude/phase detector. The tuning voltage sent to the transmitter VCO is presented in

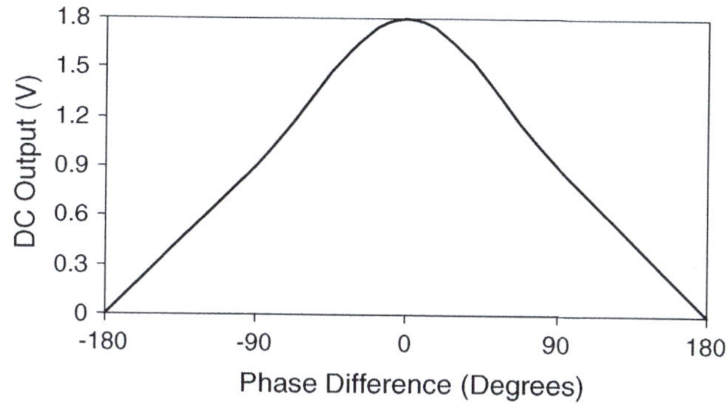

FIGURE 4.23. Measured AD8302 DC output based on phase difference.

Figure 4.24, which shows that the frequency control is very linear. Figure 4.24 shows that the transmitter output varies from 14.3 dBm at 2.1 GHz to 13.4 dBm at 2.5 GHz.

Figures 4.25, 4.26, and 4.27 show the measured spectrum of the transmitter output from 0 GHz to 10 GHz when the transmitter output signal frequency is set to 2, 2.25, and 2.5 GHz, respectively. From Figures 4.25, 4.26, and 4.27 it is clear that the spectrum is "clean" of any higher-frequency distortions due to the filtering section in the transmitter (presented in Figure 4.17).

The measured isolation between the transmitter and receiver is shown in Figure 4.28. From Figure 4.28 it is clear that the isolation is above 71 dB from 2 GHz to 2.5 GHz.

The receiver detects the amplitude and phase variations in the received tag signal by comparing it to a reference signal provided by the transmitter VCO. It is important to identify that the phase spectral signature encoding by the tag is more resilient to noise, but the design of a receiver with phase data detection is a greater challenge. The coherent transceiver architecture is characterized by the phase error that is present

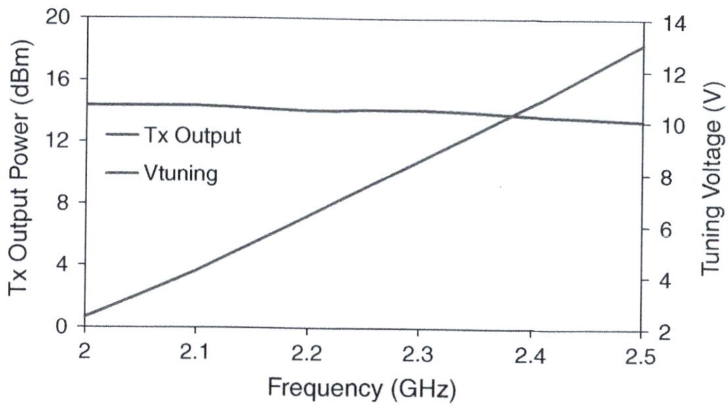

FIGURE 4.24. Measured transmitter output power and VCO tunning voltage.

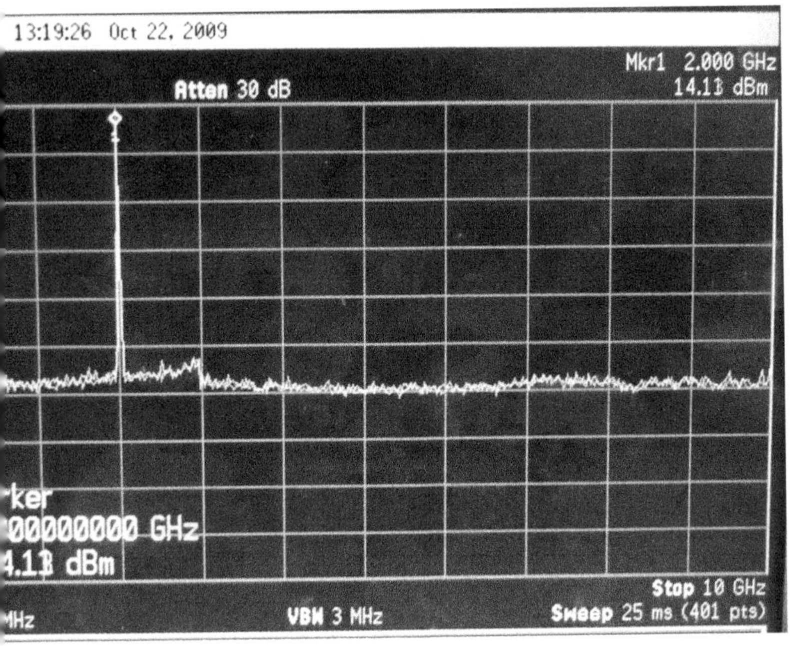

.25. Measured transmitter output spectrum from 0 GHz to 10 GHz for signal GHz.

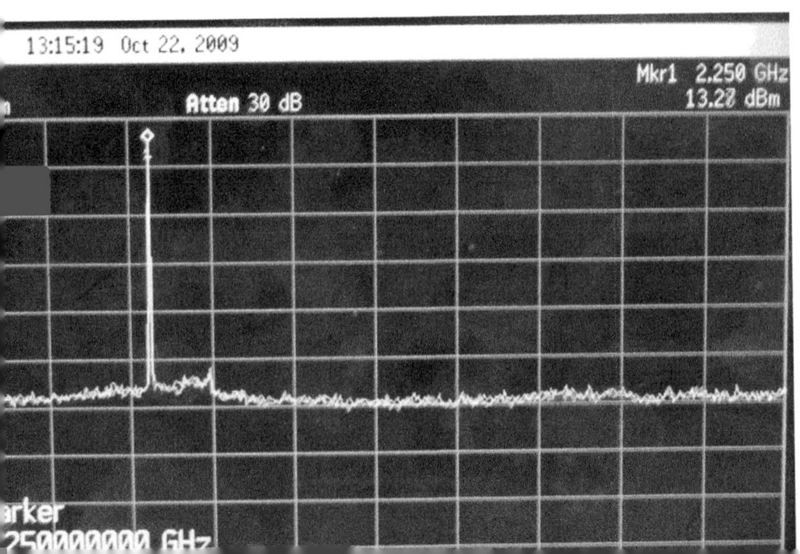

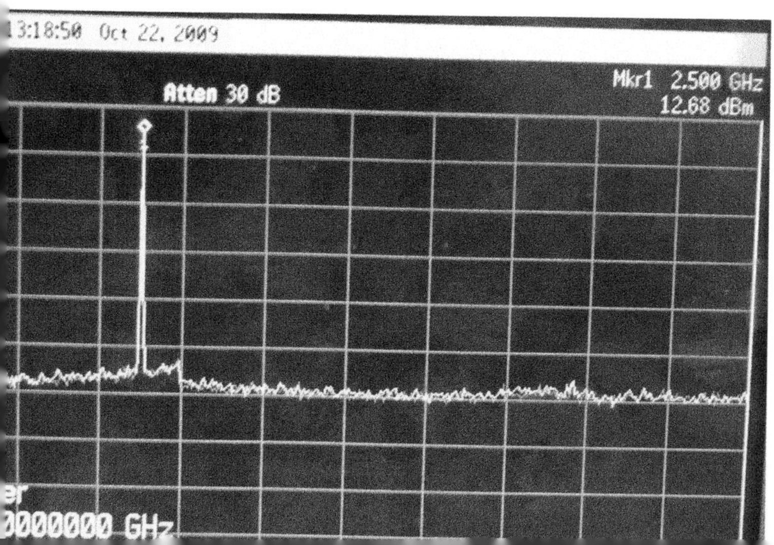
13:18:50 Oct 22, 2009
Atten 30 dB
Mkr1 2.500 GHz
12.68 dBm
er
0000000 GHz

TABLE 4.3. Transceiver Pre-design and Tested Specifications

Electrical specifications		
Specifications	Pre-design	Tested
Frequency of operation	2–2.5 GHz	1.98–2.5 GHz
Transmitting power	15 dBm	Above 13.4 dBm
Interrogation signal type	Frequency sweep	Frequency sweep
Tx/Rx isolation	60 dB	Above 71 dB
Receiver sensitivity	−35 dBm	Below −57.9 dBm
Maximum power consumption	2 W	1.6 W
Commercial		
Cost	Less than AU $200 (a guide only)	$135

transceiver. Figure 4.30 shows the measured receiver sensitivity. From Figure 4.30 it is clear that the receiver sensitivity is below −57.9 dBm.

The transceiver characteristics are summarized in Table 4.3 along with the minimum specifications set in the design section. From Table 4.3 it is clear that the transceiver has met all of the desired requirements except for the transmitter output power. Since receiver sensitivity is greater than the specified −35 dBm this will not be a problem. Thus, the design and testing of the chipless tag RFID reader transceiver has been finalized. The experimental setup of the transceiver integrated with the control section is shown in Figure 4.31.

The Agilent PNA E8361A network analyzer was used to measure the spectral signature amplitude (insertion loss) and phase (transmission phase) of the 5-bit multiresonator. Figures 4.32 and 4.33 show the measured spectral signature 00000 in amplitude and phase obtained from the PNA against the detected calibrated spectral signature using the transceiver, respectively. A tag with no resonances (11111) was used as a reference tag for calibration. Figures 4.34 and 4.35 show the detection of different amplitude and phase spectral signatures using the transceiver. As can be seen, tag 11111 gives no nulls or phase jumps as expected, whereas 10101 yields two

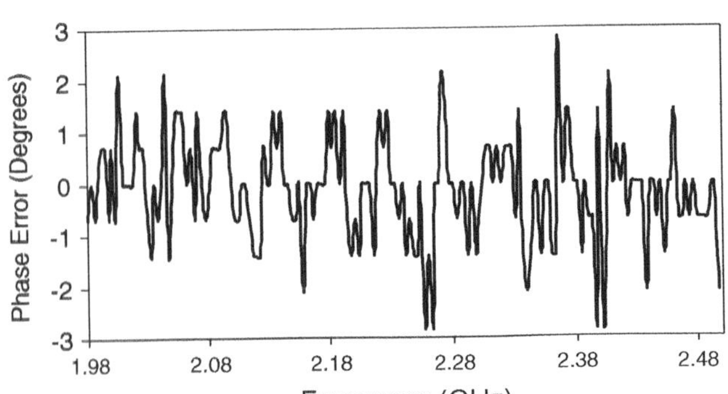

FIGURE 4.29. Measured transceiver phase error.

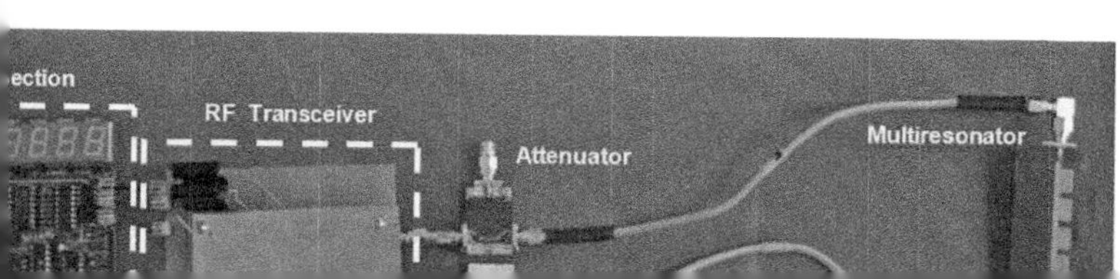

ection
RF Transceiver
Attenuator
Multiresonator

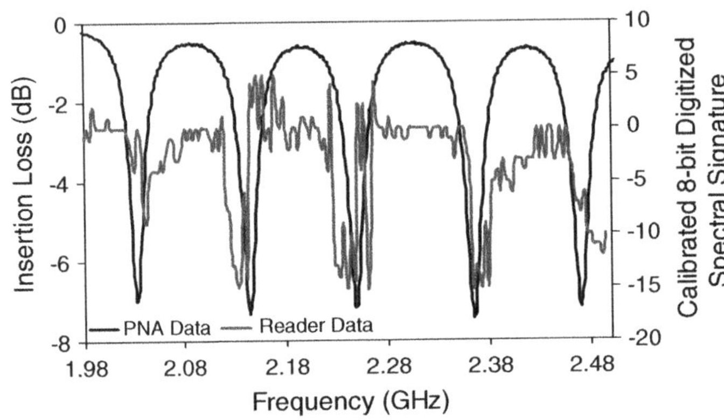

FIGURE 4.32. Measured amplitude spectral signature using Agilent's PNA and reader transceiver.

RFID reader. The chipless transponder along with the reader antennas have been mounted on fiberglass stands and placed into the anechoic chamber.

The testing of the chipless tag RFID reader (2–2.5 GHz) was conducted in an anechoic chamber using log periodic dipole (LPDA) arrays operating between 1.9 GHz and 2.7 GHz as the reader antennas, as shown in Figure 4.39. The RFID reader was tested on both amplitude and phase detection. The 6-bit chipless RFID tag was placed up to 10 cm from the reader antennas. The RFID reader in operating mode is shown in Figure 4.40.

The RFID reader interrogates the tag by sweeping the frequency spectrum in 150–180 samples (amplitude data in 180 points, phase data in 150 points). It was necessary to calibrate the RFID reader first by setting up the experiment and then determining

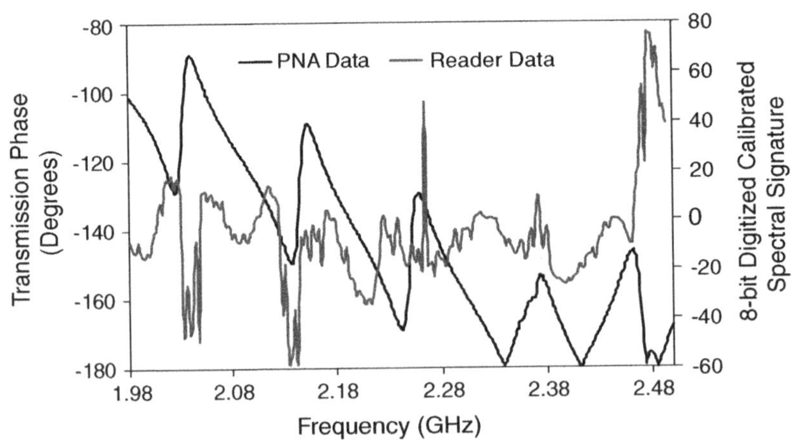

FIGURE 4.33. Measured phase spectral signature using Agilent's PNA and reader transceiver.

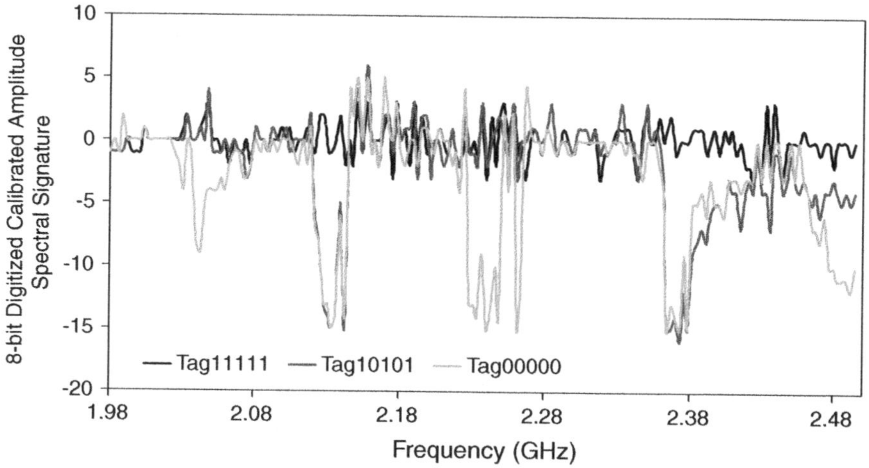

FIGURE 4.34. Different measured amplitude spectral signatures using RFID reader.

the necessary thresholds of the amplitude and phase data, which determined logics "1" and "0". This was done by interrogating a tag with all logic zeros in its ID and recording the data and then replacing the tag with all logic ones in its ID. Hence, a clear difference between what is logic "0" and what is logic "1" was created and recorded in the reader. The calibration procedure enables the detection of a tag in the reader's interrogation area by measuring the received signal strength. The tag spectral signatures ('000000' and '110111') calibrated by the reader in both amplitude and phase are presented in Figures 4.41 and 4.42, respectively.

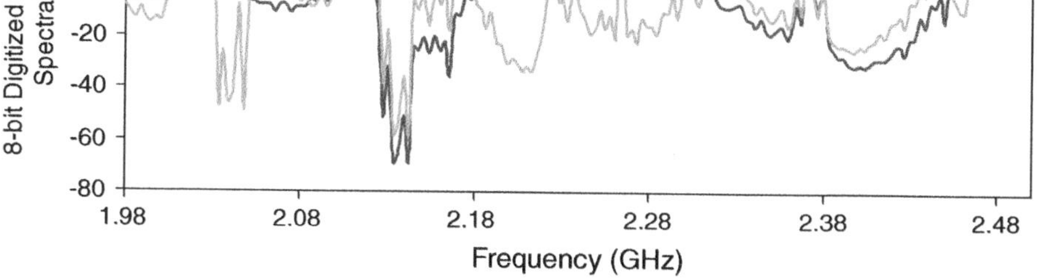

FIGURE 4.35. Different measured phase spectral signatures using RFID reader.

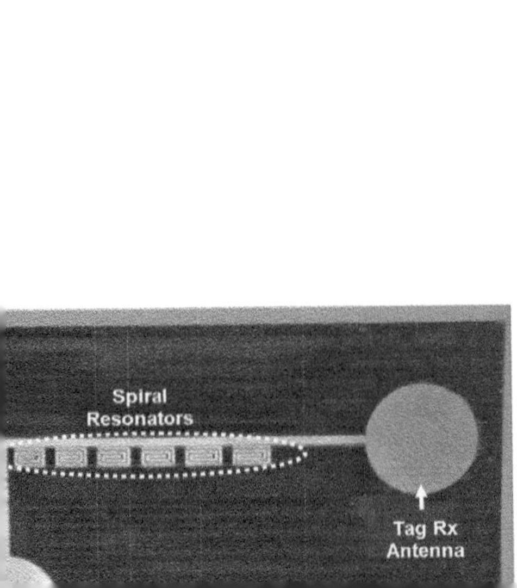
Spiral
Resonators
Tag Rx
Antenna

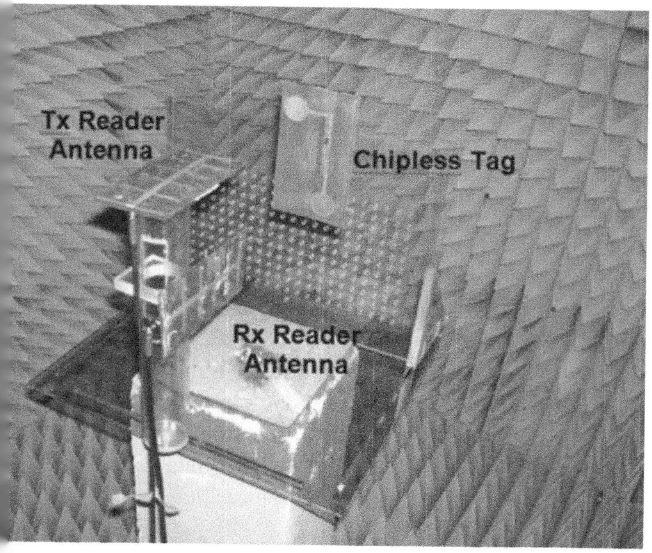

JRE 4.39. Photograph of RFID reader experimental setup.

lecoded by using the threshold values established in the calibration tag ID has been decoded, it is sent to the display and/or via RS-232 ter application. The interrogation/detection algorithm implemented readers is presented in Figure 4.43. It was necessary to calibrate the t by setting up the experiment and then determining the necessary ase data thresholds that determined logics "1" and "0". This was

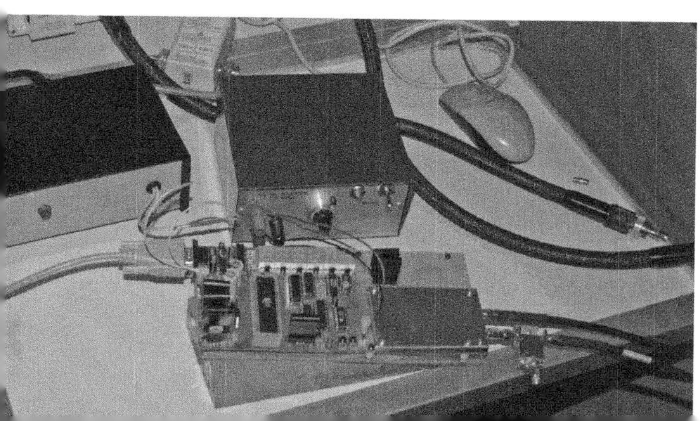

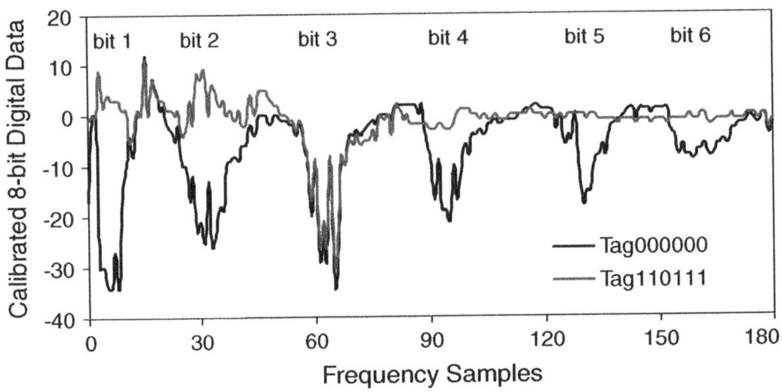

FIGURE 4.41. Recorded amplitude of tag's spectral signature after calibration by RFID reader prototype.

done by interrogating a tag with all logic zeros in its ID and recording the data and then replacing the tag with all logic ones in its ID. Hence, a clear difference between what is logic "0" and what is logic "1" was created and recorded in the reader. The calibration procedure enables the detection of a tag in the reader's interrogation area by measuring the received signal strength.

Tag interrogation starts when the reader detects the received signal strength from a tag in its interrogation zone. After setting the interrogation frequency by loading the DAC with an 8-bit digital number corresponding to the necessary analog tuning voltage for the VCO, the reader software reads data from the ADC. The data are the digitized values of the analog DC signals created by the gain/phase detector

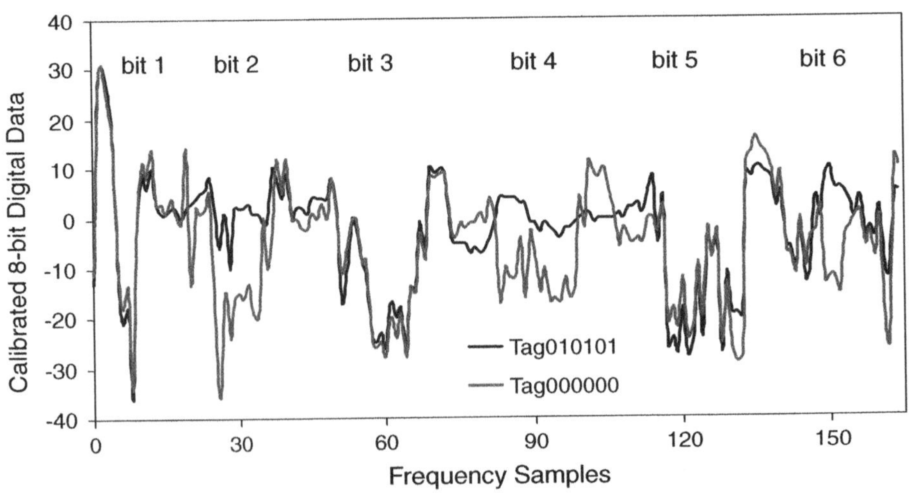

FIGURE 4.42. Recorded phase of tag's spectral signature after calibration by RFID reader prototype.

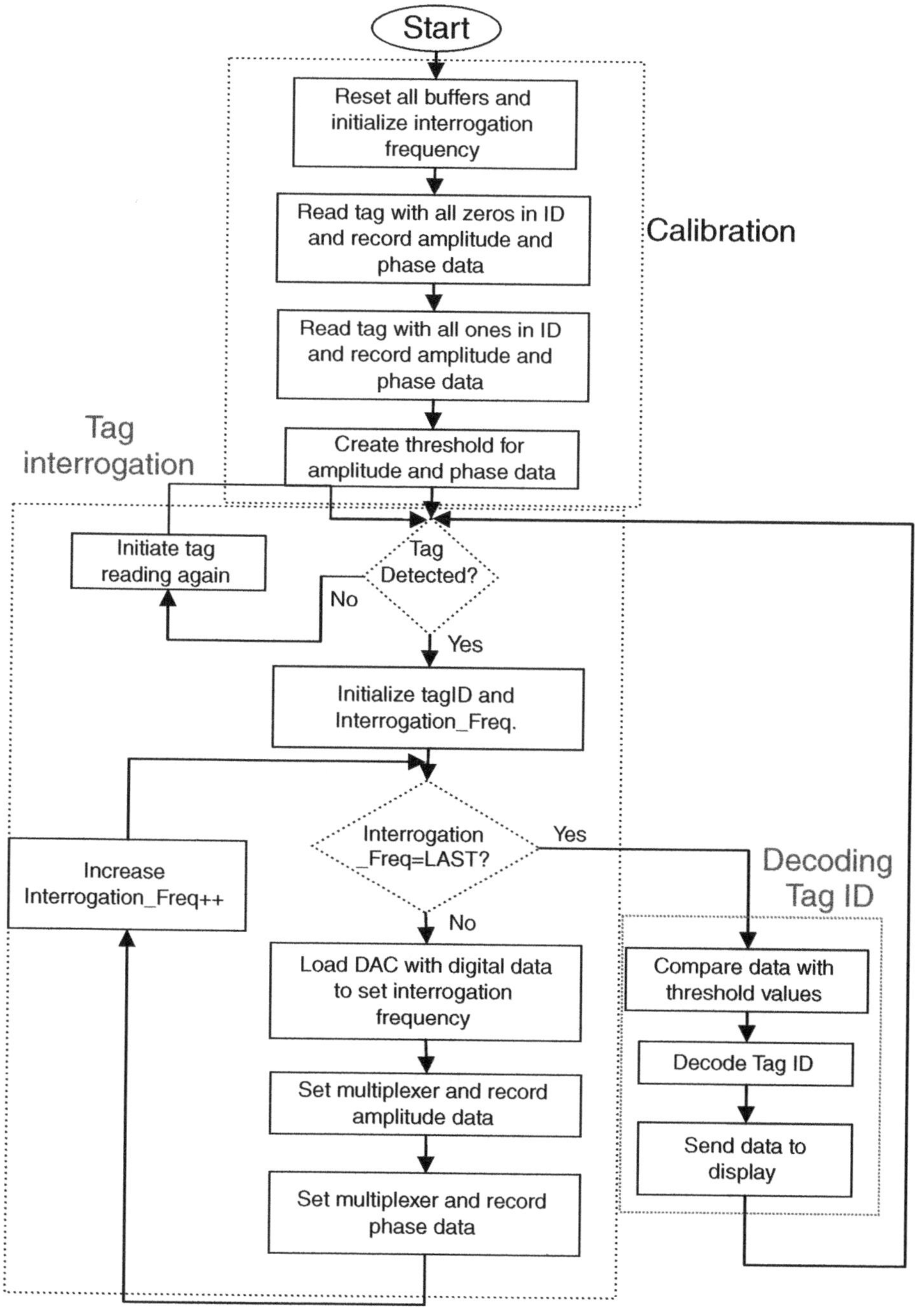

FIGURE 4.43. Flow chart of the RFID reader ID decoding algorithm.

when comparing the received signal from the tag to the reference signal. The reader continues interrogating the tag until it reaches the final frequency sample. At this point, all the necessary data are collected and the tag ID decoding commences. The tag ID is decoded by using the threshold values established in the calibration routine. It is not necessary for the tag to be in the interrogation zone of the reader in order to perform this operation. After the tag ID has been decoded, it is sent to the display and/or via RS-232 to the host computer application. The algorithm returns to Tag Interrogation mode and waits to detect the presence of the next tag.

This section has finalized the RFID reader hardware and software design.

4.10 CONCLUSIONS

RFID is an enabling technology transforming the identification, asset tracking, security surveillance, medicine, and many other sectors. This chapter has presented a comprehensive literature review of RFID readers. This comprehensive review has generated an analysis and synthesis of various RFID readers developed recently by various commercial companies. This classification is based on the unique features of RFID readers. They are power supply, communication interface, interrogation protocol, frequency response of tags, and data-encoding protocol. To the best of the authors' knowledge, such comprehensive classification based on the available literature in various resources were not presented before.

This new synthesized classification will help the end users to plan and select RFID reader according to their applications and requirements and hence will develop confidence in investment of return in using the RFID enabling technology in their manufacturing plants, supply chains, logistics, asset tracking, security, and medical diagnostics.

The current bottleneck of using RFID technology is its lack of a global standard like barcodes. Gen2 tag system will overcome this limitation. Already some manufacturers have paid attention to this issue and started designing such universal systems that can be used across the boundary of countries and applications.

A prototype RFID reader operating successfully between 1.9 and 2.5 GHz has been presented. The reader is a short-range prototype and is intended as a proof of concept circuit. An extension for operating in the UWB region will be realized in the future. The RFID reader has successfully detected and decoded a 6-bit chipless tag at 10 cm using a software core that utilizes calibration, threshold detection, and amplitude and phase data decoding for transponder ID extraction. The RFID reader can be mounted on conveyor belt applications. Future development will concentrate on the development of a long-range UWB RFID reader.

REFERENCES

1. J. Landt, *Shrouds of Time. The History of RFID*, AIM Inc. publications, October 2001 <http://www.aimglobal.org/technologies/rfid/resources/shrouds_of_time.pdf> (accessed March 2006).

2. H. Stockman, Communication by means of reflected power, *Proc. IRE*, pp. 1196–1204, October 1948.
3. D. McFarlane and Y. Sheffi, The impact of automatic identification on supply chain operations, *Int. J. Logistics Management*, Vol. 14, No. 1, pp. 1–17, 2003.
4. M. Karkkainen and T. Ala-Risku, Automatic identification—Applications and technologies, in *Logistics Research Network 8th Annual Conference*, London, September 2003.
5. K. Finkenzeller, *RFID Handbook*, 2nd edition, John Wiley & Sons, Hoboken, NJ, 2003.
6. National Livestock Identification System, official website, <www.nlis.mla.com.au>
7. Sun Microsystems, Inc., Sun and Auto-ID, Sun Microsystems white paper. <http://xml.coverpages.org/SunRFIDWP-200309.pdf#search='Sun%20and%20AutoID'>
8. EPCglobal, Inc., The EPCglobal network: Overview of design, benefits and security, EPCglobal white paper <http://www.epcglobalinc.org/> (accessed April 2006).
9. D'Hont S., The cutting edge of RFID technology and applications for manufacturing and distribution, Texas Instruments white paper, <http://www.ti.com/rfid/docs/manuals/whtPapers/manuf_dist.pdf> (accessed March 2006).
10. Mark Palmer, Build an effective RFID architecture, <http://www.rfidjournal.com/article/articleview/781/1/82/> (accessed 24 June 2007).
11. S. Preradovic and N. C. Karmakar, RFID Transponders—A review, in *Proceedings, ICECE '06 International Conference on Electrical and Computer Engineering*, Dhaka, Bangladesh, pp. 96–99, 22–26, December, 2006.
12. D. M. Pozar, *Microwave Engineering*, 3rd edition, John Wiley & Sons, Hoboken, NJ, 2005.
13. F. Kocer and M. P. Flynn, A long-range RFID IC with on-chip ADC in 0.25 /spl mu/m CMOS, in *Digest of Papers 2005 IEEE Radio Frequency Integrated Circuits (RFIC) Symposium*, 12–14 June, Long Beach, California, 2005, pp. 361–364.
14. N. C. Karmakar, S. M. Roy, S. Preradovic, T. D. Vo, and S. Jenvey, Development of Low-Cost Active RFID Tag at 2.4 GHz, in *36th European Microwave Conference*, 10–15 September, 2006, pp. 1602–1605.
15. P. Salonen, L. Sydanheimo, A 2.45 GHz digital beam-forming antenna for RFID reader, in *IEEE 55th Vehicular Technology Conference*, Vol. 4, Birmingham, Alabama, May 2002, pp. 1766–1770.
16. S. K. Padhi, N. C. Karmakar, C. L. Law, and S. Aditya, A dual polarized aperture coupled microstrip patch antenna with high isolation for RFID applications, in *2001 IEEE Antennas and Propagation Society International Symposium*, Vol. 2, Boston, July 2001, pp. 2–5.
17. G. Marocco, Gain-optimized self-resonant meander line antennas for RFID applications, *Antennas Wireless Propag. Lett.*, Vol. 2, pp. 302–305, 2003.
18. R. Hornung, ARLON FoamClad based microstrip patch antennas and arrays for RFID readers, ARLON internet white paper, <http://www.arlon-med.com/Foamclad%20for%20RFID.pdf#search='microstrip%20antennaRFID%20reader'> (accessed May 2006).
19. CHIPSILICON Pty. Ltd, Product information internet site, 2006. <http://info.chipsilicon.com:8080/chipsilicon/products/> (accessed May 2006).
20. Omron Canada Inc., Microwave RFID system V690 data sheet, 2005. <http://www.omron.com> (accessed March 2006).
21. P. Salonen, M. Keskilammi, L. Sydanheimo, and M. Kivikoski, An Intelligent 2.45 GHz beam-scanning array for modern RFID reader, in *IEEE International Conference on Phased Array Systems and Technology*, Dana Point, California, May 2000, pp. 407–410.

22. Dallas Semiconductor MAXIM, Application note 83: Fundamentals of RS-232 Serial Communications, March 2001. <www.maxim-dallas.com> (accessed March 2006).

23. Feig Electronic Gmbh, Product information internet site, 2006. <http://www.feig.de> (accessed March 2006).

24. Idesco, IR8000 reader data sheet, May 2005. <http://www.idesco.fi> (accessed March 2006).

25. Smart Code Corporation, Product information internet site, 2006. <http://www.smartcodecorp.com/products/rfid_readers.asp> (accessed April 2006).

26. Samsys Technologies Inc., UHF reader MP9320 v2.8e data sheet, 2005. <http://ww.samsys.com> (accessed March 2006).

27. Thingmagic Corporation, Product information internet site, 2006. <http://www.thingmagic.com/html/prod.htm> (accessed February 2006).

28. Symbol Corporation, XR480 RFID reader data sheet, 2006. <http://www.symbol.com/XR480> (accessed March 2006).

29. TAGSYS Corporation, Product information internet site, 2006. <www.tagsys.com/html/rfid53.html> (accessed March 2006).

30. Tagmaster, LR-3 reader product data sheet, 2006. <http://www.tagmaster.com> (accessed March 2006).

31. RF Code, Inc, Mantis II Active RFID reader data sheet, 2006. <http://www.rfcode.com/433mantis_reader.asp> (accessed April 2006).

32. IDENTEC SOLUTIONS, Inc., i-CARD CF Mobile reader data sheet, 2006. http://www.identecsolutions.com/i-cardCF.asp (accessed March 2006).

33. SAVI Technology Inc., SAVI Mobile reader SMR-650 data sheet, 2005. <http://www.savi.com/products/ds.smr650.pdf > (accessed March 2006).

34. V. Naware, G. Mergen, and L. Tong, Stability and delay of finite-user slotted ALOHA with multipacket reception, *IEEE Trans. of Information Theory*, Vol. 51, No. 7, pp. 2636–2656, July 2005.

35. C. Nak-Gwon, L. Hyuek-Jae, L. Sang-Hoon, and K. Seong-Jeen, Design of a 13.56 MHz RFID system, in *ICACT 2006 The 8th International Conference of Advanced Communication Technology*, Vol. 1, Phoenix Park, Korea, February 2006.

36. ActiveWave Inc., Product information data internet site, 2006. <http://www.activewaveinc.com/products_rfid_readers.html> (accessed March 2006).

37. R. Page, A Low Power RF ID Transponder, Grand Prize Winner 1993 RF Design Awards Contest, Wenzel Associates, 1993.

38. BALOGH RFID, HYPER X LMB-6012 RFID reader data sheet, January 2003. <http://www.balogh.cc/HyperX/Support/PDF/LMB6012-6013.pdf> (accessed March 2006).

39. RFSAW, Inc., RFSAW RFID system data sheet, 2003. <www.rfsaw.com> (accessed April 2006).

40. Impinj, The Gen 2 story: Charting the path to RFID that just works, White Paper, 2005 Impinj, Inc. <http://www.impinj.com/files/MR_GP_ED_00001_Gen2Story.pdf> (accessed June 2007).

41. EPC Global, EPC Radio-Frequency Identity Protocols Class-1, Generation-2 UHF RFID Protocol for Communications at 860 MHz–960 MHz—version 1.0.9, Internet

Article, EPCglobal Inc., January 2005. <http://www.epcglobalinc.org/standards/Class_1_Generation_2_UHF_Air_Interface_Protocol_Standard_Version_1.0.9.pdf> (accessed June 2007).

42. Alien Technology, EPCglobal Class 1 Gen 2 RFID specification, Internet white paper, Alien Technology Corporation, 2005. <http://www.alientechnology.com/docs/AT_wp_EPCGlobal_WEB.pdf> (accessed June 2007)

43. EE Times Asia, RFID Solution Enables Single Universal RFID Reader, Internet Article, January 2006. <http://www.eetasia.com/ART_8800403199__NP_798750a5.HTM> (accessed June 2007)

44. S. K. Padhi, N. C. Karmakar, and C. L. Law, Dual polarized reader antenna array for RFID application. *2003 IEEE Antennas and Propagation Society International Symposium*, Vol. 4, Columbus, Ohio, June 2003, pp. 265–268.

45. T. Nakamura and J. Seddon, Omron develops world's first antenna technology that boosts UHF RFID tag read performance, Omron Corporation press releases, http://www.omron.com/news/n_270306.htmlC (accessed June 2007).

46. Profibus-DeviceNet-Ethernet IP-Modbus-Remote I/O, http://www.rfidinc.com (accessed June 2007).

47. Texas Instruments, Inc., Series 2000 Micro Reader data sheet, 2002. <http://www.ti.com/rfid/docs/manuals/pdfSpecs/RI-STU-MRD1.pdf> (accessed June 2006).

48. Sentinel ID Systems, Inc., Product information internet site, 2006. <http://www.sentinelid.com/products.html> (accessed March 2006)

49. M. Reynolds, C. Weigand, Design considerations for embedded software-defined RFID readers, internet white paper, August 2005. <http://www.thingmagic.com/html/508EWTF2.pdf#search='508EWTF2'> (accessed April 2006).

50. Electronic Engineering Times, Anadigm Rolls Universal RFID Reader, internet site, June 2005. http://www.eetasia.com/ART_8800368056_499491_a7168a56.HTM (accessed March 2006).

51. Anadigm Inc., RangeMaster Solution for RFID Tag Readers, Anadigm product data sheet, 2005. <http://www.anadigm.com/_doc/GF041100-0001.pdf#search='anadigmrangemaster'> (accessed March 2006).

52. S. Preradovic, N. Karmakar, and I. Balbin, RFID transponders, *IEEE Microwave Mag.*, Vol. 9, No. 5, pp. 90–103, October 2008.

53. T. Osinski, Why buy EPCglobal-certified products, *RFID Journal* internet article, April 2009, Available: <http://www.rfidjournal.com/article/view/4781/1> (accessed October 2009).

54. S. Preradovic, I. Balbin, and N. C. Karmakar, Multiresonator based chipless RFID system for low-cost item tracking, *IEEE Trans. Microwave Theory Techniques*, Vol. 57, No. 5, pp. 1411–1419, May 2009.

CHAPTER 5

A DEVELOPMENT PLATFORM FOR SDR-BASED RFID READER

BEHNAM JAMALI
School of Electrical and Electronic Engineering, The University of Adelaide, Adelaide, Australia

5.1 INTRODUCTION

Radio-frequency identification (RFID) systems are demanding more data, faster logging, and higher interrogation rates, so they require a powerful system that can manage information from a variety of sources, store the data, and transmit it reliably and continuously for long periods of time to other readers or a host PC. The most common RFID readers in use today operate at ultra-high-frequency (UHF) and high-frequency (HF) industrial, scientific, and medical (ISM) bands [1]. More and more wireless devices are being designed to utilize the ISM band. As the number of these wireless devices proliferates, so does the level of destructive noise in the spectrum that can hamper the effective use of RFID [2].

To address the insatiable desire of the spectrum users for additional wireless bandwidth and burgeoning problem of spectrum scarcity, novel innovation in RFID reader technology and communication protocols are paramount. However, commercial RFID readers are like a black-box. Low-level communication layer parameters and protocol specifics are hardwired, making it difficult, if not impossible, for researchers to experiment and implement new functionality at those levels.

GNU Radio (GR) is an open-source software-defined radio project [3]. It is programmed to run on general-purpose personal computers. The Universal Software Radio Peripheral (USRP) is the hardware designed specifically for use with GNU Radio [4]. The combination of USRP and GR allows implementation of very sophisticated, yet powerful and cost-effective, software-defined radios (SDR). Since SDR and software-defined RFID are one in the same technologies, it is reasonable to use GR software and hardware to implement a software-defined RFID development platform.

Handbook of Smart Antennas for RFID Systems, Edited by Nemai Chandra Karmakar

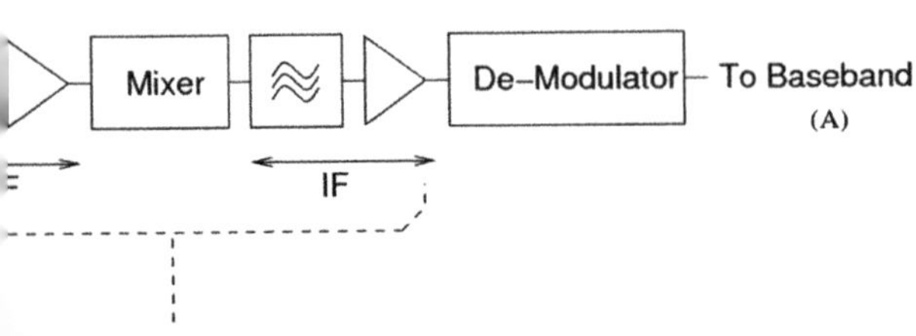

(A)

receiver requires many analog components as shown in the simplified Figure 5.1A. First the radio signal is received by the antenna and is then amplified by a low-noise amplifier (LNA). In such configuration, one receiver is required for each channel. The RF signal is down-converted to baseband by a mixer and a local oscillator. At each stage, analog filters are used to discriminate the signal outside the frequency band. At this stage the signal is decomposed into inphase (I) and quadrature (Q) components. The I and Q signals are conditioned, stabilized, and then demodulated to baseband. The result is converted to digital domain in a narrowband analog-to-digital convertor (ADC), before being fed to a micro-controller for further processing. Figure 5.1B shows how the functionality of the analog parts can be emulated in software using a SDR architecture.

5.3 NEW APPROACH

Today's continuously changing technology in RFID brings the need to build future-proof RFID readers. If the functions that were formerly carried out by hardware can be performed by software, new functionality can be deployed easily by updating the software. With the existing stringent requirements of RFID spectrum (which varies in different parts of the world), increasing traffic rates, and the need to adhere to regulations on spectrum usage, this requires even more sophisticated signal processing algorithms that can only be implemented on a software-based RFID reader system.

An RFID reader based on software will allow the addition of new functionalities with a short time-to-market. Also, user-specific functions can be configured easily to the performance required for different environments, and only the software needs to be upgraded rather than a completely new hardware design. By integrating almost everything into software, fewer external components are required, which further reduces the cost. This allows innovative new features and a rapid development cycle.

Furthermore, there is a need for flexibility in RFID reader design because the specifications are still changing, and even within a single specification, the tags can be asked to reply at different frequencies with different modulation schemes. For instance, tags working within the EPC Global Class I Gen II specifications can be asked to reply at a subcarrier frequency ranging from 40 kHz up to 620 kHz [8]. Therefore, an RFID reader must be able to be easily reconfigured to support frequency bands and protocols in different geographical regions and those that will become available in the future.

5.4 GNU RADIO

GNU Radio (GR) is designed to allow a general-purpose computer to function as high-bandwidth software radio. Using GR, a radio can be built by creating a graph where the vertices are signal processing blocks and the edges represent the data flow between them. The GR components are connected as shown in Figure 5.2. In essence, the USRP (hardware) serves as the digital baseband and IF section of a

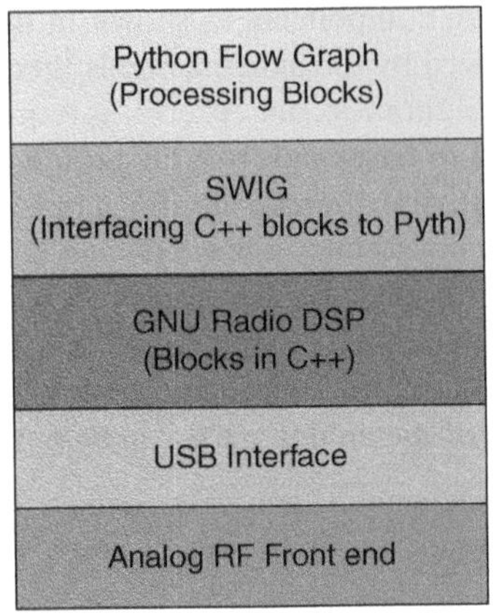

FIGURE 5.2. Hierarchy of operation of GNU Radio components.

radio communication system. The USRP has a well-defined electrical and mechanical interface to RF front-end which can translate between that IF or baseband and the RF bands of interest. The GR system is designed such that all of the waveform-specific processing, such as modulation and demodulation, should be performed on the host computer, while all of the other high-speed general-purpose operations like digital up- and down-conversion, decimation, and interpolation should be performed on the FPGA.

Figure 5.3 is a photograph of the USRP board. As shown in the picture, the USRP has two slots that accept the receiver daughterboard. These slots are labeled RxA and RxB and correspond to interfaces J666 and J668, respectively. Each of those interfaces accepts two real-valued voltage signals from the daughterboard. These analog signals are sent to two separate ADCs. The ADCs are 12 bits wide and run at a rate of 64 mega-samples per second. The collected samples are then sent to the FPGA for processing. Upon entering the FPGA, the digitized signals are routed by a multiplexer (MUX) to the appropriate digital down-convertor (DDC).

The DDC is essentially a complex mixer that expects in-phase and quadrature inputs. The inputs to DDC can be either one of the four input voltage sources or constant zeros. The DDC mixes its input signal to baseband. Once down converted, the signal is then decimated by a factor specified by the user. The decimation is performed in two stages. If a decimation of a factor of M is required, the signal is first decimated by a factor of $M/2$ using a cascaded integrator comb (CIC) filter and then decimated again by a factor of 2 when passing through a half-band filter. The DDC and decimation stages are shown in Figure 5.4. The samples are then pushed into a first-in first-out (FIFO) buffer. They are then sent over USB interface to the host PC for further processing. Every sample consists of in-phase and quadrature values, each of which is represented by a 16-bit value. Therefore each sample is 32 bits long. The USB2.0 interface can transfer data at a maximum rate of 32 megabytes per second. This limits the bandwidth of signals that can be transferred from and to the host computer.

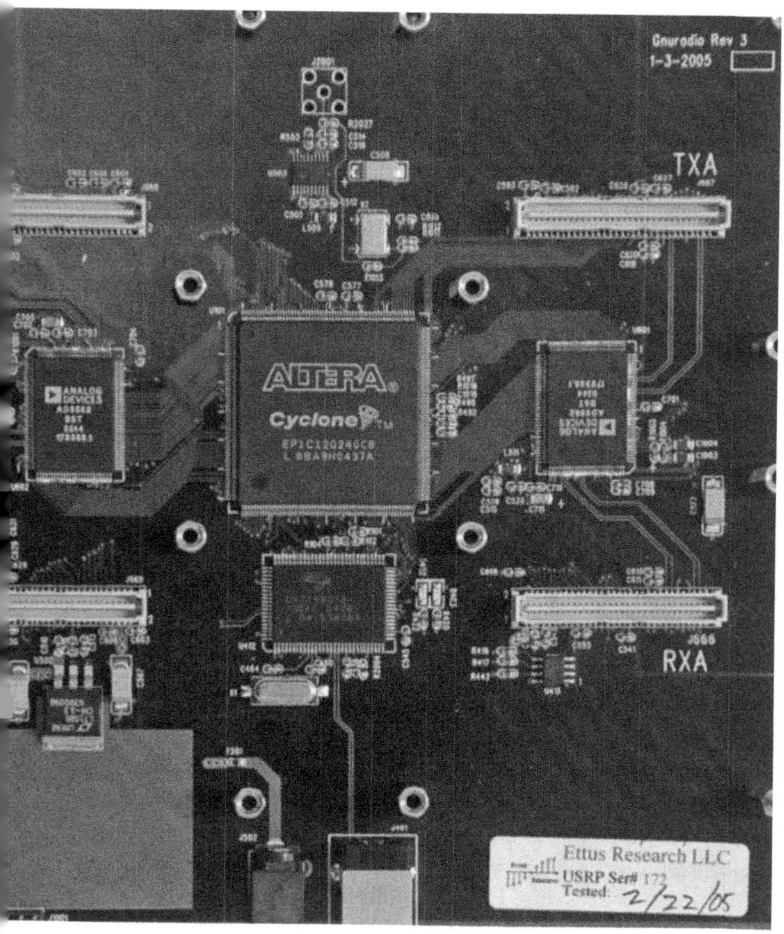

FIGURE 5.3. Photograph of USRP hardware.

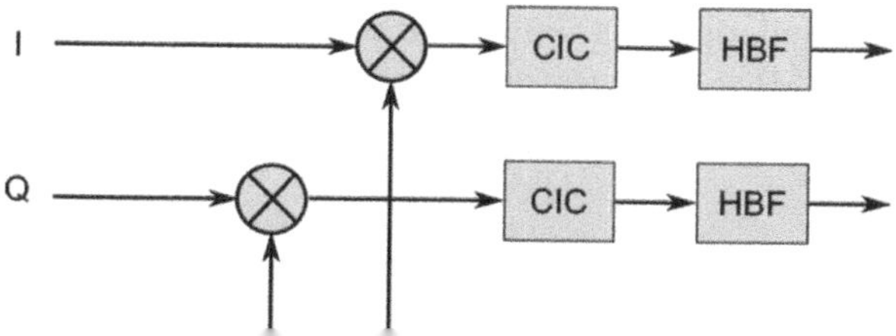

It is worth mentioning that the USRP is completely designed under an open specification project using free and open source CAD software; that is, schematics, CAD files, and all other specification details are available for download from the GNU website. gEDA and PCB [9] have been used for schematics and board layout.

Also the FPGA design is open, therefore the FPGA firmware can be easily modified. By having a closer look at the picture, we can identify four white slots that are used for connecting the daughterboards. Two slots are used for transmission (TXA and TXB) and two for reception (RXA and RXB). Signals from/to the daughterboards pass through the ADCs/DACs (the black chips between the white slots). At the center of the motherboard is the Altera Cylone FPGA, where part of the DSP is performed. Some centimeters under the FPGA, one can see the USB interface chip, used for the communication to the host computer.

5.4.1 ADC and DAC

The USRP comes with four 12-bit analog-to-digital convertors (ADC). The sampling rate of the ADCs is 64 mega-samples per second, therefore a signal with bandwidth up to 32 MHz can be digitized. Consequently, in order to digitize signals of frequency higher than 32 MHz without introducing aliasing (due to Nyquist criterion), the radio front-end needs to downconvert the incoming signal prior to being fed to the ADC. The board is also provided with a programmable gain amplifier (PGA) that is placed right before the ADCs, in order to amplify weak signals if needed. On the transmitting side, there are four 14-bit digital-to-analog convertors (DAC) with a sampling rate of 128 mega-samples per second. In this case the Nyquist frequency is 64 MHz, although oversampling should be used for better filter performance. DACs are also followed by PGA that provides up to 20-dB gain.

5.4.2 FPGA

At the heart of the system is an FPGA, which plays a central role in the USRP design. The USRP FPGA uses Verilog hardware description language, compiled by using the Quartus II web edition from Altera. This compiler is available for free; therefore a customized Verilog code can be compiled and uploaded to the FPGA firmware. The standard configuration is already suitable for a variety of applications. Alternatively, a Verilog code for particular applications can be found in the GNU Radio community website. The FPGA is connected to the ADCs, the DACs, and the USB controller. In the receiving path, the analog signal is converted by the ADC in 12-bit samples and then passed to the FPGA for further processing. Here, a multiplexer routes the signal to the appropriate Direct Down Converter, which converts the signal to the baseband and decimates it by a factor specified by the user. Data is then interleaved and passed in 16-bit samples to the USB2 controller for transmission to the host computer. The USB connection has a nominal bandwidth of 32-Mbyte/s half-duplex; that is, bandwidth needs to be partitioned between downlink and uplink. The USB bandwidth is one of the most stringent limitations when designing a software radio with USRP. Using 16-bit samples (2 bytes) results in 32 Mbytes/2 bytes, or 16 mega-samples per

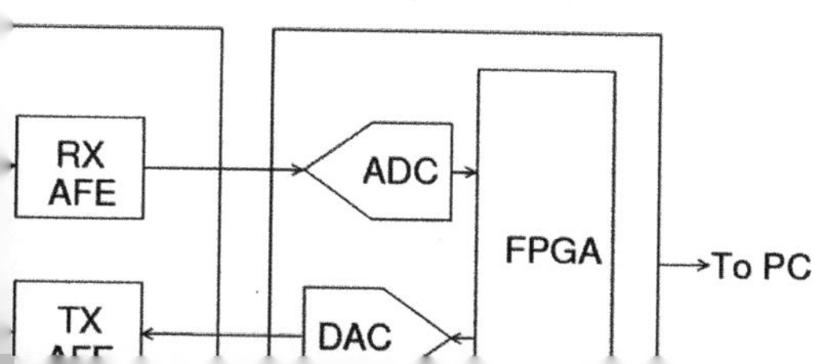

RX
AFE
ADC
FPGA
To PC
TX
DAC

[11], which is a limiting factor for the USRP because the C++ portion of GNU Radio flow graph blocks does not have access to the USRP's configurable parameters.

5.6 GR-BASED RFID READER

Wireless RFID networks are getting popular as a means of creating a pervasive wireless networking and inventory management mechanism. The central concept in RFID systems is use of multihop relaying. Such wireless links give rise to new challenges in medium access control (MAC) protocols. The challenges include interference, fading, improving network throughput, and guaranteeing fairness. In order to combat these challenges, various cross-layer design techniques must be used. Such optimization requires special measurement devices capable of measuring transient signals, bandwidth inefficient modulations, backscattered data, and passive tags. These capabilities are not commonly found in traditional test instruments, at least in real time. The GNU Radio-based RFID system (code name GRID) is a very flexible system with extraordinary real-time capabilities, supporting multistandard and multimode operation. It provides an outstanding test-bed for RFID and near-field communication (NFC) [12] characterization.

By means of very little modification to the GNU Radio framework, a very basic non-real-time RFID reader could be developed. However, because of the limitations poised by GR, the investigations undertaken by the research community in brining RFID and GNU Radio together have been very limited. In the following paragraph we list some of the more prominent projects in this area.

Several attempts have been made in building an RFID sniffer with GNU Radio [13]. In the meantime students from MIT's Department of Architecture have used GNU Radio to demonstrate and evaluate co-channel BPSK source separation algorithm. They have used passive RFID tags as signal source [14]. A professor at the University of Kansas and director of RFID Alliance Lab is proposing starting an "OpenRFID" initiative similar to that of GNU Radio [15]. However, the project has not produced any output as yet.

5.6.1 Basics of Passive RFID

A simple illustration of the concept of an RFID system is provided in Figure 5.6. An RFID system can be seen as an ad hoc network formed between a reader and the tags that are within its read range. The reader selects a target tag to establish a communication link for transfer data. The transmitter of interrogation signal, which is contained within an interrogator, communicates via electromagnetic waves with an electronically coded label to elicit from the label a reply signal containing useful data characteristic of the object to which the label is attached. The reply signal is detected by a receiver in the interrogator and made available to a control system. A passive tag is powered by harvesting energy from the reader's transmitted electromagnetic wave. The reader initiates an inventory round by sending an unmodulated carrier signal (CW) to energize the tags. The CW signal is sent out throughout the inventory round

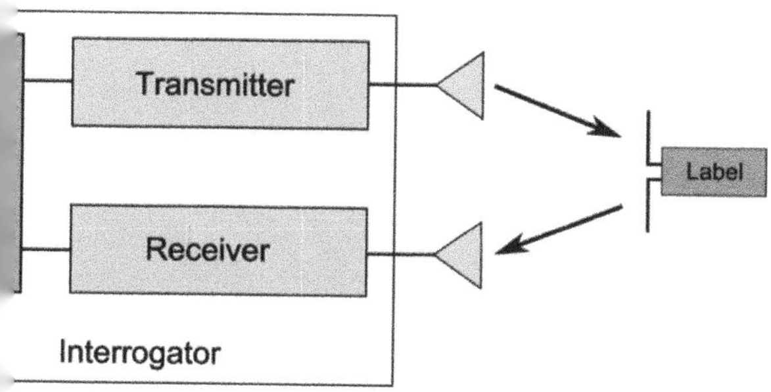

GURE 5.6. Simplified illustration of an RFID system.

ve. The reader sends a command to tags by modulating its carrier ng a command, the reader stops modulating its carrier and enters uring this period the reader still sends a CW while while listening gs respond to the reader's enquiry by backscattering the reader's ckscattering (i.e., the reflection of the incoming carrier power) is ing the state of the tag antenna impedance.

principle and operating frequency of an RFID system are driven application of the labeling system and the constraints provided by mpatibility regulations, environmental noise, and the ability of EM a scanned region of space or to penetrate intervening materials. ound in reliable and secure data collection, object or personal uthentication, and the detection of location of scanned objects. erating at ultra-high frequency (UHF) is employed when a longer (several meters) is required, and high frequency (HF) is employed etic fields that exhibit good material penetration and sharp spatial re required or low system implementation cost is desired.

tation

n implemented using the GNU Radio and USRP hardware. The composed of four basic building blocks as depicted in Figure 5.7: r, protocol processor, and network interface. In the following scuss each of these components in more detail.

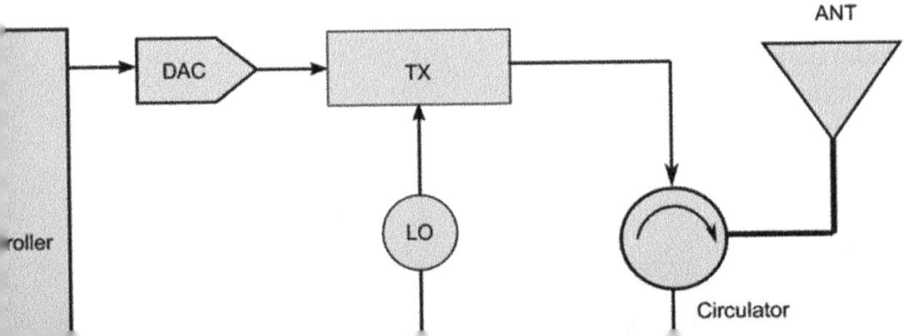

ANT
DAC
TX
roller
LO
Circulator

(ADC) channel and handed off to the FPGA for further processing before being forward to the PC for demodulation.

5.6.5 UHF Module

The air interface protocol at UHF is designed to be compliant with the EPC Class 1 Generation 2 (EPC C1G2) protocol specification as proposed by the EPC global in collaboration with its sponsoring institutions [8].

The UHF module is a UHF-to-baseband down-converter designed for frequency hopping operation. It comes with the FCC's Part 15.247 rules. These rules specify that a maximum output power of 1 W may be used in a frequency hopping system using at least 50 channels, with maximum dwell time of 400 ms at any given frequency. The module was therefore subject to the limitations of the PLL lock time and receiver's transmit-to-receive recovery time. A conscious trade-off was made between channel utilization and cost. At this stage a single synthesizer design is chosen because of its lower cost even though the synthesizer lock time would result in "dead time" during which the reader field would be off during channel transitions.

5.6.5.1 Local Oscillator. The local oscillator for the module is implemented using a phase-locked loop synthesizer module (Z-Comm Inc PSN0930A), integrating a VCO and a National Semiconductor LMX2316 PLL IC. This inexpensive module generates +3-dBm output power with phase noise specified at −100 dBc/Hz at 10 kHz. Significant harmonic energy is present at the VCO output port. A 6-dB attenuator pad is used between the VCO and the first amplifier to increase load isolation of the VCO. The amplifier's output power is approximately +8 dBm. The amplifier's output is passed through a two-pole ceramic filter with bandpass frequency centered at 915 MHz to remove the second harmonic and other spurious outputs. This is then followed by a power splitter that splits this local oscillator signal into two paths, one for transmit and one for receive.

5.6.5.2 Transmit Path. The transmit signal comprises a digitally controlled step attenuator. The attenuator provides finer control over RF output power level. The signal is then fed into final RF amplifier followed by a second surface acoustic wave (SAW) bandpass filter for harmonic and spurious output suppression. The transmitter is capable of delivering up to 30 dBm of RF power at UHF band. The modulation of the output signal is achieved through software on the host PC.

5.6.5.3 Receive Path. The majority of the receive path is implemented in software on the host PC. Therefore the analog receive section is very simple. The incoming RF signal is filtered by a SAW bandpass filter and split into two signal paths for in-phase and quadrature (IQ) demodulation. These signals are fed to two double-balanced mixers. The local oscillator signals for these mixers are generated by splitting and phase delaying the receive LO path to generate a 90° lag in the Q path as shown in Figure 5.9. There is no RF preamplifier in the receiver path before the mixers because the dynamic range of the input signal would make it difficult to

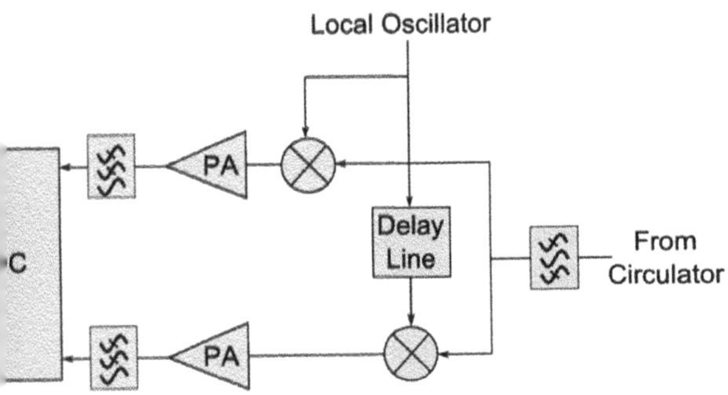

FIGURE 5.9. Block diagram of I and Q demodulator.

able signal-to-noise ratio at the input. The I and Q channel outputs d filtered by an anti-aliasing filter with a bandwidth of 450 kHz. ls are then applied to the ADC on the motherboard. Further signal complished digitally on the host PC, because this provides the most possible.

mentation. The graph in Figure 5.10 shows the data path through ion of the reader. The RF transponder signals are collected using 0 daughterboard and are passed on to the computer for processing. side, a correlator is used to find the preambles followed by a framer in he actual packet from the real-time stream. Block sets containing real- receivers can be found in the digital subsection of the GNU Radio The signal processing blocks mentioned above can be connected to form a full receiver chain.

stage, the framer block pushes the received packets into a message upper layers can pick it up to reconstruct frames, check CRC and he signaling from the payload. The signaling goes to the MAC layer while payload goes to the upper layers for EPC Class 1 Generation

rimental Results. Figure 5.3 is a photograph of the completed system was then used to read experimental transponder's replies. operating at its resonance frequency or not matched to its antenna not have enough power to complete its response to a reader command. ows a tag reply where the tag runs out of power as soon as it starts

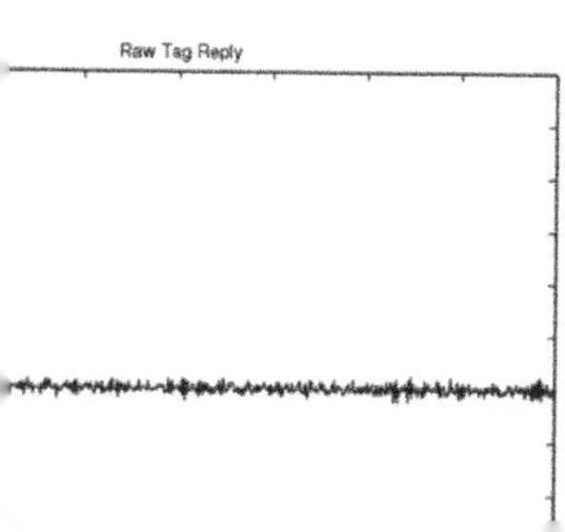

Raw Tag Reply

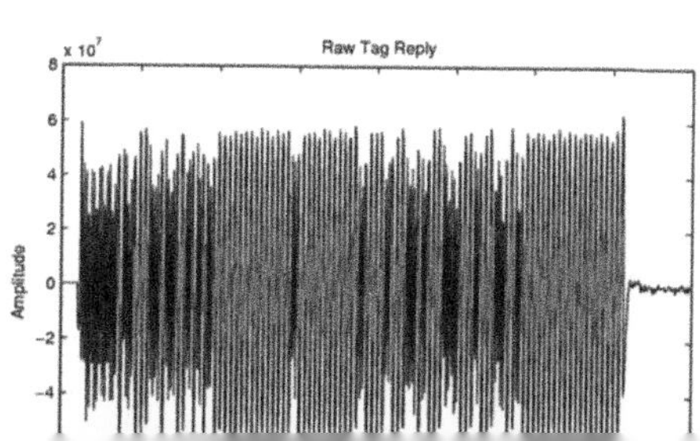

Raw Tag Reply
x 10^7
8
6
4
2
0
−2
−4
Amplitude

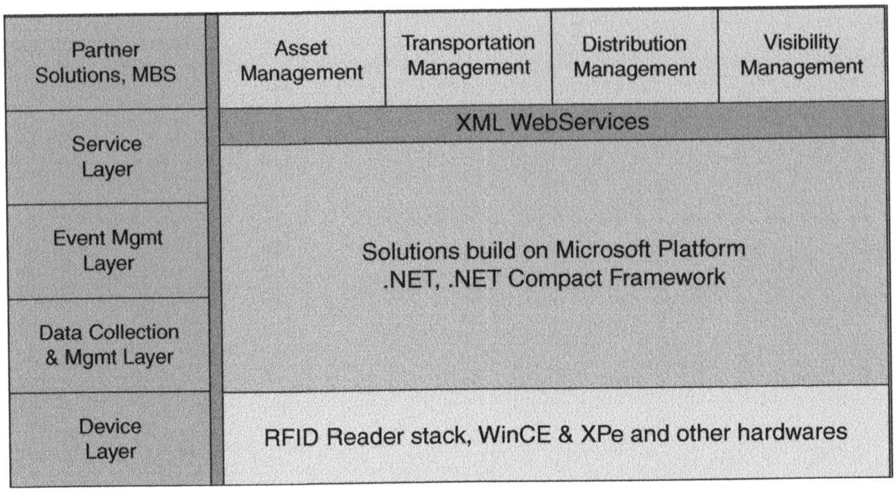

FIGURE 5.12. RFID software stack proposed by Microsoft.

reconfigurable RFID reader design using the GNU Radio. This platform provides a powerful solution for rapidly building and testing new RFID communication systems and protocols, and it forms a basis for developing highly flexible RFID infrastructures. The USRP, developed as part of the GNU Radio project, is an interface board that converts streaming baseband data to an RF output with the help of a band-specific daughterboard. Likewise, the USRP and daughterboard create a streaming data stream from the RF inputs. The RF daughterboard converts the baseband signals to and from the appropriate frequency bands. The FPGA on the USRP converts the data rate to match data rates supported by a USB 2.0 high-speed controller. In addition, the USB interface combined with the FPGA provide digital IO and low-speed analog IO signals for use by daughterboards. This approach can allow designers to build a flexible RFID reader that can be easily integrated within the existing infrastructure of a company at an affordable price. Among the market sectors it is targeting are logistics, transport, asset tracking, identification, and ticketing.

ACKNOWLEDGEMENTS

We would like to thank Professor Bevan Bates and Mr. Shivvaan Sathasilvam for their help and collaboration in setting up the hardware used in this project.

REFERENCES

1. S. Lewis, A basic introduction to RFID technology and its use in supply chain, Technical Report, January 2004.
2. M. Bhuptani and S. Moradpour, *RFID Field Guide—Deploying Radio Frequency Identification Systems*, Microsystems Press, NJ, 2005.

3. GNU Radio Wiki, USRP FPGA, GNU Radio, Free Software Foundation, 2007 [Online]. Available at http://gnuradio.org/trac/wiki/UsrpFPGA. (accessed September 2008).
4. Ettus Research LLC, The USRP, http://www.ettus.com/.
5. J. Reed, *Software Radio: A Modern Approach to Radio Engineering*. Prentice-Hall, Upper Saddle River, NJ, 2002.
6. J. Mitola, G. Maguire, Telesystems Conference, 1992, NTC-92, Washington DC, 19–20 May 1992, pp. 13/15–13/25.
7. B. Jamali and P. H. Cole, *RFID Handbook: Applications, Technology, Security, and Privacy*, CRC Press, Boca Raton, FL, 2007, pp. 760–768.
8. O. Ernst and G. Golub, EPC Class 1 Generation 2 UHF air interface protocol standard version 1.09, Technical Report.
9. Open Source EDA project homepage, http://www.gpleda.org/.
10. BBN Technologies Group, GNU Radio architectural changes, BBN Technologies, 2006 [Online]. Available: http://acert.ir.bbn.com/downloads/adroit/gnuradio-architectural-enhancements-3.pdf. (accessed September, 2008).
11. Python, an interpreted, interactive, object-oriented, extensible programming language, http://www.python.org/.
12. Near Field Communication, http://wiki.unik.no/index.php/NFC/HomePage.
13. Open PCD, RFID sniffer, Open PCD, 2005 [Online]. Available at http://www.openpcd.org/rfiddump.0.html. (accessed September 2008).
14. G. R. Woo, Demonstration and evaluation of co-channel DBPSK source separation, Master's thesis.
15. RFID Alliance Lab [Online]. Available at http://www.rfidalliancelab.org/. (accessed November 2008).
16. S. Preradovic and N. Karmakar, RFID readers—A review, in *International Conference on Electrical and Computer Engineering*, December 2006, pp. 100–103.
17. M. Reynolds and C. Weigand, Design considerations for embedded software-defined RFID readers, *Emerging Technol.*, pp. 14–15, August 2005, available at www.rfdesign.com.
18. Five factors for success—UHF Gen 2 RFID readers, Impinj technical paper, available at www.impinj.com.
19. Microsoft RFID Technology Overview [Online]. Available at http://msdn.microsoft.com/en-us/library/aa479362.aspx. (accessed November 2008).
20. K. Finkenzeller, *RFID Handbook: Radio Frequency Identification Fundamentals and Applications*, John Wiley & Sons, New York, 1999.

PART III

PHYSICAL LAYER DEVELOPMENTS OF SMART ANTENNAS FOR RFID SYSTEMS

CHAPTER 6

RFID PLANAR ANTENNA—SMART DESIGN APPROACH AT UHF BAND

SUSHIM MUKUL ROY
Department of Electrical and Computer Systems Engineering, Monash University, Clayton, Victoria, Australia

NEMAI CHANDRA KARMAKAR
Department of Electrical and Computer Systems Engineering, Monash University, Clayton, Victoria, Australia

6.1 INTRODUCTION

Numerous developments are reported globally in conventional RFID technology in the ultra-high-frequency (UHF) band. Following the standardization of ISO 18000-6C [1] by EPC global [2], RFID operations got a global boost. The band allocated to RFID technology in the UHF band is around 800 MHz in Europe and the Asian nations like Brunei, Hong Kong, Malaysia, and Singapore [3] and [2]. However, for countries like the United States, Australia, or Japan, the frequency range is around 900 MHz. To make a tag and reader work in both the United States and Europe, the solution is to make frequency-agile or dual-band systems that work in both the frequencies of 800 MHz and 900 MHz. The other option, being taken up by the government organizations, is to come to a general agreement for uniform frequency allocation.

With the different frequency allocations, developments are being reported for accessory supporting components also like Power Amplifiers working in the range of 800–1000 MHz [4]. New IP backbones are developed to provide intercompatibility between RFID systems or digital radio working at 800 MHz with an analog user [5]. The added feature of this work is the convergence of RFID systems working at an 800-MHz UHF system with the digital public address systems. Organic rectifiers [6] are reportedly developed at this frequency range to work in the RFID tags. Some advantages of using RFID in the UHF range near 800 MHz can be found in proprietary

Handbook of Smart Antennas for RFID Systems, Edited by Nemai Chandra Karmakar

Internet newsletters such as [7] or in RFID-related review articles such as [8], with special emphasis to application of RFID in retail chains.

For the RFID tags operating at this frequency, dipoles or meander-like antennas with variations in their designs are predominant. One such variation, reported in [9], enables the efficient use of the RFID tags at this frequency for use on metal containers. A variation of meander-like antenna with open ends for use in this frequency range is reported in [10], where the antenna is fabricated using conductive ink on a thin plastic substrate. Instead of using a continuous ground plane, a meander line is used as the feed with reported improvement in the bandwidth. General guidelines for development of antennas in this frequency can also be found in the proprietary documents available in the Internet like that of Ansoft Corporation® [11]. Thus it is evident that the UHF frequency range for commercial RFID deployment is a subject of current research and improvement globally.

Keeping in view the aforesaid developments in RFID technology in the UHF band, an antenna array is developed and presented in this chapter. The antenna array is a 3×2 array with six elements confined within a 400-mm $\times$ 400-mm area. The array is designed to work at 800-MHz center frequency with a (-10-dB) bandwidth of 100 MHz around the center frequency. Suitable scaling for the dimensions is shown in the later sections to make the array work at a center frequency of 915 MHz with a little bit wider bandwidth. Measured gain of 15 dBi was achieved for this array. The antenna elements are semicircular microstrip patches loaded with an L-shaped slot. By changing the slot and the patch dimensions, the array can work in two different bands of the UHF RFID system—for application in both Europe/Asia and USA/Australia/Japan.

The chapter is organized as per Figure 6.1 Initially a literature review is presented that shows the basis of the antenna design. The literature review consists of an overview of microstrip antennas with special mention of circular and slot microstrip antenna structures (MSA) and their loading effect. This is followed by an overview of techniques to enhance the bandwidth and gain and to mitigate mutual coupling between adjacent elements. This is followed by the design of the antenna where the different dimensions are shown and discussed. The results are shown along with the effect of variation of different design parameters. Different techniques are discussed related to measurement and minimizing the effect of mutual coupling, with the introduction of metallic barriers between the individual elements being the best one suited for the purpose. A six (3×2)-element array is designed and fabricated. The design, simulation, and measurement results are presented. Lastly, the effect of beam-squinting is studied categorically by incorporating a delay in the feed line by extending the length of microstrip line.

6.2 BACKGROUND LITERATURE REVIEW: BASIS OF DESIGN

Microstrip antennas were considered for their planar structure, low profile, and ease of integration with other circuit components. Basics of microstrip antenna design with their inherent shortcomings can be found in [12–17]. Different varieties of antenna,

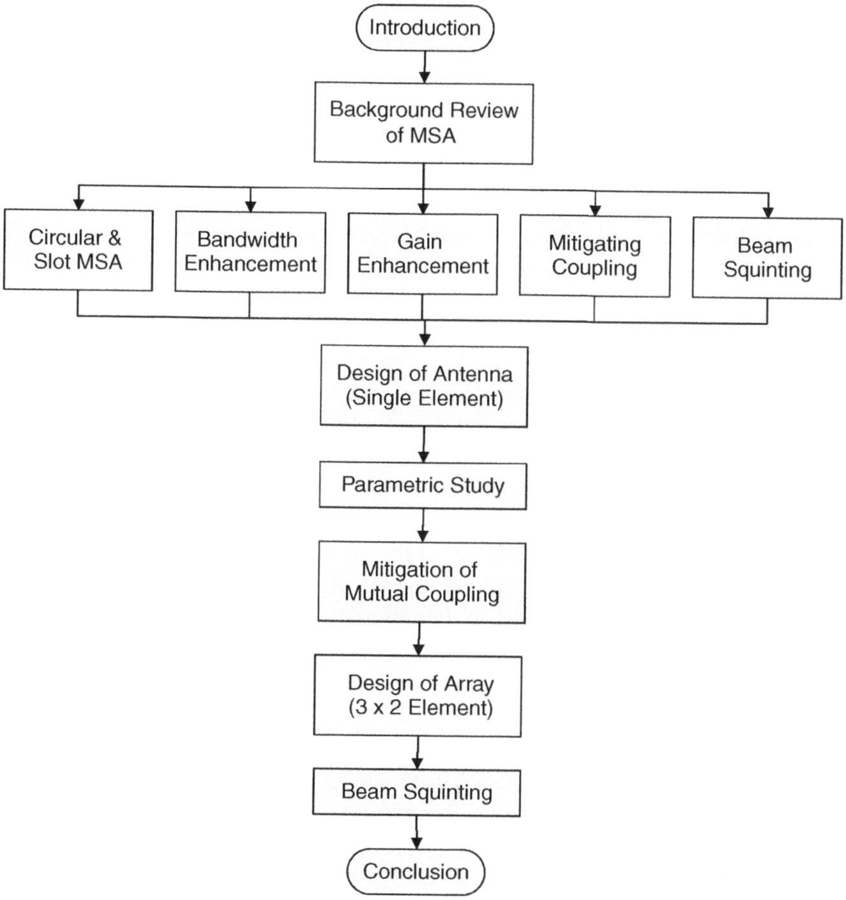

FIGURE 6.1. Flowchart of organization of the chapter.

different feed techniques, and circuit extraction are presented there. Techniques are discussed for bandwidth enhancement and design of arrays and the problems encountered in those processes.

The antenna designed in this chapter is an array of six antennas (3×2), where the individual elements are semicircular microstrip patches with an L-shaped slot. The antenna is a multilayer structure where power division takes place in one layer and the patches (for radiation) are placed in another layer. The individual patches (loaded with slot) are fed by cylindrical copper wire. The aim of this design was to produce a compact array with high gain and wide bandwidth.

Circular microstrip antennas are a part of the different available geometries of microstrip antennas. Different configurations are studied on the circular microstrip patch by shortening it [18] or cutting it into a semicircular shape. By making the microstrip antenna semicircular, the gain reduces to some extent due to the reduction of the radiating area. However, the antenna becomes half the size. The nature of the fringing fields change as a result, and this influences the capacitance [19] of the semicircular disk.

Slot antennas occupy an important part in microstrip antenna technology. In slot antennas, the radiation occurs from both edges, which makes it a good radiator within a smaller area. They are compact in size. The design guides for slots can be found in the aforesaid literature. Further analyses are shown in [20, 21], where a rectangular slot is analyzed in the ground plane with reference to aperture-coupled microstrip patch antenna. Certain shapes of slots on the patch layer can even change the polarization of the antenna; for example, placement of a cross slot on the circular microstrip patch makes it circular-polarized [22]. Studies on effects of loading with slots of different shapes are found in [23]. The work of Deshmukh and Kumar [23] is particularly helpful for this chapter because it involved the incorporation of slots in semicircular patches, thereby making the antennas broadband and compact.

When a microstrip patch antenna is loaded with a slot, two resonances occur simultaneously. The patch antenna resonates at one particular frequency determined by its geometry. The slot also resonates at a particular frequency, depending on its geometry. The frequencies of resonances of the aforementioned antenna and slot can be separated by changing the design parameters. Thus the antenna can be made to perform in dual band, which is particularly important for working at UHF frequencies for RFID near 800 MHz and 900 MHz, respectively.

6.2.1 Bandwidth and Gain Enhancement Techniques

Microstrip antennas have an inherent shortcoming of narrow bandwidth. If a microstrip antenna operates at either of the two aforesaid frequencies (800 and 900 MHz), it would perform well but exhibit narrow bandwidth, thereby limiting the operation around one of the frequencies. However, if the patch and slot are designed to resonate at the same frequency, then the antenna exhibits a wide bandwidth at the design frequency. This is a technique widely deployed in microstrip technology to produce broadband antennas [13, 15].

Different techniques for increasing the bandwidth of patch antennas are found in [24]. One of the common techniques is impedance matching at the source [25]. Multilayered structures are proposed in some of the techniques. However, designing broadband microstrip antennas in single layer is shown in [26]. Both bandwidth and gain are reported to be increased simultaneously for compact planar antennas [27]. Based on the aforesaid literature review, a semicircular microstrip patch loaded with an L-shaped slot is chosen to make the structure compact and resonate with a wide bandwidth.

The aim of this chapter is to design an antenna for usage in an RFID reader, so high gain is one of the key requirements. In addition to the aforementioned literature, gain enhancement techniques are reported in [27–29]. Most of the techniques concentrate on the gain enhancement of a single patch by incorporating a layer of thick dielectric of high dielectric constant such as ceramic, and so on, above the patch layer. Weng et al. [30] reports a novel technique of use of metamaterial structure to increase the gain. The idea behind this technique is the incorporation of microwave lenses.

Following the aforementioned literature to keep the antenna compact, experimentation was done by keeping a thick layer of ceramic slab in front the single patch. Improvement was observed in gain, but it was not so significant. However, the weight of the antenna became very high—a few kilograms. This is difficult to work with for mobile RFID readers. The concept of microwave lens [30] was also investigated. Using the calculations mentioned in the paper, incorporation of the lens resulted in some improvement of gain, but the size of the microwave lens for a single antenna element became too big to incorporate in handheld RFID readers.

Two other conclusions were drawn at this stage: First of all, if high dielectric substrate is used for the microstrip antenna (below the antenna), the size of the patch reduces, but the gain also reduces drastically. Hence the antenna is designed using air as the substrate—lowest permittivity (= 1). The thickness of air turned out to be in the range of centimeters, but the gain was high and the weight was minimal. Foam was used for mechanical support. Electrically, foam is very close to air/vacuum. There was a size constraint of 400 mm $\times$ 400 mm for handheld reader operation. So, to increase the gain within the given area, the other conclusion drawn was to go for an array of multiple elements.

6.2.2 Mitigating Mutual Coupling in Electrically Thick Array

In addition to the aforementioned literature, guidelines on microstrip arrays are found in [31]. The design of feed line for the array is based on power division using the guidelines set in [32]. In microstrip technology, like other types of antennas, mutual couplings take place when the elements of the array are placed very close to each other [33, 34]. Use of EBG structures in the array can lead to low mutual couplings in microstrip antennas [35]. The mutual couplings take place due to the interaction of the fringing fields of the adjacent elements. If this interaction can be anyhow reduced, the mutual couplings can also be reduced. Because the present design is a three-layer design with a foam layer as a sandwich between the upper and the lower layer, a different approach is taken up to reduce the mutual coupling.

Thin aluminum metal barriers are placed between the adjacent patches and also the boundary. They start from the ground plane in the bottom layer, travel all along the foam layer, and protrude a bit above the patch layer. The concept is to cage the fringing fields of the individual elements and prevent them from influencing the fields of the adjacent elements. The blinders/winkers/blinkers used on the eyes of the horses during a race can be used as an analogy of the concept: They prevent the horse from looking sideways. This concept proved to be fruitful, resulting in high gain along with narrow beamwidth and wide bandwidth. The size of the antenna was well-contained within the specified dimension constraints, and the weight of the whole system was less than a kilogram with an outer casing.

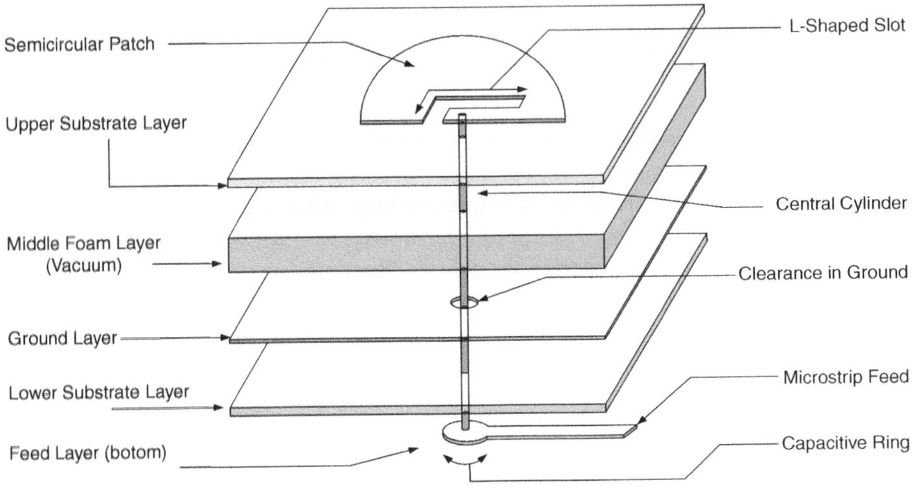

FIGURE 6.2. 3D view of the single-element semicircular patch loaded with an L-shaped slot. The lower substrate is 0.79-mm-thick Taconic TLX-0 substrate ($\varepsilon_r = 2.45$). The upper substrate layer is 1.6-mm-thick FR4 substrate ($\varepsilon_r = 4.4$). The middle layer is 25-mm-thick foam—electrically very close to vacuum. All metal parts are made of copper.

6.2.3 Beam Scan Array

Once the array was completed, one further investigation was carried out to explore if the antenna can be used as a smart antenna where the beam squinting/steering is done by incorporating delays between the elements. Information on phased-array antennas can be found in [36–38]. Beam squinting was investigated by incorporating delays using microstrip lines. Furthermore, phase shifters were designed by my colleague Saqib Iqram, who implemented them in this array and thereby making it a smart antenna.

6.3 DESIGN OF THE ANTENNA: SINGLE ELEMENT

The 3D view of the single element is shown in Figure 6.2, where the different layers are separated for clarity. The lowest layer is the microstrip feed layer. The width of the microstrip line corresponds to characteristic impedance of 50 Ω. Input power for the patch is fed through the microstrip line. The other end of the microstrip line ends in a circular pad. The radius of the circular patch is 3 mm, and above the microstrip line is the 0.79-mm-thick "lower substrate layer." The material of lower substrate layer is Taconic TLX-0 ($\varepsilon_r = 2.45$). Above the lower substrate is a 17-μm thick ground layer. From the microstrip feed line, a copper wire of 1.5-mm radius runs all the way up to the upper patch layer. In ground layer, around the central cylinder, there is a clearance of 1.5-mm width to prevent shorting of the central cylinder with the ground

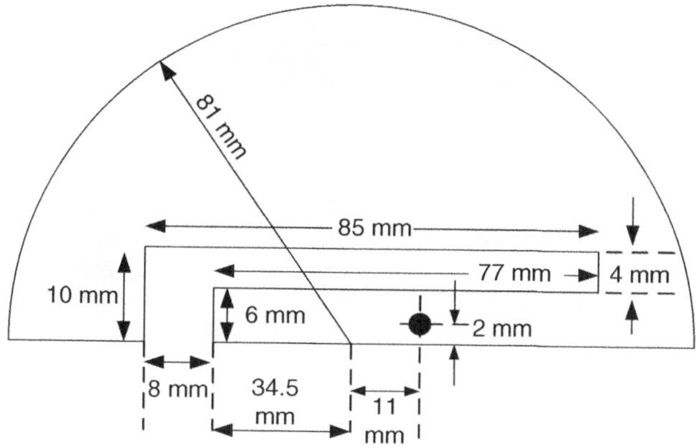

FIGURE 6.3. Top view of the semicircular patch and the slot indicating dimensions.

plane. Above the ground plane is the layer of foam. The thickness of the foam layer is 25 mm, and the dielectric constant is nearly the same as for vacuum ($\varepsilon_r = 1.02$). Above the layer of vacuum is "upper substrate layer." The upper substrate layer is 1.6-mm-thick FR4 ($\varepsilon_r = 4.4$). Above the upper substrate layer is the patch layer. The patch is a semicircular patch loaded with an L-shaped slot. The input power to excite the patch comes through the central cylinder running all the way from the bottom layer.

Figure 6.3 shows the top view of the semicircular patch loaded with the L-shaped slot. The individual dimensions are shown in the figure. The position of the central cylinder is shown by a black-filled circle with the associated dimensions indicating the center of the feed point. The center of the feed cylinder is 11 mm away from the center of the patch and 2 mm away from the edge.

The CST MWS^{TM} simulated and measured results are presented next. For simulation, the boundaries are all considered as "open-add space." A waveguide port is used between the microstrip feed and the ground plane. Three mesh planes are made to go through the lower substrate, and six mesh planes are made to go through the upper substrate. Through the foam layer, 10 mesh planes were made to go, 6 mesh planes were made to go through the central cylinder, and 2 mesh planes were made to go along the length of the L-shaped slot (85-mm length). These simulation settings conformed to the requirements of the software for maximum accuracy of simulation. In the fabricated patch, standard available copper wire of 0.125-in. (= 3.175 mm) diameter was used instead of 3-mm diameter. This might contribute to a small amount of discrepancy between the simulated and the measured return losses.

Figure 6.4a shows the comparative representation of the simulated and measured results of return loss versus frequency. As observed in the figure, the position of the resonance is same, but the magnitude of the dip of the S_{11} curve for the measured result is less than that of the simulation. Also, the bandwidth of the measured result is a bit wider, which is probably due to the lossy nature of FR4 substrate. Figure 6.4b shows the orientation of the antenna element with respect to the coordinates. The 3D

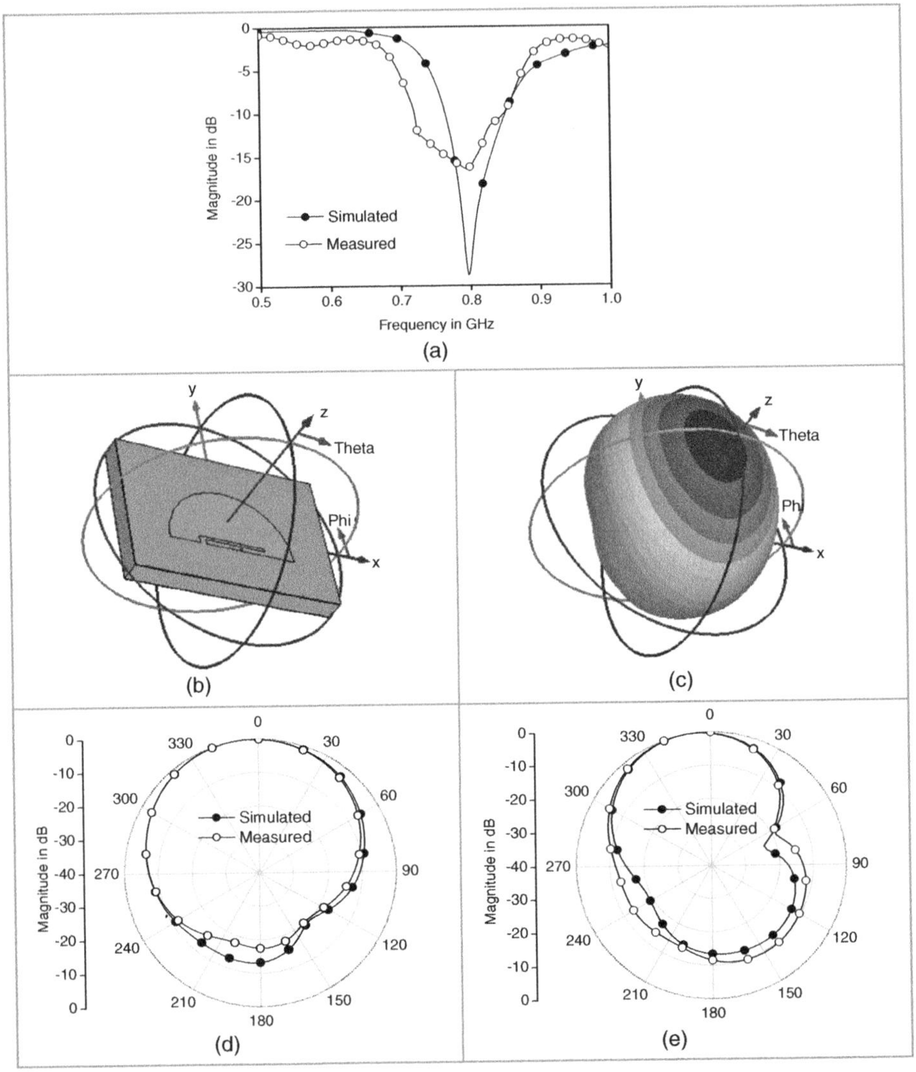

FIGURE 6.4. (**a**) Simulated and measured return loss of a unit antenna element. (**b**) Orientation of the antenna element with respect to the coordinates. (**c**) 3D view of the radiation pattern. (**d**) Normalized polar view of the radiation pattern when angle is varied along Theta keeping Phi constant at 90°. (**e**) Normalized polar view of the radiation pattern when angle is varied along Theta keeping Phi constant at 0°.

radiation pattern is shown in Figure 6.4c. The simulated gain is 7.45 dBi. The polar view of the normalized radiation patterns are shown in Figures 6.4d and 6.4e, where the angle is varied along Theta, keeping Phi constant at 90° and 0°, respectively. The orientation of the main beam is tilted 10° away from the normal. The main beam is wider along Theta when Phi is 90° compared to Phi at 0°.

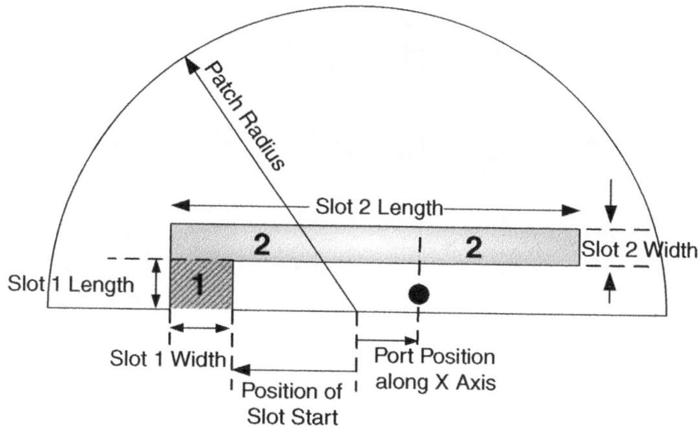

FIGURE 6.5. Nomenclature of the different design parameters and their location (top view).

6.4 STUDY OF VARIATION OF THE DESIGN PARAMETERS

Figure 6.5 shows the names of various design parameters and their positions on the L-shaped slot-loaded semicircular patch. The L-shaped slot is divided into two different segments: slot 1 and slot 2. The two segments of the slot are marked in different shades for clarity. The position of the port and starting position of the slot are offset horizontally from the center of the patch. Other design parameters include the radius of the patch, the thickness of the foam, the radius of the capacitive ring at the feed layer, and the radius of the central cylinder that runs from the feed layer to the patch layer.

Initially, for different design parameters of the slot and the patch, two different resonances were observed at a difference of 100 MHz. However, after both of them were optimized to resonate at the same frequency, the difference was less for variation of the design parameters. As and when resonance at two different frequencies are observed, it is mentioned in the following sections. All design parameters are centered in Figure 6.3, and their positions are referred to in Figure 6.5.

In Figure 6.6a the thickness of the foam is varied 20 mm centered on the design thickness 25 mm. The resonant frequency goes down in a linear way with the increase of thickness of foam. The return loss shows −30 dB dip for the design thickness. When the thicknesses of foam are 30 mm and 35 mm, second resonances are observed at 0.805 GHz and 0.796 GHz. The magnitudes of the second resonances are −22 dB and −30 dB, respectively.

In Figure 6.6b the radius of the semicircular patch is varied 20 mm centered on the design radius 81 mm. From the figure, it is observed that the resonant frequency decreases in a linear way with the increase of radius of the patch. Best matching is obtained at the design radius. For the patch radii 86 mm and 91 mm, second resonances are observed at 0.835 GHz and 0.845 GHz. However, the magnitudes of the second resonances are less than −10 dB.

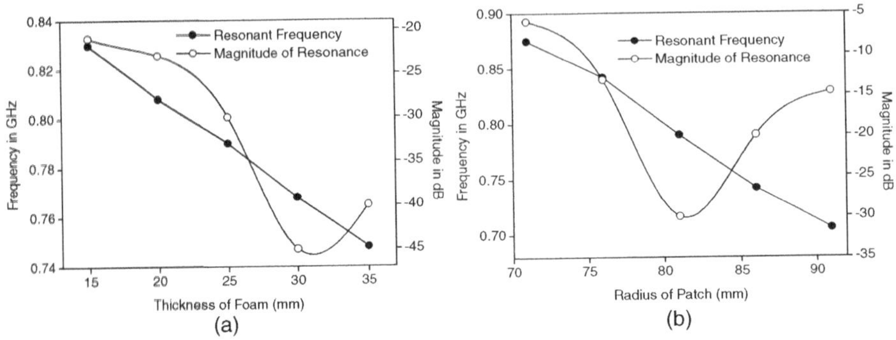

FIGURE 6.6. (**a**) Variation of frequency response for variation of thickness of foam. (**b**) Variation of frequency response for variation of radius of semicircular patch.

In Figure 6.7a the radius of the capacitive ring at the junction of the feed line and the central cylinder is varied. As observed in the figure, the resonant frequency goes up and then goes down. The best matching is obtained at the design radius of 3 mm. No second resonance is observed for the variation of this design parameter. Thus the role of the capacitive ring is proper matching.

In Figure 6.7b the radius of the central cylinder is varied. Like the variation for the capacitive ring, in this case the frequency goes up and comes down within a very small span. The magnitude at resonance shows that best matching exists for the design radius of 1.5 mm. However, a second resonance is observed when radius of central cylinder is 1 mm. The resonant frequency of the second resonance is 0.828 GHz and the magnitude of resonance is −23.5 dB.

In Figure 6.8a, the horizontal distance of the center of the port from the center of the patch is varied. With the increase of the distance, it is observed, that the frequency of resonance increases a bit, but best matching is obtained at the design offset. When

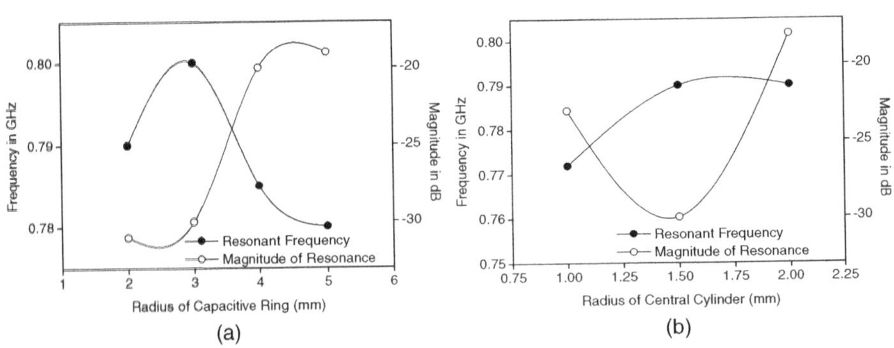

FIGURE 6.7. (**a**) Variation of frequency response for variation of radius of capacitive ring at feed line. (**b**) Variation of frequency response for variation of radius of central cylinder connecting patch with feed line.

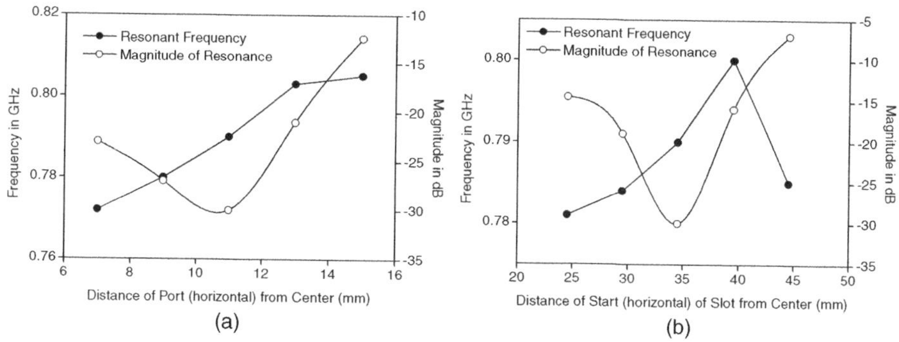

FIGURE 6.8. (**a**) Variation of frequency response for variation of horizontal distance (offset) of port from center. (**b**) Variation of frequency response for variation of horizontal distance (offset) of starting of slot from center.

the distance of the port is 7 mm away from the center, a second resonance is observed at 0.875 GHz. The magnitude of the second resonance is −8.75 dB.

In Figure 6.8b, the horizontal distance of the starting of the L-shaped slot is varied. It is observed in the figure, the frequency of resonance at first increases with the distance and then falls down. Best matching is obtained at the design offset of 34.5 mm away from the center of the patch. Second resonance is observed for the distances 25 and 30 mm. The resonant frequencies are 0.925 and 0.875 GHz. The magnitudes of resonances are around −10 dB.

In Figures 6.9a and 6.9b the dimensions of slot 1 are varied. In Figure 6.9a the length of slot 1 is varied. From the figure it is observed that a good matching exists for the design slot length, but for a slot length of 10 mm, there exists a better match. Second resonance at 0.87 GHz with magnitude less than −10 dB is observed for 2-mm length of slot 1. In Figure 6.9b, the width of the slot 1 is varied. The frequency goes down with the increase of slot width, but the matching becomes bad. A good match exists for the design width of 8 mm. For lesser widths, better matching exists,

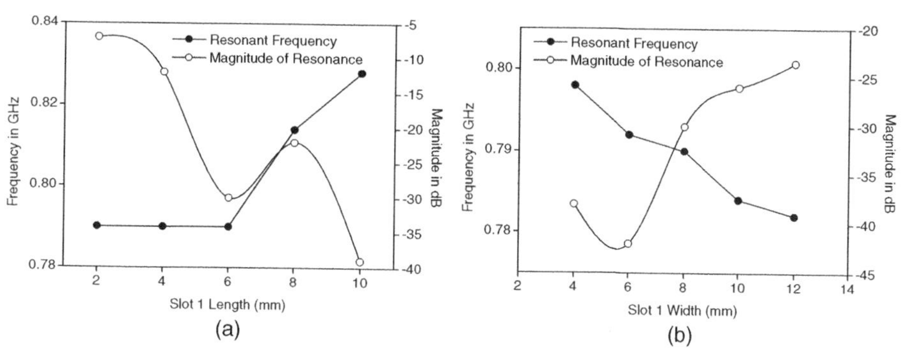

FIGURE 6.9. (**a**) Variation of frequency response for variation in length of slot 1. (**b**) Variation of frequency response for variation of width of slot 1.

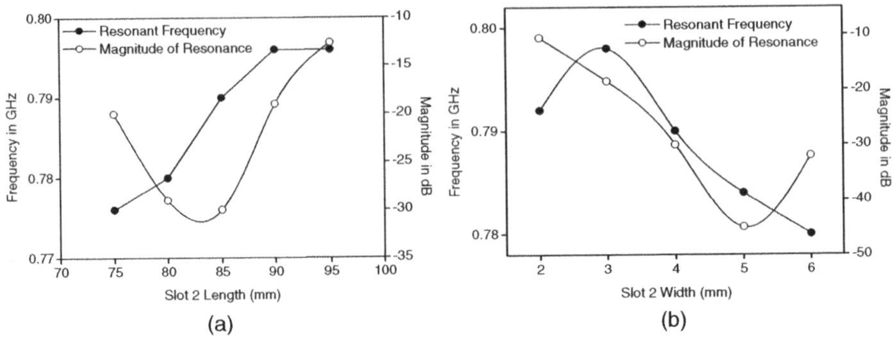

FIGURE 6.10. (**a**) Variation of frequency response for variation in length of slot 2. (**b**) Variation of frequency response for variation of width of slot 2.

but some decrease of bandwidth is observed. Second resonances are observed for slot widths of 10 and 12 mm. The second resonant frequencies are 0.85 GHz and 0.855 GHz with magnitudes around −10 dB.

In Figures 6.10a and 6.10b the dimensions of slot 2 are varied. In Figure 6.10a the length of slot 2 is varied. From the figure it is observed that a good matching exists for the design slot length of 85 mm. The resonant frequency shifts toward the higher side of the frequency spectrum with the increase in the slot length. Second resonances at 0.886 GHz and 0882 GHz with magnitudes around −11 dB are observed for 75-mm and 80-mm lengths of slot 2. In Figure 6.10b, the width of slot 2 is varied. The frequency goes down with the increase of slot width, but the matching becomes bad beyond the design width of 4 mm. A good matching exists for the design width. Second resonance is observed for slot width 6 mm. The second resonant frequency is 0.855 GHz with magnitude −10 dB.

The design parameters are varied based on the prototype mentioned in Figure 6.3. It is observed that for most of the design parameters, there exists a second resonance very close to the first resonance. The design dimensions were optimized so that that the patch and slot resonate at the same frequency. Hence the effect of the second resonance is not so prominent.

If the patch and slot are designed to resonate at two different frequencies, then for the dips of S_{11} the two frequencies become comparable in magnitude. The resonances observed at the higher frequencies are due to the resonances of the slot. On the basis of the aforesaid study, the antenna element can be designed as a dual-band antenna resonating at the two ends of the UHF band for RFID operation.

In some of the aforementioned parametric studies, it is observed that better matching exists for some other design parameters which are close to that of the prototype. They are not used in the design because in those cases the bandwidth becomes narrow.

Thus, an antenna element is designed which gives broadband performance centered on 800 MHz. All possible design parameters of the antenna are varied based on the prototype. The variations in the design parameters give an indication of the tolerance in the frequency response.

6.5 GAIN IMPROVEMENT METHOD—SUPERSTRATE (CERAMIC)

On the basis of the aforementioned literature review, an investigation was carried out to enhance the gain of the single-antenna element by the introduction of a layer of superstrate. A superstrate layer of high dielectric constant on the patch antenna offers the lensing effect of the beam, leading to enhancement of the gain.

The unit cell discussed in Section 6.3 is considered as the antenna element. Using the design guidelines and techniques for gain enhancement, a few investigations were carried out using a superstrate of high dielectric constant. In this section, one of the cases is discussed. The unit antenna element is given a layer of ceramic superstrate of dielectric constant ($\varepsilon_r = 10$). The ceramic superstrate is 28 mm thick and is placed 180 mm above the patch layer.

This thickness of the material is discussed specifically due to the commercial availability of this material. However, for a small block the weight is quite heavy (a few kilograms). The dimension of the available ceramic block on the basis of which this simulation was done is 260 mm $\times$ 180 mm $\times$ 28 mm.

The return loss for the aforementioned arrangement is shown in Figure 6.11a. As observed in the figure, the dip of S_{11} goes down to around -30 dB, but there is a shift in the center frequency. The -10-dB bandwidth is around 100 MHz. The orientation of the aforementioned arrangement with respect to the coordinates is shown in Figure 6.11b. The 3D radiation pattern is shown in Figure 6.11c. The simulated maximum gain for this arrangement is 8.1 dBi, which is 0.65 dBi higher than that for the single element. The normalized polar views of the radiation patterns are shown in Figures 6.11d and 6.11e. It is observed in the figure that the beam is a bit more directional and is more aligned with the normal to the plane of the antenna element (Z axis). However, there is development of back-lobe for this arrangement which reduces the effective gain in the front direction. Gain improvements are observed with high dielectric materials if the cross-sectional area of the dielectric slab is sufficiently large compared to the antenna element and if the slab structure is plano-concave. However, the constraint of size and weight hinders the use of dielectric lens for the current application.

Another method of gain improvement was investigated as mentioned in [30]. A 0.5-mm thin sheet of FR4 substrate was used. Scaling was done for the appropriate size of the periodic circular holes. The holes were 9.5 cm in diameter. Simulation and measurement were both done keeping the cross-sectional area of the microwave lenses within 900 $\times$ 900 mm^2. However, the gain for this arrangement also did not improve beyond 1 dB. Greater cross-sectional areas or multiple lenses are not feasible for handheld reader operation. In simulation, for a period of square lattice less than one wavelength, the periodic structure behaved as a thin-wire array and hence acted as a reflector. With the increase of the period of the square lattices, the structure starts behaving as a lens instead of a reflector.

The unit antenna element occupies a planar area of 260 mm $\times$ 180 mm with clearance on all the sides. The maximum area within which the antenna was to be designed was 400 mm $\times$ 400 mm. Hence the choice was a $3 \times 2 = 6$ element array. Each element was supposed to get equal power; hence the feed line [32] is designed by making the phase at each end the same for all the elements.

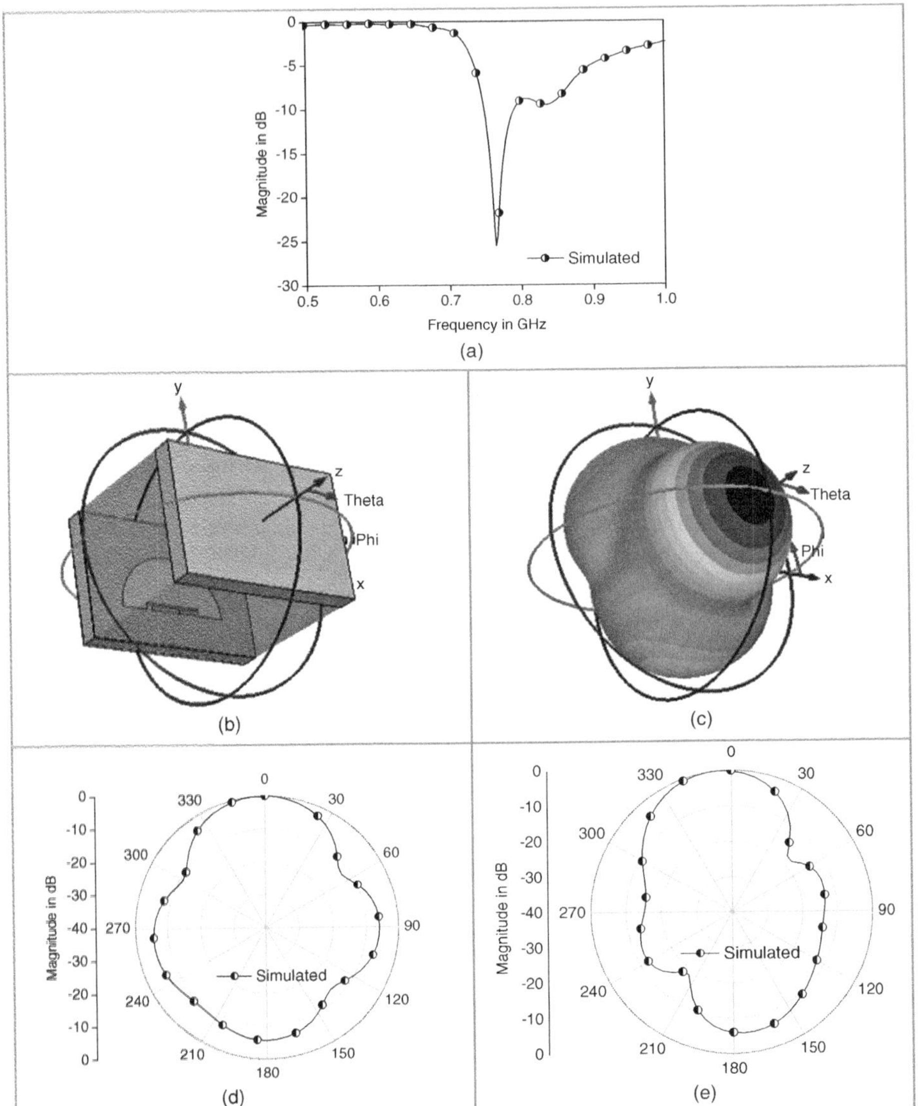

FIGURE 6.11. (**a**) Return loss of single antenna element using ceramic superstrate. (**b**) Orientation of the single antenna element with ceramic superstrate with respect to the coordinates. (**c**) 3D radiation pattern for the unit antenna element. (**d**) Normalized polar view of the radiation pattern when the angle is varied along Theta, keeping Phi constant at 90°. (**e**) Normalized polar view of the radiation pattern when the angle is varied along Theta, keeping Phi constant at 0°.

However, before designing the array, it was important to know the effect of mutual coupling among the elements. Because the effective thickness of the substrate (25 mm—foam) is electrically thick for the unit elements, the mutual coupling is supposed to be very high. The distance between the elements needed to be adjusted accordingly to minimize this coupling. This was done in several steps following investigation of the actual couplings among the antenna elements for different spacing. This investigation is discussed in the following section.

6.6 ARRAY—MUTUAL COUPLING BETWEEN ELEMENTS

In the aforementioned section, dielectric constant of the superstrate was varied up to 300 (ε_r) with associated variation of the distance of superstrate above the patch. However, maximum gain achievable was only 10 dB for a dielectric constant of 100 for the superstrate. So the next investigation carried out was to make an array of the unit elements.

Before going into the design of the array, the mutual couplings between the antenna elements were studied for their different orientations as shown in Figure 6.12a–d.

Figure 6.12a shows the placement of three patches (slot loaded) in vertical manner so that they are equidistant from each other and also from the external boundary of the substrate. The patches are numbered 1, 2, and 3 as shown in the figure. The length of the substrate is 400 mm. The patches are given excitations individually (separately excited), and the frequency response hence obtained is shown in Figure 6.12b.

It is observed in the figure that the return losses (S_{11}) of the individual patches are fairly similar, but the isolations (S_{21}) are low between the patches 1 & 2 and 2 & 3. The S_{21} and S_{23} parameters went up to -10 dB, indicating that 10% of the power was absorbed by the adjacent patch directly. The isolation between patches 1 & 3 was acceptable and in the range of -20 dB. Thus it was clear that placing patches equidistant in a vertical manner gives rise to a good amount of mutual coupling between the patches.

The other placement that was investigated along with the former one is shown in Figure 6.12c. Because the maximum planar area is 400 mm $\times$ 400 mm, it was possible to incorporate another column of three patches. For that reason, the patches were placed as shown in Figure 6.12c. Two investigations were done simultaneously for this placement. The frequency response of the second arrangement is shown in Figure 6.12d. The return losses (S_{11}) of all the patches are fairly similar. The isolation between patches 1 & 3 is comparatively better, being in the range of around -16 dB. However, the isolation (S_{21}) between patches 1 & 2 and 2 & 3 goes up to around -8 dB. This indicates that the orientation of the patch 3 cannot be used in the array.

From the results of Figures 6.12a–d it is evident that a good amount of mutual coupling exists between the patches if they are simply placed equidistant in the given area. The effects of mutual couplings are further studied in Figures 6.13a–f, where the distances are insertion losses are plotted for different distances between the adjacent

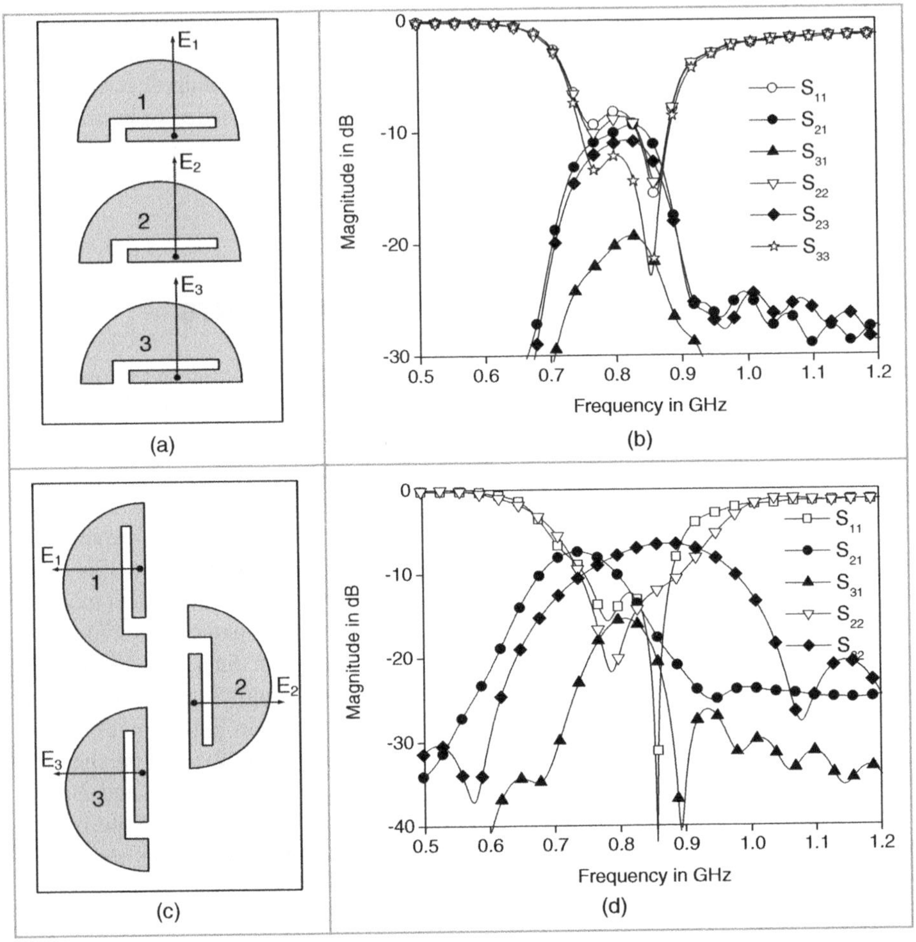

FIGURE 6.12. (**a**) Placement of three patches vertically where patches are in same orientation. (**b**) S_{11} and S_{21} parameters observed for the vertical orientation of the patches. (**c**) Placement of two patches (1 and 3) horizontally in the same orientation while the middle patch (2) is placed in different orientation. (**d**) S_{11} and S_{21} parameters observed for the placement of the patches with mixed orientation (part c).

patches. It is observed that with an increase of distance between the elements, the power absorbed by the adjacent elements decrease in a predictable manner. The different S parameters plotted in Figure 6.12 are all centered at 800 MHz.

To minimize the mutual coupling between the patches, some techniques were investigated, including the placement of EBG structures [35]. But as per the technique showed in the paper, the space consumed becomes large due to placement of periodic EBG structures between the antenna elements. The isolation was lowered from −10 dB to around −15 dB, which was not enough.

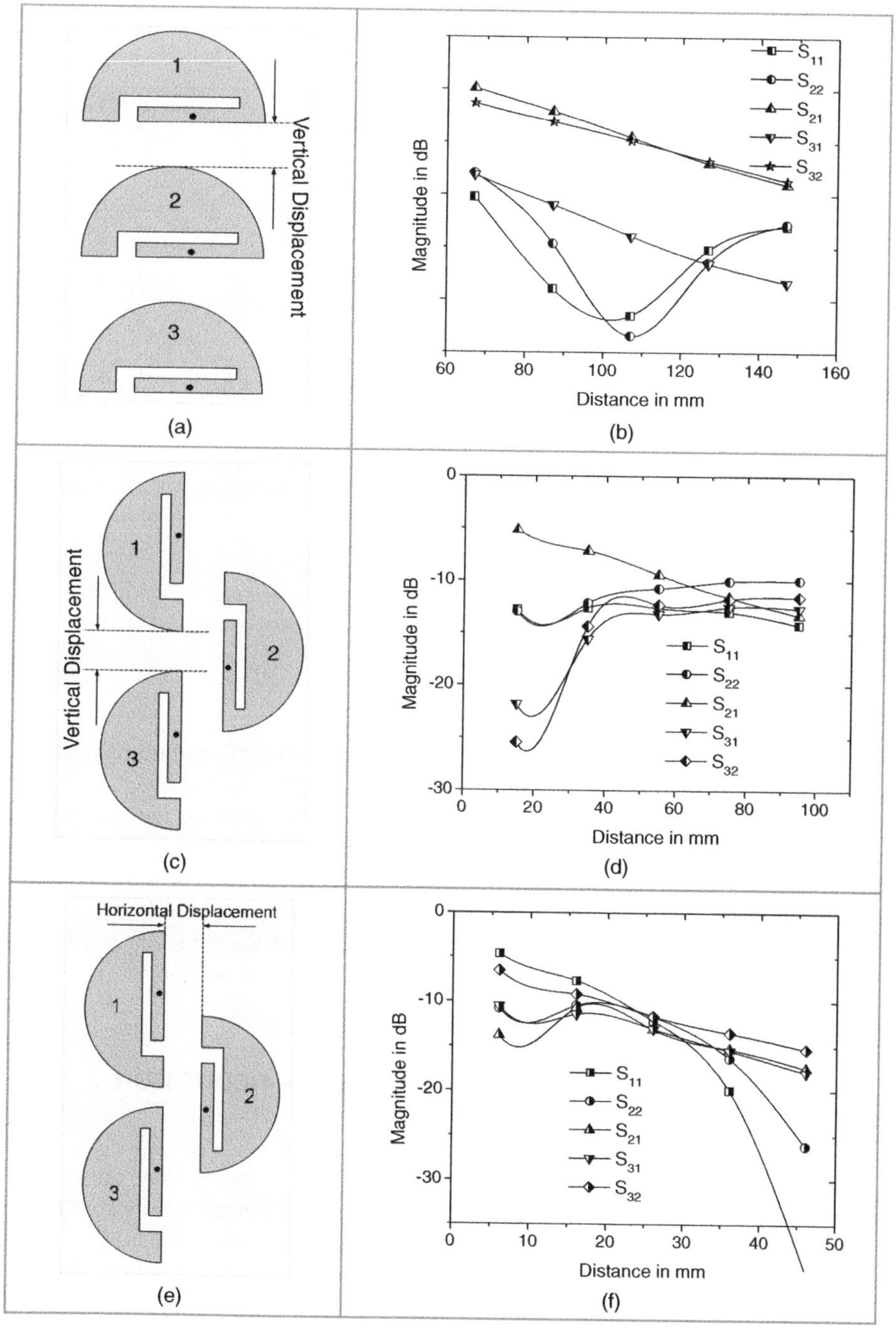

FIGURE 6.13. (**a**) Variation in vertical direction for similar patch orientation. (**b**) *S* parameter curves for variation in vertical direction for similar patch orientation. (**c**) Variation in vertical direction for dissimilar patch orientation. (**d**) *S* parameter curves for variation in vertical direction for dissimilar patch orientation. (**e**) Variation in horizontal direction for dissimilar patch orientation. (**f**) *S* parameter curves for variation in horizontal direction for dissimilar patch orientation. All the plots are centered at 800 MHz.

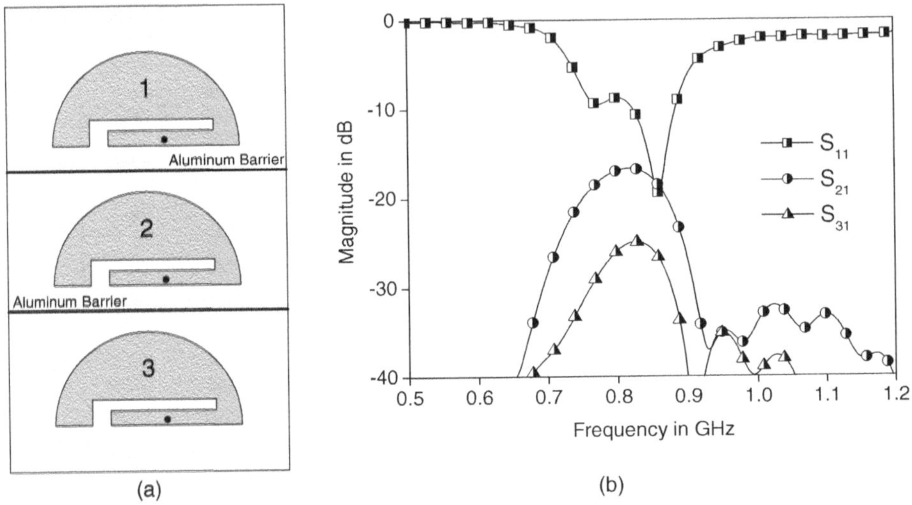

FIGURE 6.14. (**a**) Vertical equidistant placement of three antenna elements with horizontal barrier of 0.5-mm-thick aluminum plate. (**b**) Frequency response observed for the given arrangement of the patches.

Thus a different technique was investigated where 0.5-mm-thick aluminum plates were introduced between the individual elements from the ground layer to 2 mm above the patch layer. The aim of this technique was to minimize the interaction between the transverse components of the fringing fields. The idea was to cage the transverse fields. The transverse components would get reflected from the aluminum barriers but would not influence the adjacent element. The aluminum barriers were placed between the antenna elements, but the four ends of the substrate were kept open. Investigation of this arrangement is discussed in the next section.

6.7 INTRODUCTION OF ALUMINUM BARRIERS—REDUCTION OF MUTUAL COUPLING

Figure 6.14a shows the arrangement of the three patches when they are placed equidistant to each other forming a single column. The aluminum barriers are shown in blue. The frequency response of the arrangement is shown in Figure 6.14b.

In Figure 6.14b, because the return losses of the individual elements is similar, only the S_{11} curve is shown. Wide bandwidth centered on 800 MHz is observed for S_{11}. From the figure, considerable improvement is observed in the isolation between the adjacent elements.

The S_{21} (and S_{23}) parameters came down to less than -18 dB from the previous -10 dB. The S_{31} parameters came down to less than -30 dB. Investigations were also done for the horizontal arrangement of two adjacent patches. In that case also, the isolation came down to less than −20 dB.

Thus the use of the aluminum barriers proved to be a good technique in bringing down the mutual coupling between the adjacent patches. The distance between the outer edge of the individual patches and the metal barrier was 26 mm along the *Y* axis (vertical). The distance between the outer edge of the patches and the outer edge of the substrate along the *Y* axis was also 26 mm. The distance between the outer edge of individual patches and the outer edge of substrate was 19 mm along the *X* axis (horizontal).

The array of $3 \times 1 = 3$ elements was next simulated. A comparison was done between the array with aluminum barriers and the array without barriers. This is presented in the next section.

6.7.1 3 x 1 Element Array—With and Without Aluminum Barriers

Figure 6.15a shows the orientation of the 3×1 element array without any barrier in between the antenna elements. All three elements are simultaneously excited and are at the same phase. The three-dimensional view of the radiation pattern is shown in Figure 6.15b. The simulated gain for the 3×1 element array is 8.01 dBi, whereas the simulated gain for the single element is 7.45 dBi. Thus it is evident that mutual coupling plays a big role when the elements are placed 52 mm away from each other. The radiation pattern shows a wider beamwidth when the angle is varied along Theta, keeping Phi constant at 90° (*H*-plane), thereby resembling the radiation pattern of the single element shown in Figure 6.4c.

Figure 6.15b shows the orientation of the 3×1 element array when 0.5-mm-thick aluminum barriers are introduced between the adjacent elements. The four sides of the substrate are kept open. The three dimensional view of the radiation pattern for the resulting arrangement is shown in Figure 6.15d. The simulated gain in this case rises to 10.4 dBi, which is almost a 3-dB improvement over the single element. Thus it is observed that the introduction of the aluminum barrier plays a significant role in reducing the mutual coupling between the adjacent elements and thereby enhances the gain.

Another feature is notable in Figure 6.15d: The beamwidth of radiation pattern, when varied along Theta and keeping Phi constant at 90° (*H*-plane), reduces considerably with the introduction of the aluminum barriers.

The radiation patterns for the two arrangements are shown in Polar plots showing directivity in Figures 6.15e and 6.15f, where Theta is varied keeping Phi constant at 90° (*H*-plane) and 0° (*E*-plane), respectively. From the figures it is evident that the beam gets narrower and more directive with the introduction of the aluminum barriers. Gain improves significantly with the resultant reduction of mutual couplings.

This investigation is the basis of the next design, where $3 \times 2 = 6$ element array is designed using the same antenna elements. Aluminum barriers are used between the adjacent elements and also at the four open sides.

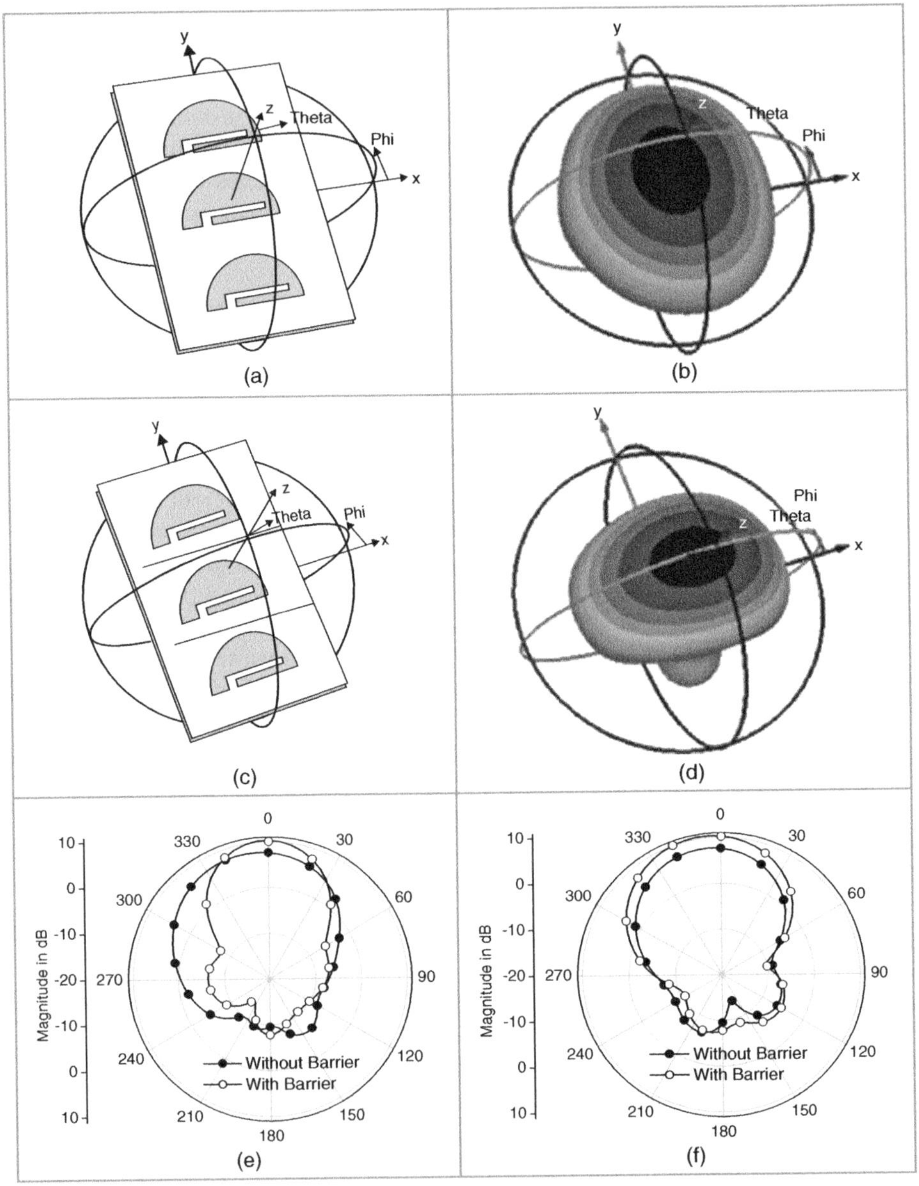

FIGURE 6.15. (**a**) Orientation of the 3 × 1 element array without metal barriers between elements. (**b**) Three-dimensional radiation pattern for the array without aluminum barriers. (**c**) Orientation of the 3 × 1 element array with horizontal aluminum barriers between the individual elements. (**d**) Three-dimensional radiation pattern for the array with aluminum barriers. (**e**) Comparative polar graph (directivity) when angle is varied along Theta, keeping Phi constant at 90°. (**f**) Comparative polar graph (directivity) when angle is varied along Theta, keeping Phi constant at 0°.

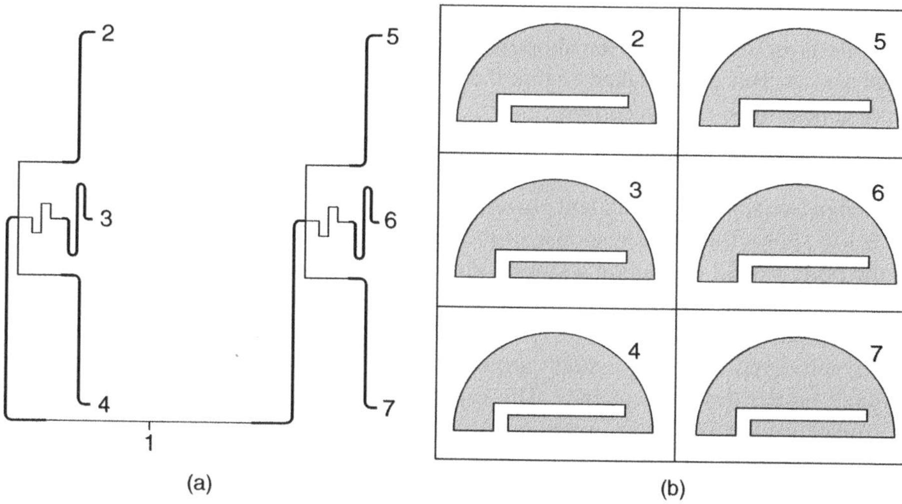

FIGURE 6.16. (**a**) Top view of the feed layer showing the number and position of the individual elements. (**b**) Top view of the patch layer showing the number and placement of the individual elements; 0.5-mm-thick aluminum barriers exist between the adjacent elements and at the four boundaries of the array.

6.7.2 3 x 2 Element Array—With Aluminum Barrier at All Sides and Between Elements

On the basis of the aforementioned findings, a six-element array is designed. The array consists of three rows and two columns. The top views of the patch layer and the feed layer are shown in Figures 6.16a and 6.16b.

Figure 6.16a shows the top view of the feed line. The number 1 denotes the position of the input port, which can be an SMA connector or an N-type connector depending on the cable attached to the array. The antenna elements are named from 2 to 7, and their positions are shown in the figure. After coming through Port 1, the input power divides into two parts. A symmetric T (Tee) junction is used for the division of power. Quarter-wavelength matching is done at the two ends of the T junction so that the width of the microstrip line can correspond to 50-Ω characteristic impedance following the quarter-wavelength. The microstrip line again divides into three branches of equal division to feed the elements of each column. Quarter-wavelength transformation is done again so that the width of the microstrip line at the feed of each element corresponds to 50-Ω characteristic impedance. The phases are matched exactly so that each element gets the input power in the same magnitude and phase.

The top view of the patch layer is shown in Figure 6.16b. The numbering of the patches corresponds to the numbering of the ends of the feed line. The individual elements are spaced equidistant from the adjacent elements and from the outer boundary. All the outer boundaries and the midway between the individual elements are aluminum barriers. The barriers are 0.5 mm thick and extend from the ground plane to 2 mm above the patch layer. The barriers run through the foam layer and the upper substrate layer. The barriers are shorted to the ground and among themselves using metal screws. But care is taken so that the barriers or the metal screws do not touch or come near the bottom feed line.

In Figure 6.17a, the measured return loss (magnitude) of the feed network is shown. It is observed from the figure, the magnitude of S_{11} shows a wide band response at the design frequency. The incident power is divided equally among the six feeds. Figure 6.17b shows the phase response of the feed network. As observed in the figure, all the ports from 2 to 7 are at a similar phase with maximum 2° variation of phase among them. The comparative representation of the simulated and measured return losses are shown in Figure 6.17c.

The measured return loss shows around 100-MHz bandwidth of −10 dB or less centered on the design frequency. The orientation of the array with respect to the coordinate axes is shown in Figure 6.17d. The 3D radiation pattern for the array is shown in Figure 6.17e. The simulated gain for the array is 12.4 dBi and the measured gain is 11.8 dBi. Because the simulated gain of the unit cell is 7.25 dBi, the maximum theoretically possible gain for the six-element array is 15.2 dBi.

Being confined within the given area of 400 mm × 400 mm, a gain of 12.2 dBi is a good achievement for the designed array. The normalized polar plot for the radiation pattern for the simulation is shown in Figure 6.17f, where the angle is varied along Theta while keeping Phi constant at 90° and 0°, respectively. The corresponding measured results are shown in Figure 6.17g. There is good agreement between the simulated and the measured results. The orientation of the main beam is along the Z axis (normal), and the half-power (−3 dB) beamwidth is 42.6° (simulated) and 48.8° (measured).

Figure 6.18a is a photograph of the feed network, and Figure 6.18b is a photograph of the patch layer. The weight was less than a kilogram.

6.7.3 Beam Squinting—Introduction of Delay Between Feed Line

The radiation patterns of the array mentioned above was measured from 700 MHz to 900 MHz. Consistency was there in the radiation pattern. Beyond 900 MHz, nulls started developing. After the design, fabrication, and measurement of the aforementioned six-element array, investigation was carried out to observe the performance of the array as a smart antenna, where the orientation of the main beam can be shifted angularly. On the basis of the aforementioned literature with special attention to [31] and [38], the effect of introduction of delay in the feed line on the orientation of the main beam was investigated.

Figure 6.19a shows the incorporation of the delay line in the feed arrangement. The elements 5, 6, and 7 are given a delay together with respect to the elements 2, 3, and 4. The delay is incorporated by extension of the length of the microstrip line. To increase the length of the microstrip line suitable to the feed with minimum mutual coupling, the minimum increase in length that needed to be considered was 240 mm as shown in the figure. After that, the length of the microstrip line was extended along the Y axis (vertical) and a comparison was done.

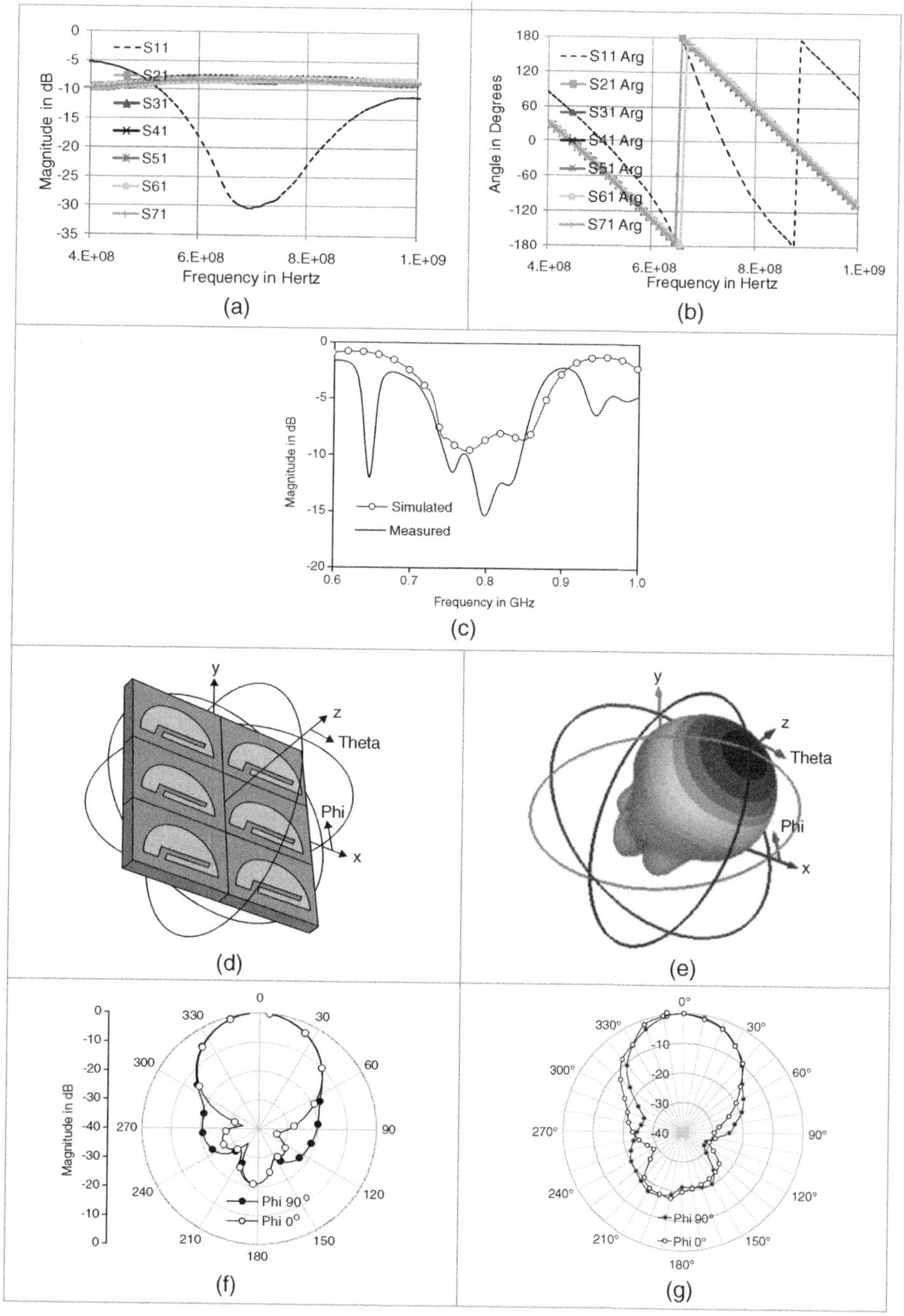

FIGURE 6.17. (**a**) Measured return loss (magnitude) of feed network. (**b**) Measured return loss (phase) of feed network. (**c**) Simulated and measured return loss (magnitude) of final 3×2 element array. (**d**) Orientation of array with respect to the coordinates. (**e**) Simulated 3D radiation pattern of array. (**f**) Simulated normalized polar plot for radiation pattern for variation of angle along Theta. (**g**) Measured normalized polar plot for radiation pattern for variation of angle along Theta. Phi in both the cases kept constant at 90° and 0°, respectively.

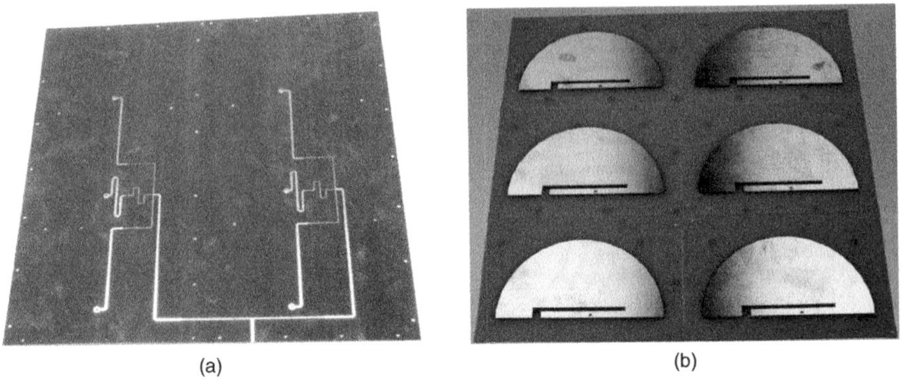

FIGURE 6.18. (**a**) Photograph of the feed network. (**b**) Photograph of the patch layer.

Figure 6.19b shows the comparative representation of the increase of the microstrip line and the delay observed. The curve shows the change in phase in degrees with increase in length of microstrip line as an isolated case. The other three curves show the change in phase as observed in the actual feed line in the array. In the case of the actual feed line, the curve does not follow a straight path. It follows a sinusoidal pattern, keeping the isolated line as the central axis. The reason behind this might be the different junctions and power division involved in the actual feed line.

With the above set of data, a delay was incorporated in the feed line for the 3×2 array. The observations are presented in a comparative format in Table 6.1.

From the table it is observed that beam can be steered within an angular width of 65°. If the orientation of the main beam is along the Z axis (normal), then increase or decrease in the length of the delay line steers the orientation of the main beam linearly. However, if the orientation of the main beam reaches one of the extreme angles (−30° or +35°), then for further increase of the delay line, the beam slowly

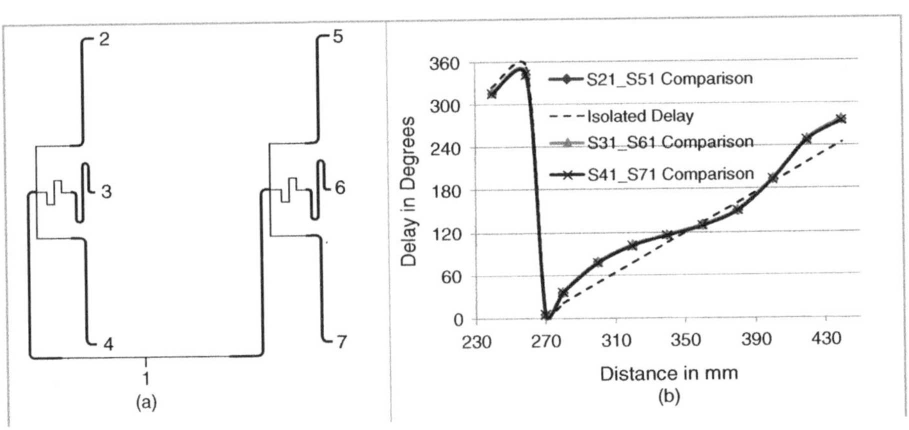

FIGURE 6.19. (**a**) Top view of the feed line with incorporation of delay between the two columns. (**b**) Comparative representation of the phase shift versus increase in length of the microstrip line.

TABLE 6.1. Comparative Representation of the Effect of Increase in Length of Microstrip Line and Change in Orientation of the Main Beam, Phase, and Half-Power Beamwidth

Phase (Degree)	Distance (mm)	Orientation of Main Beam (Degree)	−3-dB beamwidth (Degree)
326°	240	5°	50.4°
354°	260	0°	43.8°
8°	270	−10°	48.3°
22°	280	−20°	47.4°
50°	***300***	***−30°***	47.6°
78°	320	−30°	49.2°
106°	340	−30°	49.3°
134°	360	−30°	48.2°
162°	***380***	***35°***	45.6°
190°	400	30°	42.3°
218°	420	17°	46.5°
246°	440	12°	49°

breaks up into two lobes: the main lobe and the grating lobe. This goes on until the main lobe develops absolutely on the other extreme angle. Again the beam can be steered angularly with increase in length of microstrip line (delay).

An example of this would make it clear. As per the data provided in Table 6.1, following the first few rows, the orientation of the main beam follows a linear relationship with the increase in length of the microstrip line. This goes on until the increase in length of microstrip line is 300 mm when the orientation of the main beam is −30°. Beyond that length, for the subsequent increase in length of the microstrip line, the side lobes develop incrementally toward the other extreme, that is, +35°. The orientation of the main beam remains constant at −30° throughout this phase. This goes on until the increase in length of the microstrip line is 380 mm when the previous side lobe changes fully to become the main lobe at +35°. Beyond this length, the orientation of the main beam again has a linear relationship with the increase of microstrip line until the other extreme of −30° is reached.

In all the aforementioned cases, the Return Loss of the array remains below −10 dB at the design frequency. The maximum gain in the direction of the main beam varies between 11.8 dBi and 12.4 dBi. When the beam pattern breaks into two lobes at the extreme angles, then the gain decreases due to proportional division among the two lobes. The half-power (−3-dB) beamwidth remains contained within 43° to 50° except in the cases when the beam breaks into two similar lobes.

Figures 6.20a–d show the three-dimensional view of the radiation pattern when the beam is steered from 0° to −30°. The increase in the length of the microstrip line for this purpose was from 260 to 300 mm. It is observed in the figure that a side lobe develops along with the main lobe which becomes more prominent with the increase of the angle. Beyond −30°, for subsequent increase of 80-mm length of the microstrip line, the side lobe becomes bigger whereas the main lobe becomes smaller.

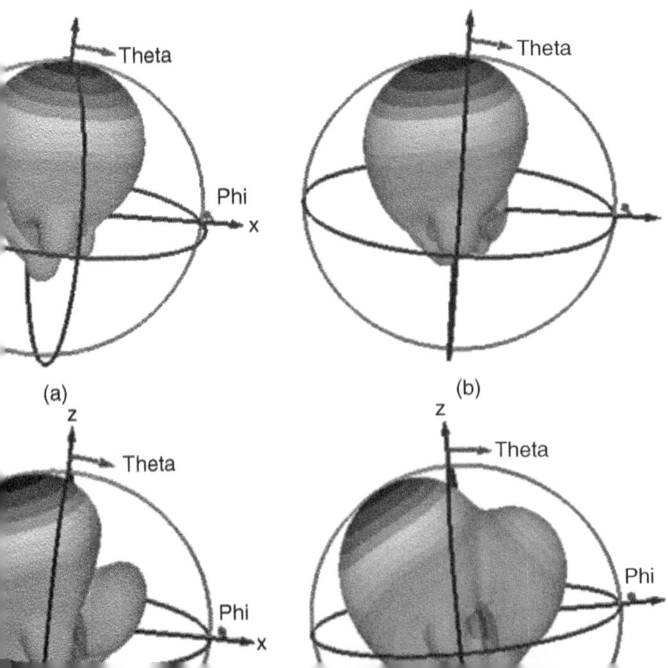

Theta
Phi
x
Theta
(a)
(b)
z
Theta
Phi
x
z
Theta
Phi

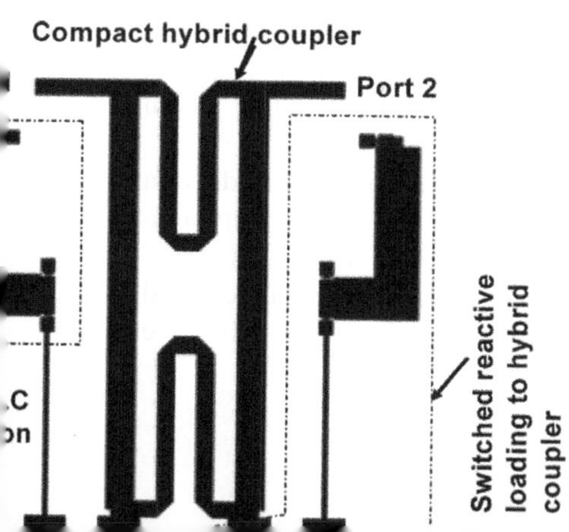
Compact hybrid coupler
Port 2
Switched reactive loading to hybrid coupler

6.8.2 Use as Dual-Band Antenna at UHF Range—Proposed Application

Each antenna element of the aforementioned array consists of a semicircular patch loaded with an L-shaped slot. The design parameters chosen in this antenna element and array were optimized so that the slot and the patch resonate together at the same frequency, thereby producing a wide bandwidth. However, if the slot and the patch in the aforementioned antenna array are designed in such a way that one of them resonates centered on 800 MHz and the other resonates centered on 900 MHz, then the antenna array can be efficiently used in all parts of the world. The feed line can be designed to be broadband by the incorporation of power dividers such as Wilkinson Power Divider, and so on.

If, for example, a container carries goods from Europe (868 MHz) to the United States (928 MHz) and RFID is incorporated in the logistics, then for purposes of cross checking of tag data at both the continents, this type of dual antenna can prove to be very useful. However, this is a proposed area of application and was not actually investigated.

6.9 CONCLUSION

A 3×2 element antenna array is designed in this chapter to act as an antenna for an RFID reader operating at the UHF band. Size of the antenna was a constraint for nature of its usage, but maximum gain and directed beam was the requirement of the design.

After reviewing some available literature regarding planar antennas, a semicircular microstrip patch antenna loaded with an L-shaped slot was chosen. The dimensions were optimized according to the design requirements. The single element was simulated and measured. Simulated gain observed was 7.45 dBi. The antenna was a semicircular patch to make the design more compact. The patch was loaded with a slot designed to resonate at the same frequency. The reason behind this was to exhibit broadband resonance.

Various techniques of increasing the gain of the antenna element were investigated, including the introduction of the superstrate or the use of microwave lenses. However, the gain did not improve over 10 dBi for these arrangements, and the weight or size turned out to be massive.

To improve the gain within the given constraint of the area and weight, the idea of an array was contemplated. The mutual couplings of the elements were investigated within the constraints of the dimensions. It was observed that strong mutual coupling exists between the elements if they are simply placed equidistant to each other within the given area.

The problem of mutual coupling was reduced to a good extent by the introduction of the aluminum barriers between the elements and the four open sides. The aluminum barriers acted to some extent as blinders/winkers in the eyes of the horses which prevent them from looking at the sides. The simulated and measured gain increased to 12.4 dBi for six elements where the individual gain was 7.45 dBi. Thus the introduction of the aluminum barriers proved to be a good option to reduce mutual couplings.

Investigations were carried out to improve the return loss of the array by incorporation of quarter-wavelength transformer [32] directly and also by Chebyshev distribution. The return loss improved drastically to around −35 dB at the design frequency, but the bandwidth narrowed down drastically too. Hence this was not incorporated in the design.

An investigation was carried out to explore the extent to which the orientation of the main beam of the designed array could be steered. It was observed that the main beam of the array could be steered up to 65° by incorporation of suitable phase shifters. The phase shifters [39] were designed and incorporated in the array. The design parameters of the array were scaled and investigated [39] for use in the 900-MHz region of the UHF band. The applications of the antenna array were discussed at the end of the chapter. The antenna array can be used in the UHF band of RFID.

REFERENCES

1. International Organization for Standardization—Internet archive www.iso.org/iso/home.htm.
2. EPC Global—Internet archive www.epcglobalinc.org/home.
3. L. Y. Min, Interim Report on the work on RFID, *in The APT Forum Interim meeting*, Document AWF-IM1/15, Bangkok, Thailand, March 2005.
4. A. Mendelson, 2-W PA serves 800–1000 MHz RFID applications, *RFID* World—Internet archive at www.rfid-world.com/201805357.
5. D. Jackson, Betting on IP, *Mobile Radio Technol.*, pp. 14–17, April 2004—Internet archive searchable in www.iwce-mrt.com.
6. J. Collins, IMEC announces organic rectifier for tags, *RFID Journal*—Internet archive at www.rfidjournal.com/articleprint/1804/-1/1.
7. RFID Technologies—Internet archive www.rfid.com.ru/en-technology.htm.
8. G. Roussos, Enabling RFID in retail, *IEEE Computer Soc.*, pp. 25–30, March 2006.
9. H. W. Son, G. Y. Choi, and C. S. Pyo, Design of wideband RFID antenna for metallic surfaces, *Electron. Lett.*, Vol. 42, No. 5, pp. 263–265, March 2006.
10. H. W. Son, G. Y. Choi, and C. S. Pyo, Open-ended two-strip meander-line antenna for RFID tags, *ETRI J.*, Vol. 28, No. 3, pp. 383–385, June 2006.
11. M. Laudien, Radio frequency identification (RFID) antenna and system design, in *Converge—An Applications Workshop for High-Performance Design*, Ansoft Corporation, USA, pp. 1–19, 2006.
12. D. M. Pozar, Microstrip antenna, *Proc. IEEE*, Vol. 80, No. 1, pp. 79–91, January 1992.
13. Garg, P. Bhartia, I. Bahl, and A. Ittipiboon, *Microstrip Antenna Design Handbook*, Artech House, Boston, 2001.
14. C. A. Balanis, *Antenna Theory: Analysis and Design*, John Wiley & Sons, Hoboken, NJ, 2005.
15. K. L. Wong, *Compact and Broadband Microstrip Antenna*, John Wiley & Sons, NY 2002—Online ISBN 9780471221111.

16. K. R. Carver and J. W. Mink, Microstrip antenna technology, *IEEE Trans. Antennas Propag.*, Vol. AP-29, pp. 2–24, January 1981.
17. Y. T. Lo, D. Solomon, and W. F. Richards, Theory and experiment on microstrip antennas, *IEEE Trans. Antennas Propag.*, Vol. AP-27, No. 2, pp. 137–145, March 1979.
18. S. K. Satpathy, K. P. Ray, and G. Kumar, Compact shorted variations of circular microstrip antennas, *Electron. Lett.*, Vol. 34, No. 2, pp. 137–138, January 1998.
19. W. C. Chew, and J. A. Kong, Effects of fringing fields on the capacitance of circular microstrip disk, *IEEE Trans. Microwave Theory Tech.*, Vol. MTT 28, No. 2, pp. 98–104, February 1980.
20. D. M. Pozar, A reciprocity method of analysis for printed slot and slot-coupled microstrip antennas, *IEEE Trans. Antennas Propag.*, Vol. AP-34, No. 12, pp. 1439–1446, December 1986.
21. P. R. Haddad, and D. M. Pozar, Characterization of aperture coupled microstrip patch antenna with thick ground plane, *Electron. Lett.*, Vol. 30, No. 14, pp. 1106–1107, July 1994.
22. H. Iwasaki, A circularly polarized small-size microstrip antenna with a cross slot, *IEEE Trans. Antennas Propag.*, Vol. 44, No. 10, 1399–1401, October 1996.
23. A. A. Deshmukh, and G. Kumar, Various slot loaded broadband and compact circular microstrip antennas, *Microwave Opt. Technol. Lett.*, Vol. 48, No. 3, pp. 435–439, 2006.
24. A. Henderson, J. R. James, and C. M. Hall, Bandwidth extension techniques in printed conformal antennas, *Military Microwaves*, MM 86, pp. 329–334, 1986.
25. H. F. Pues, and A. R. Van de Capelle, An Impedance matching technique for increasing the bandwidth of microstrip antennas, *IEEE Trans. Antennas Propag.*, Vol. AP 37, pp. 1345–1354, November 1989.
26. T. Huynh, and K. F. Lee, Single-layer single-patch wideband microstrip antenna, *Electron. Lett.*, Vol. 31, No. 16, pp. 1310–1312, August 1995.
27. C. Y. Huang, J. Y. Wu, C. F. Yang, and K. L. Wong, Gain-enhanced compact broadband microstrip antenna, *Electron. Lett.*, Vol. 34, No. 2, pp. 138–139, January 1998.
28. X. H. Shen, G. A. E. Vandenbosch, and A. R. Van de Capelle, Study of gain enhancement method for microstrip antennas using moment method, *IEEE Trans. Antennas Propag.*, Vol. 43, No. 3, pp. 227–231, March 1995.
29. D. R. Jackson, and N. G. Alexopoulos, Gain enhancement methods for printed circuit antennas, *IEEE Trans. Antennas Propag.*, Vol. AP-33, No. 9, pp. 976–987, September 1985.
30. Z. Weng, N. Wang, Y. Jiao, and F. Zhang, A directive patch antenna with metamaterial structure, Microwave and Optical Technology Letters, Vol. 49, No. 2, pp. 456–459, February 2007.
31. R. J. Mailloux, J. F. Mcllvenna, and N. P. Kernwels, Microstrip array technology, *IEEE Trans. Antennas Propag.*, Vol. AP-29, pp. 25–37, January 1981.
32. D. M. Pozar, *Microwave Engineering*, New York, John Wiley & Sons, New York, 1998.
33. W. Y. Tam, A. K. Y. Lai, and K. M. Luk, Mutual coupling between cylindrical rectangular microstrip antennas, *IEEE Trans. Antennas Propag.*, Vol. 43, No. 8, pp. 897–899, August 1995.
34. R. P. Jedlicka, M. T. Poe, and K. R. Carver, Measured mutual coupling between microstrip antennas, *IEEE Trans. Antennas Propag.*, Vol. AP-29, No. 1, pp. 147–149, January 1981.

35. F. Yang, and Y. R. Samii, Microstrip antennas integrated with electromagnetic band gap (EBG) structures: A low mutual coupling design for array applications, *IEEE Trans. Antennas Propag.*, Vol. 51, No. 10, pp. 2936–2946, October 2003.

36. J. J. Schuss, J. D. Hanfling, and R. L. Bauer, Design of wideband patch radiator phased arrays, *IEEE Antennas Propag. Symp. Dig.*, pp. 1220–1223, 1989.

37. R. E. Munson, Conformal microstrip antennas and microstrip phased arrays, *IEEE Trans. Antennas Propag.*, pp. 74–78, January 1974.

38. M. Y. Li, S. Kanamaluru, and K. Chang, Aperture coupled beam steering microstrip antenna array fed by dielectric image line, *Electron. Lett.*, Vol. 30, No. 14, pp. 1105–1106, July 1994.

39. N. C. Karmakar, S. M. Roy, and M. S. Iqram, Development of smart antenna for RFID reader, in *IEEE RFID Symposium*, Las Vegas, NV, April 2008, CD ROM.

CHAPTER 7

HANDHELD READER ANTENNA AT 5.8 GHz

SUSHIM MUKUL ROY, ISAAC BALBIN
Department of Electrical and Computer Systems Engineering, Monash University, Clayton, Victoria, Australia

NEMAI CHANDRA KARMAKAR
Department of Electrical and Computer Systems Engineering, Clayton, Victoria, Australia

7.1 INTRODUCTION

Similar to the handheld optical barcode reader, there is a significant market segment for the handheld readers. The examples include the wand for library database management system, apparel tagging and checking of hanging tags, animal tracking, dispatch and inventory control at the manufacturing premises, courier services and deliveries, and so on. The existing systems work in LF and HF band around 132 kHz and 13.56 MHz, respectively, and the handheld readers use a loop antenna (magnetic dipole also called wand as used by magicians) to read the tags. These antennas have omnidirectional patterns to cover maximum reading zone as possible. However, due to the magnetic coupling between the reader and the tag, the distance of the reader is very short, usually less than a few tens of centimeters.

Figure 7.1a shows the configuration of a handheld reader used in the inventory check of books on a library book shelf. As can be seen, the RFID reader consists of a handheld loop antenna and an RFID reader (with battery pack, display, etc.), which are connected by an RF cable for data transmission and reception. The handheld antenna (also called wand) is carried by the hand of a person who reads the tags attached to the item. The scanning is done by hand sweeping the tagged items (books on a library book shelf) with the wand at a close proximity. These antennas are LF and HF magnetic loop antennas. Sometimes the antennas are covered with a plastic radome. The reader electronics with a battery pack is bulky with the RF transceiver

Handbook of Smart Antennas for RFID Systems, Edited by Nemai Chandra Karmakar

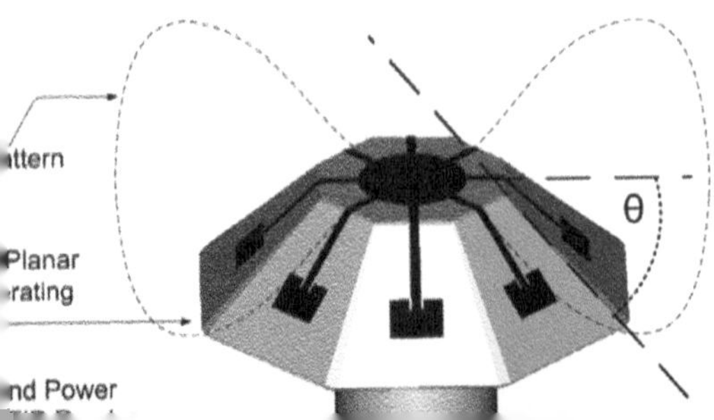
attern
Planar
rating
nd Power
θ

electronics, a digital data control section, and a display—all are integrated in a single case. The reader is carried on the shoulder, with a strap connected to the reader like a bag or attached to a belt to be fixed on a person's waist. The conventional system is bulky and needs separate parts like the wand, the cable assembly, and the reader electronics—hence it is not convenient to carry and use.

The current demand of the market is to improve the situation by pushing the operating frequency in S and C bands by exploiting the far-field EM radiation. The objectives are to achieve compact design, long reading range (exploiting the far-field radiation), high reading speed, and provision to read multiple tags in a very fast sequence.

To achieve the objective, the reader needs to cover as many tags as possible in its reading zone but with high-gain beams to leverage the link budget. The objective can be fulfilled by designing antenna arrays that are compact in design and moderately high gain. The wireless local area network (WLAN) uses antennas with difference patterns to cover as many terminals as possible in an office. The approach in the design discussed in this chapter has adopted the same principle of WLAN to generate the difference patterns and/or wide-beam patterns. To fulfill the design goal, a comprehensive design exercise is performed in the chapter.

Figure 7.1b shows an IP4™ RFID Reader from Intermec Technologies™ operating in UHF band. The reader operates as a complete standalone module. In a similar way, in the systems proposed in this chapter, all components will be integrated into a single module as shown in Figures 7.1c–e. The antenna is a fully planar microstrip patch antenna array with a compact circular power divider and antenna elements in a circular grid. Based on the operational need, the antenna can be designed to produce sum (directional) patterns and difference patterns (omnidirectional) in order to maximize the reading distance and zone, respectively. With a switching electronics and control algorithm, the antenna's beam patterns can be controlled adaptively and switched between the two modes of operation to maximize the signal-to-interference-plus-noise ratio and system capacity. As can be seen in the conceptual design of the integrated handheld RFID reader, the reader electronics can be designed as a baton and the planar antenna can be mounted on a side (Figure 7.1c) on the top of the baton with adjustable angles as an angle grinder (Figure 7.1d), as well as on a side of the baton as a conformal antenna array (Figure 7.1e). In Figures 7.1d and 7.1e, the angle θ can be adjusted manually to optimize the reading distance and zone. In Figure 7.1e, the elements can be energized simultaneously to produce a low-gain omni-pattern to cover maximum zone or a more directional beam with a sequential switching algorithm using switched beamforming electronics [1]. To make the antenna fully conformal to the body of revolution, the angle θ should be 90° so that the most compact handheld design of the integrated reader is achieved.

To achieve the goal of a compact design for a handheld reader, a novel compact power divider is designed to replace the conventional corporate feed power divider. The power divider comprises a central circular disc as the distributive element with an input power connecting at the center of the disc with a coaxial-to-microstrip transition, and the output ports are the peripheral microstrip lines with 50-Ω impedance. This design not only offers a compact configuration, but also allows us to use only 50-Ω output ports instead of very high impedance output ports as found in conventional corporate feed power dividers. The conventional power dividers create a huge problem of (a) designing the tracks in the available space and (b) broadband matching and power distributions. The high-impedance track for a large number of power divisions is not possible to realize in practice due to its very thin track width.

A set of eight-element circular array geometries with different orientations of patch antenna elements has been investigated to generate difference as well as sum patterns. Spatial orientation and phase compensation of the antenna elements provide different beamwidth and shapes of main beam patterns which are very useful in handheld application for maximum coverage. Next, to maximize the reading capabilities for arbitrarily oriented tags, a circularly polarized (CP) antenna array with sequential feeding and phase compensation is designed. The phase excitation is performed in two ways: A set of delay lines are used sequentially to the antenna elements to compensate for their physical locations and orientations in the circular grid of the array. However, design of the delay lines makes the interelement spacing of antenna array large, hence the total antenna size. A new solution is to use a novel negative refractive index (NRI) meta-material phase shifter in a compact design to replace the long uneven delay lines. The second approach generates the exact phase shifts to create an axial ratio of less than 2.5 dB over the frequency band. Both designs are implemented and their input impedance return loss, the radiation and gain patterns, and axial ratio bandwidth are investigated. The design yield required broad beamwidth characteristics, which are suitable for handheld reader application.

The latter design (NRI metamaterial) offers a more compact design with a uniform array design, while the former design needs more real estate to accommodate the long delay lines. Therefore, the delay-line-based approach is less favorable for compact handheld reader design. Irrespective of the tag antenna's polarization, the CP reader antenna receives at least half the power from the tag and makes the reading more efficient with respect to linearly polarized antennas (tag and reader) where polarization mismatch severely degrades the reading process.

The planar antenna arrays are fed in two ways: (i) A corporate feed network is a replica of stages of equal power dividers and transmission carrying the power equally to each antenna element in an array from the RF input port. The feed network ensures precise equal power division and transmission phase to generate a collimated beam pattern from the antenna array. However, to produce matched transmission lines and equi-phase distances, the configuration becomes very large with the number of the antenna elements. Therefore, the corporate feed network is not suitable for the proposed handheld application. (ii) A series feed network progressively leaks the power to the antenna elements in the array as the the power progresses to the end of the array. Usually the end of the feed network is terminated to a match load; hence there is unwanted power wastage of the residual power after the distribution of power to the element. Though the series feed network is more compact, it also needs phase balance between elements progressively and hence needs extra length of lines. To solve the real estate problems of the two conventional approaches of beamforming networks, a novel circular power divider is proposed in the chapter. The power divider is a circular resonator at 5.8 GHz. The power is input via a central coaxial probe, and the outputs are tapped from the periphery of the circular disc. All ports are inherently matched to 50 Ω; therefore, the requirements of large matching sections in the conventional power dividers are relinquished. The chapter is organized as shown in Figure 7.2.

7.2 BACKGROUND REVIEW OF RFID SYSTEMS AT 5.8 GHz

Griffith and Durgin [2] proposed multi-antennas for RFID tags to reduce the fading in an RFID channel, consequently increasing the throughput and system capacity in an RFID system. To accommodate multiple antennas in a tag, they advocated pushing the operating frequency to 5.8-GHz ISM band. Therefore there is a pressing requirement to develop reader antennas at that frequency band. The advantages of operating in the upper frequency band are [3]: (i) compact (handheld) reader electronics and antennas, (ii) small tag dimensions with multiple antennas, (iii) large bandwidth, (iv) higher reading rate, (v) improved capacity, (vi) implementation of multiple-input multiple-output (MIMO) antenna systems [4], and (vii) improved anti-collision features.

This chapter addresses these issues and envisages the design and development of a handheld reader antenna at 5.8 GHz. This antenna can also incorporate beam scanning by NRI phase shifters and switching electronics. As per the existing standards, RFID systems operate centered on 134.2 kHz, 13.56 MHz, 433 MHz, 915 MHz, 2.4-GHz, and 5.8-GHz frequencies. However, a few developments are reported in 5.8 GHz which promise miniaturization of the circuits and improved noise performance.

Certain limitations exist at the 5.8-GHz region: The antennas have very narrow bandwidth and become very directive at this high frequency [5], and few microchips are available to modulate or process the signal. The problem of bandwidth can be resolved by use of CPW-fed slot antennas [6] and other conventional bandwidth enhancement techniques. But the miniaturization of the tags open up new areas of application in RFID technology, such as their usage in pipes used to drill oil [7]. Because the tags are small, they can withstand the pressure of the oil flow through the pipes with minimum obstruction.

In the Internet-archived newsletter of June 2005 of Georgia Tech, Atlanta, GA, Nikolau et al. [8] reported the benefits of using compact multiband antennas with reconfigurable features. They developed annular slot antennas that operate at three different frequencies centered at 5.8 GHz with reconfigurable radiation patterns. According to the authors, this system provides higher efficiency and noise immunity. These antennas are cheap and can be used in RFID transponders/tags with easier anti-collision algorithms, thereby adhering to the benchmarks of RFID systems [1, 9]. The authors also incorporated the use of multiple antennas.

In any RFID system, the readers have to comply with the benchmarks of high gain and directional beam with optional wide bandwidth. For systems mentioned in reference 8, the readers ought to provide higher bandwidth along with gain. Reader antennas also require miniaturization and low profile. Microstrip technology comes in very handy for this purpose. However, microstrip antennas generally provide low

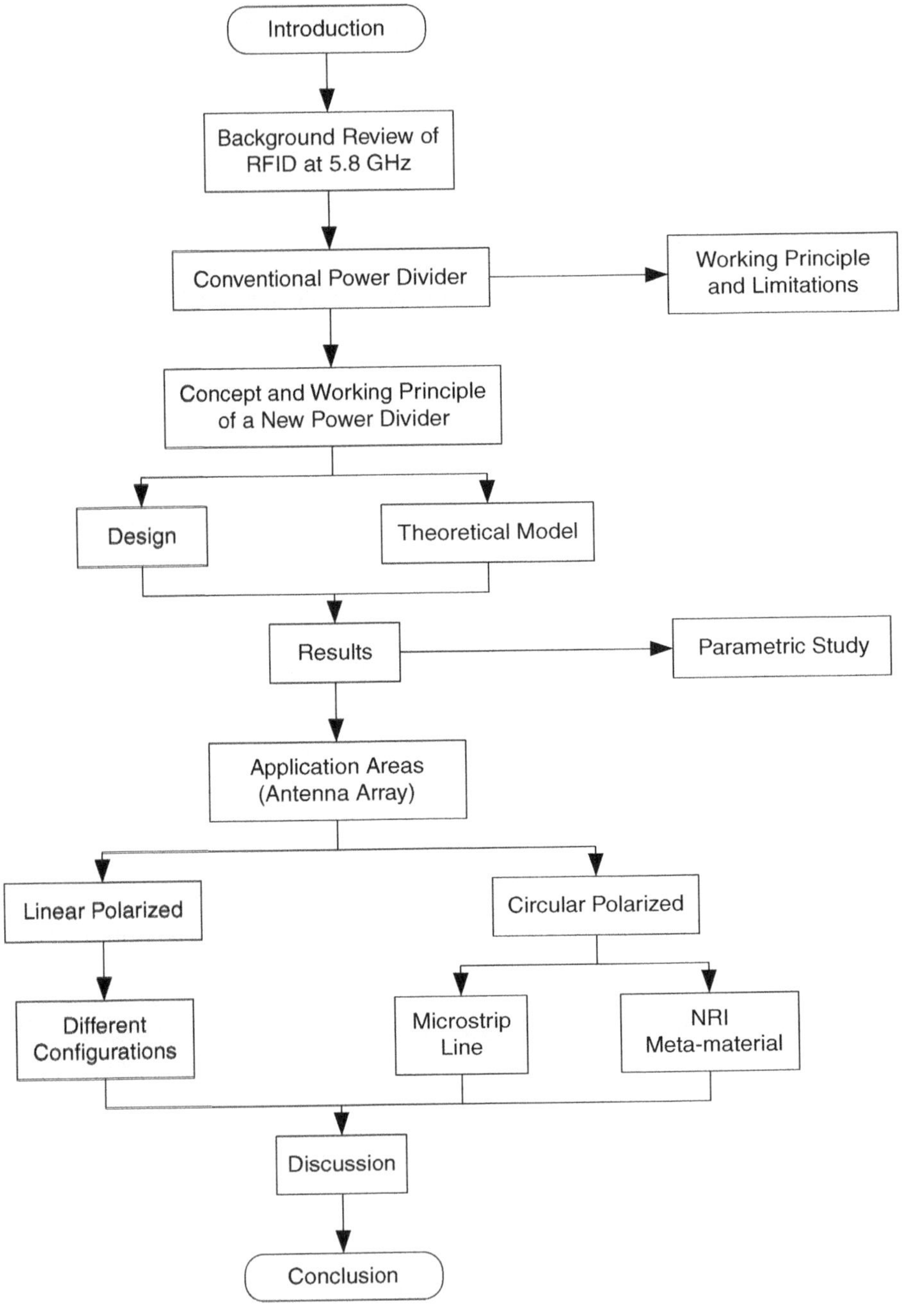

FIGURE 7.2. Organization of the chapter.

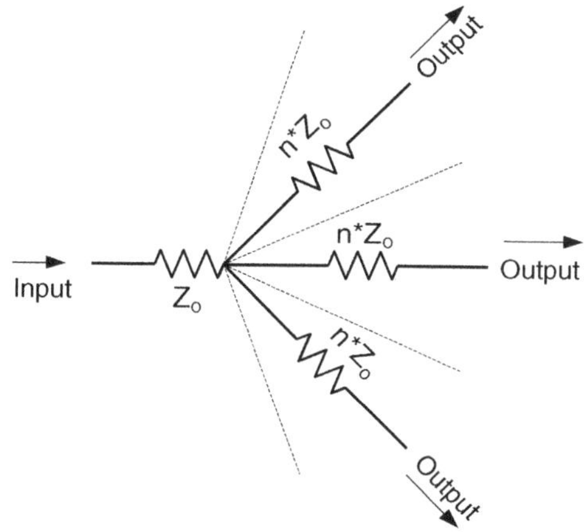

FIGURE 7.3. Principle of operation for a conventional power divider.

gain. This problem to a certain extent can be solved by configurations such as aperture-coupled microstrip patch antenna (ACMPA) [10] or microstrip patch antennas with cylindrical metal cavity enclosures [11]. However, the gain does not improve more than 6–7 dB, which is insufficient for an RFID reader. Thus the necessity arises for designing suitable arrays in reader antennas using microstrip technology.

In the process of designing an antenna array, the designers face a big challenge of efficient power division. The conventional power dividers like T-junction, Wilkinson power divider [12], or Bagley power dividers [13] work well with substrates of low dielectric constant (ε_r) and for two- to three-way power division. For substrates with higher dielectric constants and higher-order power division, even with quarter-wavelength matching, the width of the microstrip lines becomes difficult to realize practically and the cost of precision shoots up.

Based on the aforementioned observations, a novel power divider is designed in this chapter. The power divider can accommodate any number of power divisions from a central circular microstrip patch. Input is given centrally by coaxial probe feed, and output is taken out by microstrip lines in a radial manner. The designed power divider is broadband in nature and useful for array configurations which are also novel in design.

7.3 CONVENTIONAL POWER DIVIDER

Conventional power dividers work according to the principle shown in Figure 7.3, where power division takes place in a 1 to N way. For conventional power dividers, using microstrip technology, the characteristic impedance Z_0 is calculated using Eqs. (7.1) and (7.2), where ε_r is the relative dielectric constant and ε_e is the effective dielectric constant of the substrate. For thin microstrip lines, W is the width of the microstrip line and d is the thickness of the substrate.

$$\varepsilon_{\text{eff}} = \frac{\varepsilon_r + 1}{2} + \frac{\varepsilon_r - 1}{2}\left[\frac{1}{\sqrt{1 + \frac{12d}{W}}}\right] \tag{7.1}$$

$$Z_0 = \begin{cases} \frac{60}{\sqrt{\varepsilon_{\text{eff}}}} \ln\left[\frac{8d}{W} + \frac{W}{4d}\right] & \text{for } W/d \leq 1 \\ \frac{120\pi}{\sqrt{\varepsilon_{\text{eff}}}\left[\frac{W}{d} + 1.393 + 0.667\ln\left(\frac{W}{d} + 1.444\right)\right]} & \text{for } W/d \geq 1 \end{cases} \tag{7.2}$$

Taking an example of a readily available and inexpensive microwave laminate FR4, of dielectric constant (ε_r) 4.4 and thickness (d) 1.6 mm, the width of the 50-Ω microstrip line is 3.1 mm. If there is an eight-way power division, each of the microstrip lines coming out of the power divider should have characteristic impedance corresponding to $8 \times 50\ \Omega = 400\ \Omega$. This corresponds to thickness 0.0002 mm, which is very difficult to realize. Even if the designer goes for a quarter-wavelength match, the characteristic impedance of the quarter-wavelength (λ/4) lines turn out to be $\sqrt{(50\times 400)}\ \Omega = 141.42\ \Omega$. This impedance corresponds to an 0.225-mm-wide microstrip line, which is still very hard to realize practically and is expensive to fabricate. Moreover, dependency of λ/4 transformers on a particular frequency makes the design very narrowband and may not be able to fulfill the requirement of the high data rate communication.

7.4 DESIGN APPROACH AND WORKING PRINCIPLE: ANALOGY WITH FLOW OF WATER

Using cavity model analysis, a circular microstrip patch antenna is supposedly a cavity where the side walls of the cavity are magnetic walls and the upper and lower walls are electric walls [14, 15]. When a high-frequency electrical signal is fed to the patch antenna, electromagnetic energy is radiated from the edges of the patch in the form of fringing fields. In the power divider designed in this chapter, instead of allowing the energy to radiate out into space, it is tapped out using microstrip lines from the outer edge of the circular disc.

If an analogy is drawn between the working principles of the conventional power dividers and that of the designed power divider, then the conventional power dividers can be compared with the flow of water in a river through its system of tributaries and distributaries. The designed power divider can be compared with the flow of water in a reservoir in a public water distribution system where flow of water into the reservoir takes place through one pipeline whereas flow of water out of the reservoir takes place through another set of pipelines. All the pipes through which water flows out of the reservoir have the same pressure at the source—that is, the reservoir. Figure 7.4 shows the analogy of the working principle of the designed power divider.

The energy is fed at the bottom through the input port while energy is tapped out at the top through the output ports, which are radial to the central circular patch. In

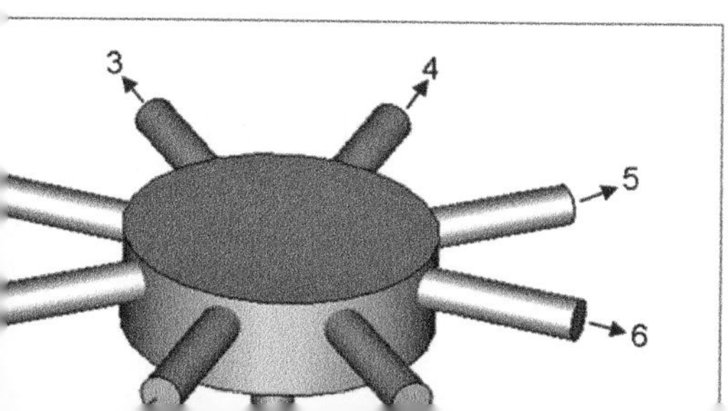

3
4
5
6

this number can range anywhere between 1 and 16 or more, depending on the application. For measurement by a standard vector network analyzer (VNA), the width of the output microstrip lines correspond to 50-Ω characteristic impedance. The output ports must be at equiangular separations to ensure uniform power distribution in each of the ports.

The second layer is the upper substrate layer. It can be made of any nonconducting dielectric material. The surface area of this layer has to be bigger than the patch layer to minimize the fringing fields from the edge of the upper patch layer. The third layer is the ground made of thin copper. This layer is the common ground for both the input and output ports. The ground layer has same cross-sectional area as the upper patch layer. To prevent shorting of the central cylinder, there is a clearance (circular gap) in the ground layer around the central cylinder. The central cylinder runs vertically through all the layers and transfers energy from the input port to the output ports.

The fourth layer is the lower substrate layer. This layer is of same cross-sectional area as the ground layer and is made of nonconducting dielectric material. The material of the lower substrate layer may be the same as or different from that of the upper substrate layer. The lowermost layer is the feed layer. The layer consists of a 50-Ω microstrip line. One end of the microstrip line is connected to the central cylinder with a capacitive ring. The other end is connected to the input port, which might be a standard connector such as an SMA or N-type. The capacitive ring along with the clearance in the ground layer is necessary for proper matching of the power divider.

The power divider is fabricated in two parts from standard double-sided microwave laminates. The upper part consists of the circular patch, upper substrate, and ground. The lower part consists of the ground, lower substrate, and the feed layer. The two grounds are aligned and soldered without keeping any gap between them to ensure uniform electrical conductivity in the ground layer.

In this chapter, a power divider with eight output ports is designed and fabricated. The results presented in the later sections correspond to the eight-way power divider. The radius of the upper patch is 16.77 mm. The substrate considered is Flame Retardant-4 (FR4) due to its ready availability and low price. The dielectric constant (ε_r) of the substrate is 4.4 and the thickness (d) is 1.6 mm. Copper thickness is 17 μm. The diameter of the center conductor is 0.5 mm and the radius of clearance around the center conductor in the ground layer is 1.5 mm. The radius of the capacitive ring at the feed layer is 2.5 mm. The widths of input and output ports correspond to 50 Ω.

7.6 THEORY AND CIRCUIT CONFIGURATION

The initial radius of circular patch was taken as 16.77 mm. The TE_{10} resonance of the patch for the aforementioned dimension is 2.4 GHz. This resonance is suppressed by feeding the patch at the center. As a result, resonance at the next harmonic of 5 GHz is obtained with the patch only (no output port). Depending on the number of output ports, the resonant frequency shifts toward the right side along with associated change in impedance.

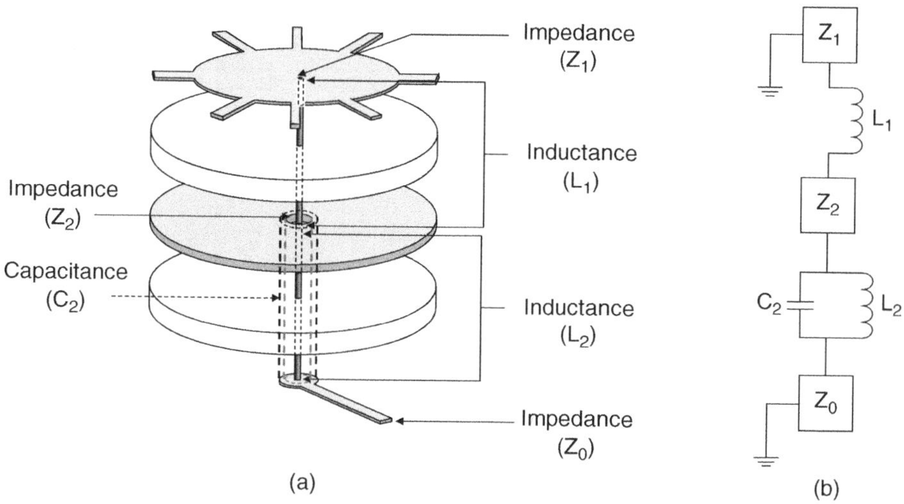

FIGURE 7.6. (**a**) Impedance, Inductance, and capacitance in the designed power divider, depending on the geometry. (**b**) Interconnection between the impedance, inductance, and capacitances in circuit form.

In Figure 7.6a, impedance Z_1 is observed by a commercial EM solver (CST Microwave Studio^{TM}) using "discrete ports." Z_1 is the impedance at the center of the upper cylinder with respect to the common ground. The other inductances, capacitances, and impedances are connected in series and parallel shown in Figure 7.6b. This makes the equivalent circuit of the designed power divider. When the power divider has only the input port and no output port, resonance is observed at 5 GHz. Impedance Z_1 is $(43.33-j39.83)\ \Omega$ at 5 GHz. When eight output ports are connected, the power divider resonates at 5.2 GHz. Impedance Z_1 is $(37.98 + j24.47)\ \Omega$ at 5.2 GHz.

The central cylinder gives an inductance L_1 when it passes through the upper substrate layer. When the central cylinder crosses the ground plane, due to the circular clearance it behaves as a coaxial cable of characteristic impedance is Z_2. However, the thickness of the ground plane being too small (17 μm), the influence of this impedance can be neglected.

When the central cylinder crosses the lower substrate, it gives an inductance L_2 at the given frequency. But in case of the lower substrate, depending on the dimension of the clearance in the ground layer and dimension of the capacitive ring in the feed layer, a cylindrical capacitance C_2 develops which is in parallel to the inductance L_2. This is followed by characteristic impedance Z_0 of the microstrip line. The inductances L_1 and L_2 and capacitance C_2 along with Z_1 lead to resonance at 5.8 GHz.

In the aforementioned circuit analysis, the impedance (Z_1) at the center of the upper patch was measured using a discrete port of the EM solver. If the upper patch is considered to be an octagon and it is assumed that there are four output ports connected to the upper patch given by the numbers 1, 2, 3, and 4 in Figure 7.7a, then

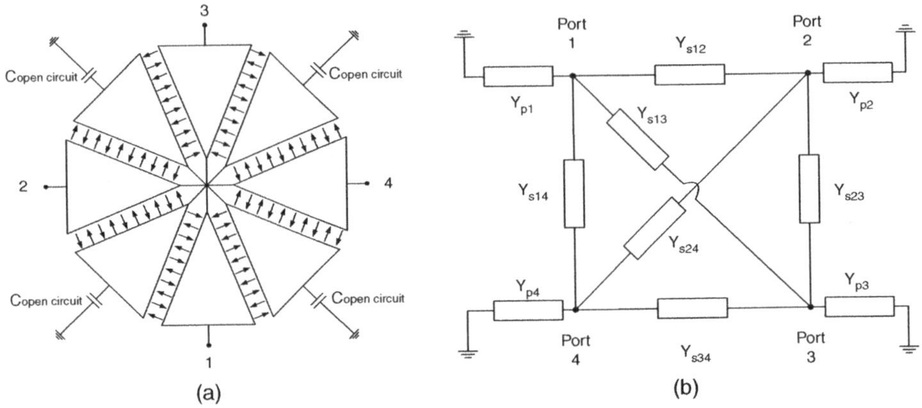

FIGURE 7.7. (**a**) Multiport network representation of the upper patch of Figure 7.6a when only four output ports are considered. (**b**) Schematic representation of a multiport network model when only four ports are concerned [18].

the upper patch can be broken up into eight triangular sections, which are connected to each other along their two edges. This arrangement becomes a multiport network where the end effects can no longer be explained by Quasi-Static Analysis using microstrip discontinuities.

The circuit explanation of this type of structure is done by full–wave analysis [19]. Additionally, Figure 7.7b shows the complexity involved in the multiport network [18]. With four ports only, six interconnections are involved. With eight output ports, eight open-ended triangular stubs, and an input port in the center, the calculation becomes very complex. To avoid that, the impedance at the center of the upper patch (Z_1) was measured using a discrete port of the EM solver.

7.7 RESULTS FOR THE PERFORMANCE OF THE EIGHT-WAY POWER DIVIDER

Figure 7.8 shows the photographs of the top and bottom view of the fabricated power divider. The left-hand picture shows the upper patch and the output ports. The right-hand picture shows the lower feed line and the input port. In both the pictures, the central cylinder is soldered at the central position relative to the upper patch.

Figure 7.9 shows the simulated and measured results of the designed power divider. Simulation was done by a commercial MoM solver, CST Microwave Studio™. Measurements are done in an Agilent E8361A Vector Network Analyzer. The simulated and measured results show good agreement. The S_{21} curves for the ports 2, 4, and 6 are given only to avoid congestion. In both simulation and measurement, maximum deviation of 0.5 dB is observed among the receiving ports. All the output ports are at same phase. The S_{21}, S_{41}, and S_{61} curves for the measured results show

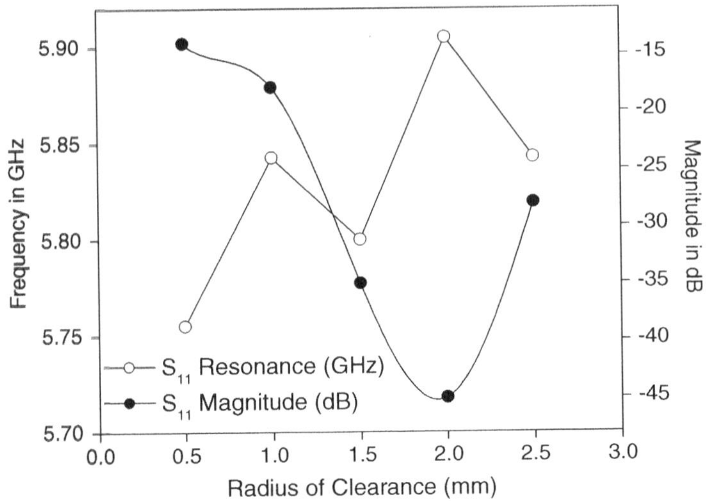

FIGURE 7.10. Variation of resonant frequency and magnitude of S_{11} for variation in radius of clearance in ground.

7.8.1 Variation of Radius of Clearance in the Ground Plane Around the Central Cylinder

The effects of variation in radius of the clearance in ground plane are presented in Figure 7.10. The resonant frequency is observed to oscillate, whereas the magnitude at resonance shows a clear dip. The S_{21}–S_{91} parameters vary within a range of 0.5 dB for the entire variation. Thus it is observed that clearance in the ground is related to proper matching in the designed power divider.

7.8.2 Variation of the Radius of the Central Cylinder

The effects of variation in radius of the central cylinder are presented in Figure 7.11. Radius of the central cylinder is observed to play an important role in matching as well as determining the resonant frequency of power divider. For a variation of 0.5-mm radius, the resonant frequency of the power divider varies up to around 1 GHz. Thus this parameter plays an important role in fine tuning the power divider.

7.8.3 Variation of Radius of Capacitive Ring at Feed

The effects of variation in radius of the capacitive ring are presented in Figure 7.12. The resonant frequency is observed to oscillate with increase in radius of the capacitive ring, but the magnitude of S_{11} at resonance shows a sharp dip. For radii less than 2 mm, the dip at resonance is very low, which is also the case with radii 3 mm and beyond. The S_{21}–S_{91} parameters vary within a range of around 1 dB for the entire variation.

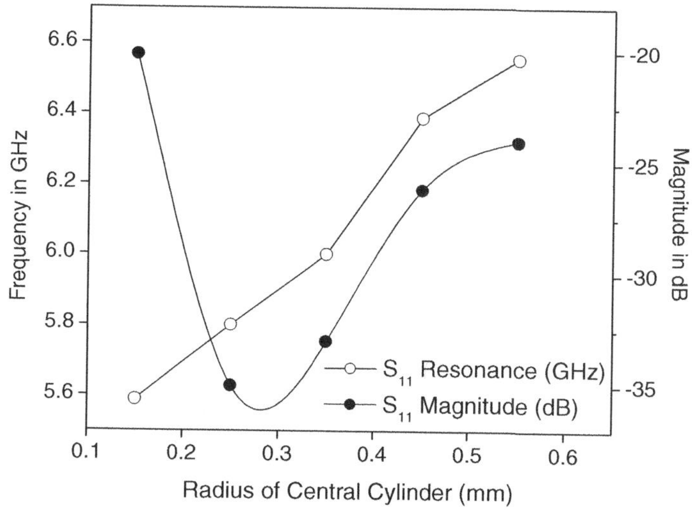

FIGURE 7.11. Variation of resonant frequency and magnitude of S_{11} for variation in radius of central cylinder.

7.8.4 Variation of Radius of the Upper Patch

The effects of variation in radius of the upper patch are presented in Figure 7.13. The radius of the patch is observed to determine the resonant frequency of the power divider. For smaller patch size, a bit lowering in the level of the S_{21}–S_{91} curves is

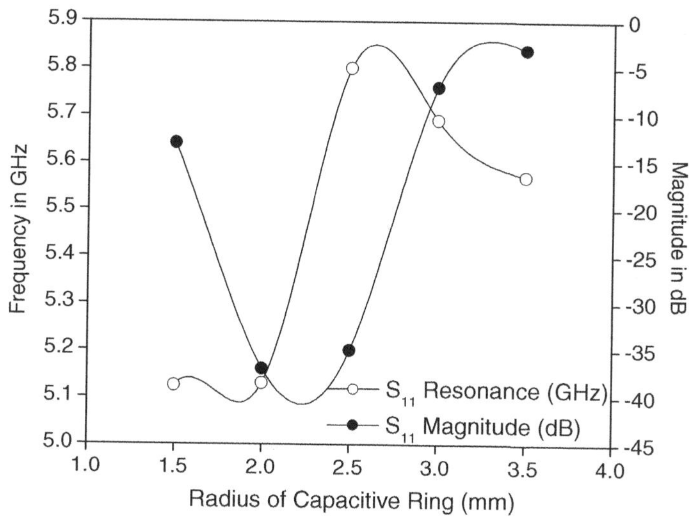

FIGURE 7.12. Variation of resonant frequency and magnitude (S_{11}) for variation in radius of capacitive ring at feed.

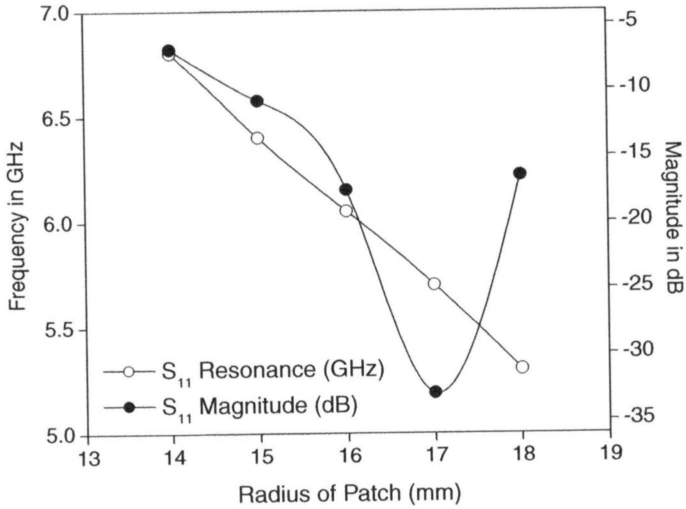

FIGURE 7.13. Variation of resonant frequency and magnitude (S_{11}) with variation in radius of upper patch.

observed. The reason may be due to the mutual coupling between the adjacent output ports for smaller size of the patch. However, for any given size, the S_{21}–S_{91} curves have same magnitude and phase.

With proper choice of the design parameters, the power divider can give a −10-dB bandwidth of more than 1 GHz centered on the design frequency of 5.8 GHz, where the input ports and output ports are exactly matched at 50 Ω. When designing the power divider for any given frequency, the dimension of the patch is to be determined first, followed by variation of the other matching parameters to achieve optimum result.

7.8.5 Variation in the Angular Distribution of Output Ports

In the prototype and the aforementioned design variations, the output ports were placed at equal angles from each other for uniform power distribution to each port. If the ports are not placed at uniform angular distances, discrepancies are observed in the $S_{21}, S_{31}, \ldots, S_{n1}$ curves.

7.9 POTENTIAL APPLICATIONS OF THE POWER DIVIDER

Microstrip reader antenna arrays can efficiently utilize this power divider. In this section, certain configurations are discussed where the power divider is used at the center of the array. Some observed limitations of the power divider are also discussed. A rectangular microstrip patch is used as the antenna element for the discussed arrays.

7.9.1 Unit Element Microstrip Patch Antenna

A microstrip line-fed rectangular patch antenna with inset cut is designed in this section. The inset cut is required for proper matching. The patch antenna is designed on the 1.6-mm-thick FR4 of dielectric constant 4.4. The guidelines for rectangular patch antenna [14] and matching [12] are used for design.

The effective dielectric constant (ε_{eff}) was calculated using Eq. (7.3). The width (w) was approximated using Eq. (7.4). The effective length (L_{eff}) of the antenna was calculated using Eq. (7.5), and the increase in length (ΔL) was calculated using Eq. (7.6). The actual length (L) as the difference of the two lengths is shown in Eq. (7.7), and the conformity of the substrate height with the limiting conditions of the aforesaid equations is shown in Eq. (7.8). In the equations, c corresponds to velocity of light in vacuum. The equations are based on the transmission line model of the rectangular patch antenna.

$$\varepsilon_{\text{reff}} = \frac{\varepsilon_r + 1}{2} + \frac{\varepsilon_r - 1}{2}\left(1 + 12\frac{h}{w}\right)^{-1/2} \tag{7.3}$$

$$w = \frac{c}{2f_0\sqrt{\frac{\varepsilon_r+1}{2}}} \tag{7.4}$$

$$L_{\text{eff}} = \frac{c}{2f_0\sqrt{\varepsilon_{\text{reff}}}} \tag{7.5}$$

$$\Delta L = (0.412h)\left[\frac{(\varepsilon_{\text{reff}} + 0.3)\left(\frac{w}{h} + 2.64\right)}{(\varepsilon_{\text{reff}} - 0.258)\left(\frac{w}{h} + 0.8\right)}\right] \tag{7.6}$$

$$L = L_{\text{eff}} - 2\Delta L \tag{7.7}$$

$$h \le \frac{0.3c}{2\pi f_0\sqrt{\varepsilon r}} \tag{7.8}$$

With the above-mentioned calculations, the approximated dimensions of the antenna were obtained. Based on that, simulation was done in EM solver CST Microwave Studio^{TM}. The length of the patch is 11.6 mm, while the width of the patch is 16 mm. The dimensions of the cut at the inset are 3 mm × 1 mm. The antenna is fed with a microstrip line corresponding to 50-Ω characteristic impedance.

Figure 7.14a is a photograph of the fabricated patch antenna. Figure 7.14b shows the simulated and measured return losses. Good matching is observed for the resonant frequencies, but the measured result shows wider bandwidth. This is due to the higher losses of the FR4 substrate. The return losses are compared in a Smith chart in Figure 7.14c. Figure 7.14d shows the orientation of the antenna with respect to the coordinate axes. Figure 7.14e shows the 3D radiation pattern of the unit element rectangular patch. The patch is linearly polarized. Figures 7.14f and 7.14g show the normalized simulated and measured radiation patterns. Half-power beamwidth is 60°. The simulated gain of the antenna is 6.5 dBi, and the measured gain is 4.5 dB.

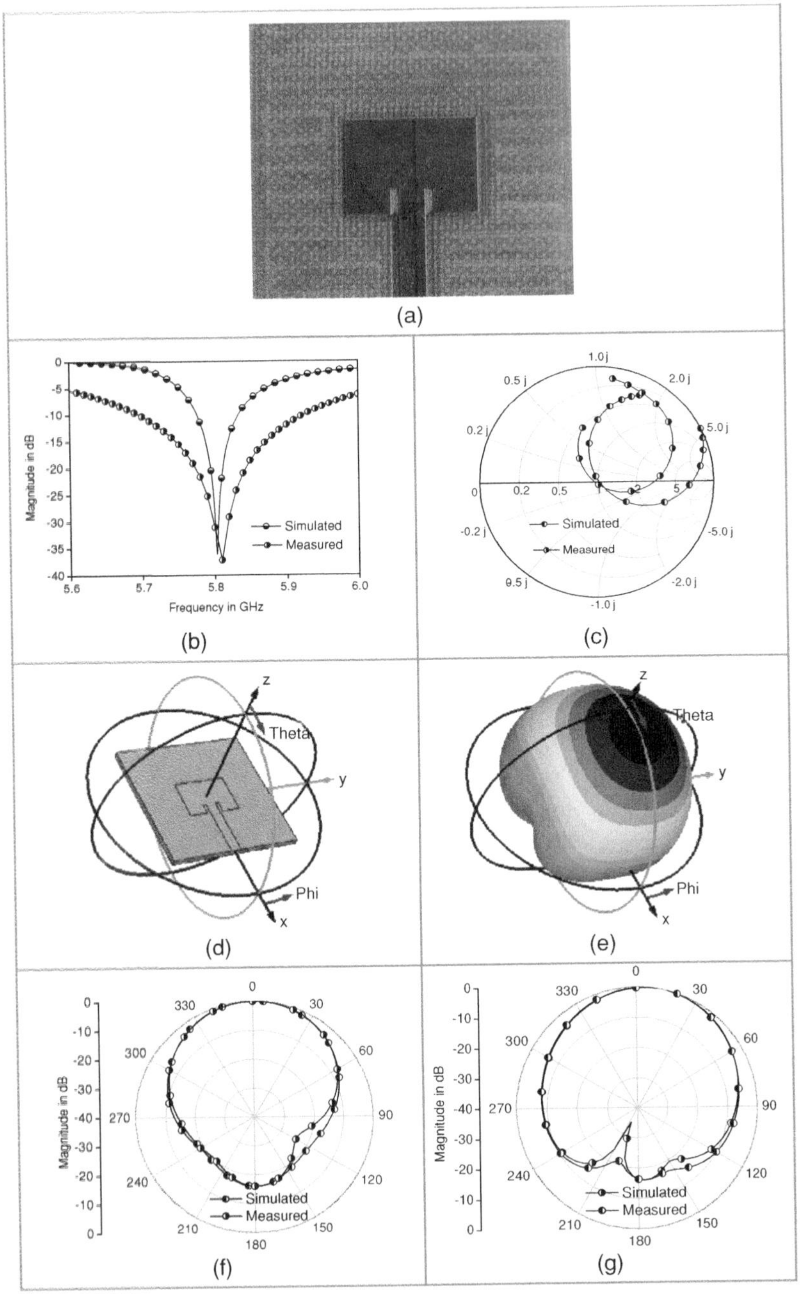

FIGURE 7.14. (**a**) Unit element rectangular microstrip patch antenna. (**b**) Simulated and measured return loss. (**c**) Simulated and measured return loss in a Smith chart. (**d**) Orientation of the rectangular patch with respect to the coordinates. (**e**) 3D radiation pattern of the rectangular patch. (**f**) Normalized radiation pattern for varying Theta, keeping Phi constant at 90°. (**g**) Normalized radiation pattern for varying Theta, keeping Phi constant at 0°.

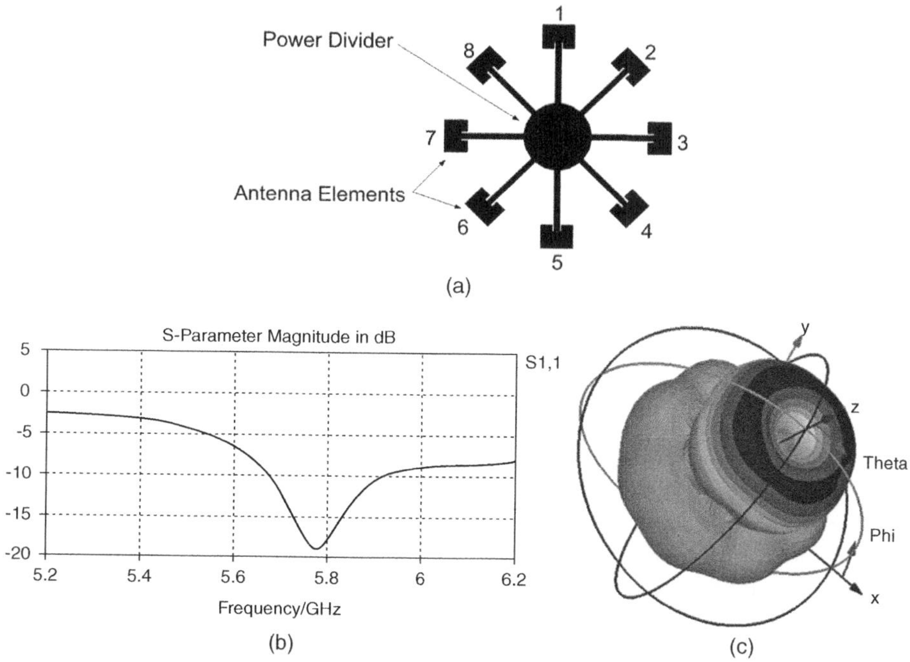

FIGURE 7.15. (**a**) Top view of the linear array for configuration 1. (**b**) Simulated return loss of the linear array of configuration 1. (**c**) 3D view of the radiation pattern.

7.9.2 Linear Array—Configuration 1

Following the design of rectangular patch antenna and the power divider, the first array configuration that was investigated is shown in Figure 7.15a. The designed power divider was used at the center, and the rectangular patches were used as the antenna elements. The simulated return loss of the array shown in Figure 7.15b shows a -10 dB return loss for around 200 MHz. The 3D radiation pattern is shown in Figure 7.15c, where Z axis is normal to the plane of the array through the center. The antenna elements are placed 70 mm away from the center. This distance was an optimization between mutual coupling and the side lobes. The simulated gain is 9.4 dBi. However, from the radiation pattern it is evident that the pattern is differential with a deep null at the center—comparable to a volcanic crater. This radiation pattern is not applicable for RFID readers.

7.9.3 Linear Array—Configuration 2

The next investigation carried out to solve the problem of differential radiation pattern was to make the feedline of each element come out at an angle of 45° with the normal to the circle in the center. Considering the electric fields to be vector, the idea behind this configuration was to create a torque where the electric fields would enhance each other instead of canceling each other. The antenna elements are still radial in their spatial orientation.

The top view of the array for this configuration is shown in Figure 7.16a, where the antenna elements are placed 80 mm away from the center of the circular power divider. The return loss (S_{11}) for this array configuration is given in Figure 7.16b. The return loss is shown in a Smith chart in Figure 7.16c. The orientation of the simulated array with respect to the coordinates is shown in Figure 7.16d, and the 3D radiation pattern is shown in Figure 7.16e.

The simulated gain is 12 dBi. However, it is observed that a null still exists in the center of the array with the main beam concentrated around that null. The elements in this array are placed 80 mm away from the center, leading to the formation of side lobes. The half-power beam width is quite narrow.

It is observed from the simulation results that this configuration cannot solve the problem of null in the center. The reason behind this central null was assumed to be the mutually canceling directions of electric fields coming out the power divider. However, the electrical fields reaching the antenna elements were still radial. To solve the problem of the central null, the next investigation carried out was to make the electrical fields come out of the power divider in radial manner, but reach the antenna elements in a different angle so as to enhance each other. The idea is discussed in configuration 3.

7.9.4 Initial Assumption, Corresponding Limitations, and First Solution

The distribution of the electric fields in the linear array configuration is shown in Figure 7.17a. This is independent of the angle or shape of the feedline, reaching each of the elements as long as they are in same phase. The central circle is the power divider with the radial black elements emerging out of it as the output ports. Connected to it are the individual antenna elements with their respective numbering.

The *E*-fields of every antenna element of the power divider are denoted by the suffix of that individual element and are shown in the form of arrows emerging out of the individual elements. If the *E*-fields are treated as vectors, then they are positioned in such a way that they cancel each other at the center for the opposite element; that is, E_1 cancels with E_5, E_2 cancels with E_6, and so on, thereby radiating no power. This canceling effect decreases as the distance from the center increases. This gives rise to a differential radiation pattern.

To overcome this problem, a different configuration was investigated as shown in Figure 7.17b. The power divider and the antenna elements are left unchanged; only the length of the output port is changed for each of the elements to arrive to the given shape. The elements are tilted 45° compared to the elements in array of configuration 1 or 2. The idea behind changing the angular orientation of each of the elements was to produce a kind of torque if the electric fields are thought to be vectors. It was hypothesized that for this configuration they should enhance each other and make a circular orientation of the electric fields.

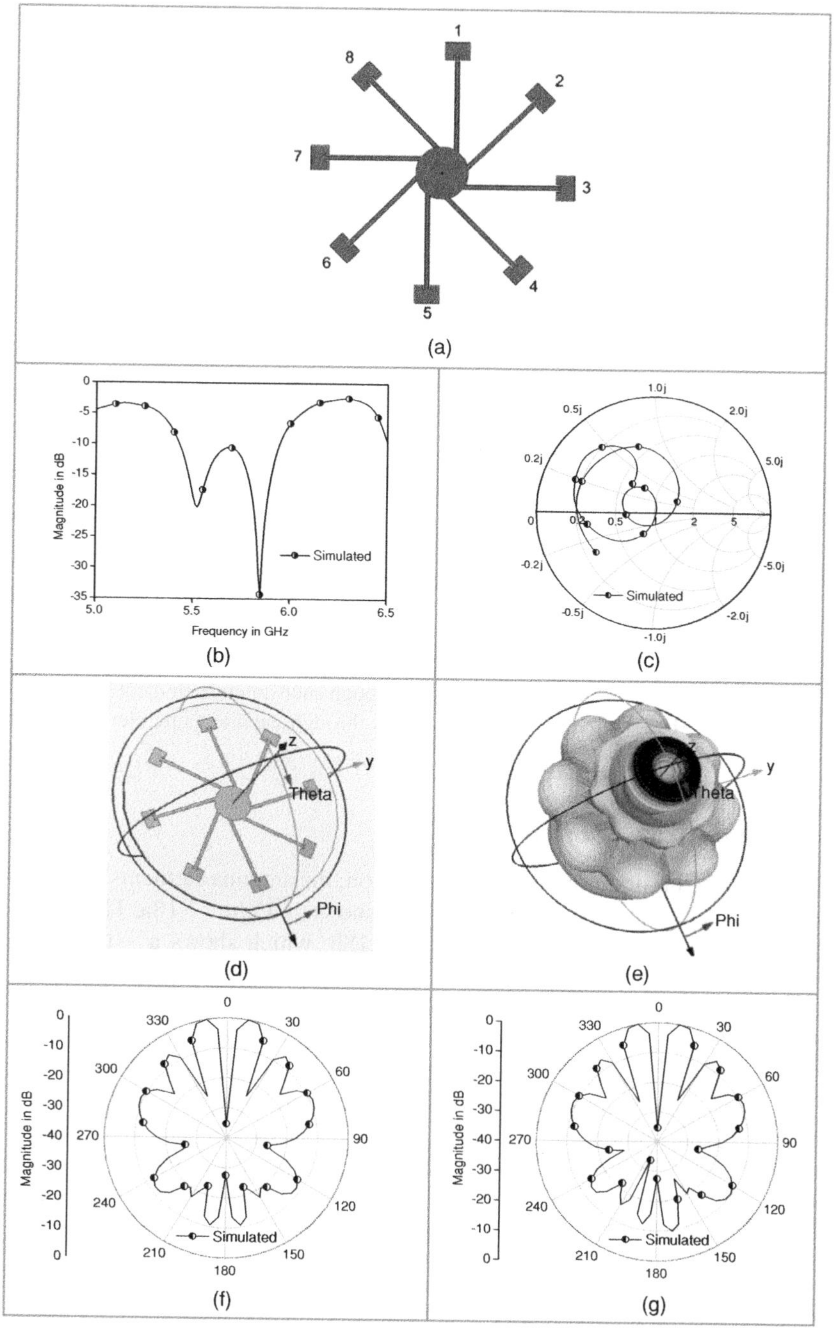

FIGURE 7.16. (**a**) Top view of the linear array with radial orientation of patch but angular orientation of feed—configuration 2. (**b**) Simulated return loss of the array. (**c**) Simulated return loss as observed in a Smith chart. (**d**) Orientation of the array with respect to the coordinates. (**e**) 3D radiation pattern of the array. (**f**) Normalized radiation pattern for varying Theta, keeping Phi constant at 90°. (**g**) Normalized radiation pattern for varying Theta, keeping Phi constant at 0°.

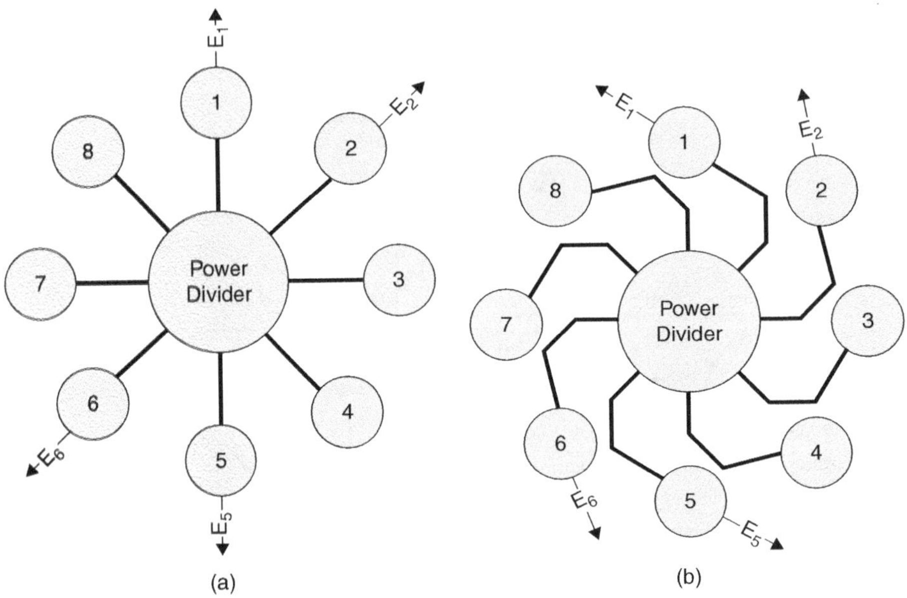

FIGURE 7.17. (**a**) Distribution of Electric field through each antenna element in linear array of configuration 1. (**b**) Distribution of electric field through each antenna element in linear array of proposed configuration 2.

7.9.5 Linear Array—Configuration 3

In the array configuration presented in this section, the antenna elements are bent 45° instead of the feed. The top view of the array is shown in Figure 7.18a. The simulated return loss of the array is shown in Figure 7.18b, which shows a −10-dB return loss over a 300-MHz bandwidth. The 3D radiation pattern is shown in Figure 7.18c, where the Z axis is normal to the plane of the array through the center. The antenna elements are placed at a distance of 68 mm away from the center. The simulated gain is 12.8 dBi, which shows an improvement of 0.8 dBi over configuration 2; also, the half-power beamwidth is wider than the previous configuration. Due to the change of the orientation, the distance between adjacent elements increased, leading to development of side lobes. The cancellation of electric fields is lesser. This is evident from the enhanced gain and wider half-power beamwidth. However, a differential radiation pattern is still observed with a deep null at the center.

7.9.6 Linear Array—Configuration 4

Following the investigations of the aforementioned three linear configurations, it was observed that the direction of the electric fields cannot be changed at the center of each antenna element if they are kept radial with respect to the power divider. If the direction of the electric fields cannot be changed, then the problem of null in the center cannot be solved.

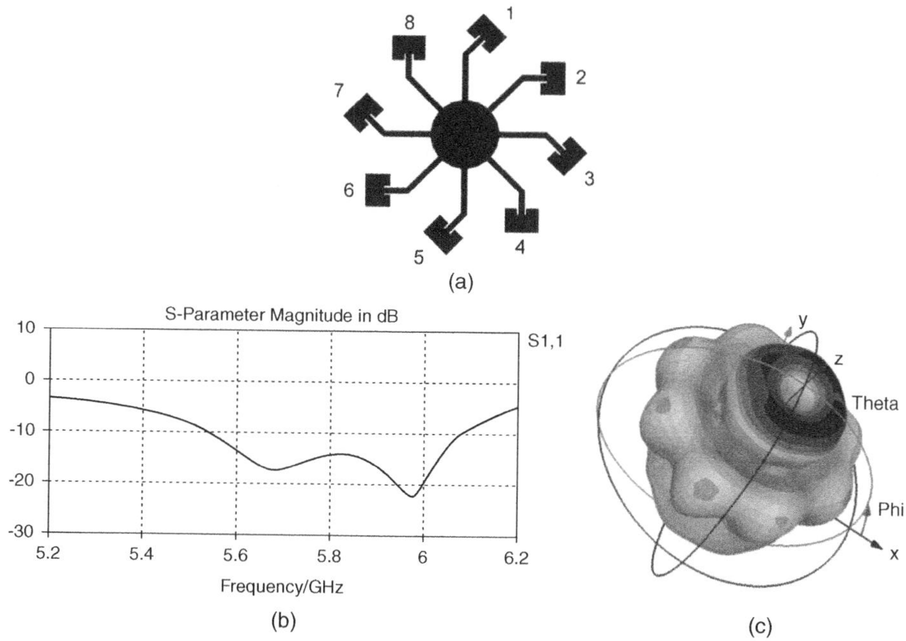

FIGURE 7.18. (**a**) Top view of the linear array for configuration 3. (**b**) Simulated return loss of the linear array of configuration 3. (**c**) 3D view of the radiation pattern for the linear array of configuration 3.

Thus a final configuration was investigated where each antenna element has similar angular orientation and spatial distributions from the center. The top view of this configuration is shown in Figure 7.19a. The distance from center of each patch to the center of the array is same. Each feedline is exactly at the same phase. The elements are placed 80 mm away from the center of the patch. Figure 7.19b shows the measured and simulated return losses of the linear array for this orientation. The return losses are further shown in a Smith chart in Figure 7.19c. Figure 7.19d shows the photograph of the fabricated array, and Figure 7.19e shows the 3D radiation pattern.

The central null that was observed in the aforementioned three configurations, is totally eliminated in the present design. A continuous narrow beam is developed and the simulated gain is 13.5 dBi, where the maximum possible theoretical gain in 15.5 dBi with eight elements. (Each element has a gain of 6.5 dBi.) The comparative normalized radiation patterns are further shown in polar plots in Figures 7.19f and 7.19g.

In the radiation pattern, a narrow main beam is observed along Theta $= 0°$ (Z axis). But along with that, quite a good amount of side lobes are also observed in the figures of the radiation patterns. The reason behind this is the mutual coupling between the feed elements, which led to uneven power distribution to some of the elements. In Figure 7.19a, it is observed that for the elements 1 & 2 and elements 4 & 5, the feed lines come quite near to each other. This leads to the formation of coupled microstrip

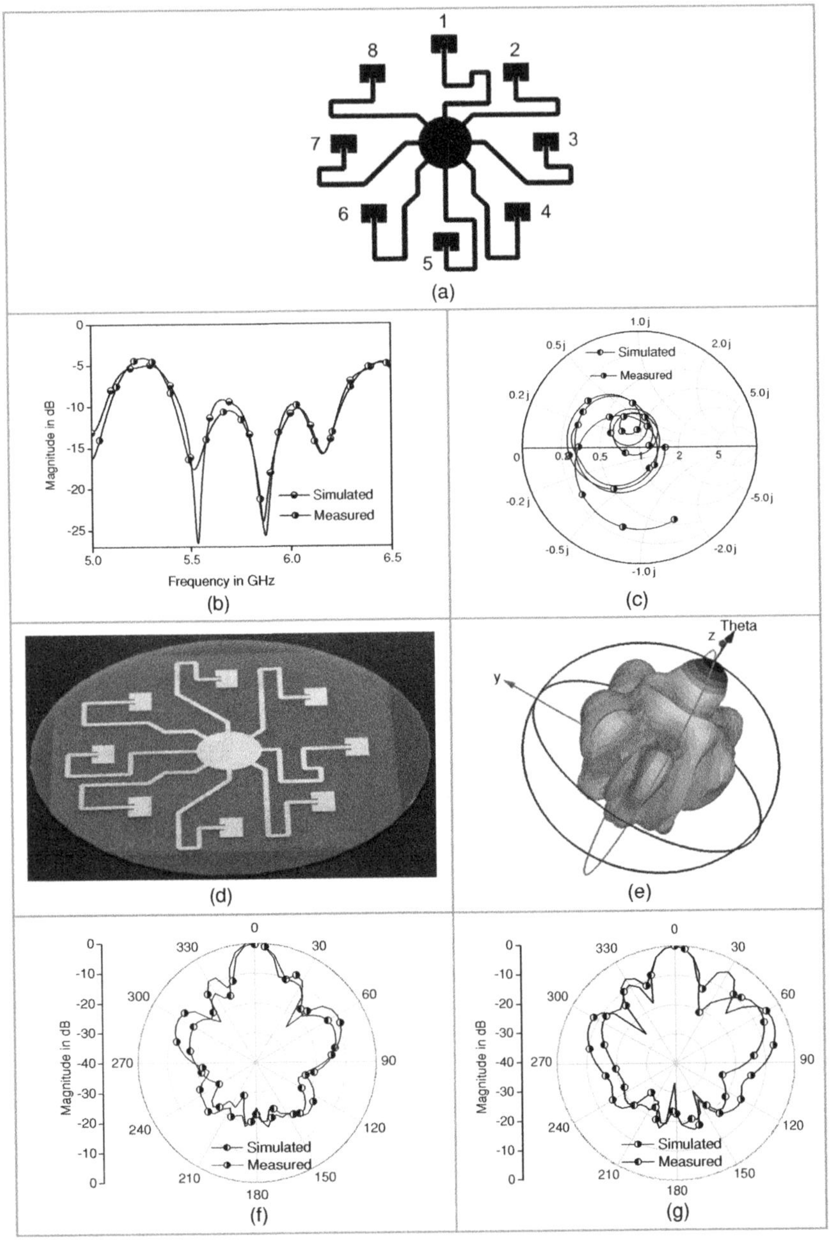

FIGURE 7.19. (**a**) Top view of the linear array with similar orientation and spatial distribution of the antenna elements—configuration 4. (**b**) Simulated and measured return loss of the array. (**c**) Simulated and measured return loss as observed in a Smith chart. (**d**) Photograph of the fabricated array. (**e**) 3D radiation pattern of the array. (**f**) Normalized radiation pattern for varying Theta, keeping Phi constant at 90°. (**g**) Normalized radiation pattern for varying Theta, keeping Phi constant at 0°.

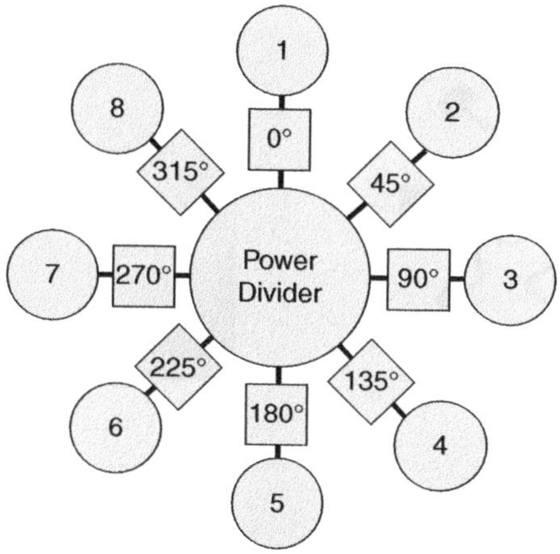

FIGURE 7.20. Concept of circular polarization.

lines, resulting in reduced power reaching the antenna elements. The phases remain the same, but there is imbalance in the magnitude of power reaching the individual antenna elements. This problem can be sorted out by more efficient design of feed lines.

The achievement of this configuration was designing the linear high gain array without any central null. The problem of mutual cancellation of the electric fields is solved. Refinements over this design can be done to reduce the side lobes. The power divider is successfully used to make a linear array with a substrate radius of 100 mm and 12 dBi measured gain at 5.8 GHz for RFID reader antenna.

7.9.7 Circular Polarized Array—Concept

Next, the power divider is placed in the center of a circular polarized array for an RFID reader antenna. The block diagram of the array is shown in Figure 7.20. Each element is radial in placement and has 45° incremental phase difference with respect to its adjacent element (clockwise). The element number 1 is considered at phase 0°. The maximum nullifying influence for element 1 comes from the electric field of element 8. In this design, the electric field of element 8 is in opposite direction to element 1, but the phase of element 8 is reversed simultaneously by incorporating a 180° delay. Thus the electric fields of elements 1 and 8 are in the same phase. However, fields of the other adjacent elements have some influence in determining the ultimate orientation and magnitude of the field. The resultant array is circularly polarized.

7.9.8 Circular Polarized Array—Configuration 1

Figure 7.21a shows the top view of the circular polarized array. The phase differences are introduced by incorporating incremental length of microstrip lines. Figure 7.21b

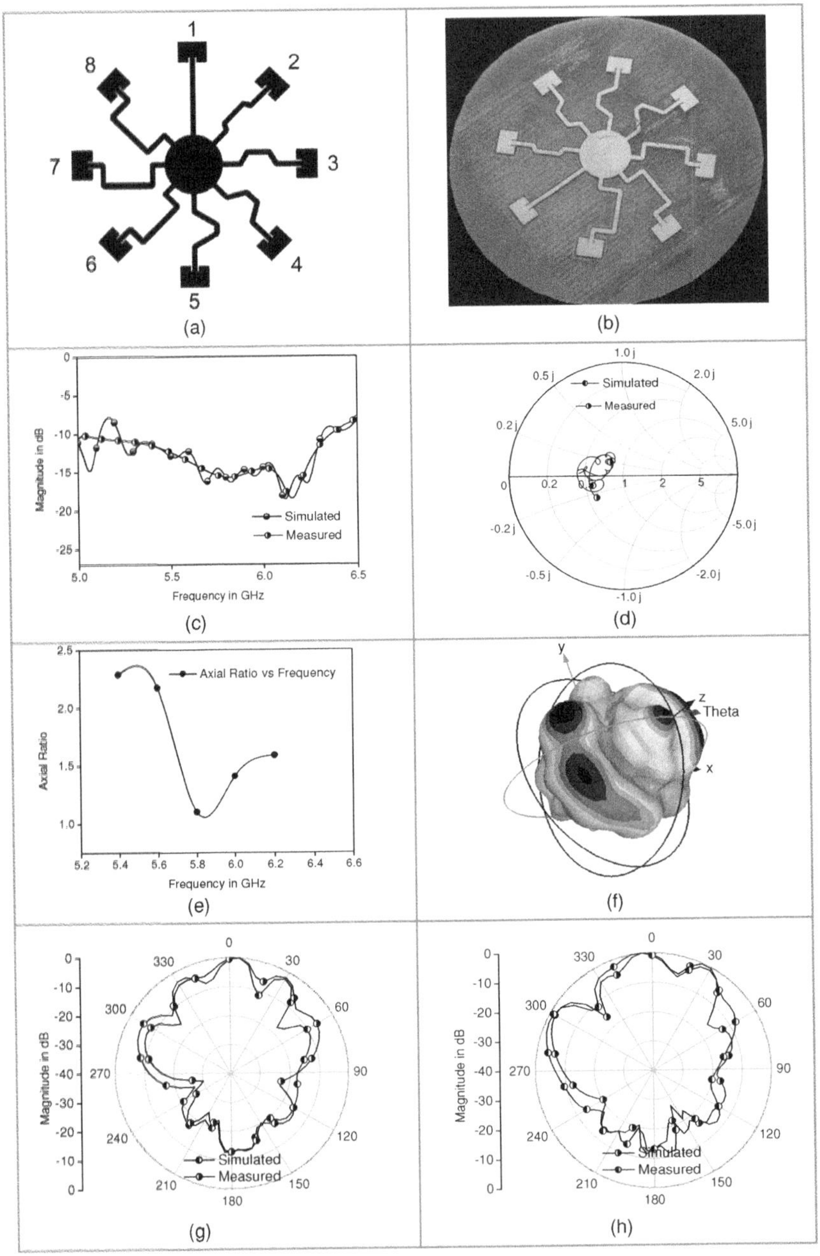

FIGURE 7.21. (**a**) Top view of the CP array with radial orientation of the antenna elements. (**b**) Photograph of the fabricated array. (**c**) Simulated and measured return loss of the array. (**d**) Simulated and measured return loss as observed in a Smith chart. (**e**) Axial ratio versus frequency for the array. (**f**) 3D radiation pattern of the array. (**g**) Normalized radiation pattern for varying Theta, keeping Phi constant at 90°. (**h**) Normalized radiation pattern for varying Theta, keeping Phi constant at 0°.

shows the photograph of the fabricated circular array. The simulated and measure return losses are shown in Figure 7.21c. The results are shown in a Smith chart in Figure 7.21d to have an idea of both magnitude and phase. Figure 7.21e shows the graph of axial ratio versus frequency. The axial ratio is observed to be 1.22 at the design frequency. Measured gain is 9.104 dBi. Figure 7.21f shows the 3D radiation pattern. Figures 7.21g and 7.21h show the comparative radiation patterns in polar graphs. A good amount of side lobes is observed. The reason behind this is the mutual coupling between the different feed lines. This leads to imbalance in the amount of power reaching the antenna elements.

To mitigate the problem of mutual coupling of feed lines, a different approach is adopted in the next section where phase shift of the individual antenna elements are given by negative refractive index (NRI) meta-materials. Additionally, this approach saves space.

7.9.9 Design of NRI Phase Shifter

NRI meta-materials are based on the theoretical work Veselago [19] and the experimental verification of Pendry et al. [20]. One special property of an NRI meta-material is to create a reverse-phase velocity for an incident signal. In the case of a conventional material, if electromagnetic wave enters the material from vacuum at some angle, then on entering the material, the path of the wave turns toward the normal at the separation of the two media. If it is a NRI meta-material, then the path of the wave, on entering the material, deviates away from the normal.

This is the reason why they are called "negative refractive index" materials. In effect, this means that as the signal moves forward, a constant phase point will travel in reverse direction. This is a particularly useful effect that can be applied to design compact delay lines.

The common way that 1D NRI meta-materials are realized in planar microstrip format is to periodically load a transmission line. A standard transmission has an equivalent circuit with a series inductance and a shunt capacitance. This is referred to as a proper refractive index (PRI) material with a positive refractive index. To create a negative refractive index, an equivalent series capacitance and a shunt inductance are needed. However, it is not possible to remove the PRI components of the transmission line. So, when a transmission line is loaded to create an equivalent NRI meta-material, actually a composite between PRI and NRI is created. This is known as a "composite right/left-handed material" (CRLM) with an equivalent circuit shown in Figure 7.22a [21].

An example of dispersion relation of a realized CRLM is shown in Figure 7.22b [22]. The band structure has a fundamental band that is bound by frequencies f_b and f_{c1} followed by a stopband bound by frequencies f_{c1} and f_{c2}. The phase shift per unit cell changes through the band.

If a design frequency is selected in the passband and the passband is shifted up and down ensuring that the design frequency has suitable transmission characteristics, a series of phase-shifters can be designed for the design frequency. The frequency

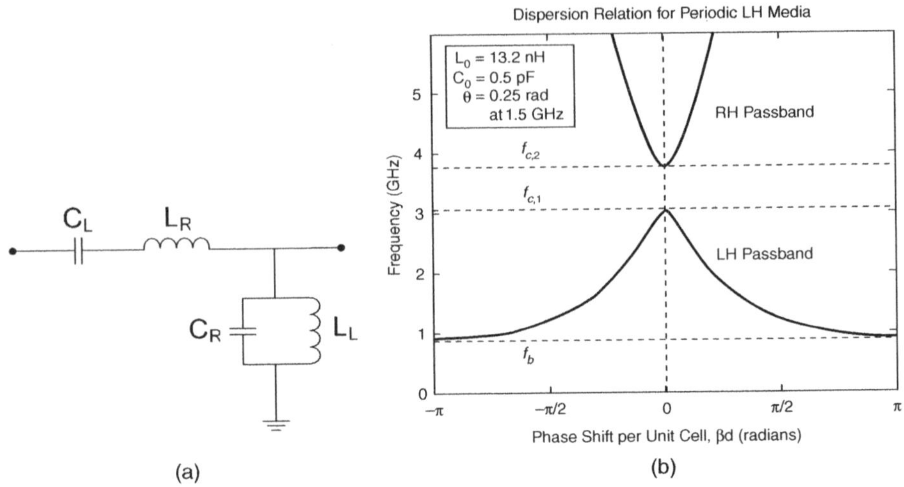

FIGURE 7.22. (**a**) Equivalent circuit model of a one-dimensional CRLH TL meta-material [21]. (**b**) Dispersion relation for periodically *L*–*C* loaded LHM using $L = 13.2$ nH, $C = 0.5$ pF, and electrical length $\theta = 0.25$ radians at 1.5 GHz, indicating Bragg frequency f_b and stopband limits f_{c1} and f_{c2} [22] (Copyright IEEE).

bounds are shown in Eqs. (7.9–7.11) along with the simulated results.

$$f_b = \frac{1}{4\pi\sqrt{L_0C_0}} \tag{7.9}$$

$$f_{c1} = \frac{1}{2\pi}\sqrt{\frac{v}{C_0Z_0d}} \tag{7.10}$$

$$f_{c2} = \frac{1}{2\pi}\sqrt{\frac{v}{L_0Y_0d}} \tag{7.11}$$

In this section, an NRI is designed for $f_b = 5$ GHz, $f_{c1} = 7$ GHz, and $f_{c2} = 8$ GHz. The basic parameters are as follows: $L_0 = 0.796$ nH, $C_0 = 0.318$ pF, length of unit cell $= 1.1$ mm (all based on a 50-Ω Z_0 system). Number of cells $= 6$ in series.

With the aforementioned parameters, simulation is done for six-unit NRI cells in series. The magnitude and phase response for the S_{11} and S_{21} curves are shown in Figures 7.23a and 7.23b. From the magnitude curves it is observed that there exists a passband for a wide bandwidth. From the phase of the curves it is evident that a phase shift takes place in the given arrangement. This shows that the design parameters are suitable but eight different variations are needed with an incremental phase shift of 45° to use in the array.

To start with, a section of microstrip line is considered—"reference line." The reference line is a microstrip line with no additions or subtractions. It was observed that it is possible to significantly affect the phase shift by simply altering the lumped

TABLE 7.1. Comparative Overview of the Values of L and C Used and the Corresponding S_{11} and S_{21} Values

	Angular Shift	S_{11} mag (dB)	S_{21} arg Simulated	S_{21} arg Ideal
0°	Reference line	−31.5	57.48°	57.48°
45°	$L = 1.04$	−16.06	11.4°	12.48°
90°	$L = 1.5$	−16.53	−34.64°	−32.52°
135°	$L = 2$	−14.12	−78.43°	−77.52°
180°	$L = 0.15$	−10.341	−120.5°	−122.52°
225°	$L = 0.25$	−10.49	−168.3°	−167.52°
270°	$L = 0.36$	−31.91	148.5°	147.48°
315°	$L = 0.5$	−16.72	97.79°	102.48°

Capacitance used = 0.636 pF, unit of L = 0.1 nH.

shunt inductance. The results are shown in Table 7.1 along with magnitude of the return loss to ensure that enough power reaches the relevant antenna element.

Table 7.1 gives the comparative overview of the design parameters used and the frequency and phase responses produced due to their implementation. It is observed that for the phase differences of 180° and 225°, the magnitude of S_{11} comes down to around −10 dB. However, the space consumed is the same for all the phase shifts: Six unit cells are required in series, each of them being 1.1 mm in length. Thus the problem of space and mutual coupling can be avoided by this method. However, the availability of proper lumped components for the L and C is important to design the array.

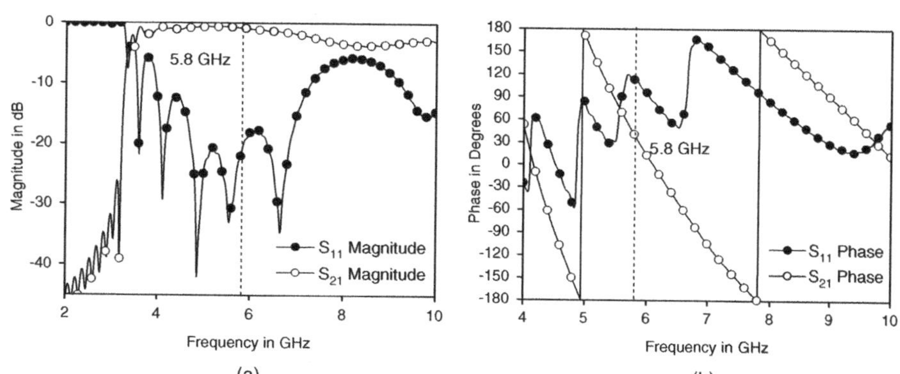

FIGURE 7.23. (**a**) Frequency response showing magnitude of S_{11} and S_{22} when 6 NRI cells are used in series with $L_0 = 0.796$ nH, $C_0 = 0.318$ pF, and length of unit cell = 1.1 mm. Design was based on 50-Ω system. The other design parameters considered were $f_b = 5$ GHz, $f_{c1} = 7$ GHz, and $f_{c2} = 8$ GHz. (**b**) Phase of S_{11} and S_{21} showing the phase shift for the design parameters mentioned in part a.

7.9.10 Circular Polarized Array—Configuration 2

Figure 7.24a shows the circular polarized array using NRI phase shifters. The little white rectangle marks in each feed line are a set of six unit cells placed in series. The antenna elements are placed in radial orientation at a distance of 70 mm from the center of the power divider. The magnified view of the six cells in series is shown in Figure 7.24b, where each cell is 1.1 mm in length.

Each cell consists of a lumped capacitance followed by a section of microstrip line of length 0.7 mm. The microstrip line corresponds to 50-Ω characteristic impedance. The microstrip line is followed in series with another capacitance of the same value, while it has a shunt inductance connected to the ground. The space between two consecutive microstrip lines is 0.4 mm.

With the fabrication facility present in the university, it was not possible to realize the array with such minute precision. The lumped components were also not available. But, these limitations can be overcome by changing the design suitably. The importance of use of NRI phase shifters lies in the fact that in the whole process, there is no scope of any mutual coupling between the feed lines. Thus maximum output can be expected out of the antenna elements when they are placed in radial orientation. Only simulation results are shown and discussed for this array configuration.

The simulated return loss is shown in Figure 7.24d. Good matching of −10 dB or more is observed at the design frequency. The 3D radiation pattern is shown in Figure 7.24e. The maximum simulated gain is 10.65 dBi and the axial ratio at the design frequency is around 1.23. The radiation pattern shows the development of side lobes along with the main lobe. The reason behind this is the imbalance of power reaching the elements 5 and 6 (second data column in Table 7.1). The normalized radiation patterns are plotted in Figures 7.24f and 7.24g. It is observed that the gain improves by around 1.5 dBi by using the NRI phase shifters. If the oscillations in the beam pattern are minimized by proper design of the phase shifters, then a radiation pattern with a wide beamwidth can be achieved by using this array configuration.

7.10 DISCUSSION

This section discusses the achievement and limitations of two different array patterns discussed in this chapter followed by an overview of the performance of two proposed systems. In Figure 7.19 it was observed that there exists a good amount of side lobes in addition to the main lobe. To investigate the reason, two different kinds of simulations were carried out with CST MWS™. In the first simulation, the antenna elements were kept intact as per their orientation and distribution in Figure 7.19a, but there were no feed lines. The antenna elements were excited with eight discrete ports; hence there was no distortion in the magnitude and phase of the input power reaching the antenna elements.

The mutual coupling between the elements as a result of this arrangement was observed. It was observed that the mutual coupling ($S_{ij\,(i \neq j)}$) was −24 dB or lower. The mutual coupling was observed to be greater between the adjacent elements

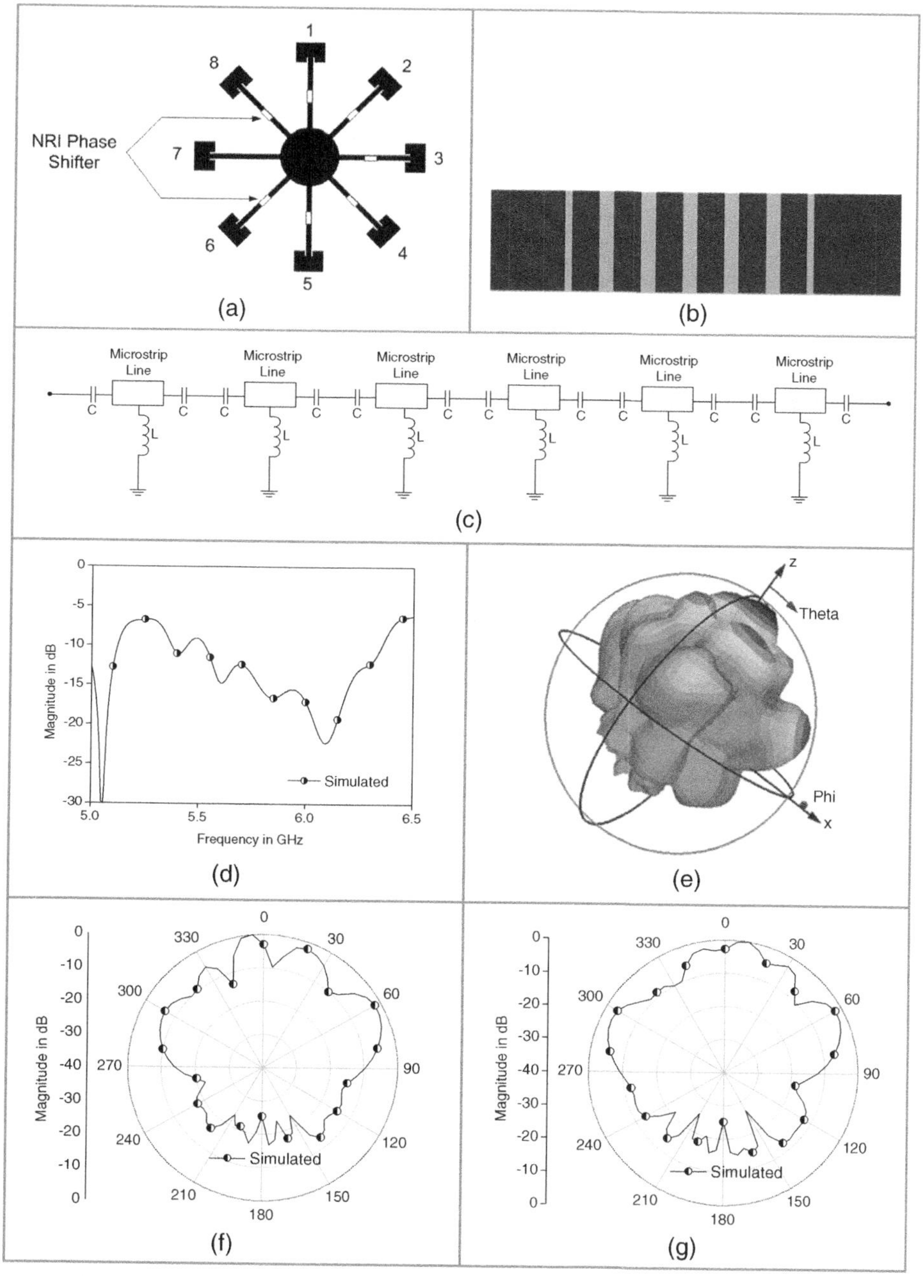

FIGURE 7.24. (**a**) Top view of the CP array with radial orientation of the antenna elements. The rectangles show the position of the NRI cells. (**b**) Magnified view of the six NRI cells in series. (**c**) Schematic of the six cells in series. The dotted oval encloses a single cell. (**d**) Simulated return loss. (**e**) 3D radiation pattern of the array. (**f**) Normalized radiation pattern for varying Theta, keeping Phi constant at 90°. (**g**) Normalized radiation pattern for varying Theta, keeping Phi constant at 0°.

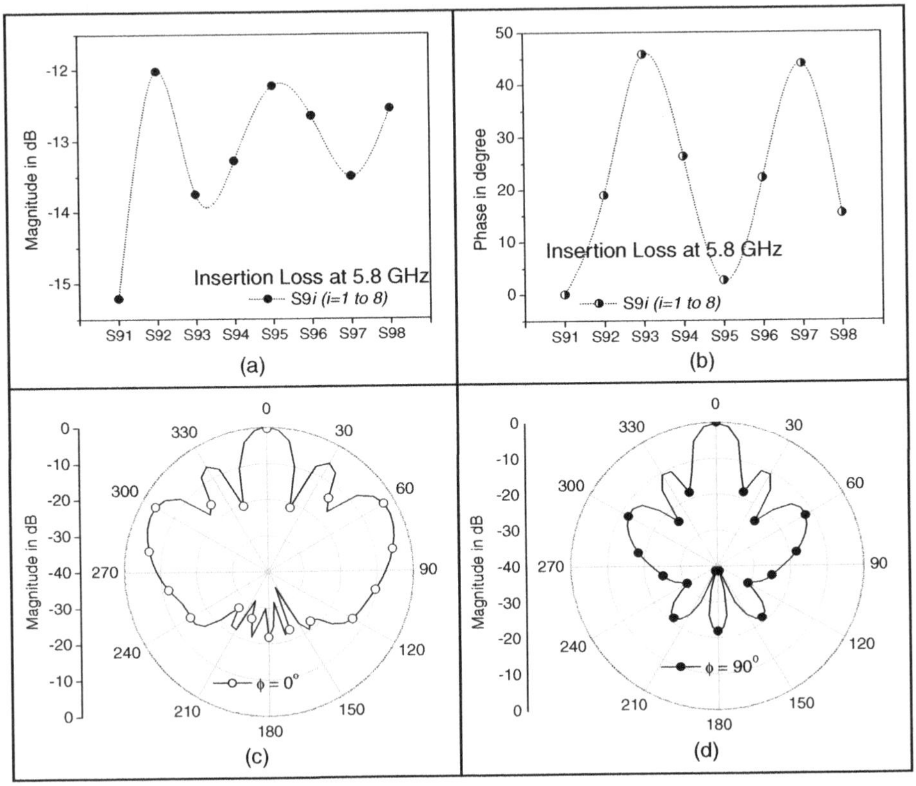

FIGURE 7.25. $S_{9j\,(j=1\text{–}8)}$ curves for the feed lines of the linear array (configuration 4). (**a**) Magnitude curves of S_{9j}. (**b**) Phase curves of S_{9j}; the phases shown are relative with respect to feed line 1. Normalized radiation patterns for the linear array. (**c**) Phi = 0°. (**d**) Phi = 90°.

(e.g., between elements 1 & 2 and 1 & 3). However, because the maximum power going to the other elements is less than −24 dB, the feed lines were investigated next.

In this investigation, the antenna elements shown in Figure 7.19a were removed, only keeping the central power divider and the feed lines. The feed lines were terminated by discrete ports of 50-Ω characteristic impedance. It was observed that due to the orientations and different distances between parts of adjacent feed lines, both magnitude and phase of the input power reaching the different antenna elements got distorted. The power reaching the element 1 suffered maximum distortion in magnitude where the phase was distorted in a sinusoidal manner over all the elements. Figures 7.25a and 7.25b shows the magnitude and phase of the input signal at the different ports. The central excitation port is numbered 9.

In addition to the aforementioned simulations, only the antenna elements of Figure 7.19a were excited by eight discrete ports; the feed lines were removed to avoid any distortion in magnitude and phase of the input signals. The discrete ports were excited simultaneously keeping their phases and amplitudes same. The normalized radiation pattern hence obtained are shown in Figures 7.25c and 7.25d. The maximum gain was observed to be 13.6 dBi with a half-power beamwidth of 16°, which is the same as observed in configuration 4 (Section 7.9.6). However, a good amount of side lobes were observed even in this ideal condition of excitation when the angle Theta (θ) was varied keeping Phi (f) constant at 0°. Thus, it can be concluded that efficient design of feed lines can improve the performance of the antenna to a certain extent only from the current ones.

Similar investigations were carried out for the circular polarized antenna presented in Figure 7.21. The mutual couplings between the elements were observed where the feed lines of Figure 7.21a were removed and each antenna element was excited by a discrete port. It was observed that the mutual coupling ($S_{ij\,(i \neq j)}$) was −21 dB or lower. The mutual coupling was observed to be greater between the opposite elements (e.g., between elements 1 & 5 or 2 & 6) instead of the adjacent elements. However, the maximum power going to the other elements being lesser than −24 dB, the feed lines were investigated next.

In this investigation, the antenna elements shown in Figure 7.21a were removed, only keeping the central power divider and the feed lines. The feed lines were terminated by discrete ports of 50-Ω characteristic impedance. In this investigation also, both magnitude and phase of the input power reaching the different antenna elements got distorted. The power reaching element 8 suffered maximum distortion in magnitude where the phase was distorted to a good amount between the elements 1 & 2. Figure 7.26a and 7.26b shows the magnitude and phase of the input signal at the different ports. The central excitation port is numbered 9.

In addition to the aforementioned simulations, only the antenna elements of Figure 7.21a were excited by eight discrete ports; the feed lines were removed to avoid any distortion in magnitude and phase of the input signals. The discrete ports were excited simultaneously keeping their amplitudes same. However, the phase of each adjacent element was incremented by 45° in a clockwise manner. The normalized radiation pattern hence obtained are shown in Figures 7.26c and 7.26d. The maximum gain was observed to be 10.4 dBi, which is the same as observed in configuration 4 (Section 7.9.6). However, in this ideal condition the radiation pattern was very symmetrical with minimum back lobes. Thus, it can be concluded that efficient design of feed lines can improve the performance of the antenna to a good extent.

The NRI array could not be fabricated due to nonavailability of very precise etching facility and, more importantly, the nonavailability of the lumped components of the specified values. Further investigation is required to make the values of lumped elements similar to the values available in the market.

Figures 7.27a–d show the performance of the proposed use of the designed antenna array in Figure 7.1d. The antenna elements are at 45° inclination with respect to the plane of the power divider in Figure 7.27a. The radiation pattern obtained from this arrangement of the antenna elements is shown in Figure 7.27b. The power distribution shows a difference pattern that is distributed with equal proportions at four angles with adjacent nulls between them. The maximum gain observed for this arrangement is 7.219 dBi. Another version of the hat top antenna is shown in Figure 7.27c, where the antenna elements are perpendicular to the plane of the power divider. In

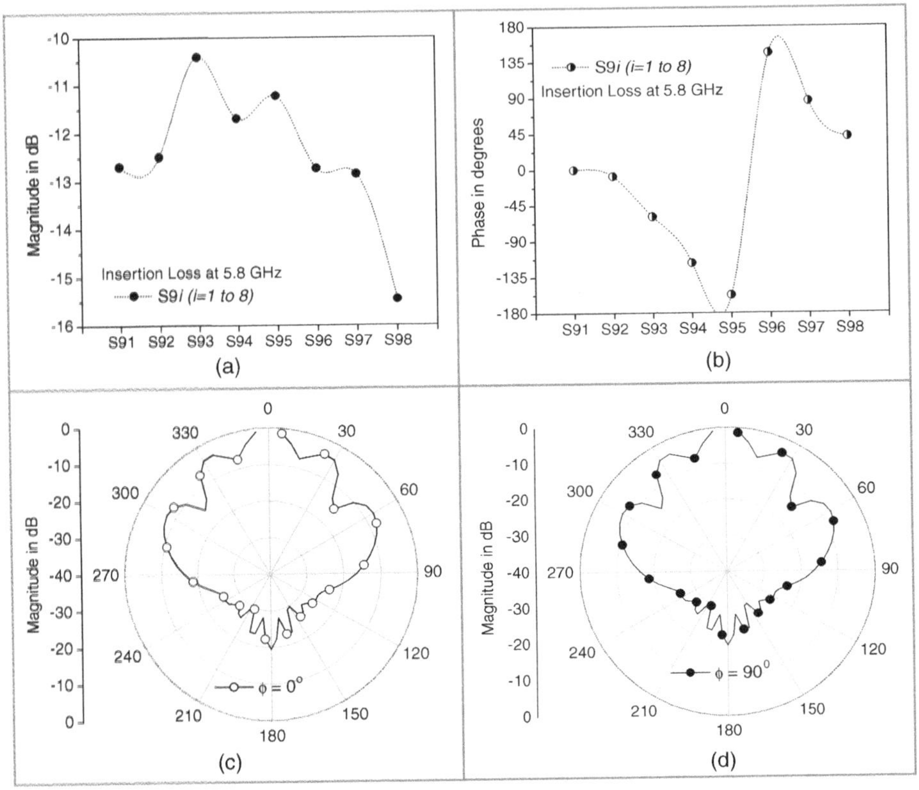

FIGURE 7.26. $S_{9j\,(j=1\text{–}8)}$ curves for the feed lines of the circular array (configuration 1). (**a**) Magnitude curves of S_{9j}. (**b**) Phase curves of S_{9j}; the phases shown are relative with respect to feed line 1. Normalized radiation patterns for the circular array. (**c**) Phi = 0°. (**d**) Phi = 90°.

this situation also, the radiation pattern observed in Figure 7.27d is differential in nature. However, for this arrangement, the power distribution is more concentrated perpendicular to the plane of the power divider compared to the situation observed in Figure 7.27b. To get the radiation pattern as proposed in Figure 7.1d, the feed lines should probably be modified similar to Figure 7.19a but in three-dimensional space. This remains to be an area of further design and investigation.

7.11 CONCLUSION

In this chapter, antenna arrays are proposed that are centered on 5.8 GHz and are suitable for RFID readers. The antenna arrays consist of a novel power divider and different orientation and distribution of the individual antenna elements. The frequency band of 5.8 GHz is chosen because that range is unlicensed (hence can be allocated for RFID systems) and is relatively free. The sizes of the components are

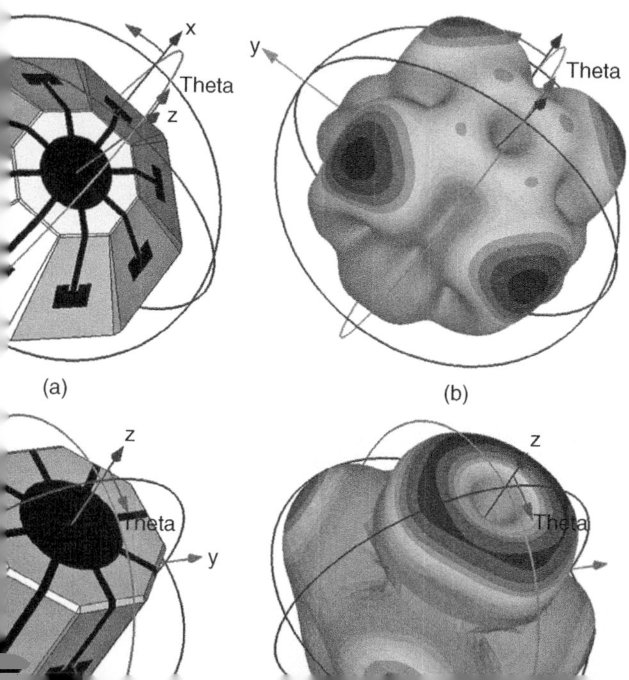
x
Theta
z
y
Theta
(a)
(b)
z
Theta
y
z
Theta

results were very good where 13.5 dBi gain was achieved with eight elements. The maximum theoretical gain achievable is 15.5 dBi.

Next, a circular polarized array was designed using the designed power divider and the rectangular microstrip patches. Incremental lengths of microstrip lines were used to change the phase between the successive patches. The results were good, but imbalance was observed in the power reaching the individual elements, leading to formation of side lobes. For handheld operations where the accuracy of position is not a critical requirement, this designed antenna array can serve well as an RFID reader antenna. The reason for imbalance was the mutual couplings between the feed lines. This imbalance led to the formation of side lobes.

To resolve this problem, NRI phase shifters were investigated and designed to replace the microstrip lines. The NRI phase shifters use a very little amount of space along the length of the microstrip line without allowing any situation to arise for mutual couplings. However, the design of the NRI phase shifters showed that there was some imbalance of the power received for three of the elements compared to the other elements of the array. This imbalance can be resolved with proper design of the lumped components and width of the microstrip line. The importance of this series of phase shifters is the absence of any mutual coupling between the adjacent feed lines.

The results obtained in two different array configurations are analyzed. It is observed that the feed lines need to be designed in a better way to improve the radiation patterns. The maximum gains were not affected, but the symmetry in the radiation patterns was disturbed with the creation of side lobes and back lobes. For the linear configuration, maximum mutual couplings were observed between the adjacent elements whereas for the circular configuration, maximum mutual couplings were observed between the opposite elements. The radiation patterns for the ideal cases (no feed line—excitation by multiple discrete ports) showed the presence of side lobes too. Thus the array was analyzed in different ways to investigate the design flaws and to what extent the performances can be improved.

The arrays can be used in some other three-dimensional configurations also, including the hat-top configuration. CST MWS™ simulation results for the two hat-top configurations are shown. The designed power divider can also be used in real-time location of objects. The delay lines of Figure 7.24a can be replaced by single-pole double-throw (SPDT) switches. Each SPDT switch would be connected to the corresponding antenna element or a 50-Ω lumped resistance. At any time, any one of the output ports would feed power to the corresponding antenna element while the other output ports would drain their power to the 50-Ω resistances. The SPDT switches can be controlled by selector switches. With appropriate switching [23], the arrangement can be efficiently used for tracking items in real time.

REFERENCES

1. N. C. Karmakar and M. E. Bialkowski, A beam-forming network for a circular switched-beam phased array antenna, *IEEE Microwave Guided Wave Lett.*, Vol. 11, No. 1, pp. 1–3, January 2001.

2. J. D. Griffin and G. D. Durgin, Reduced fading for RFID tags with multiple antennas, in *IEEE Antenna Propagation Society International Symposium Digest*, Honololu, HI, July 2007, pp. 1201–1204.
3. J. Lee, S. K. Das, and K. A. Kim, Analysis of RFID anti-collision algorithms using smart antennas, in *Sensys '04*, November 3–5, 2004, Baltimore, MD, pp. 265–266.
4. J. Lee, Wireless networks characterizations: Interference, collision, and localization, unpublished doctoral dissertation, Seoul University, Korea.
5. P. R. Foster and R. A. Burberry, Antenna problems in RFID systems, in *IEE Colloquium on RFID Technology: Microwave and Antenna Systems*, Malvern, UK, 1999, pp. 3/1–3/5.
6. S. Y. Chen and P. Hsu, CPW-fed folded-slot antenna for 5.8 GHz RFID tags, *Electron. Lett.*, Vol. 40, No. 24, pp. 1516–1517, November 2004.
7. B. Strassner and K. Chang, Passive 5.8 GHz radio-frequency identification tag for monitoring oil drill pipe, *IEEE Trans. Microwave Theory Tech.*, Vol. 51, No. 2, pp. 356–363, February 2003.
8. S. Nikolau, J. Papapolymerou, and M. M. Tentzeris, Reconfigurable smart antenna for mobile electronics, Internet archive of Packaging Research Center, Georgia Tech newsletter, June 2005, archived at http://www.prc.gatech.edu/newsletter/june2005.htm.
9. M. A. Ingram, Smart reflection antenna system and method, US Patent No. US 6,509,836, B1, January 21, 2003.
10. D. M. Pozar, A review of aperture coupled microstrip antennas: History, operation, development, and applications, Internet archive of University of Massachusetts at Amherst, archived at http://www.ecs.umass.edu/ece/pozar/aperture.pdf, May 1996.
11. N. C. Karmakar, Investigations into a cavity-backed circular-patch antenna, *IEEE Trans. Antennas Propag.*, Vol. 50, No. 12, pp. 1706–1715, December 2002.
12. D. M. Pozar, *Microwave Engineering*, John Wiley & Sons, New York, 1998.
13. T. Wuren, K. Taniya, I. Sakagami, and M. Tahara, Miniaturization of 3-way and 5-way Bagley polygon power dividers, in *Proceedings IEEE APMC 2005*, Vol. 4, 4 pp.
14. C. A. Balanis, *Antenna Theory: Analysis and Design*, John Wiley & Sons, Hoboken, NJ, 2005.
15. P. Bhartia, I. Bahl, R. Garg, and A. Ittipiboon, *Microstrip Antenna Design Handbook*, Artech House, Norwood, MA, 2001.
16. S. M. Roy, I. Balbin, and N. C. Karmakar, Novel N-way power divider and array configuration for RFID readers operating at 5.8 GHz, in *IEEE RFID Conference*, March 2008, Las Vegas, NV (CD ROM).
17. R. P. Parrikar and K. C. Gupta, Multiport network model for CAD of electromagnetically coupled microstrip patch antennas, *IEEE Trans. Antennas Propag.*, Vol. 46, No. 4, pp. 475–483, April 1998.
18. J. Kim, H. T. Kim, K. S. Kim, J. S. Lim, and D. Ahn, An equivalent circuit model for multi-port networks, in *Proceedings of the 37th European Microwave Conference*, Munich, Germany, 2007, pp. 901–904.
19. V. G. Veselago, The electrodynamics of substances with simultaneously negative values of ε and μ, *Soviet Phys. Uspekhi*, Vol. 10, No. 4, pp. 509–514, January–February 1968.
20. J. B. Pendry, A. J. Holden, D. J. Robbins, and W. J. Stewart, Magnetism from conductors and enhanced nonlinear phenomena, *IEEE Trans. Microwave Theory Tech.*, Vol. 47, No. 11, pp. 2075–2084, November 1999.

21. A. Lai and T. Itoh, Microwave composite right/left-handed metamaterials and devices, in *Microwave Conference Proceedings, IEEE Asia-Pacific Conference Proceedings*, Vol. 1, December 2005, 4 pp.
22. G. Eleftheriades, Planar negative refractive index media using periodically *L–C* loaded transmission lines, *IEEE Trans. Microwave Theory Tech.*, Vol. 50, No. 12, pp. 2702–2712, December 2002.
23. N. C. Karmakar, L. S. Firmansyah, and P. Hendro, Tunable PIFA using low cost band switch diodes, in *Proceedings APS IEEE Conference, 2002*, pp. 516–519.

CHAPTER 8

FPGA-CONTROLLED PHASED ARRAY ANTENNA DEVELOPMENT FOR UHF RFID READER

NEMAI CHANDRA KARMAKAR, PARISA ZAKAVI,
and MANEESHA KAMBUKAGE
Department of Electrical and Computer Systems Engineering,
Monash University, Clayton, Victoria, Australia

8.1 INTRODUCTION

Radio-frequency identification (RFID) has been classified based on its frequency designations: low-frequency (LF), high-frequency (HF), ultra-high-frequency (UHF), and microwave frequency (MW) tags. Traditionally, LF and HF tags dominated approximated 90% of the RFID tag and reader market due to their mass deployment as animal tags and item tagging [1]. However, after the mandates of using UHF RFID by the biggest retail chain Wal-Mart, UHF RFID has gotten new momentum [2]. The motto is to comply with the UHF regulation by thousands of vendors to tag their manufactured goods and items. The US Department of Defense (DoD) has mandated UHF tags for their operations. UHF RFID system has many advantages over the other types. They are as follows [3]:

1. UHF tags have a maximum reading distance of up to 20 m.
2. Contrary to the microwave tags, UHF tags have the advantage of non-line-of-sight reading capability and little attenuation by the contents and features of items, therefore, reading of UF tags is not blocked by human presence (not attached to body), water, and so on.
3. It has already Gen1 and Gen 2 tags and is universally used in various parts of the world under the mandates of EPC Global, GATG, and CEPT.
4. Contrary to the LF and HF short-distance inductive coupling, UHF tags enjoy long-distance electromagnetic capture and work on the principle of backscatter radiation from the tag.

Handbook of Smart Antennas for RFID Systems, Edited by Nemai Chandra Karmakar

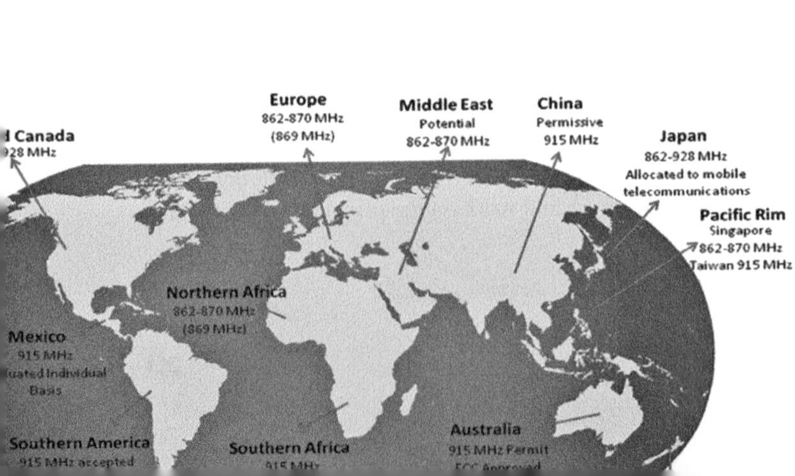

Canada
Europe
862-870 MHz
(869 MHz)
Middle East
Potential
862-870 MHz
China
Permissive
915 MHz
Japan
862-928 MHz
Allocated to mobile
telecommunications
Pacific Rim
Singapore
862-870 MHz
Taiwan 915 MHz
Northern Africa
862-870 MHz
(869 MHz)
Mexico
915 MHz
Individual
Basis
Southern America
Southern Africa
Australia
915 MHz Permit

TABLE 8.1. Specifications of Global UHF RFID Reader Phased Scanned-Array Antenna

Design frequency	910 MHz
10-dB RL Bandwidth	100 MHz
Beam scanning	3D azimuth and elevation beam scanning
Beamwidth	42°
Peak gain	12 dBi
Weight	Less than 5 kg
Dimensions	45 cm × 58 cm × 2.9 cm

on the order of 12 dBi, the antenna should have large aperture dimension. Sometimes patch antennas on high dielectric constant substrate materials are designed. However, due to the high concentration of electric energy in the dielectric materials with high dielectric constant, the radiation efficiency degrades significantly. Also, significant loss incurs in the form of surface wave propagation. Therefore, in our design we use moderately low dielectric constant materials for the antenna design. Also, to enhance the operational bandwidth of the antenna, height is increased substantially. Conventional materials cannot be used because the materials are very expensive. To reduce the cost and weight, the antennas are made on composite materials with polystyrene foam and low-cost FR4 material. All these electromagnetic, mechanical, and cost factors make the phased array antenna design a daunting task. Figure 8.2 illustrates the design challenges and progression to develop a phased array antenna for the reader, which relinquishes the requirement of multiple antennas for different zones and can serve the global reach with a single design.

Phased array antennas are defined as scanned-array antennas where each antenna element is fed with prescribed amplitude and phase weights to make a scanning beam in the 3D space. The phased array antenna can also steer nulls toward the interferers if the weights are made agile in both amplitude and phase. Such antenna arrays are reported in the open literature. However, the phased array antennas are a complex system with a combination of multidisciplinary designs encompassing array synthesis and analysis, design of RF feed network, design of phase shifters biasing network, and electronic switching control and programming of the control algorithm with either a microcontroller or a field programmable gate array (FPGA). With the improved capacity of FPGA and dropping cost, FPGAs are quickly replacing the

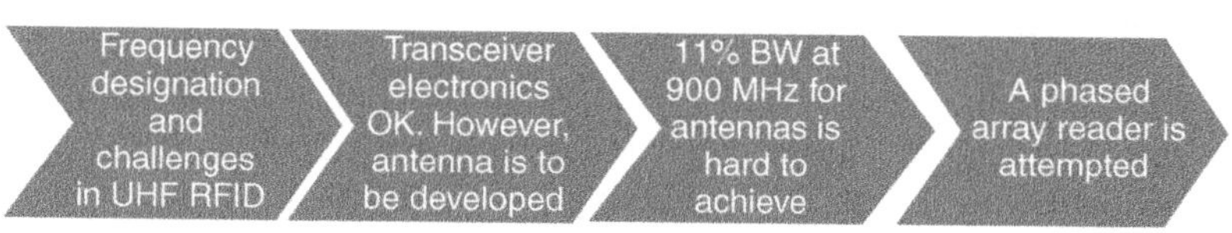

FIGURE 8.2. Challenges in UHF antenna design and proposed solution.

microcontrollers that were used in almost every system a few years ago. Nowadays in every university undergraduate course, programming with FPGA is being taught in design classes. Another advantage of the use of FPGA as the control element of the smart antennas for the RFID system is that FPGA is being used in the RFID reader electronics. Therefore, the same FPGA can be used to control the switching elements of the phased array antenna without any extra hardware implementation and hence cost. Figure 8.2 shows the multidisciplinary exercise to design and develop a FPGA controlled-phase scanned smart-array antenna from scratch. As can be seen in the first stage of the design and development, the design specifications as stated in Table 8.1 are used as the design objective of the phased array antenna. Whole gamuts of hardware and software resources in various disciplines are essential to achieve the design and development goals. The first phase of the design calculation to meet the radiation performances such as directivity, beamwidth, gain and steering vectors, and so on, is the array analysis and synthesis. From the knowledge of antenna array theory and with the help of related software, the physical parameters such as the number of antenna elements in the array, the inter-element spacing between the elements, overall dimension of the array, the current/power distributions for individual antenna elements to attain the required beamwidth, direction in the 3D space and subsequently, the beamforming feed network, and the phase shifter sequence can be determined. The whole exercise of the array synthesis and analysis is defined as design warehouse in Figure 8.3. In the following section the design phases of the phased array antenna development as shown in the figure is presented.

8.2 DESIGN OF PHASED ARRAY ANTENNA

The design phase of the phased array antenna can be categorized into: numerical analysis; passive microwave design such as patch antenna elements and feed networks, active microwave design such as digital phase shifters and the bias network; analog electronic design such as BJT switching circuit and the opto-coupler design; digital electronics design; and programming of FPGA. Since the development of the phased array antenna is done from scratch, each step involves comprehensive exercises of theoretical basis and the practical design and finally validation of the design with prototyping phase. Each stage of the design is time-consuming and needs to go through iterative process even in the age of highly efficient computer power and sophisticated software tools.

8.2.1 Array Analysis and Synthesis

The array analysis and synthesis need appropriate knowledge of antenna array theory. A good resource is any antenna handbook or textbook in the graduate level [5]. There are various topologies for the antenna array. In our present development, we have selected a 3 × 2-element rectangular array antenna. Assuming that the broadband patch antenna element will offer around 8-dBi gain [6] at 900-MHz frequency, an

array of 3 × 2 elements should give sufficient gain to achieve 12 dBi at boresight of the antenna plane. Once the assumption is made, a comprehensive investigation is made to calculate the weight vectors for the individual antenna elements for the beam pattern to be directive in various angles in 3D space. The normalized array factor (to its maximum amplitude) of a planar array of M × N elements is represented [5] by

$$AF(\theta, \phi) = \left\{ \frac{1}{M} \frac{\sin\left(\frac{M}{2}\psi_x\right)}{\sin\left(\frac{\psi_x}{2}\right)} \right\} \left\{ \frac{1}{N} \frac{\sin\left(\frac{M}{2}\psi_y\right)}{\sin\left(\frac{\psi_y}{2}\right)} \right\} \tag{8.1}$$

where

$$\begin{aligned} \psi_x &= kd_x \sin\theta \cos\phi + \beta_x \\ \psi_y &= kd_y \sin\theta \cos\phi + \beta_y \end{aligned} \tag{8.2}$$

where

d_x is the inter-element spacing along the x axis
d_y is the inter-element spacing along the y axis
k is the free-space wave number ($2\pi/\lambda$)
β_x is the progressive phase difference between antenna elements along the x axis
β_y is the progressive phase difference between antenna elements along y axis

The total radiation pattern for an antenna array is obtained by multiplying the radiation pattern of individual element with the array factor as

$$E_{\text{Tot}} = E_{\text{single element}} \times \text{array factor} \tag{8.3}$$

As can be seen in Eqs. (8.1)–(8.3), the design complexity increases with the following variables:

1. Number of antenna elements and
2. Beamforming weights ψ_x and ψ_y

The above equations are for the uniform distribution only. For the consideration of side-lobe levels, position of grating lobes and the variations of the input impedance matching and side lobes with the scan angle and mutual coupling between elements, the design becomes even more complex [5]. However, some readily available software tools ease the array analysis and synthesis. In our design we have taken simplistic approach of using uniform amplitude distribution. Also, we have not considered the mutual coupling in our design calculation. Figure 8.4 shows the radiation pattern of the 3 × 2-element antenna array using personal-computer-aided antenna design

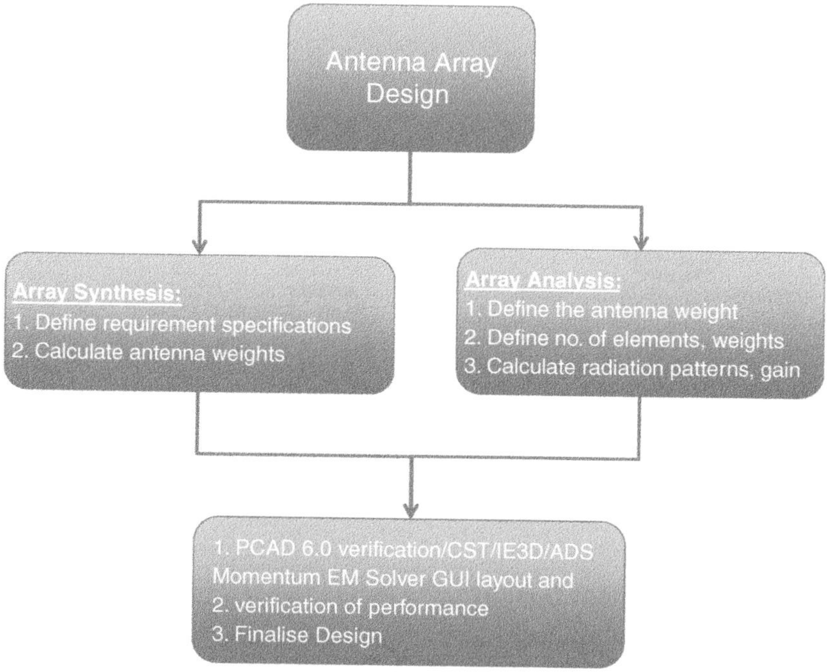

FIGURE 8.4. Antenna array design synthesis and analysis for antenna element weight calculation and performance check.

(PCAAD 6.0) [7]. The important design guidelines are achieved from PCAAD 6.0 are as follows:

1. The required number of antenna elements and inter-element distance to obtain the required beamwidth
2. The progressive phase shifts between elements to obtain beam scanning in 3D space

The numerical calculation can also be extended to design with prescribed side-lobe reduction in a non-uniform array with approximated current distribution based on some polynomial distributions. Popular polynomial approximations for antenna array designs are binomial, Chebyshev and Taylor distributions. Figure 8.4 shows the flow chart from the antenna array design synthesis and analysis to extract the necessary physical data and antenna elements' current amplitude and phase distributions. The design can be further extended in physical layout in the graphical user interface (GUI) in advanced software such as CST Microwave Studio, ADS Momentum, and so on.

Figure 8.5 shows the 3D radiation patterns of the 3 $\times$ 2-element array as obtained from running PCAAD 6.0. All the phase data for different beam positions are produced in Table 8.2 in Section 8.4.1.

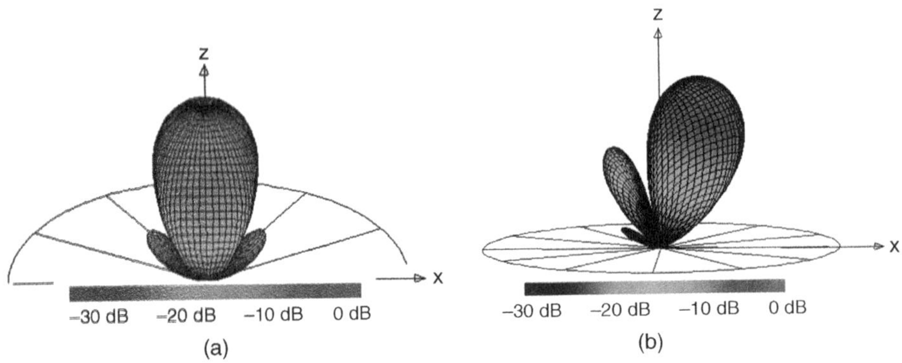

FIGURE 8.5. PCAAD 6.0-generated radiation pattern of 3 × 2-element array at elevation angles: (**a**) $\theta = 0^\circ$ and (**b**) 20° for $\phi = 0^\circ$.

The data are used in the phase shifter design and the phased array tracking electronics design using FPGA.

8.2.2 Antenna Element Design

After performing the antenna array analysis, the next decision is to select an appropriate antenna element. Patch antenna array is selected as the most suitable candidate due to its most important salient features as follows:

1. Fully printable planar configuration
2. Compatibility with monolithic microwave integrated circuit (MMIC) and coplanar waveguide (CPW) structures
3. Low cost and ease of fabrication.

As mentioned before, the antenna element needs to satisfy 11% bandwidth at 910 MHz. It is not a trivial task to improve the bandwidth of the patch antenna element. Conventional probe feed and direct feed microstrip patches are narrowband devices, and the bandwidth is on the order of within 1–3%. There are a few techniques to improve the bandwidth. They are: using very low dielectric constant thick substrate materials; use of aperture-coupled and proximity-coupled feed so that the coupling aperture and the capacitive reactance of the gap coupling contributes to the resonance to increase the bandwidth; using slot loading to the patch antenna; use of LC matching section with the coaxial probe feed; and so on. In the current design the patch dimension is on the order of half the wavelength of the operating frequency. This constitutes approximately 164 mm for the patch side dimension. This is a very large aperture when the element is being used in a 3 × 2-element array. To achieve a compact design, a semicircular patch antenna is used. The dominant mode of a circular patch antenna is TM_{11} and the design frequency [5] is:

$$f_{11} = \frac{1.8411c}{2\pi a_e \sqrt{\varepsilon_r}} \tag{8.4}$$

where c is the velocity of light in free space, a_e is the effective radius of circular patch, and ε_r is the relative permittivity of substrate on which the patch is designed. Incorporating the fringing field effect, the empirical formula of the effective radius is given [5] by:

$$a_e = a\sqrt{1 + \frac{2h}{\pi a \varepsilon_r}\left[\ln\left(\frac{\pi a}{2h}\right) + 1.7726\right]} \tag{8.5}$$

where h is the dielectric thickness and a is the physical radius of the circular patch antenna. A transcendental code can be written to calculate the physical dimension of the circular patch using Eqs. (8.4) and (8.5) firstly knowing the design frequency and substrate parameters and then iteratively solved for a_e, the effective radius.

Based on the above equations, the approximate radius of the patch antenna with a relative dielectric constant close to air (foam is used to emulate air dielectric, which has relative permittivity close to 1). The radius is approximately 94 mm. This will consume 188 mm of space for each element. In a 3×2-element array with an optimum spacing of 0.7λ (234 mm) to avoid grating lobe, the total size of the array antenna would have been 750 mm by 522 mm (to keep the fringe of approximately 50 mm around the patches). Therefore, a prudent and expert design choice should be envisaged reducing the dimensions of the antenna array. Figure 8.6 shows the flow chart of the design procedure of the patch antenna element. As can be seen, the design is initiated with the design specifications as stated in Table 8.1, but with a redefined design requirement for antenna element which can satisfy the goal of the array gain and beamwidth. The operational bandwidth is defined by the input return loss bandwidth for the patch antenna element. This is because the inherent radiation bandwidth for the patch antenna is much larger than the input impedance bandwidth.

That is why in designing a patch antenna element, the first step is to design the antenna element that will meet the input return loss bandwidth. The gain and radiation pattern are selected so that in an array the elements will provide the required combined gain and pattern meeting the specification requirements. Next the substrate is selected for the patch antenna design. As mentioned above, the patch should be designed on a thick but low dielectric constant substrate. The coaxial feed patch antenna is energized with the feed probe that pierces through the ground plane and then connects to the patch metallization. The patch is usually designed on a low-cost substrate to be supported only by the laminate layer and stuffed with foam or hung on the air with a shorting post. Such a design can be found in the previous work by the lead author [8, 9]. A planar LC compensation technique is used in the form of an L-shaped slot on the patch antenna as shown in Figure 8.7. The next phase in the design exercise is selecting a layout that can yield the required bandwidth [10]. A comprehensive parametric study on the variables of the patch antenna was performed. Finally, the design is optimized based on the performance parameters such as input return loss bandwidth, radiation patterns, directivity, and gain bandwidth. Once the design is finalized, it is translated to the prototyping and design validation process through experimental procedure. In the experiment, the first set of data extracted is the S_{11} (in dB) versus frequency. The S_{11} parameter (in dB) is the input return loss. Figure 8.7a

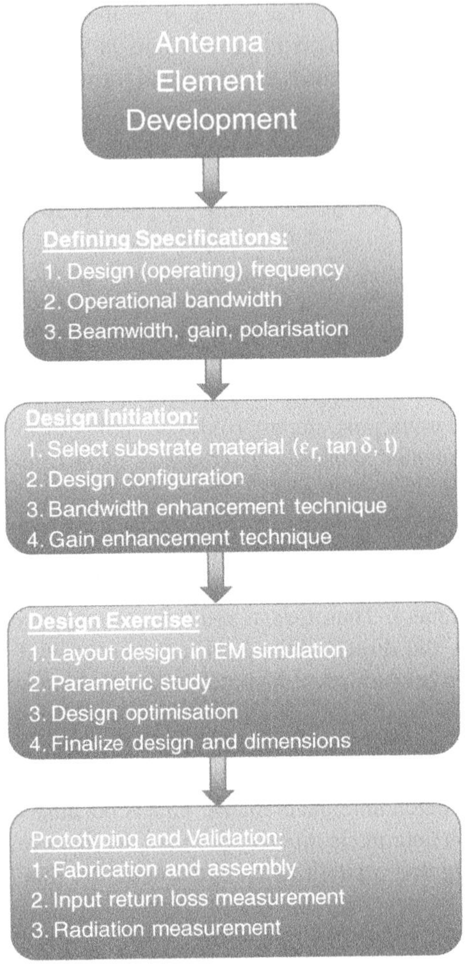

FIGURE 8.6. Design procedure of antenna element.

shows the 3D layout as generated in CST Microwave Studio, and Figure 8.7b shows the input return loss versus frequency of the optimized patch antenna. As can be seen in the figure, the 10-dB input return loss bandwidth is 860–960 MHz, meeting the required design specification for the global UHF RFID frequency designation.

Once the input return loss bandwidth meets the design specification, the radiation characteristic is investigated. Figure 8.7c shows the *E*- and *H*-plane radiation patterns of the patch antenna. As can be seen in the figure, the antenna has a 3-dB *E*-plane radiation pattern bandwidth of approximately 70° and the boresight gain of the antenna is approximately 7 dBi. From the results it can be concluded that the designed antenna element has met the required specification set for the single antenna element. The next phase is the phase shifter design as explained in the following section.

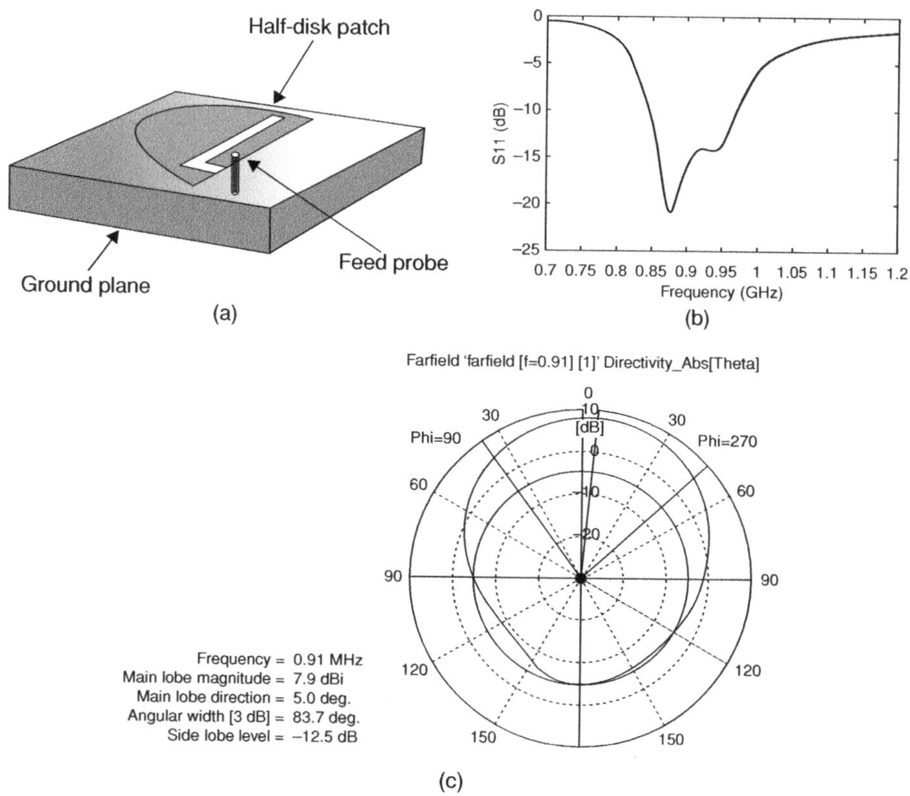

FIGURE 8.7. CST Microwave Studio-generated (**a**) layout, (**b**) input return loss versus frequency of the half-disk patch antenna, and (**c**) radiation patterns at 910 MHz. FR4 laminate $\varepsilon_{r1} = 4.2$, thickness 1.6 mm, and airgap height 17 mm (Copyright © IEEE, reprinted from [14]).

8.2.3 Phase Shifter Design

There are many varieties of digital-phase shifters used for the phased array antennas. However, they are mainly classified into two main categories: *reflection-type* and *transmission-type phase shifters*. A comprehensive coverage on the development of phase shifters can be found in [11–13]. We need a low-cost, lightweight, and compact phased array design that can cover the 11% bandwidth in UHF band. The most suitable candidate is the hybrid-coupled reflection-type phase shifter due to its following salient features:

1. Hybrid-coupled phase shifters are inherently broadband designs, comprised of a central element—a quadrature hybrid coupler.
2. Irrespective of its phase bits in degrees, it needs only two diodes. As an active device, the diode is the most expensive and involves components in the design.

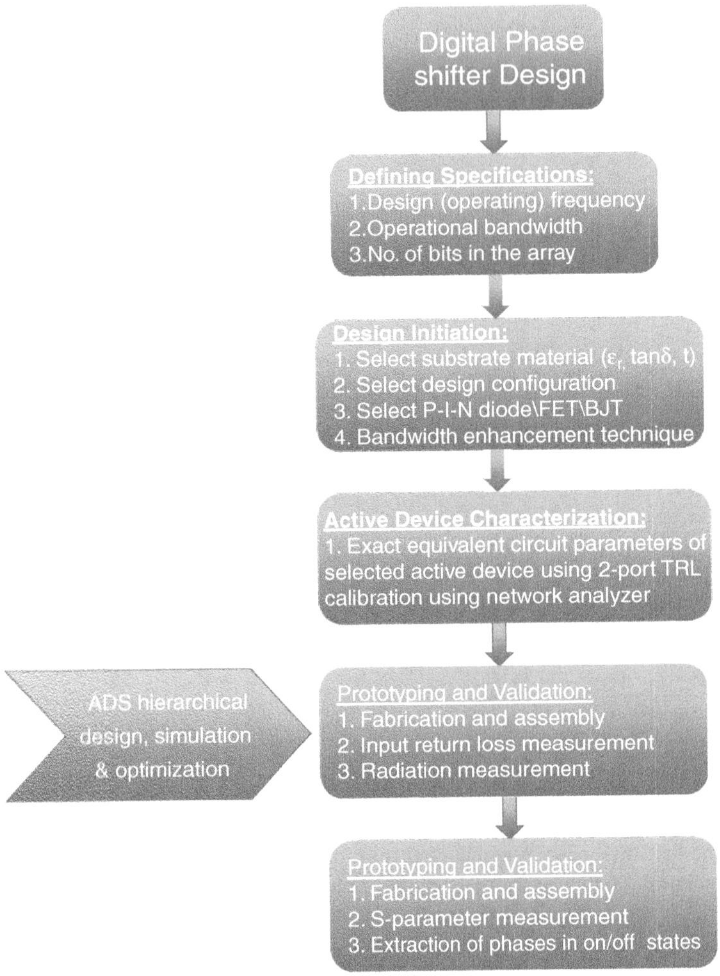

FIGURE 8.8. Procedure of designing a digital phase shifter from scratch.

3. It also needs a biasing network and control electronics for operation. Therefore, the least number of diodes in a design is the best for cost consideration, compactness, and light weight.

However, the hybrid-coupled phase shifter consumes significant real estate. In the design the coupler is compacted with the bends of the 50-Ω lines. The design phases of a digital phase shifter using active devices such as diodes, BJTs, and FETs are shown in Figure 8.8. The design phase starts with the objective of meeting the specification requirements such as operational frequency and bandwidth, number of phase shifter bits, and losses. The transmission loss of the phase shifter is an important consideration as the signal flows through the passive and active devices. To minimize the loss, a high-quality microwave material called Taconic TLX-0 of thickness

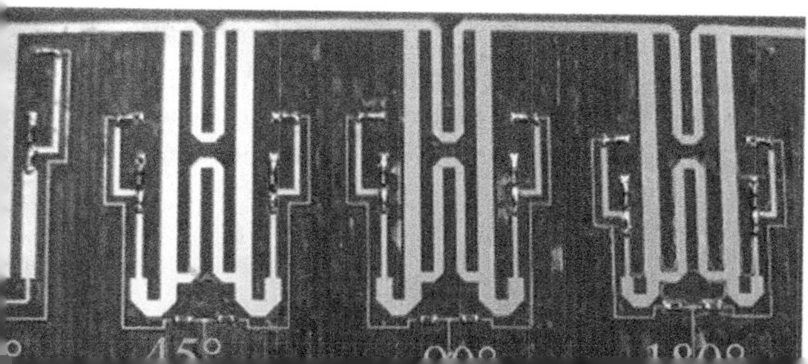
45°

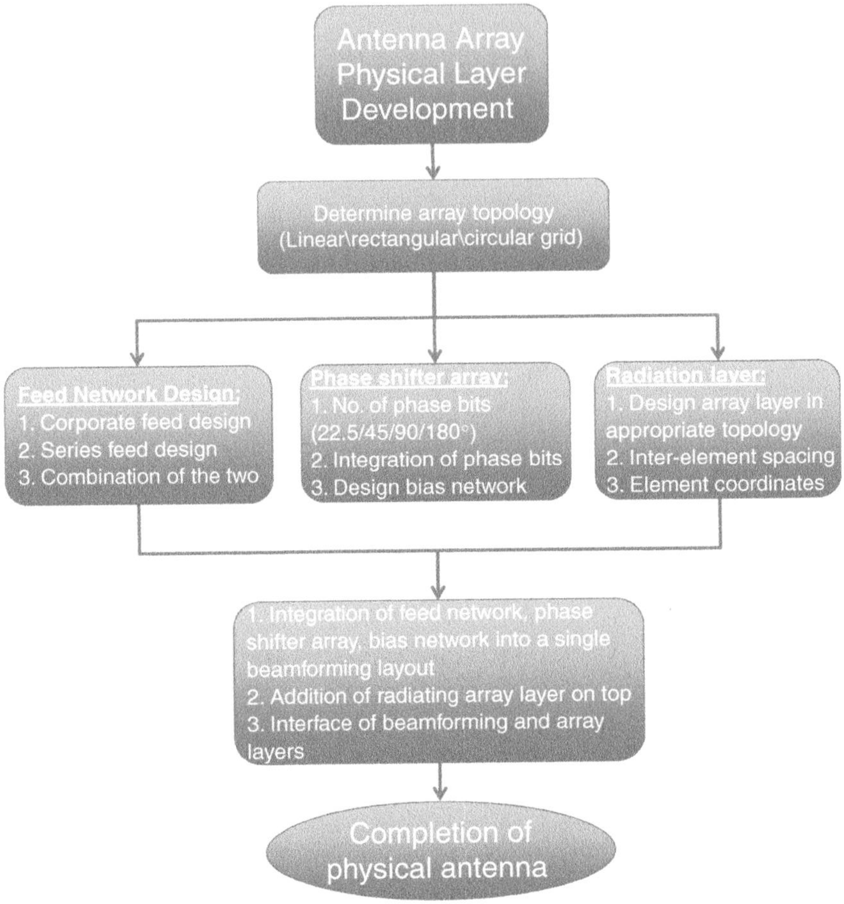

FIGURE 8.10. Procedure of phased array antenna physical layer development.

phase shifter design presented in the preceding section needs enormous effort to extract diode parameters, design of complete phase shifters, testing, and integration of the phase shifter array in the beamforming module. The physical realization of the phased-array antenna and examining its performance is really an expensive process. Modern computer-aided design tools have eased the process a bit and created the facility to test the phased array antenna before making and testing it. First the multilayer phased-array design is performed in a sequence as shown in Figure 8.10. As can be seen in the figure, the antenna array topology is determined first based on the array analysis and synthesis as discussed above and illustrated in Figure 8.4. Once the physical configuration is fixed, the elements are placed in the appropriate coordinate systems so that the array radiation layer provides optimum performance. It is essential to have inter-element distance on the order of half-wavelength to avoid grating lobes when the beams are squinted in a 3D plane. Based on the inter-element

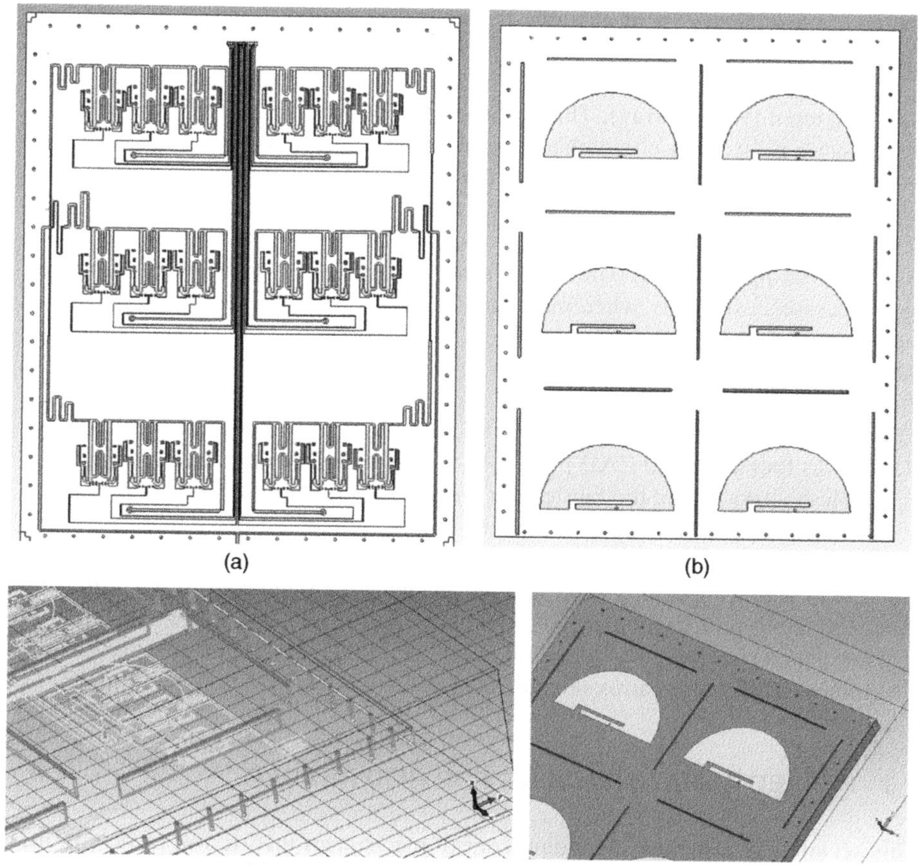

FIGURE 8.11. CST Microwave Studio-generated (**a**) beam-forming network, (**b**) array layout, (**c**) interleaved antenna array, and (**d**) complete antenna array. (Radiating layer laminate: FR4 $\varepsilon_{r1} = 4.2$, thickness 1.6 mm, and airgap height 17 mm; beamforming layer laminate Taconic TLX0, $\varepsilon_{r1} = 4.2$, thickness 0.787 mm).

spacing, the beamforming network comprised of the feed network design and the phase shifter array design are delicately positioned and the bias lines for the phase shifters are run from a central D-connector to each phase shifter. The task is the most challenging one to fit all modules in the beamforming array. Moreover, a compact beamforming network is desirable to save material cost because the microwave materials are very expensive items. The integration of the feed network, the phase shifters with their biasing network, and the radiating layer completes the array physical layer design. The interface between the beamforming layer and the radiating array antenna layer is done through feed wires. This completes the integrated array development. Figure 8.11 shows the complete beamforming network (Figure 8.11a), array radiating layer (Figure 8.11b) and interleaved complete antenna array (Figures 8.11c, d). The antennas are separated using metallic fences so that mutual coupling is minimized [14]. Total antenna size is 45 cm × 58 cm × 2.9 cm.

The next phase is to check the performance of the antenna array without physically making it (virtual array). This software prediction of microwave and radiation performance is also very comprehensive and, hence, time-consuming. However, it is worthwhile to check the performance of the total antenna array before fabricating the whole array. This intermediate stage of checking the performance in CAD tools helps not only in saving thousands of dollars but also in developing confidence before going to making the array. The whole array assembly and testing itself are complex processes, especially when many phase shifters are switched in correct on/off sequences to develop the collimated beam in the prescribed direction. Figure 8.12 shows a virtual antenna array design using the extracted *S*-parameters of the phase shifters in their on and off states as the black boxes in the simulation software. A comprehensive reporting of CAD design of the phased array antenna can be found in [15]. It was found that the CAD prediction give significant confidence and methods of troubleshooting before the fabrication process and field trials. Some of the results of the beam patterns are shown in Chapter 6. The next phase is the fabrication and assembly of the interleaved phased array antenna. Figure 8.13 is a photograph of the PCB of the beamforming layer and the antenna radiating layer. Along with fully characterization and testing of the antenna array, the FPGA-controlled section is also developed concurrently. In the following section the development of the control electronics of the phased array antenna is presented.

8.3 PHASED ARRAY TRACKING FOR RFID READER

This section describes the electronic control mechanism involved in switching the beam of the phased array antenna. Figure 8.14 shows the flow chart of the development of the electronics. Each phase shifter in the antenna beamforming network switches on at 5 V DC and switches off at −18 V DC. Thus, in order to obtain the required beam direction, specific phase shifters must be turned-on while leaving the rest turned off. This process is conducted through a microcontroller combined with a specifically designed switching interface module, often referred to as the digital control system, which drives the beam according the user-specified inputs.

Figure 8.15 shows the block diagram of the earlier-mentioned digital control section of the RFID phased array antenna. As can be seen in this figure, the control section comprises an Altera field programmable gate array (FPGA) Cyclone II microprocessor and an optically isolated analog bipolar junction transistor (BJT) switching interface biased with an external DC power supply. The outputs of the switching interface are +5 V and −18 V DC bias voltages to individual phase shifters of the 3 × 2-element phased array antenna module.

In the initial plan, a 4-bit phase shifter array comprised of 22.5°, 45°, 90°, and 180° phase shifters were designed (Figure 8.9). However, due to the space limitation and fabrication constraints of the RF electronics, a 3-bit phase shifter array comprised of 45°, 90°, and 180° phase shifts was selected (Figure 8.13a). However, the original

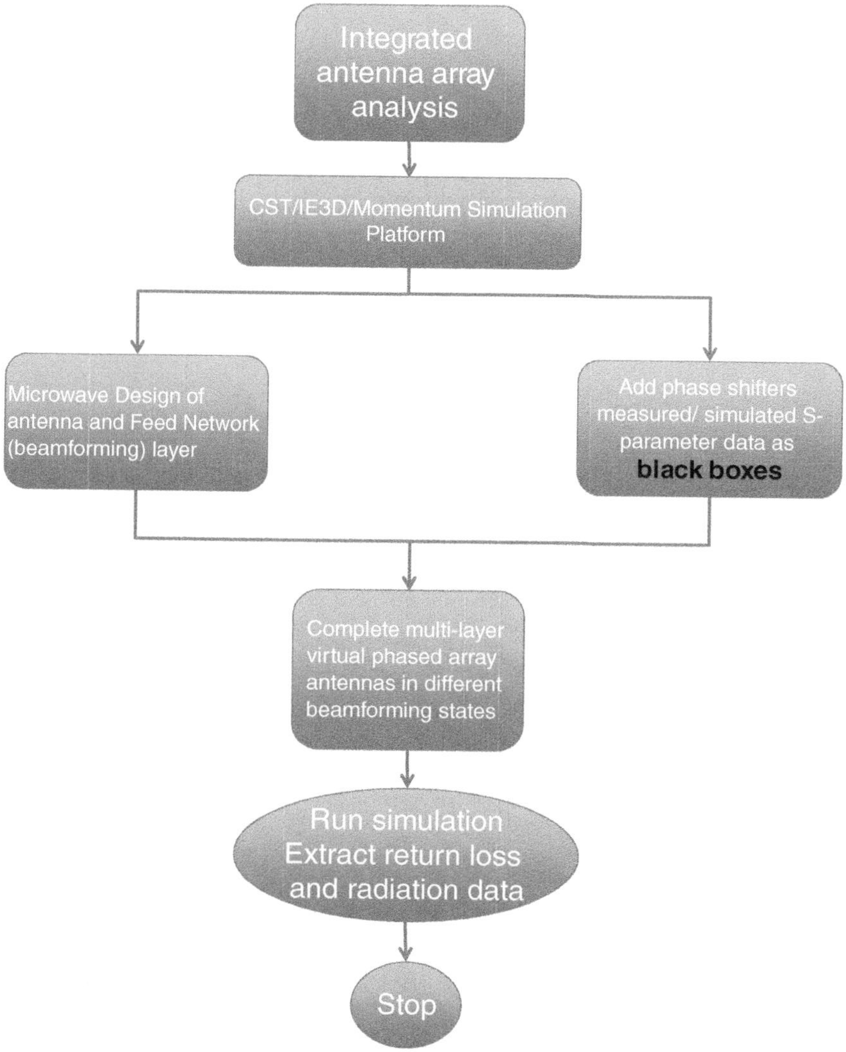

FIGURE 8.12. Computer-aided smart antenna performance prediction before prototyping and field trial.

design of the digital control section still has the provision of 24 outputs for 24 phase shifters. The operational analysis of each component of the above control system will be described in the sections to follow.

8.3.1 Microcontroller

The purpose of the microcontroller in this system is to identify which phase shifters must be switched on or off according to the user-specified beam direction. Once

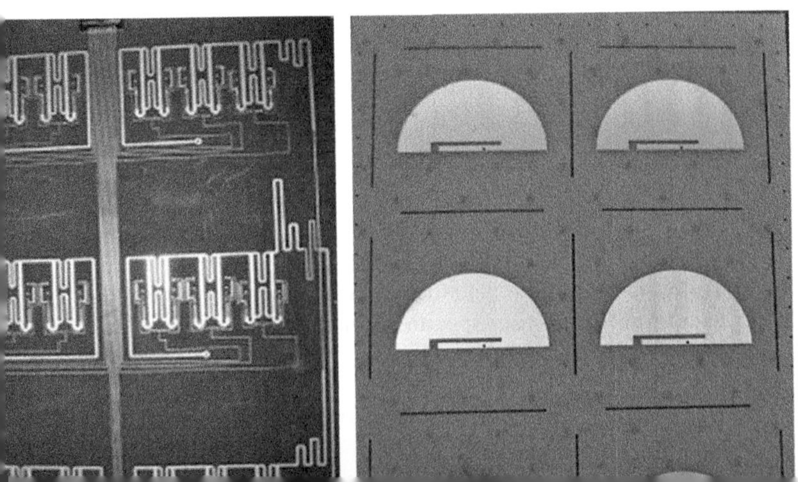

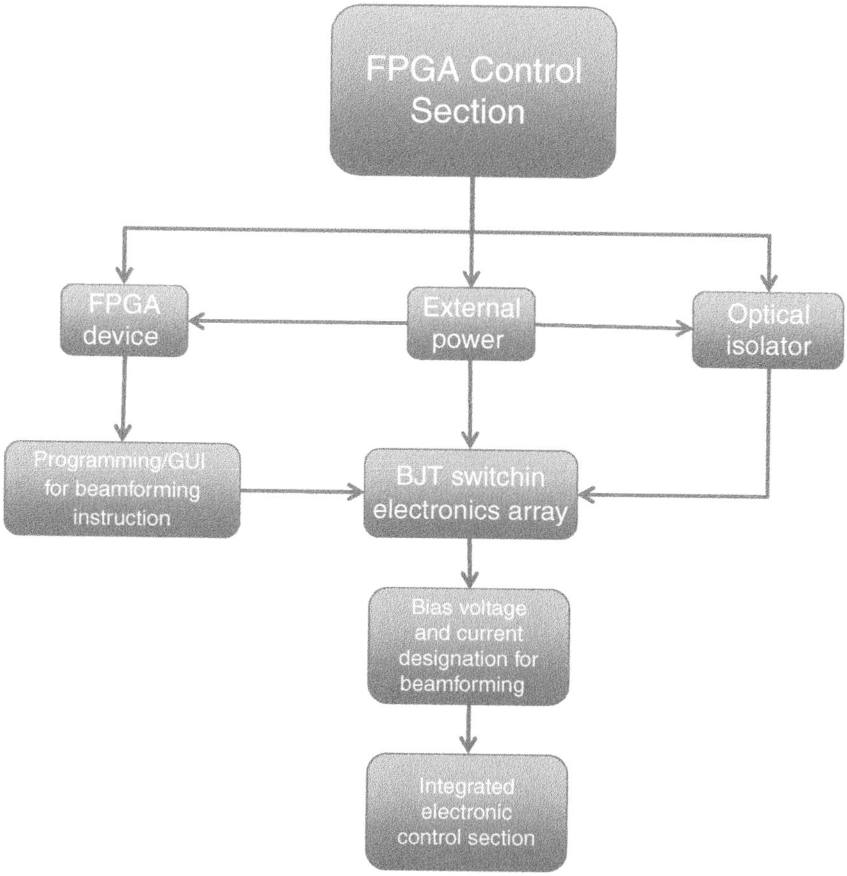

FIGURE 8.14. Flow chart of the control mechanism of FPGA-based switching electronics.

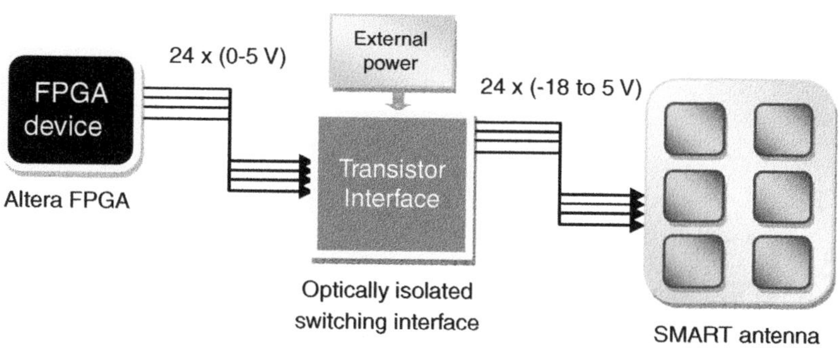

FIGURE 8.15. Block diagram of digital control section of RFID phased array antenna.

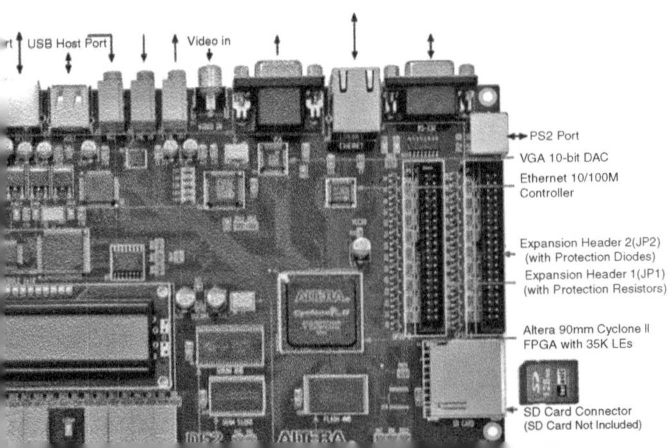
USB Host Port
Video in
PS2 Port
VGA 10-bit DAC
Ethernet 10/100M
Controller
Expansion Header 2(JP2)
(with Protection Diodes)
Expansion Header 1(JP1)
(with Protection Resistors)
Altera 90mm Cyclone II
FPGA with 35K LEs
SD Card Connector
(SD Card Not Included)

it would drive a total current of $24 \times I_{ON}$ Amperes from the prototyping board, where I_{ON} is the turn-on current per optocoupler. In order to limit this current, a current limiting resistor R_{series} is used as shown in Figure 8.18. The upper threshold for this current was set to 2 mA or less per optocoupler, giving a total maximum limit of 48 mA. Since standard optocouplers require much more current, a low-current optocoupler HCPL-070A was used. Its characteristic features are:

- Ultra-low input current capability (40 μA)
- Typical power consumption <1 mW
- Small SOIC-8 footprint
- High current transfer ratio

8.3.3.3 Final Design Simulation. Once the basic circuit was successfully simulated, it was implemented. In the final design, the measured oscilloscope trace in Figure 8.19b shows the output waveform at 1 kHz switching speed for the phase shifters, which is sufficient for this application. This result makes sure the output is smooth at 1-ms response rate.

8.4 SOFTWARE DESIGN

The FPGA device was programmed using Altera QUARTUS II and NIOS II programming environments. This development package consists of a schematic editing module and a programming/coding environment, among other things [16, 17]. Thus the development of the software involves three stages:

1. Building the NIOS microprocessor (using SOPC builder module)
2. Editing the schematic diagram of the complete processor system with input/output ports
3. Programming the microprocessor using a programming language (C/C++)

Once the NIOS processor is built with the selection of cache memory and other debug utilities, necessary memory, input/output connections, and other essential modules to complete the soft-core microcontroller named "Antenna_controller" can be added.

Once this is completed using the SOPC builder, the microprocessor block diagram can be imported to the Quartus Schematic editing environment (Figure 8.20) in which the processor I/O connections are linked to the actual I/O pins. "antenna_controller" is the NIOS soft-core processor designed specifically for our application. The block "LPM" is an LPM modulus counter that acts as a clocking or a select signal for the two multiplexors named "muxer" and "selec". These multiplexors are used to select between normal operation and a "testing operation" in which a special test routine will be carried out to verify that the circuitry is running fine. This processor can then be programmed using the NIOS programming environment with C/C++ programming language.

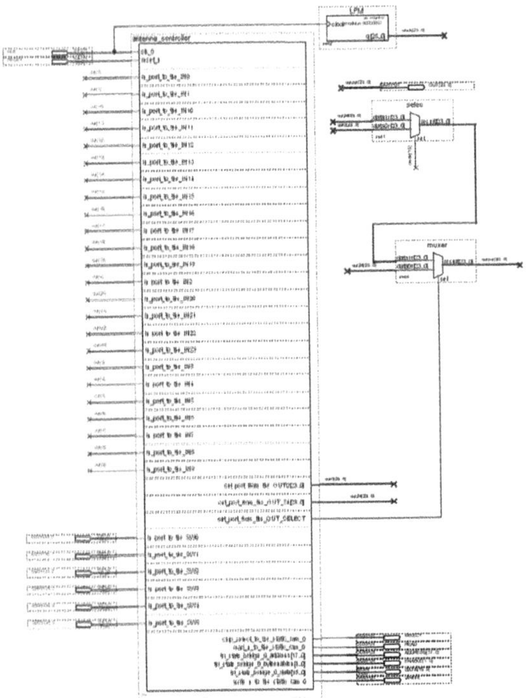

(a)

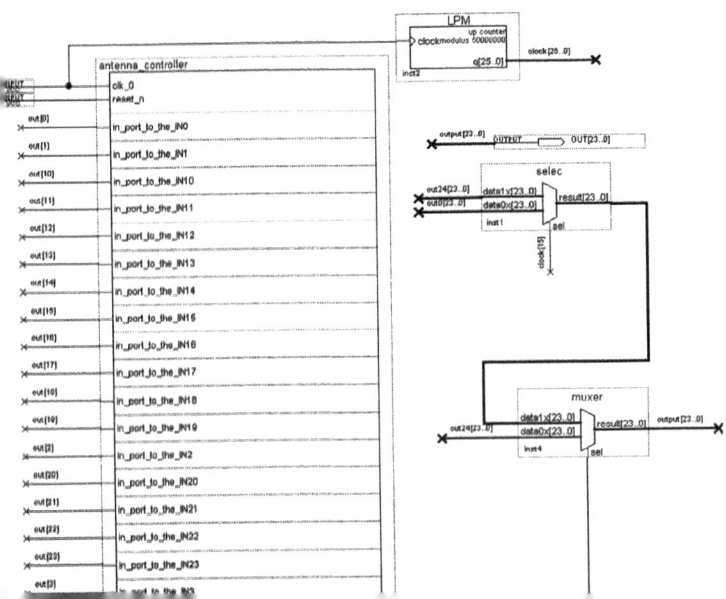
LPM
up counter
clock
modulus 50000000
q[25..0]
clock[25..0]
inst2
antenna_controller
clk_0
reset_n
out[0]
in_port_to_the_IN0
out[1]
in_port_to_the_IN1
out[10]
in_port_to_the_IN10
out[11]
in_port_to_the_IN11
out[12]
in_port_to_the_IN12
out[13]
in_port_to_the_IN13
out[14]
in_port_to_the_IN14
out[15]
in_port_to_the_IN15
out[16]
in_port_to_the_IN16
out[17]
in_port_to_the_IN17
out[18]
in_port_to_the_IN18
out[19]
in_port_to_the_IN19
out[2]
in_port_to_the_IN2
out[20]
in_port_to_the_IN20
out[21]
in_port_to_the_IN21
out[22]
in_port_to_the_IN22
out[23]
in_port_to_the_IN23
output[23..0]
OUTPUT
OUT[23..0]
selec
out24[23..0]
data1x[23..0]
out0[23..0]
data0x[23..0]
result[23..0]
inst1
sel
clock[15]
muxer
data1x[23..0]
data0x[23..0]
result[23..0]
output[23..0]
inst4
sel

8.4.1 Functional Description of the Program

This section describes the functionality of the program written to instruct the processor. In summary, the program requires the user to input the desired beam direction as an angle in degrees. It then uses this information to compare with an existing pre-coded reference table using a function called **retrv**. This function contains data as to which phase shifters need to be turned-on in order to obtain a specific beam direction. Theoretical information (using PCAAD 6.0) about these angles and their relevant phase shifter values are listed in Table 8.2. The leftmost columns theta (θ) and phi (ϕ) are the user inputs to the program, and the rightmost six columns show the phase weights (binary data: $1 = \text{on}, 0 = \text{off}$) for each of the patch antennas in the 3×2-element array. Please note that this table was constructed to control 24 phase shifters. However, the antenna was designed with a 3-bit phase shifter array consisting of 18 phase shifters.

Table 8.3 shows the comparison of the theoretical phase shifts and the actual phase shift by exciting 3-bit phase shifters (45°, 90°, and 180°). As can be seen in the table, the real phase-shifts yielded by the physical phase shifters are very close to the theoretical values predicted running PCAAD 6.0. Based on the above values, relevant phase shifters must be turned on to give a combination phase shift necessary for the relevant beam angle as shown in Figure 8.5.

The software program and its major functions are as follows: It begins the process by requesting the user to input the desired "phi" and "theta" angles. Then it calls the function "**retrv**" in order to obtain the specific phase shifters necessary to produce the desired outcome. Once the information is retrieved, the program automatically calls the function "**pulse**" which outputs the sequence of ON/OFF bits to the BJT switching module causing them to turn on/off the respective phase shifters. Once this process is completed, the ***pulse*** function will collect the output bit sequence bit by bit to conform that the output is correct. The reader may refer to the program and the Quartus project files to understand this pulsing method.

Figure 8.21 illustrates an image capture of the input terminal to further explain the above function. Due to the reduction of the number of the phase shifters in the final design as shown in Figure 8.13a, program functionality had to be changed to adjust to this new design. This could be done in a few minutes, and this is the major advantage of using an FPGA device such as the Altera Cyclone II device. This also allows a knowledgeable user to modify the software quickly to implement new routines or simply change the existing code. Thus this electronics circuitry was designed to have the capability to run various testing methods as well as to be able to constantly change throughout the development process of the phased array antenna.

8.5 FIELD TRIALS

After the successful integration of the antenna array, the FPGA control electronics, and software programming, field trials were performed using the antenna as well as Omron's reader electronics v750, antennas, and associated cable assemblies. Figure 8.22 shows the overall signal flow graph of the field trials. As can be seen, the reader

TABLE 8.2. Theoretical Calculations of Beam Angles and Corresponding Phase-Shifter Operation

User Input		Main Beam		3-dB												
θ	φ	Angle	$\Delta\theta$	Beamwidth	1	2	3	4	5	6	1	2	3	4	5	6
30	0	24	19	32	0	−97.2	−194.4	0	−97.2	−194.4	0000	0100	1000	0000	0100	1000
30	30	21	18	31.69	0	−84.2	−168.4	−24.3	−108.5	−192.7	0000	0100	0111	0001	0100	1000
30	45	17	18	31.27	0	−68.7	−137.5	−34.4	−103.1	−171.8	0000	0100	0110	0001	0100	1000
30	60	12	18	30.88	0	−48.6	−97.2	−42.1	−90.7	−139.3	0000	0010	0100	0010	0100	0110
30	90	0	14	30.48	0	0	0	−48.6	−48.6	−48.6	0000	0000	0000	0010	0010	0010
45	0	34	19	33.93	0	−137.5	−274.9	0	−137.5	−274.9	0000	0110	1100	0000	0110	1100
45	30	29	18	37.9	0	−119	−238.1	−34.4	−153.4	−272.5	0000	0101	1100	0001	0111	1100
45	45	24	18	32.09	0	−97.2	−194.4	−48.6	−145.8	−243	0000	0100	1000	0010	0110	1100
45	60	17	18	31.27	0	−68.7	−137.5	−59.5	−128.3	−197	0000	0011	0110	0010	0110	1000
45	90	0	12	30.48	0	0	0	−68.7	−68.7	−68.7	0000	0000	0000	0011	0011	0011
60	0	43	20	36.36	0	−168.4	−336.7	0	−168.4	−336.7	0000	0111	1111	0000	0111	1111
60	30	36	19	34.47	0	−145.8	−291.6	−42.1	−187.9	−333.7	0000	0111	1100	0010	1000	1111
60	45	29	18	32.98	0	−119	−238.1	−59.5	−178.6	−297.6	0000	0101	1100	0010	1000	1100
60	60	21	18	31.69	0	−84.2	−168.4	−72.9	−157.1	−241.3	0000	0100	0111	0011	0111	1100
60	90	0	6	30.48	0	0	0	−84.2	−84.2	−84.2	0000	0000	0000	0100	0100	0100
90	0	−42	20	36.07	0	−194.4	−388.8	0	−194.4	−388.8	0000	1000	1111	0000	1000	1111
90	30	43	19	36.36	0	−168.4	−336.7	−48.6	−217	−385.3	0000	0111	1111	0010	1100	1111
90	45	34	19	33.93	0	−137.5	−274.9	−68.7	−206.2	−343.7	0000	0110	1100	0011	1001	1111
90	60	24	17	32.09	0	−97.2	−194.4	−84.2	−181.4	−278.6	0000	0100	1000	0100	1000	1100
90	90	0	17	30.48	0	0	0	−97.2	−97.2	−97.2	0000	0000	0000	0100	0100	0100
0	0	0	18	30.48	0	0	0	0	0	0	0000	0000	0000	0000	0000	0000
0	30	0	18	30.48	0	0	0	0	0	0	0000	0000	0000	0000	0000	0000
0	45	0	18	30.48	0	0	0	0	0	0	0000	0000	0000	0000	0000	0000
0	60	0	18	30.48	0	0	0	0	0	0	0000	0000	0000	0000	0000	0000
0	90	0	18	30.48	0	0	0	0	0	0	0000	0000	0000	0000	0000	0000
75	0	−44	23	36.79	0	−187.8	−375.6	0	−187.8	−375.6	0000	1000	1111	0000	1000	1111
75	30	41	20	35.76	0	−162.6	−325.2	−46.9	−209.6	−372.2	0000	0111	1111	0010	1001	1111
75	45	33	20	33.66	0	−132.8	−256.6	−66.4	−199.2	−331.9	0000	0110	1100	0011	1001	1111
75	60	23	19	31.97	0	−93.9	−187.8	−81.3	−175.2	−269.1	0000	0010	1000	0100	1000	1100
75	90	0	18	30.48	0	0	0	−93.9	−93.9	−93.9	0000	0000	0000	0100	0100	0100

```
witch 0 in the DE2 board to initiate the program

ie THETA : 45

ie PHI : 30

rify the following beam data before proceeding:
______
 PHI  |
------|
 30   |
______|

y/n)? y

........

 output stream: 0000-0000 0000
```

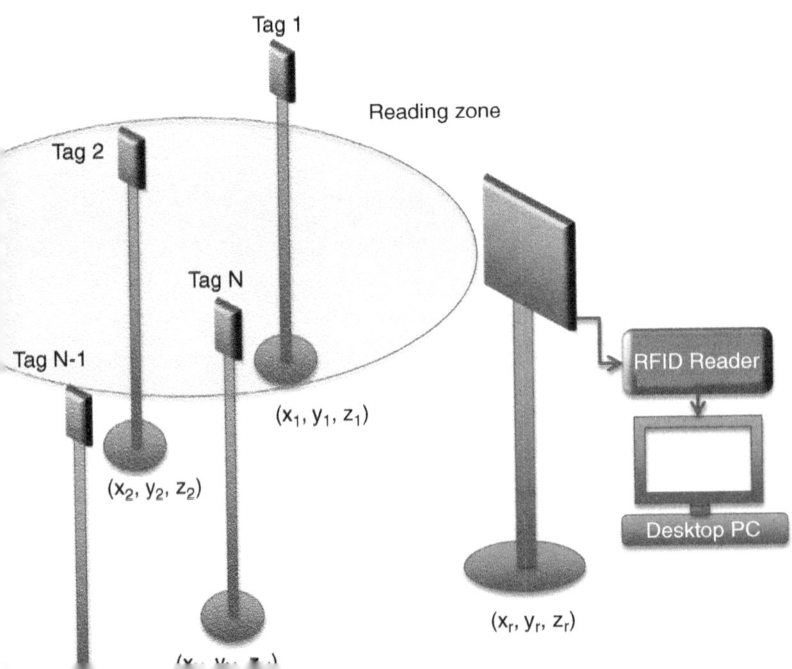

Tag 1
Reading zone
Tag 2
Tag N
Tag N-1
RFID Reader
(x_1, y_1, z_1)
(x_2, y_2, z_2)
Desktop PC
(x_r, y_r, z_r)

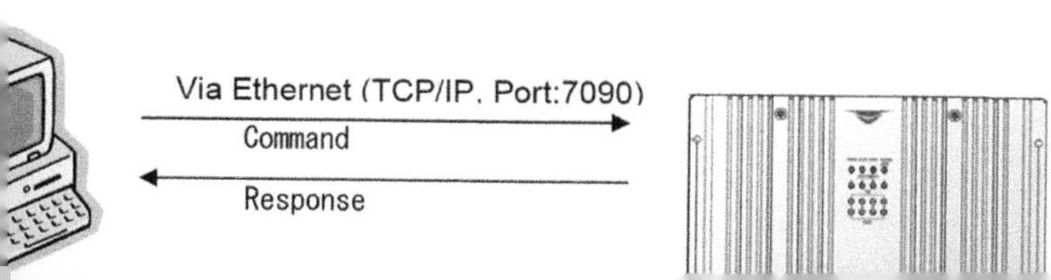
Via Ethernet (TCP/IP, Port:7090)
Command
Response

was observed even with low transmission power detection. With the beam scanning, the reader detected all tags within the read range.

8.6 CONCLUSION

RFID is an enabling technology that is transforming the identification, asset tracking, security surveillance, medicine, and many other industries at an accelerated pace. With the mandates of the retail giant Wal-Mart and US DoD to comply with the UHF tag standard, there is a global demand for the UHF RFID tag system. To attract the mass global market of RFID system, a broadband UHF FPGA controlled phase scanned antennas was developed from scratch. The design involved comprehensive multidisciplinary design. Implementation started with antenna element design, phase shifter design, beamforming network design, and integration of the antenna into a planar phased array antenna. The exercise has resulted in a 3×2-element low-cost, lightweight, and compact planar-phase scanned antenna array for RFID readers. A detailed design guideline for the component level physical layer development of the smart antenna has been presented. The antenna is designed at 910 MHz with 100-MHz bandwidth to cover a plethora of applications in this frequency band. The development certainly has a potential international market. The FPGA programming and switching electronic design is an important development to control the phased-array antenna. The field trail of the tags using an Omron reader has signified the usefulness of the developed phased array antenna.

ACKNOWLEDGMENTS

This work was supported by Australian Research Council's Discovery Project Grant # DP665523: Chipless RFID for Barcode Replacement. The software supports from Agilent and CST are also acknowledged. Acknowledgments also go to Panfilo Tarulli of Omron Australia Pty Ltd for the RFID reader, Agilent for ADS, and CST for Microwave Studio. Special thanks to Tony Brosinsky, Ray Chapman, Ian Reynolds, and Ray Cooper for mechanical construction, electronics, and purchasing components, respectively. Technical support of Sushim Mukul Roy and Mohammad Saqib Ikram are also acknowledged.

REFERENCES

1. R. Das and P. Harrop, RFID Forecasts, Players & Opportunities 2006–2016, IDTechEx, London, UK. Retrieved from http://www.idtechex.com/products/en/view.asp?productcategoryid=93 on 11 September 2007.
2. M. A. Ingram, Smart reflection antenna system and method, US Patent No. US 6,509,836, B1, January 21, 2003.

3. T. Nakamura, and J. Seddon, Omron Develops World's First Antenna Technology that Boosts UHF RFID Tag Read Performance, Omron Corporation press releases, 2006. Retrieved from http://www.omron.com/news /n_270306.html in June 2007.

4. J. Chamberlain, C. Blanchard, S. Burlingame, S. Chandramohan, E. Forestier, G. Griffith, M. L. Mazzara, S. Musti, S. I. Son, G. Stump, C. Weis (2006, May). IBM WebSphere RFID Handbook A Solution Guide [Online], Available at: http://www.redbooks.ibm.com/redbooks/pdfs/sg247147.pdf.

5. C. A., Balanis, *Antenna Theory: Analysis and Design*, 2nd edition, John Wiley & Sons, New York, 1982.

6. A. A. Deshmukh and G. Kumar, Various slot loaded broadband and compact circular microstrip antennas *Microwave Optical Technol. Lett.*, Vol. 48, No. 3, pp. 435–439, 2006.

7. M. H. Kori and S. Mahapatra, Integral analysis of hybrid coupler semiconductor phase shifters, *Proc. Inst. Electr. Eng.*, Vol. 134, pp. 156–162, 1987.

8. S. K. Koul and B. Bhat, *Microwave and Millimetre Wave Phase Shifters*, Vol. II, Artech House, Norwood, MA, 1991.

9. R. V. Garver, *Microwave Diode Control Devices*, Artech House, Norwood, MA, 1975, Chapter 10.

10. R. W. Burns, R. L. Holden, and R. Tang, Low cost design techniques for semiconductor phase shifters, *IEEE Trans Microwave Theory Tech.*, Vol. MTT-22, 675–688, 1974.

11. N. C. Karmakar and M. E. Bialkowski, Designing a low cost L-band phased array antenna for mobile communications, in *Proceedings, 1997 Asia-Pacific Microwave Conference*, Hong Kong December 1997.

12. N. C. Karmakar and M. E. Bialkowski, An L-band 90° hybrid-coupled phase shifter using UHF band PIN diodes, *Microwave Optical Technol. Lett.*, Vol. 21, No. 1, pp. 51–54, January, 1999.

13. N. C. Karmakar, Antennas for Mobile Satellite Communications, Unpublished doctoral dissertation, The University of Queensland, Australia, 1999.

14. N. C. Karmakar, A low sidelobe sub-array of portable VSATS in *Proceedings, 2003 IEEE International Antennas and Propagation Symposium and URSI North American Radio Science Meeting*, Columbus, Ohio, June 22–27, 2003.

15. J. F. White, Diode phase shifters for array antennas, *IEEE Trans. Microwave Theory Tech.*, Vol. MTT-22, pp. 658–674, 1974.

16. Altera DE2 prototyping board [Online], Available at http://www.terasic.com (accessed December 1, 2009).

17. Altera Development & Education Board—User Manual, Available at http://www.terasic.com.

CHAPTER 9

OPTICAL BEAMFORMING PHASED ARRAYS FOR UWB CHIPLESS RFID READER

AROKIASWAMI ALPHONES and PHAM QUANG THAI
School of Electrical and Electronic Engineering, Nanyang Technological University, Singapore

NEMAI CHANDRA KARMAKAR
Department of Electrical and Computer Systems Engineering, Monash University, Clayton, Victoria, Australia

9.1 INTRODUCTION

The current market volume of the RFID industry is worth approximately $5.25 billion in 2009. The industry will grow exponentially with the invention of a very low cost tag, with unit cost in the range of less than a cent. RFID technology can be classified into: (i) systems where the tag contains a microcontroller and (ii) systems where the transponder is "chipless." Das and Harrop [1] argue that the largest market in RFID lies with the chipless RFID as opposed to any type of chipped RFID. The cost of chipless tags is generally much lower than that of other tag types due to the absence of the microcontroller. To make significant market impact, chipless RFID systems need to have reasonable read range (>30 cm), with tags that are small in size, flexible, and printable for low cost to compete with the optical barcode. With the exception of surface acoustic wave (SAW)-based RFID [2], no other chipless technology has made a significant impact on the market. However, SAW tags are rigid and bulky, making them unsuitable for some applications. There are a few printable chipless tags reported in the open literature [3–9]. They are printed stripline reflectors [3], 11-element capacitive gap-coupled printed dipoles [4], 5-bit Hilbert and Peano curves [5], a spiral mil-resonator-based printed tag [6], reactive loaded patch antennas [7], and distributive capacitive loaded transmission lines [8,9].

Handbook of Smart Antennas for RFID Systems, Edited by Nemai Chandra Karmakar

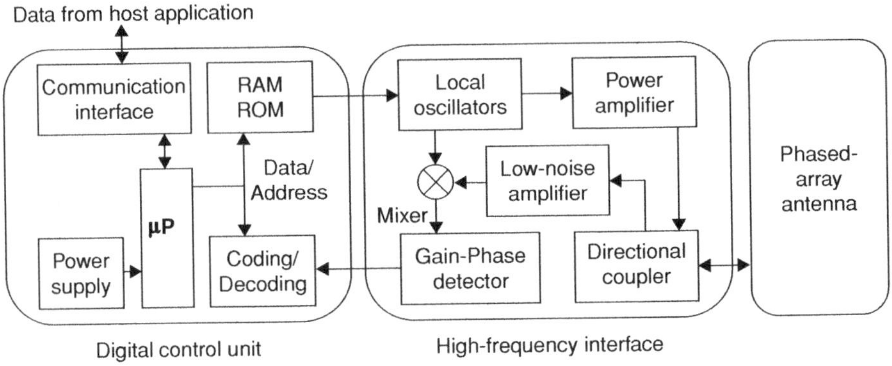

FIGURE 9.1. Block diagram of UWB chipless RFID reader.

With the progress on the chipless printable RFID tags [3–9], they will offer the most likely solution to a significant reduction in cost despite the fact that existing developments are still in their infancy and in the laboratory experimentation stage. The limitations are: (i) The information-carrying capacity and reading efficiency are still maturing, and (ii) the proposed methods do not support the tag-driven reading of multiple RFID tags, which is required in a library application. The main reason is to make the RFID tag printable and without an information carrying microchip; the conventional interrogation system is not compatible with the tag. The information-carrying capability is very limited (only 8-bit in time domain [3, 8, 9] and 34-bit in frequency domain [6] are reported) and cannot accommodate a discrimination protocol due to the absence of the microchip. Therefore, to compensate for the limited capacity of information in the chipless RFID tag, the bulk of processing is to be maintained in the RFID reader. The reader must be advanced enough to read multiple chipless tags in the close vicinity and capable of discriminating the tagged items almost instantly. However, such an RFID reader has not been reported in the open literature yet to the best of the investigators' knowledge. In conclusion of the review, while many strides are being taken in reduction in the price of the RFID tag, there are no commercial chipless RFID systems in the market to cater for the need for the chipless tags with a required amount of data bits and reading distance that have the potential to replace the optical barcode. The only possibility to compete with the optical barcode is to develop a chipless RFID system that can provide 124 bits of data. The Monash Universisty RFID research team has developed a UWB chipless RFID system [10] that has the promise to yield a large number of data bits using frequency signatures in the 3 to 10-GHz frequency band. Figure 9.1 shows the reader system.

An RFID reader [11] consists of three main parts shown in Figure 9.1: (i) control section; (ii) high-frequency (HF) interface capable of providing UWB signals ranging from 3 to 10 GHz; and (iii) a UWB phased-array antenna. At the user end, the reader connects the host application such as enterprise software. The control section does digital signal processing and procedures over the received data from the RFID transponder. Also, the control section enables the reader to communicate with the transponders wirelessly by performing modulation, anti-collision procedures and decoding the received data from the transponders.

A smart antenna operating over the UWB band can leverage the performance of the reader. With the capabilities of spatial, temporal, and polarization diversities and enhancement of signal-to-noise ratio (SNR), smart antennas can offer many functionalities that lead to anti-collision procedures and capacity improvement of the RFID system. However, being a UWB system, conventional electronic beamforming cannot offer uniform functionality over the band of operation. It is well known that electronics phase shifters, biasing network, matching network, and power dividers are inherently narrowband devices. Hence, the only solution to the UWB operation with uniform functionality is the optically controlled beamforming network.

Reader Smart Antenna

Being a UWB reader to cover the UWB frequency spectra, the proposed smart antenna is an optical beamforming phased-array antenna. The phase-scanned-array antenna is comprised of a number of antenna elements (up to 20 elements reported in the chapter) and associated beamforming modules that are presented in detail in this chapter. The beamforming module is a hybrid optical beamformer with an eight-element antenna array. The optical phase-shifter arrays have the capabilities of generating multiple beams (here two beams are investigated) and controlling the beams in 3D plane. Even several of the antennas in various positions can be used to exploit special diversity. For example, in an active three-antenna configuration, a single-pole three-throw (SP3T) switch activates three adjacent-array modules to collimate the in-phase beams even in further fine resolution. Thus the smart antenna produces high-gain scan coverage in 360° azimuth and elevation plane patterns. The antenna will detect individual chipless RFID tags. The exact bearing will be calculated from the beam position of the antenna.

Being very directive, the smart antenna filters out site interferences. An alternative is to select frequencies with a reconfigurable antenna. Prior study and knowledge of types of site interferences and their frequency signature are useful in this regard. Following is a detailed overview of the optical beamforming phased-array antennas for the UWB chipless RFID system. The antenna can also be implemented in UWB radars and other emerging technologies.

The chapter is organized as follows: Section 9.1 highlights the importance of UWB operation for the multi-bit chipless RFID tags and reader systems followed by the optically controlled smart antenna. Section 9.2 presents an overview of the optically beamforming phased-array antennas. Section 9.3 presents the effect of group delay ripples in beamforming, and Section 9.4 presents the optical beamforing using the dispersive delay and the nondispersive delay as a hybrid approach. Finally, conclusions are presented in Section 9.5.

9.2 OPTICAL BEAMFORMING PHASED ARRAYS

Activities on array theory and experimentation were blooming in the two decades before 1940. Much of the array work was performed in the United States and Great Britain during World War II, and then the interest in arrays returned in the early

1960s. The first extensive coverage of phased array was in the three-volume book *Microwave Scanning Antennas* [12] in 1964.

The main beam of a phased-array antenna can be steered to a desired direction by varying the relative phases between elements. For conventional beamformers, the difference in relative phases of antenna elements is introduced by phase-shifter modules connected to individual radiating elements of the array antenna. There are many kinds of phase shifters [13], which can be classified as electrical or mechanical, analog or digital, feed-through or reflex, active or passive. All in all, their operation is usually based on the variation of the electrical length of a transmission line, connection of a reactive element to a transmission line, or vector addition of several signals. The most commonly used phase shifters are made of microwave lenses, waveguides, transmission lines, and printed microwave circuits.

Initially, phased-array antennas were designed to work with narrow bandwidth. The distinguished ability of the phased-array antenna, which is the control of the beam position in 3D space, is quite a blessing. This ability is the essence for their functions and provides considerable performance benefits. However, for recent communication technologies such as ultra-wideband and worldwide interoperability for microwave access, the operating frequency and bandwidth are higher and higher. Also, short-pulse and dual-band radars reveal another weakness of the conventional beamformer.

The phase difference between elements is related to the scanning angle and the working wavelength of the array. In principle, this does not cause any difficulty, since a typical phase array should have instantaneous bandwidth of about 1% of the signal frequency. However, if the signal is wide band as in the case for the UWB chipless RFID system, the direction of the beam will change across the bandwidth. This phenomenon, called "beam squint" [14], is the largest obstacle a phased-array antenna has to face.

In conventional phased-array antennas, this problem is solved by introducing time delay, instead of phase shift, into the feed signal. The solution, although promising in theory, is almost impossible in reality. For example, in a 400-element array with half-wavelength spacing working at 3 GHz, steering to 60° will require a 12-m-long coaxial feed line for the farthest element. Moreover, the time delay has to be varied according to the beam direction.

The feed system of a phased-array antenna is already expensive and heavy with all its phase shifters, connecting lines, and power splitters. Trying to apply time delay using conventional means such as transmission lines or waveguides definitely burdens the system with more complexity and inflexibility. In addition, isolation issues and electromagnetic interference are always there to cause more troubles.

Optical-controlled techniques have been extensively studied in an effort to address those drawbacks. In an optical domain, there is virtually no leakage of the signal, which is the main problem in electrical realizations. Electromagnetic interference is not a problem anymore. The size and weight of the feed network is significantly reduced. Thanks to the inherent large bandwidth and low loss of optical fiber, there is almost no limitation in bandwidth and length of the feeding lines. The true-time delay characteristic inherited with optical beamformers really caught the eyes of many researchers.

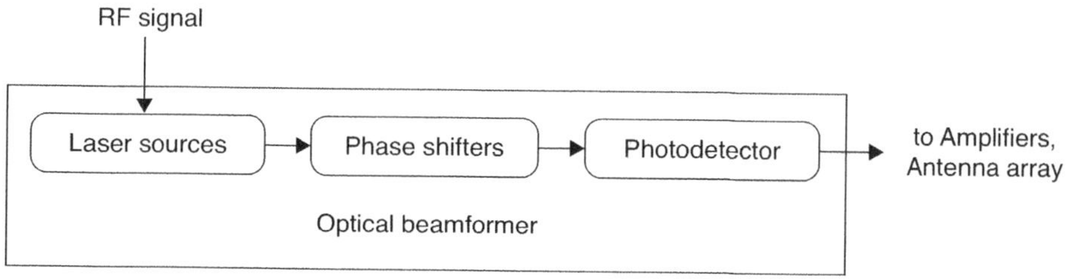

FIGURE 9.2. Simplified block diagram of an optical beamformer.

Early photonic approaches for array antennas have seen considerable reduction in hardware complexity, weight, and size in the systems. However, they did not have much impact at the beginning. It was mainly because of the material and fabrication technologies at that time, which could not provide enough sophisticated and affordable photonic devices. Fortunately, those excellent research efforts did not go to waste. Recently, optical beamforming has once again received increasingly interest.

Optical beamforming techniques can be divided into three categories in general. The first approach is the optical implementation of time-integral correlation techniques, which is so difficult in practice that this approach has received considerably less effort. The second approach provides narrowband phase-only control of a microwave signal at each element. The idea was to reduce size, weight, and system complexity while being immune to electromagnetic interference. This idea is further improved in the last category, which includes the optical implementations of large instantaneous bandwidth with true time delay.

One of the most important things in a phased-array antenna system is to find a phase-shift element that can show features such as small size, low losses, continuous variability, and true time delay. A nice and informative collection of original papers, reprinted in reference 15, provides an insightful summary about the various attempts and efforts. Photonic feed systems can be implemented using many approaches. Over the last decade, this field has been rapidly developed. There are optical beamformers using Fourier-optic [16], fiber-optic [17], integrated-optic [18], acousto-optic liquid-crystal [19], optically injection locking of microwave oscillators [20], optical switching [21], delayed fibers [22], optical polarization switches [23], and wavelength multiplexed systems [24]. It is not possible to cover every last optical beamforming method in this small section. Only the main approaches using dispersive and nondispersive delay are discussed, since those approaches have received more consideration and/or been applied in recent reports.

Generally speaking, an optical beamformer can be broken down into three modules, as shown in Figure 9.2. The first part is the optical source. The RF signal is usually modulated into the optical signal at this point. The next part is used to introduce delays into the modulated optical signal. In most cases, this part characterizes the optical beamformer. Being optical delay lines, delay matrices, dispersive fiber delay lines, or Bragg fiber gratings, the mean to achieve time delays blesses the optical beamformer with its advantages but weakens the system with its drawbacks. The pros and cons of each approach are discussed later in this section. The last part is commonly constructed from photodiodes, since the role of this part is to provide electrical signals to the array.

To steer the main beam, the optical beamformer has to be able to vary the time delay. There are only two possible ways to satisfy that requirement:

- Use nondispersive optical fiber, and then change the length of the fiber (true time delay).
- Use dispersive fiber, and then change the wavelength of the signal passing through it (reactive loading).

Stability, power budget, loss, and system complexity are some of the key issues involving the optical beamformers using those techniques.

For the first approach, one of the earliest schemes was reported back in 1990. Goutzoulis and Davies designed an optical beamformer based on switched binary fiber delay [24]. In that scheme, photonic delay lines were used to introduce delay. Along a binary delay chain, semiconductor optical switches were used to switch in or out delay segments that were binary multiply of a minimum delay. This setup encountered several problems such as uniformity, stability, drift, and IP3 (third-order interception point—applied to the amplifiers to offset link loss). Photonic delay lines also has high loss, so there have to be some means to compensate and control the amplitude. Later papers using optical delay lines and optical switches tried to deal with the stability and the response time. However, since the structure of the phase shifters were still a combination of optical delay lines and optical switches, those optical beamformers still faced the same problems.

In other to achieve true time delay using optical devices, fiber dispersion prisms are usually used [25]. Although both transmit and receive models of this approach were demonstrated, due to the limited dispersion currently available, a rather long length of high-dispersion fiber was required to steer an array. For example, a highly dispersive fiber should have a dispersion coefficient of about -70 to -100 ps/km · nm. It means a modulated optical signal with 1-nm spectrum width acquires 70- to 100-ps delay after 1 km of fiber length. This drawback not only increased the size of the beamformers, but also introduced significant signal latency. The beamformers were also susceptible to temperature and might need additional temperature-controlled devices.

In one of the newest papers following this direction [26], photonic crystal fiber was used as a substitute for a dispersive fiber. Since a photonic crystal fiber has a much higher dispersion, the optical delay line's length was largely reduced. In that paper, the dispersion of the photonic crystal fiber was around 600 ps/nm/km. As a result, the optical beamformer became even more compact and lightweight. However, a photonic crystal fiber usually has high loss. An attenuation of about 10 dB/m is not uncommon. Since the fed signals traveled through photonic crystal fibers at different length, their amplitudes were largely different. This problem would raise the side lobe of the radiation pattern. Moreover, many tunable lasers had to be used to steer the array's main beam.

The second approach was considered as a more compact alternative for dispersion-based phased arrays. Chirp fiber grating as a phase shifter was first introduced in reference 27. The time delay provided by a 2-cm-long chirped fiber grating is equivalent to that provided by a 200-m high-dispersion fiber over a 10-nm bandwidth. Since the delay characteristic of a linear chirped fiber grating is continuous and nearly linear, continuously beam steering at higher frequencies is possible. Instead of a set of chirped gratings, one wideband chirped grating can be used to further reduce the system size [28]. Although those approaches were more compact, the grating length still remained a major problem, because wideband chirped fiber grating was not an easy task. With a large array, it should become very hard to fabricate. In addition, there was always some initial time delay between elements, since the wavelengths that controlled each element occupied different grating regions in a common grating. Therefore, other means, which usually was optical delay line, had to be used to compensate those initial delays.

Optical beamforming using multichannel chirped fiber grating (MCFG) was introduced in reference 29. This approach addressed the two drawbacks of the chirped fiber grating mentioned above. The fiber was several times shorter, and there was no longer a need to compensate the initial delays, since each wavelength operated in its own channel.

Optical beamformers using dispersive lines, Bragg gratings, and chirped gratings usually require one tunable laser for each array element. Tuning the laser's wavelength introduces time delay into the corresponding optical signal. However, this simple operational scheme faces serious problems when the antenna array size is increased. At this time, a tunable semiconductor laser is not affordable enough to be classified as a low-cost component. Therefore, reports about optical beamformers using fiber grating are usually stopped at their initial level or mere laboratory demonstration with a very small size array.

Recently, an attempt to address this issue was proposed in reference 30. In that paper, broadband laser was used instead of several tunable lasers. However, that scheme still faces the same problems as the one in reference 28 since they both used a long chirped grating to introduce delay.

9.3 THE EFFECT OF GROUP DELAY RIPPLES

9.3.1 Introduction

Phased-array antennas combining with beamforming techniques provide many advantages such as better directivity, higher gain, and 3D beam-steering. However, it is very difficult for an electrical beamformer to support wideband RF signals at high frequencies, since the beam direction of the array will change across the RF signal's bandwidth. On the other hand, an optical beamformer has the ability to operate squint-free with wideband, high-frequency signals, thus avoiding the most significant drawback of the traditional electrical beamformers.

As mentioned in the previous section, there are two mayor approaches in order to achieve photonic beamforming. Optical beamformers utilizing devices capable of providing dispersive delay usually employ chirped grating. Such devices allow more compact systems with simpler controlling schemes. Reports about optical beamformers using chirped grating, more often than most, state that the beamformers are true time delay, squint-free, and suitable for signal at high frequency with larger bandwidth [28–30].

In those optical beamformers, the time delay between array elements is introduced through the chirped grating. The amount of delay is the most important factor in controlling the array's main beam. Thus, the group delay characteristic of the grating should be as linear as possible. However, there are always fluctuations, called group delay ripple (GDR), across the grating region [31]. Those inevitable errors occur during the fabrication of every chirped grating.

Since chirped grating is usually used as a dispersion compensator, the negative effect of GDR on dispersion compensation has been rigorously studied in many papers. However, when chirped grating is utilized in optical beamformers, there is rarely any report about the impact of GDR on the system performance. In some recent proposed optical beamformers using chirped grating [30, 32], the negative impact of GDR was mentioned. The errors were measured, but no analysis was done. An interesting study about GDR and true time-delay systems was carried out in reference 33. However, that study emphasized more on the distortion when a chirped pulse is used in a general true time-delay system than when it is used in a specific optical beamformer system. Another paper [34] concluded that the wavelength spacing relates to errors in the main beam direction.

In a recent paper [35], the impact of GDR on the most advantageous points of an optical beamformer using chirped grating, which are squint-free, suitable for higher signal frequencies, easy to work with large signal bandwidth, and being able to support large size arrays, was rigorously studied. The mathematical analysis and simulation results in that paper have indicated that those strong points have to be reconsidered.

For example, the mathematical model in reference 35 has shown that GDR causes additional distortion in the RF signal. More importantly, GDR also makes the main beam-squint across the RF signal's bandwidth. The simulations using the measurement of a commercially available product chirped grating further strengthen the mathematical analysis. In addition, the results have pointed out that GDR also limits the operation frequency, the RF signal's bandwidth, and the number of elements. The detailed analysis is in the next section.

9.3.2 Mathematical Analysis of Group Delay Ripples

A delay in time is necessary in order to steer the main beam of an array toward a particular direction without beam-squint. There are many approaches to introduce time delay into the RF signal using optical devices. However, in order to study the GDR that is present in chirped gratings, only methods using such gratings are considered. A typical optical beamformer using chirped fiber grating is shown in Figure 9.3.

Using the basic block diagram in the previous section, the typical optical beamformer using chirped grating can be broken down into several fundamental blocks: multiple tunable laser sources to provide optical carrier signals, an electro-optic

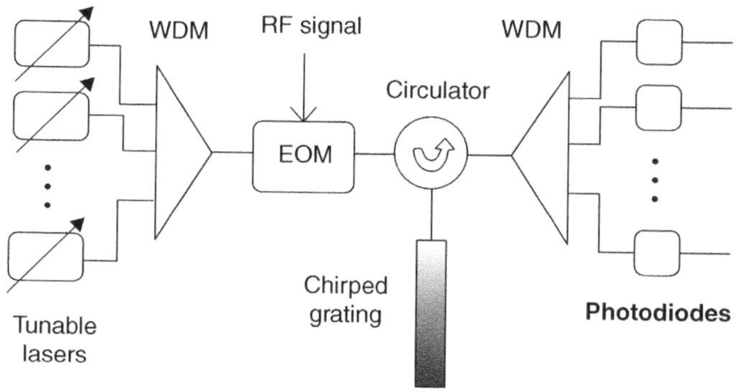

FIGURE 9.3. Schematic of a typical optical beamformer using chirped fiber grating.

modulator to modulate the RF signal with the carriers, a chirped grating to introduce time delays into the modulated signals, and finally several photodiodes to convert the signal from optical domain back to electrical domain. Since a chirped grating causes different delays to different wavelengths, the required delays can be achieved by tuning the wavelength of the lasers.

If the time-delay characteristic of the chirped grating is linear, the difference in time delay between elements at each frequency across the RF bandwidth should be the same. However, in practice, there are always some random errors, called GDR, across the grating period.

In general, an optical beamformer using chirped grating can be implemented in two ways. In the first approach, several chirped gratings are used. Each tunable laser's wavelength requires one grating. By tuning the wavelength, one can vary the delay in each optical carrier signal. An example of this approach can be found in reference 27.

Another approach uses one long chirped grating instead of several chirped gratings. This type of scheme can be found in reference 28. In this scheme, the optical carrier signals occupy different grating periods in one common long chirped grating.

Recently, multichannel chirped fiber grating (MCFG) has been used as the alternative solution for both approaches. The MCFG has several nearly identical channels. So, in a sense, it is similar to several chirped gratings connected together. In each channel, the response is the same as a chirped grating. The average group delay ripple in each channel of the MCFG is less than 13 ps, which is a good value in comparison with a typical chirped grating written using a phased mask. Therefore, in comparison with the first approach mentioned above, each channel of the MCFG can be considered as one good chirped grating for one optical carrier.

In comparison with the second approach, the MCFG is even better. Since in the second approach all optical carriers are inside the long chirped grating, most of the optical carriers are positioned farther away from the center wavelength of the grating. In those farther regions, the GDR is larger than the GDR near the center wavelength and is usually larger than 13 ps. Therefore, using the MCFG, the carrier signals suffer less from GDR.

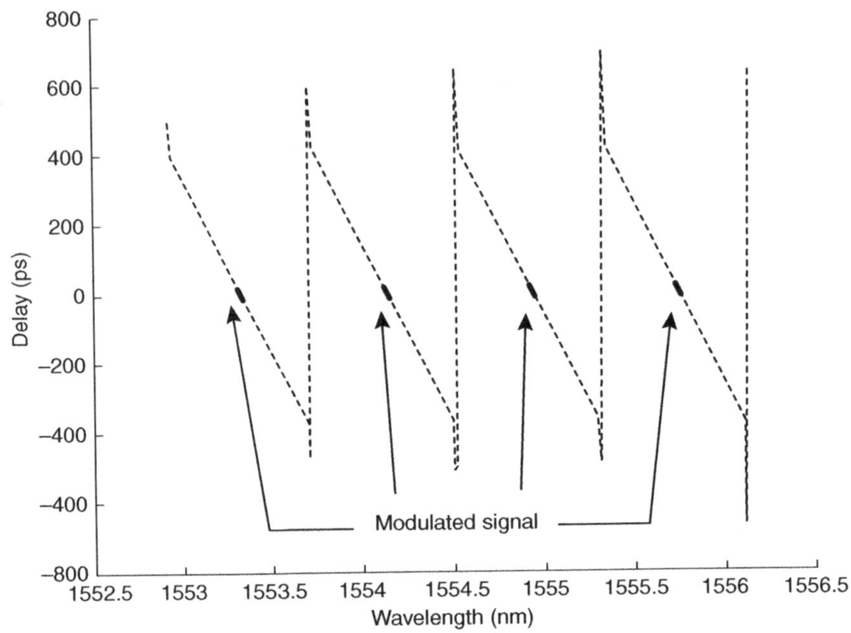

FIGURE 9.4. Measured delay characteristic of the MCFG from channel 27 to 30 (dotted lines), along with the regions occupied by the modulated signals (solid lines).

Since the GDR value at each optical carrier using MCFG is better or at least equal to the GDR value at each carrier in other approach using chirped grating, it should be adequate to use MCFG instead of long chirped grating for this study.

The wavelength of each tunable laser would occupy one channel as shown in Figure 9.4. At first glance, the group delay responses of the channels are more or less the same. However, across grating period within the modulated signal bandwidth, there are fluctuations in time delay as shown in Figure 9.5. The reflectivity of the MCFG also has ripples as shown in Figures 9.6 and 9.7. However, within each channel, the peak-to-peak values of the ripples on the reflectivity spectrum are only around 0.1–0.2 dB. The reflectivity's histograms of some channels are shown in the Figures 9.8 and 9.9. There were around 1000 samples in each plot. The measured regions were the regions occupied by the modulated signal. A Gaussian curve fit is used for each histogram. Since the variances are small and the differences in amplitude won't change the main beam's direction, the reflectivity values can be approximated by the mean values of the reflectivity.

The signal before the photodiode in Figure 9.3 is derived in the following section. The RF signal, denoted as x_{RF}, is modulated with the optical signal by the modulator. In most cases, the modulator is set to operate in single-side-band-plus-carrier mode to avoid signal fading. For better clarity, only the upper side band of the modulated

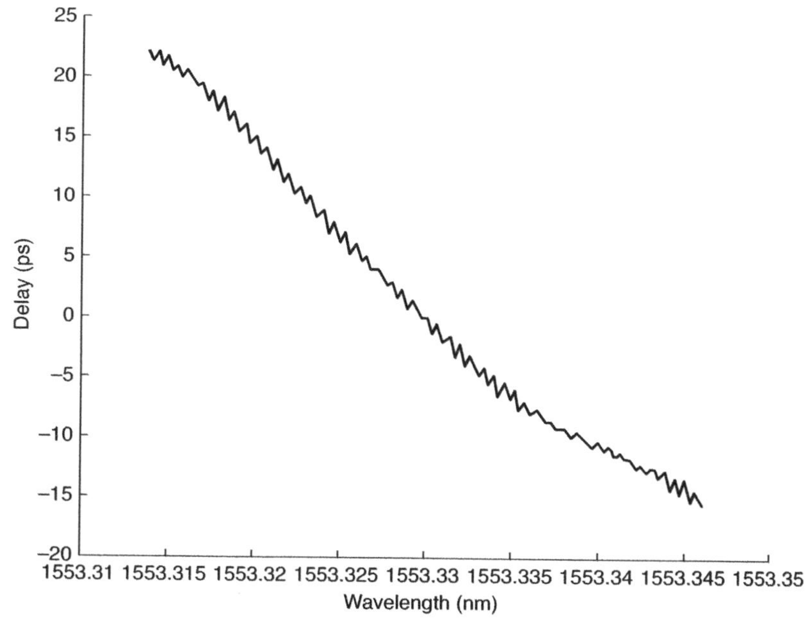

FIGURE 9.5. Fluctuations in time delay across the modulated RF signal bandwidth within channel 30.

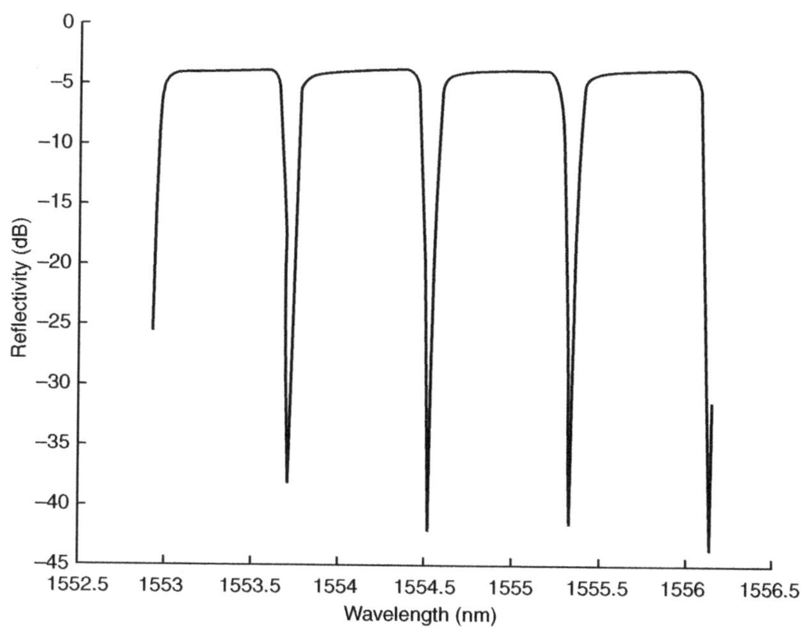

FIGURE 9.6. Measured reflectivity characteristic of the MCFG from channel 27 to channel 30.

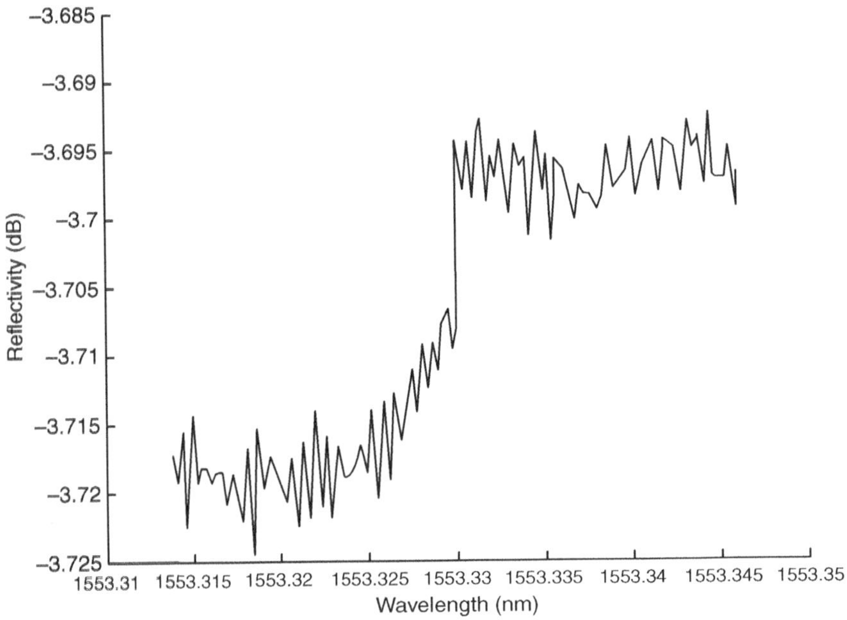

FIGURE 9.7. Fluctuations in reflectivity across the modulated RF signal bandwidth within channel 30.

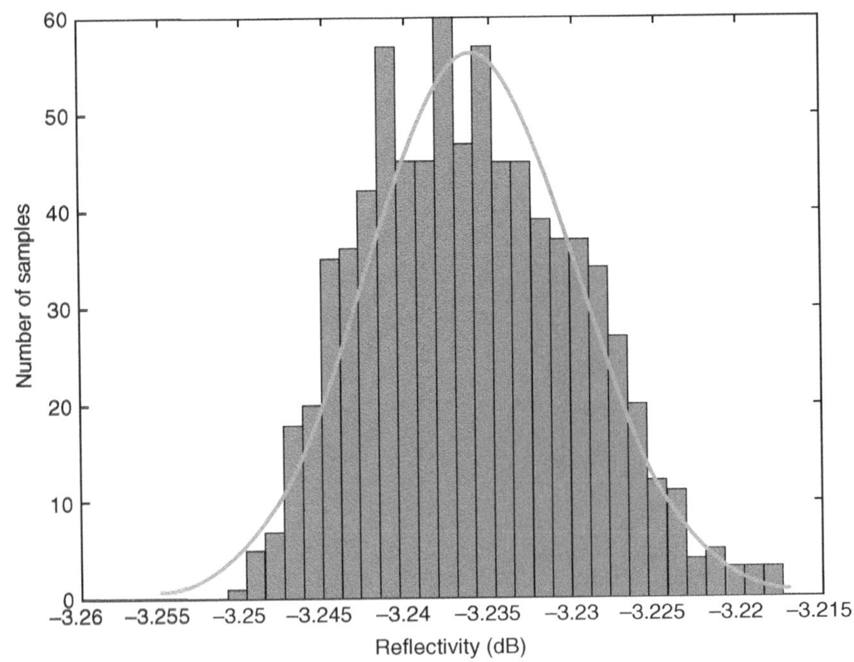

FIGURE 9.8. Histogram of the reflectivity of channel 40. Mean = −3.2361 dB, variance = 0.0063 dB.

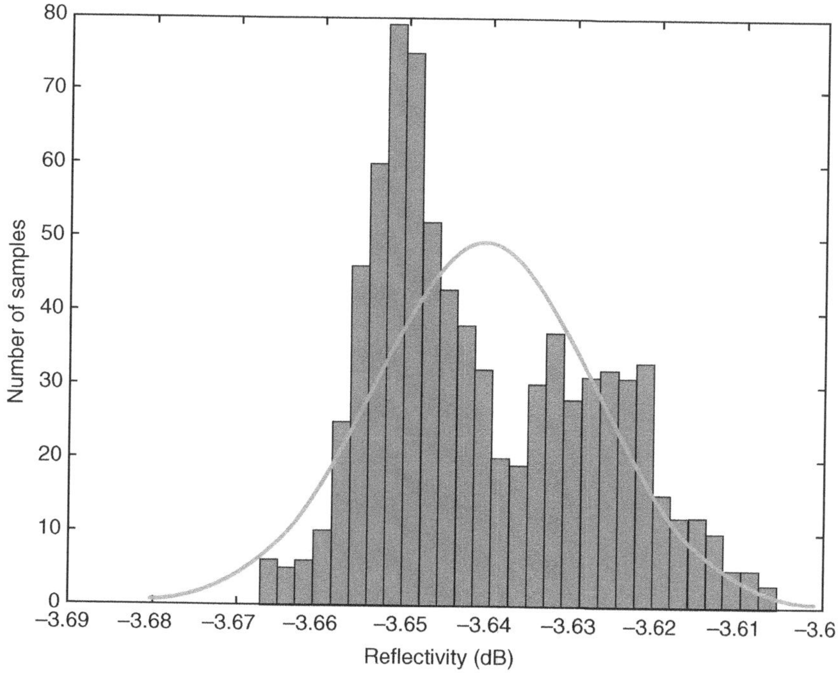

FIGURE 9.9. Histogram of the reflectivity of channel 41. Mean = −3.6407 dB, variance = 0.0131.

signal is considered in the analysis. The inverse Fourier transform of the upper side band after the modulator is

$$x_{\text{upper}}(t) = \frac{M}{2\pi} \int_{-\infty}^{\infty} X_{\text{upper}}(j\omega) \exp(j\omega t)\, d\omega \tag{9.1}$$

where M is the modulation index, and $X_{\text{upper}}(\omega)$ is the Fourier transform of the upper side-band signal $x_{\text{upper}}(t)$. After the chirped grating, the delayed upper-side-band signal becomes

$$x_{\text{upper-delay}}(t) = \frac{M}{2\pi} \int_{-\infty}^{\infty} X_{\text{upper}}(j\omega) \exp(j\omega t) \exp(-jD\omega) \exp[-j\varphi(\omega)]\, d\omega \tag{9.2}$$

where D is the constant representing the linear dispersion delay of the chirped grating. The GDR causes random delay at each harmonic frequency of the signal, which leads to a complex mathematical model. However, by considering it as random shifting in time domain, a random value $\varphi(\omega)$ can be used to represent its effect as seen on the right-hand side of Eq. (9.2). $\varphi(\omega)$ has random dimensionless value at different frequencies.

The function $\varphi(\omega)$ representing the GDR's effect can be expressed by its half-range Fourier series as

$$\varphi(\omega) = \sum_{n=1}^{\infty} b_n \sin(kn\omega) \tag{9.3}$$

where b_n is the half-range Fourier series' coefficient, and k is a constant.

Since $\varphi(\omega) \ll 1$, it is acceptable to consider $b_n \ll 1$. Since $b_n \ll 1$ and $n\omega \gg 1$, the exponential term can be approximated as follows:

$$\begin{aligned}\exp[-j\varphi(\omega)] \\ = \exp\left[-j\sum_{n=1}^{\infty} b_n \sin(kn\omega)\right] \approx 1 - j\sum_{n=1}^{\infty} b_n \sin(kn\omega)\end{aligned} \tag{9.4}$$

Substituting Eq. (9.4) into Eq. (9.2), the delayed upper-side-band signal can be approximated as

$$\begin{aligned}x_{\text{upper-delay}}(t) \\ \approx \frac{M}{2\pi}\int_{-\infty}^{\infty} X_{\text{upper}}(j\omega)\exp(j\omega t)\exp(-jD\omega)\,d\omega \\ - j\frac{M}{2\pi}\int_{-\infty}^{\infty} X_{\text{upper}}(j\omega)\exp(j\omega t)\exp(-jD\omega)\sum_{n=1}^{\infty} b_n \sin(kn\omega)\,d\omega\end{aligned} \tag{9.5}$$

The inversed Fourier transform of the right-hand side of Eq. (9.5) is

$$\begin{aligned}x_{\text{upper-delay}}(t) = M x_{\text{upper}}(t - D) \\ - \frac{M}{4\pi}\sum_{n=1}^{\infty} b_n[x_{\text{upper}}(t - D - kn\omega) + x_{\text{upper}}(t - D + kn\omega)]\end{aligned} \tag{9.6}$$

The photodiode in Figure 9.3 acts as a demodulator. Assuming that the other devices in the system have flat phase response, which is usually the case, the signal after the photodiode is

$$\begin{aligned}x(t) = \alpha M x_{\text{RF}}(t - D) \\ - \frac{\alpha M}{4\pi}\sum_{n=1}^{\infty} b_n[x_{\text{RF}}(t - D - kn\omega) + x_{\text{RF}}(t - D + kn\omega)]\end{aligned} \tag{9.7}$$

where $\alpha < 1$ is a constant represent the loss after the demodulation process.

From Eq. (9.7), it is clearly shown that the delay signal consists of two terms. The first term is the original RF signal delayed by a fixed amount of time. This is the ideal result when there is no time-delay error. The second term is the additional distorted signal caused by GDR. Although the GDR is random, its value is fixed after the fabrication of the grating. Therefore, the GDR can be measured and the Fourier coefficients b_n can be calculated beforehand. Given a particular signal, the amount of distortion can be estimated. If that amount of additive error can be subtracted from the output signal of the beamformer, the GDR's error is avoided.

To understand the effect of GDR on the main beam direction, the differences in time delay between two elements should be calculated. Taken from Eq. (9.2), the delay of a single frequency at the ith element is

$$t_i = D(\omega_{\mathrm{RF}} + \omega_{ci}) + \sum_{n=1}^{\infty} b_{ni} \sin[kn(\omega_{\mathrm{RF}} + \omega_{ci})] \tag{9.8}$$

where ω_{RF} is the particular frequency of interest within the bandwidth of the RF signal, and ω_{ci} is the optical carrier frequency of the ith element. The difference in time delay between two elements therefore equals

$$\begin{aligned}\Delta t_{21} &= D(\omega_{\mathrm{RF}} + \omega_{c2}) + \sum_{n=1}^{\infty} b_{n2} \sin[kn(\omega_{\mathrm{RF}} + \omega_{c2})] \\ &\quad - D(\omega_{\mathrm{RF}} + \omega_{c1}) - \sum_{n=1}^{\infty} b_{n1} \sin[kn(\omega_{\mathrm{RF}} + \omega_{c1})] \\ &= D(\omega_{c2} - \omega_{c1}) \\ &\quad + \sum_{n=1}^{\infty} \{b_{n2} \sin[kn(\omega_{\mathrm{RF}} + \omega_{c2})] - b_{n1} \sin[kn(\omega_{\mathrm{RF}} + \omega_{c1})]\}\end{aligned} \tag{9.9}$$

Once again, the expression has two terms. The first term only depends on the optical carrier frequency, and thus, it remains unchanged across the RF signal's bandwidth. However, the second term, which represents the effect of GDR, has the RF signal frequency component. As a result, if there is no GDR, the time delays between elements are fixed across the RF signal bandwidth, and the main beam direction should stay at the desired direction. On the other hand, if GDR is present, the time delays between elements at each RF signal frequency are different.

In other words, the main beam direction of the array is randomly changed across the RF signal bandwidth. Therefore, in contradiction to the statement about "squint-free operation" that usually appears in papers about optical beamformer using chirped grating, the beamformer actually makes the beam squint to random direction across the signal bandwidth. The simulation results in the next section further strengthen the above conclusion. Moreover, the results also show many more limitations that an optical beamformer using chirped grating has to face.

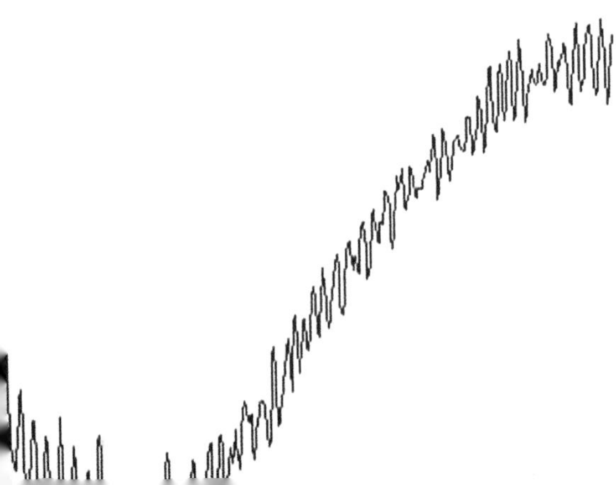

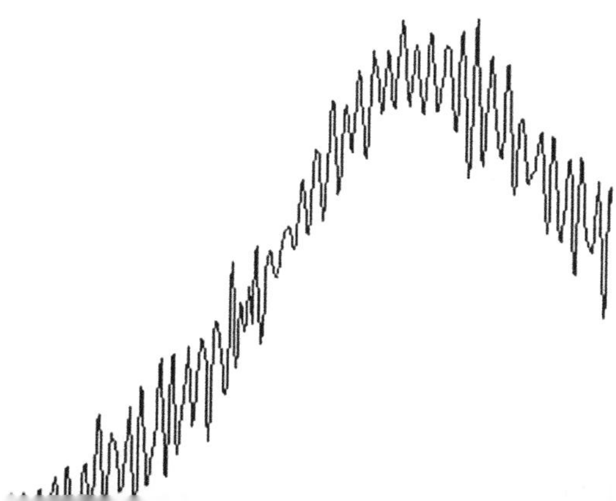

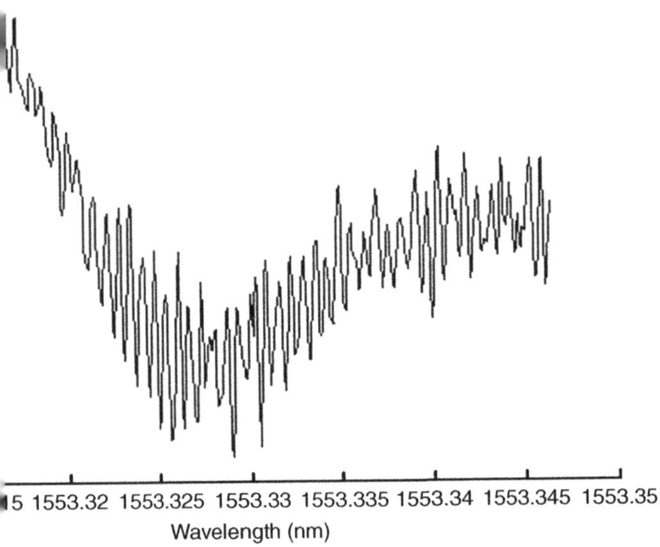

erences in time delay across the RF signal's bandwidth between channel

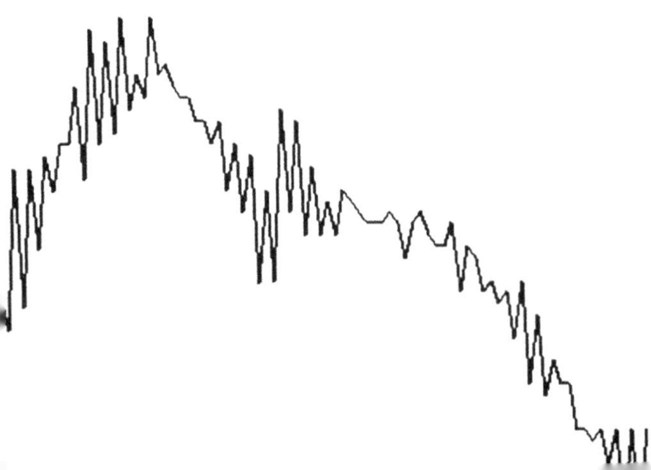

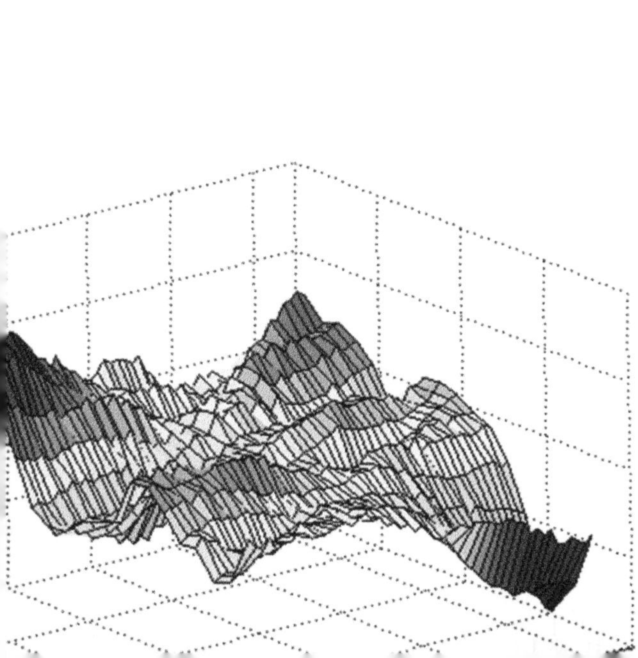

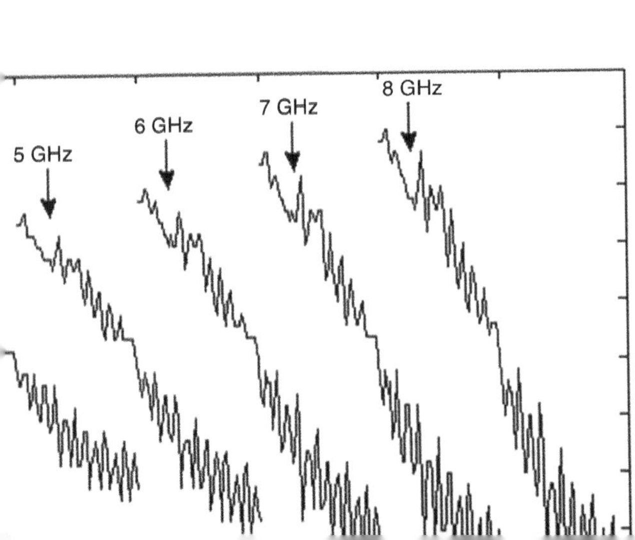

5 GHz
6 GHz
7 GHz
8 GHz

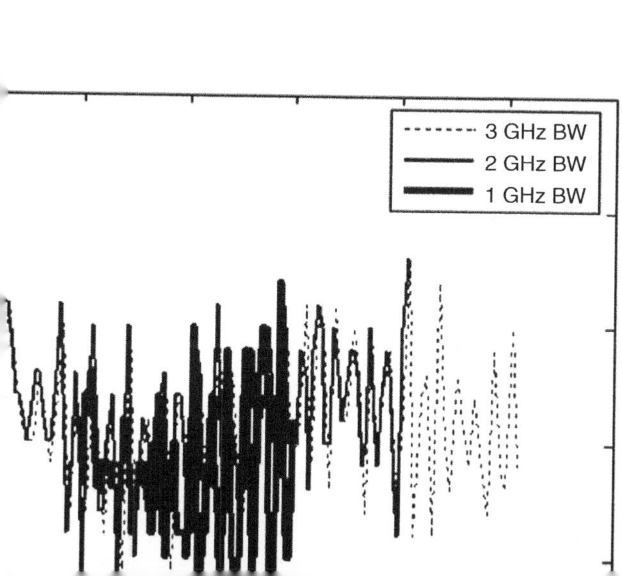
3 GHz BW
2 GHz BW
1 GHz BW

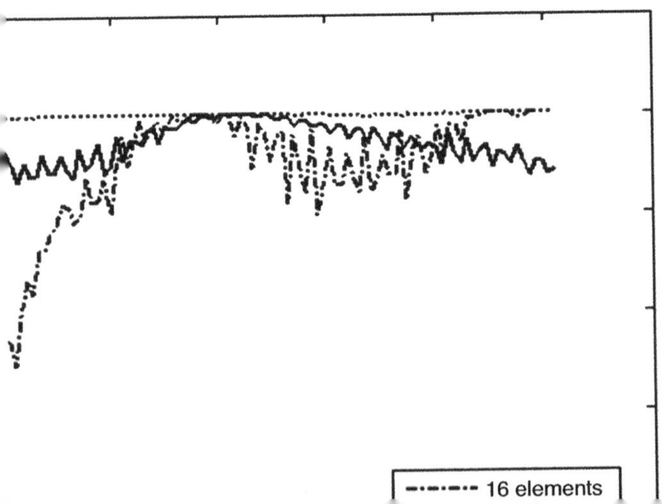
16 elements

effect of GDR should be considered and overcome in all optical beamformer systems using chirped grating.

9.4 OPTICAL BEAMFORMING USING THE DISPERSIVE DELAY AND NONDISPERSIVE DELAY HYBRID APPROACH

9.4.1 Introduction

As mentioned before, the two major approaches for a true time-delay photonic beamformer are dispersive delay and nondispersive delay. Each solution has its own strengths and weaknesses. For beamformers using nondispersive delay devices, the length and the loss of the optical signal are the largest obstacles. On the other hand, beamformers employing dispersive devices depend on tunable laser sources to operate. The cost of those tunable lasers severely limits the array size.

Wavelength division multiplexing technique has been explored to greatly reduce the system complexity and hardware requirement for photonic beamformers [36]. Furthermore, this technique can be utilized to form multiple beams. One of the first hybrid approaches between dispersive and nondispersive delay was first introduced in reference 24. However, given the technology of that time, the system was too complex and had to give way for other approaches.

Recently, the hybrid systems have been proposed again in references 37 and 38. The system in reference 37 simply used a set of gratings and a set of tunable optical delay lines for dispersive and nondispersive delay, respectively. Although none of the weakness from the two approaches was addressed in that system, at least the combination allowed 2D beam steering. The other system [38] was more aware of the advantages and disadvantages from each approach. That hybrid system has dramatically reduced the number of required tunable laser sources and the required delay length of the optical lines at the same time.

That achievement was obtained with the multichannel chirped grating. The combination of multichannel chirped fiber grating (MCFG) and optical delay line has not only addressed the hardware issue but has also supported multiband multibeam arrays and 2D arrays with independently and simultaneously steering ability. Moreover, the system is more robust against the error cause by group delay ripple that has riddled any optical beamformer using chirped grating. The simple controlling scheme and expandable capability make it suitable for phased-array system requirements in terms of beam scanning and reconfiguration.

9.4.2 Principles

Optical beamformers using dispersive lines, Bragg gratings and chirped gratings usually require one tunable laser for each array element. Tuning the laser's wavelength introduces time delay into the corresponding optical signal. However, this simple operational scheme faces serious problems when the antenna array size is increased, since too many expensive optical devices are needed.

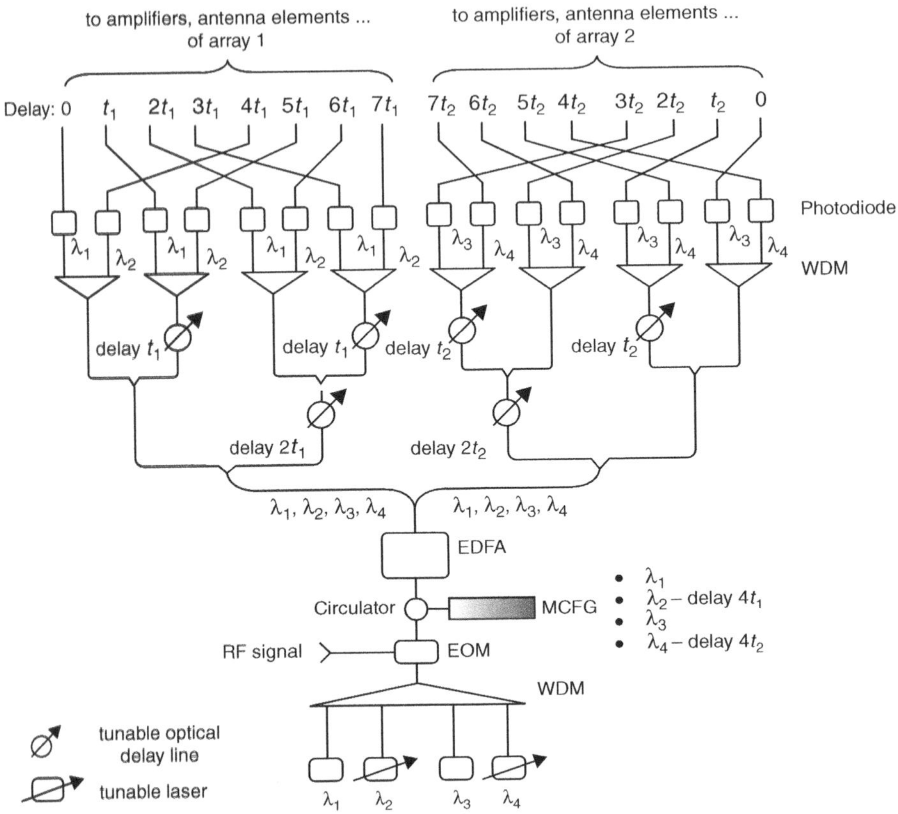

FIGURE 9.19. Topology of the hybrid beamformer, with array size $N = 8$.

Fortunately, by reusing the optical carriers, it is possible to dramatically reduce the number of required components. The proposed optical beamformer is depicted in Figure 9.19. In the schematic, the size of the array is limited to two sub-arrays, and the number of elements N is equal to 8 for better clarity. However, a much larger array size can be supported as discussed later in this section.

The MCFG has been used to provide dispersive time delays in optical transmission systems. Multiple channels with nearly identical characteristics are created using a sampled grating technique. The MCFG is now a commercially off-the-shelf product, which is mainly used in wavelength division multiplexing networks as a dispersion compensator. The channel spacing follows the ITU-T standard, and the time delay can be as long as 3000 ps, and the total number of channels can be as many as 50.

The optical source arrangement consists of two fixed-wavelength lasers and two tunable lasers. Their wavelengths are denoted as λ_1, λ_2, λ_3, and λ_4 as shown in Figure 9.19. The four wavelengths fall within four separated channels of the MCFG.

A multiplexer combines the waves of the lasers into one optical fiber. The optical signals are modulated with the RF signal via an external modulator. In order to avoid signal fading, the modulator is adjusted to generate a single side band plus carrier signal.

These optical signals are then passed through the MCFG using an optical circulator. Initially, the tunable lasers are tuned so that there are no differences in the time delay between λ_1 and λ_2 and between λ_3 and λ_4 in the reflected signals. In Figure 9.19, it means that t_1 and t_2 are equal to zero.

The array consists of two sub-arrays. Each array forms its beam in a different direction. This is one of the several approaches to achieve photonic multiple beamforming as studied in reference 36. Moreover, for this particular scheme, it is intended to adapt the proposed beamformer into a dual-beam and dual-band system. That system may require the beamformer to support two different sub-arrays. Dividing or segmenting the array into several subsets also allows beam-switching, which will increase the coverage of the system.

It is assumed that a delay t_1 is required between two consecutive elements for the first array, and a delay t_2 is required between two consecutive elements for the second array. The first tunable laser is tuned in such a way that, after the MCFG, λ_2 is delayed by $Nt_1/2$ in comparison with λ_1. Similarly, the second tunable laser is tuned so that λ_4 is delayed by $Nt_2/2$ in comparison with λ_3.

The reflected signals after the MCFG are divided into several branches using 3-dB power splitters. Tunable optical delay lines then further delay the signals in those branches before they are all de-multiplexed. The tunable optical delay lines are nondispersive, which means that all the wavelengths experience the same amount of delay in the delay lines.

Commercial optical delay lines introduce a time delay by several ways. The most common method is a mechanically tunable delay line. However, for the proposed beamformer, another type of tunable optical delay line is more suitable. It routes the optical signals through several segments whose lengths increase successively by a power of 2. An electrical signal is used to control the device. Commercial products of this type are also available and can offer up to micro/nanosecond delays. Unfortunately, such devices provide discrete delays. Commercial optical delay lines usually have 4-bit resolution, which is acceptable for this beamformer.

There are $\log_2 N - 1$ layers of tunable optical delay lines. The tunable optical delay lines in the first layer delay the signals by $Nt_i/4$. The tunable optical delay lines in the second layer delay the signals by $Nt_i/8$ so the tunable optical delay lines which are placed right before the de-multiplexers delay the signals by $Nt_i/2^{\log_2 N} = t_i$, $i = 1, 2$.

The layout of the optical delay lines when $N = 8$ is shown in Figure 9.19. There are two layers of optical delay lines. The corresponding delay for each optical delay line is also noted in Figure 9.19. All four wavelengths are present in all branches. However, the wavelengths λ_1 and λ_2 are used to control the first array, while the wavelengths λ_3 and λ_4 are used to control the second array. Therefore, the de-multiplexers only select the necessary wavelengths as shown in Figure 9.19.

After the de-multiplexers, there are $2N$ signals, with time delays $0, t_1, 2t_1, \ldots, Nt_1$ for the first array and $0, t_2, 2t_2, \ldots, Nt_2$ for the second array. Photodiodes are used to convert the signals back into an electrical domain. Power amplifiers may be used to raise the RF signal power to an acceptable threshold before the signals are fed to the array's elements.

As a result, by adjusting the tunable lasers and the tunable optical delay lines, the desired time delay is introduced into the feed signal of the corresponding element. Thus, the main beam of each array should point to the required direction. The electrical signals are arranged so that the first array has increasing delays between elements and the second array has decreasing delays between elements. This setting allows the beams to scan different sides.

Using this method, the N-element sub-array can be controlled using one tunable laser. In the other reported optical beamformers using chirped fiber grating [28] and MCFG [29], three tunable lasers are needed to control a small four-element array. Considering the cost of a tunable laser, this scheme clearly has an advantage in the system's cost.

The number of tunable delay lines is also kept to a minimum. Since the scheme is a hybrid between dispersion delay line and optical delay line approaches, it is wise to compare it with other optical beamformers that only use tunable optical delay lines. In a typical beamformer with such an approach, each array's element required a tunable optical delay line to operate, which means that an N-element array will need N optical delay lines. In this case, the number of required tunable optical delay lines is reduced to $(N/2) - 1$.

The required tuning ranges for both tunable laser and tunable optical delay line are small. Since the time delay is contributed by both the MCFG and the optical delay line, the requirement for them is eased. For example, in an eight-element array, the longest delay can be as large as $7t_{max}$, where t_{max} is the delay between two consecutive elements at the maximum scanning angle. Using the proposed scheme, the maximum delays are $4t_{max}$ for the MCFG, $2t_{max}$ for the tunable optical delay lines at the first layer, and t_{max} for the tunable optical delay lines at the second layer. This feature enables a wider scanning range.

Furthermore, with the MCFG, multiple beams can be formed and steered independently and simultaneously with little adjustment. This is a significant achievement of the approach. Reports about optical multiple beamformers [39, 40] are rare. Optical multiple beamformers that can freely steer their beams are far more uncommon. Since each array works independently, different data streams can be transmitted at the same time. The beamformer is also transparent to data type. Since the size of the array is no longer limited by the number of tunable lasers, this scheme can be used in radar applications, where large arrays are usually required.

Theoretically, the scheme in Figure 9.19 can be further expanded. One more fixed-wavelength laser source and one more tunable laser source are required to support an additional sub-array. Their wavelengths should fall within two unoccupied channels of the MCFG. The optical signal after the amplifier is divided into one more branch. The optical signal in that branch will go through a similar feeding arrangement to reach the additional sub-array. In order to increase the size of the sub-array, the optical signal can be further divided into more branches, and additional layers of tunable optical delay lines can be used.

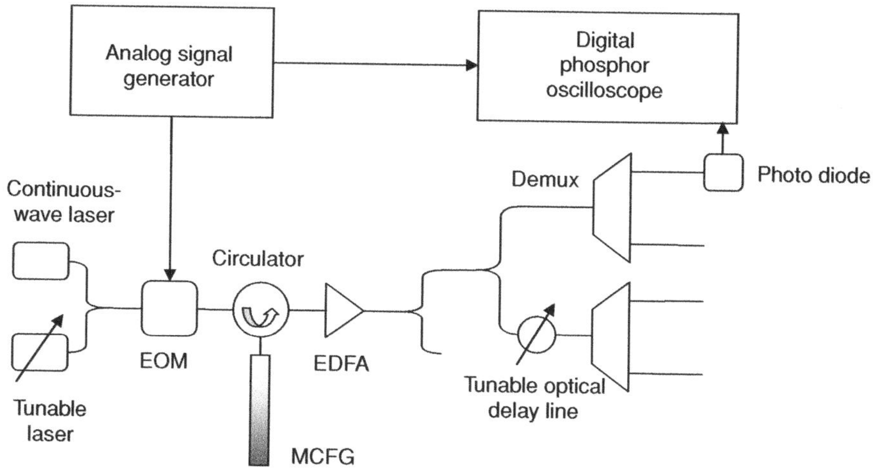

FIGURE 9.20. Experiment setup.

9.4.3 Experimental Results

In order to test the hybrid system, experiments have been conducted. The experiment system setup is shown in Figure 9.20. Each branch of the system shown in Figure 9.19 was tested at a time. The setup in Figure 9.20 was for the first branch in the left of Figure 9.19. Tunable lasers sources were used at all optical inputs for easier initial fine-tuning. However, for the tunable lasers representing the continuous-wave sources, the wavelengths were fixed during the experiments. The wavelengths for four sources were 1547.503, 1548.312, 1549.112, and 1549.917 nm, respectively. The wavelengths followed the ITU-T spacing but did not actually fall in the center wavelength grid, since this setting allowed the longest delay time within the system 3-dB bandwidth. The couplers were used instead of multiplexers because there were only two wavelengths for each array.

The photodiode was use to convert the signal back into electrical domain at each output of the de-multiplexer. The output RF signals were then compared with the reference input RF signals. The tuning ranges of the tunable sources were from 1547.373 nm to 1547.643 nm for the first wavelength and from 1548.982 nm to 1549.252 nm for the third wavelength. The second wavelength and the fourth wavelength were kept at their initial values. The optical delay lines were tuned accordingly to the absolute delay between the wavelengths as explained in the previous section. The resulting delay between the RF output signals after the photodiodes for the first array are shown in Figure 9.21.

In Figure 9.22, the frequency of the input RF signal was changed from 2 GHz to 8 GHz. The linearity of the delay in both Figure 9.21 and Figure 9.22 are acceptable. Considering the optical system as a black box, those results indicated that the system is able to support signals for L-, S-, C-, and X-band arrays. Furthermore, the RF signal bandwidth can be quite large. For optical beamformers with such capability, an

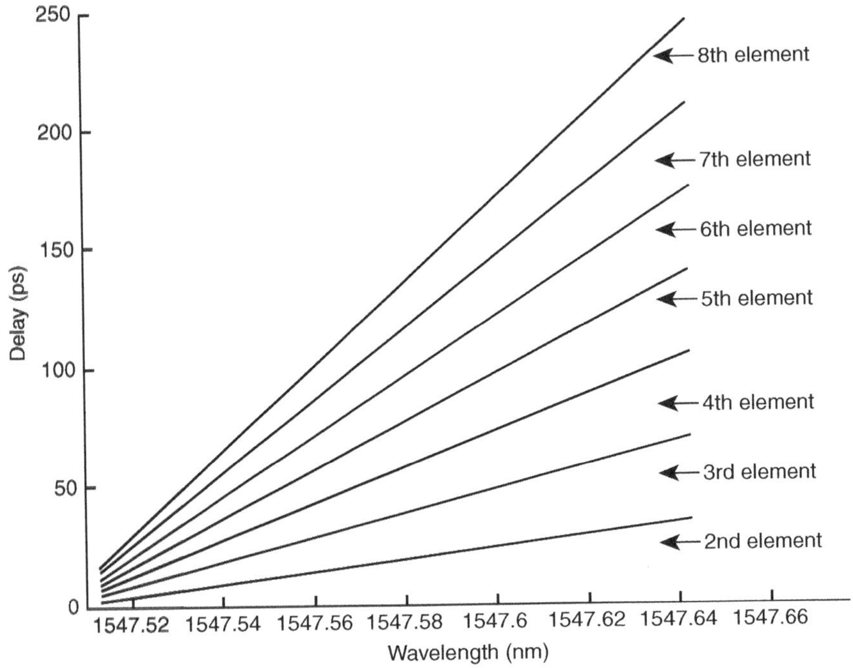

FIGURE 9.21. Time delay between elements for the first sub-array.

ultra-wideband system is a suitable application. A good example is the UWB chipless RFID system and ground penetrating radar, where the range of radiated frequency varies across several frequency bands up to 10 GHz and the signal bandwidth reaches several octaves. Moreover, the system provides two sets of independent delays. This attribution means that the system can be lightly adjusted to support two beams at the same time or one 2D beam.

In the next experiment, the system was tested with different RF frequencies at different frequency bands. The first wavelength and the second wavelength were modulated with RF signal at 2 GHz, while the third wavelength and fourth wavelength were modulated with RF signal at 8 GHz. The bandwidths of the RF signals were 0.2 GHz and 0.8 GHz, respectively. The delays between output signals after the photodiodes are shown in Figures 9.23 and 9.24.

The delays are mostly linear within the RF signal bandwidth. Since the beamformer can support two independent sets of delay for two RF signal at different frequency bands, the system can be used to provide a fed signal to dual-band arrays. Such dual-band arrays, working at L band and X band, can be used in multifunction naval radars, with better searching and tracking functions. The different sets of delay can be adjusted at the same time, which means that the two beams working at different frequency bands for different tasks can be scanned and steered independently

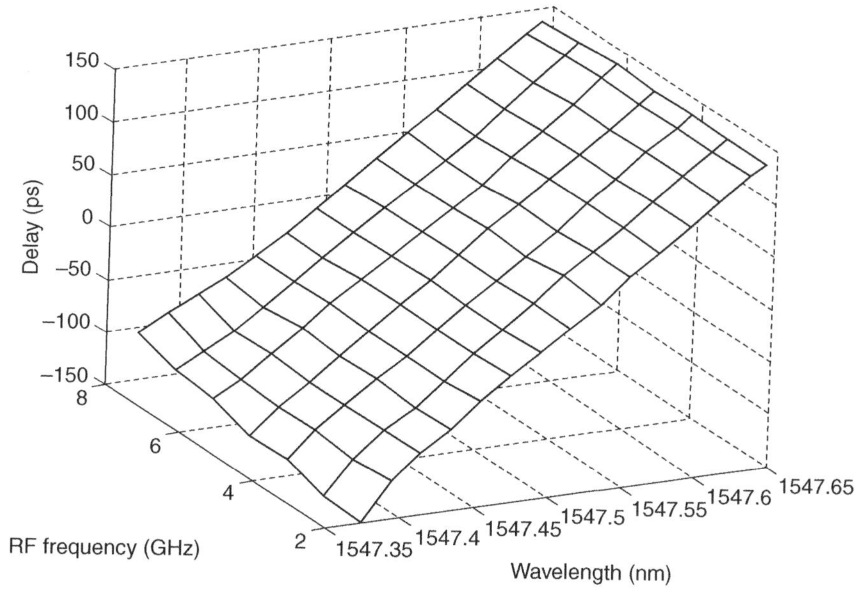

FIGURE 9.22. Time delay within the tuning range of the first array, at different RF frequencies.

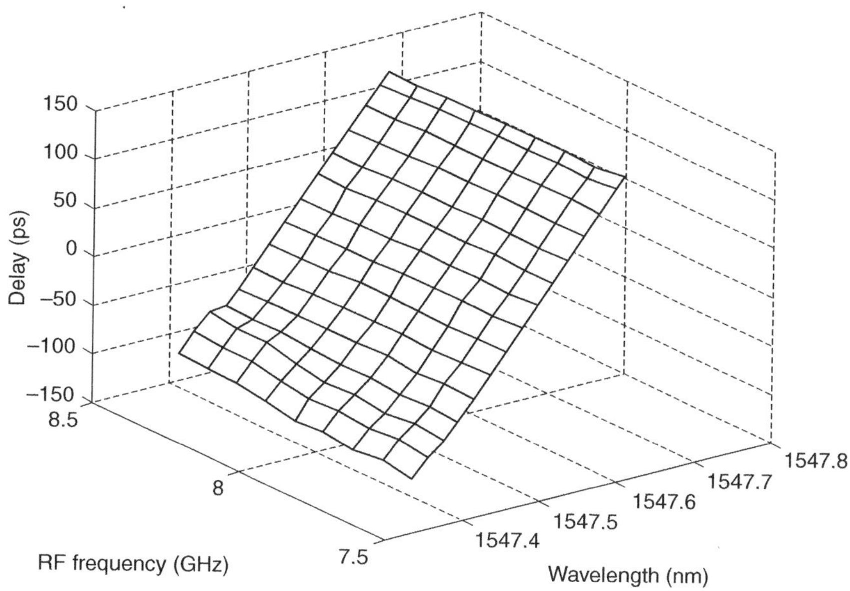

FIGURE 9.23. Time delay within the tuning range of the first array, working with a 0.8-GHz bandwidth signal.

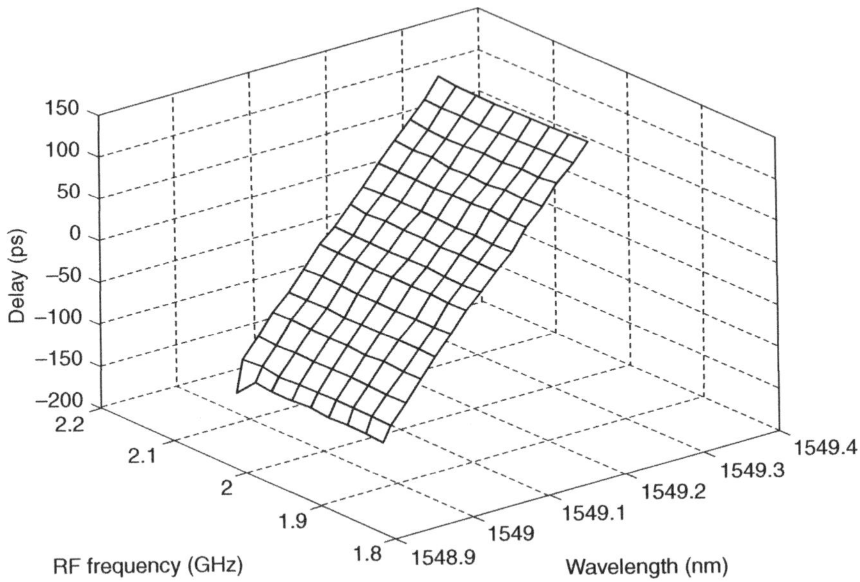

FIGURE 9.24. Time delay within the tuning range of the second array, working with a 0.2-GHz bandwidth signal.

and simultaneously. This capability of independent tuning of multiple beams helps tremendously mitigating collisions between RFID tags which are almost collocated.

The measured data after the photodiodes were imported into a simulation program in Matlab in order to simulate to radiation pattern. The radiation patterns for both cases, dual-beam dual-band and 2D beam, are shown in Figures 9.25 to 9.30. The maximum tuning range of the lasers was only 0.27 nm, providing around 245-ps delay. However, thanks to the hybrid approach as discussed in the previous section, the system can provide wide-range steering for signals above 7 GHz in dual-beam dual-band case and wide-range azimuth steering for signals above 3.5 GHz for the 2D beam case. Steering the main beam only barely raised the side lobe. However, the null level was much higher at a farther scanning angle. The reason for this drawback is explained in the following experiment.

Unfortunately, the size of the supported array, is limited by the power loss. It is a trade-off between the size of each sub-array, the number of beams, and the acceptable output power after the photodiode. The loss is highly related to the device in used. In those particular experiments, the optical losses of the devices are listed below:

- The 3-dB polarization maintaining a 2×1 coupler after the laser sources: -5 dB
- The Mach 10 external modulator from Covega: -4 to -8 dB, depending on the RF frequency and the bias voltage
- The circulator and the multichannel chirped fiber grating from Teraxion: -4 dB

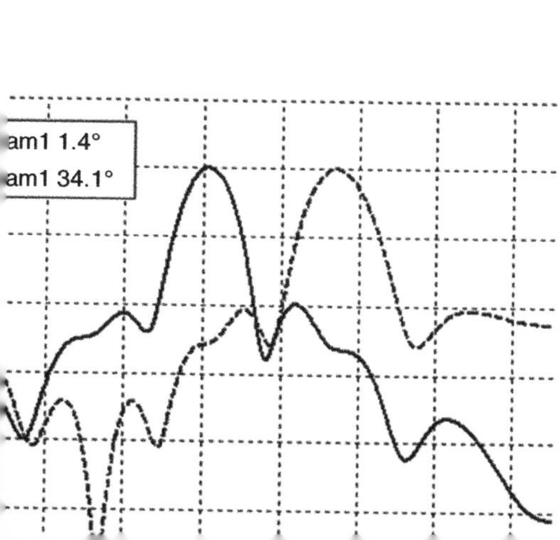

am1 1.4°
am1 34.1°

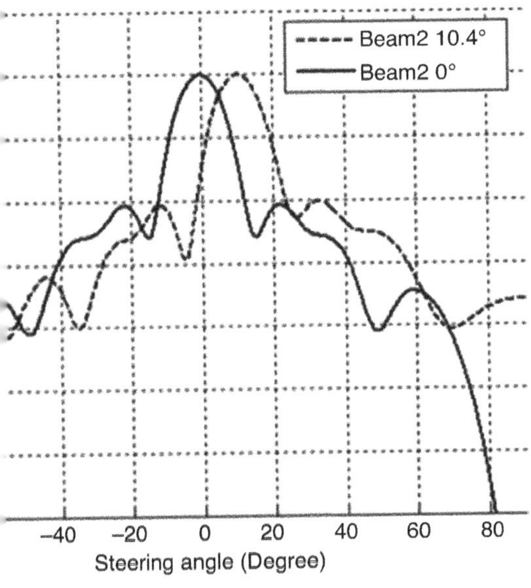

Radiation patterns for the second beam at 2 GHz.

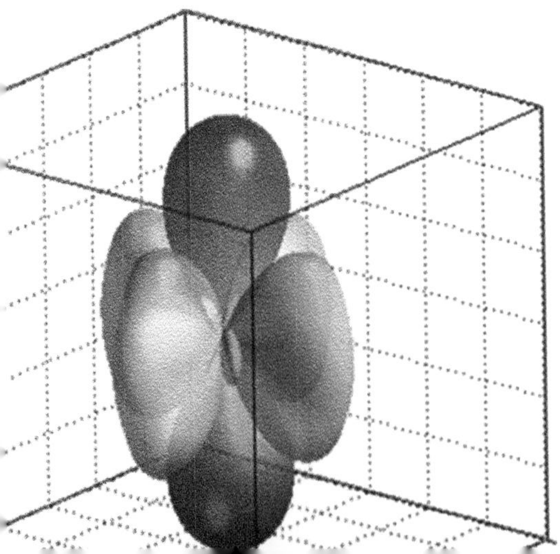

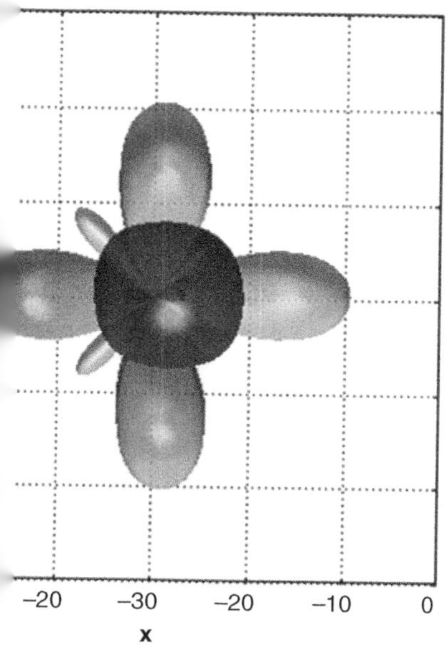

) radiation pattern at 5° in x–y planes.

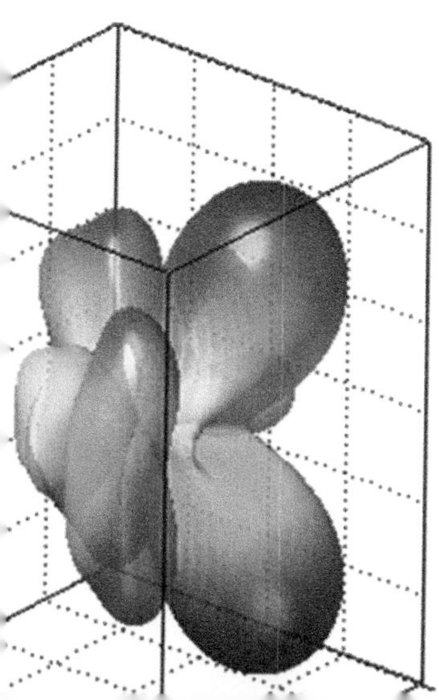

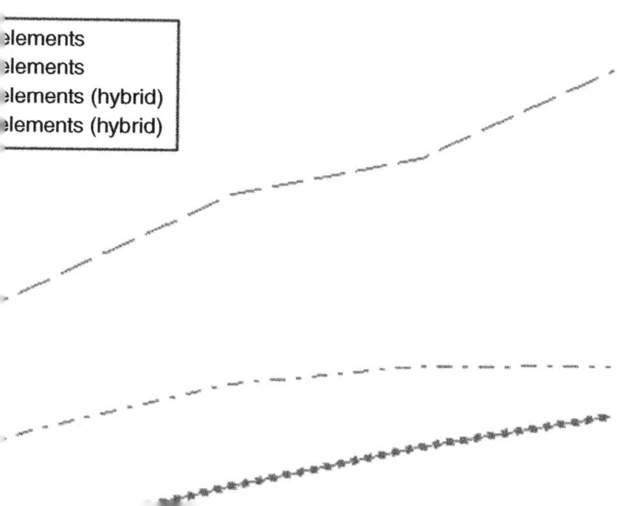

elements
elements
elements (hybrid)
elements (hybrid)

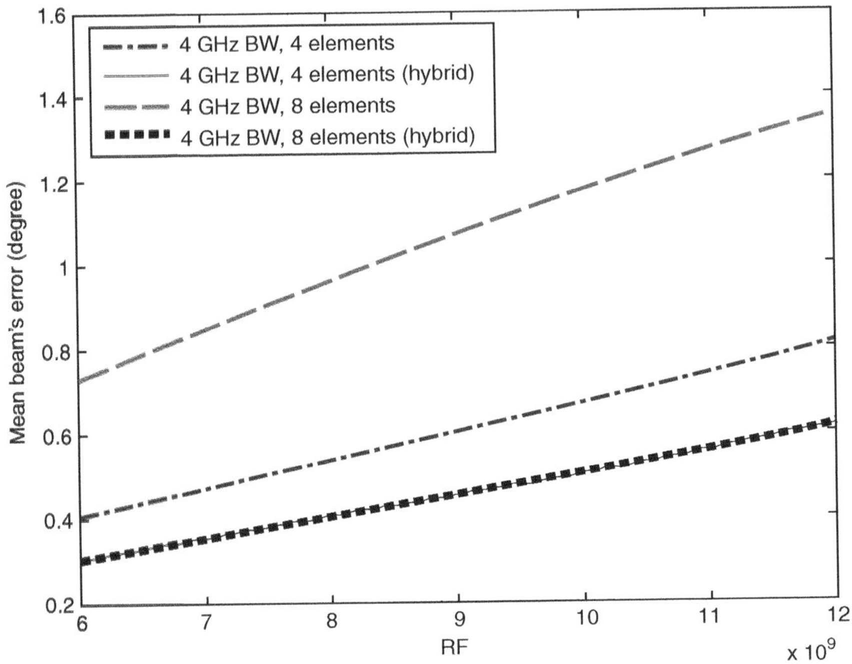

FIGURE 9.33. Main beam direction error at different RF bandwidths and different array sizes.

the drawbacks. However, more researches and studies are still required to overcome limitations such as high loss and the ability to fully control both signal's amplitude and signal's delay.

9.5 CONCLUSION

Although only a part of an RFID reader, the beamformer has a vital role in improving the system performance. In addition to spatial, temporal, and polarization diversities, smart antennas in cooperation with beamforming techniques promise higher gain, better signal-to-noise ratio, and anti-collision procedures.

Unfortunately, traditional electrical controlled beamformers do not fare well with the UWB reader system. The demand in signal bandwidth and signal frequency poses a great challenge for an electrical beamformer initially designed as a narrowband system. Many limitations arise such as the high loss, the bulky devices, and especially the beam-squint error.

An Optical beamformer, on the other hand, is a bright candidate. Being lightweight and compact, an optical beamformer inherits the large instantaneous bandwidth of photonic devices and provides squint-free operation for UWB signals. Among the many approaches to achieve true time-delay optical beamforming, beamformers using dispersive and nondispersive delay have received better favor for their simplicity and compactness. However, those advantages come with new obstacles.

For an optical beamformer using nondispersive delay, the high loss of the delay line remains as a disadvantage. Besides distorting the radiation pattern, the long delay line also suffers from latency and temperature dependence. An optical beamformer employing dispersive delay avoids those drawbacks, but suffers from other issues. In such beamformers, controlling one element in the array requires one tunable laser. For large arrays, the expensive price of the hardware overshadows the system's advantages. Moreover, the GDR errors in the grating severely cripple the system performance. Those inevitable errors cause the beam to randomly squint to different directions, denying the true time-delay characteristic of the beamformer. The signal frequency, the signal bandwidth, and the array size are also limited. The negative effect of GDR places many threshold on the design of the system. Those drawbacks should be considered not only in optical beamformers but also in other true time-delay systems.

A possible solution to subdue those drawbacks is the hybrid between dispersive and nondispersive delay. Clever placement of dispersive and nondispersive devices as well as re-using the optical carrier can greatly reduce the hardware requirement. The proposed system using such a hybrid approach in this chapter has been dramatically reduced in the number of required devices. The beamformer can support a signal with a wide RF bandwidth from L band to X band. Moreover, the scheme can be adjusted and expanded to provide beam-steering for multiple beams in both 2D and 3D planes. All the beams are independent and able to operate in different RF bands at the same time. The beamformer is also more robust against GDR. Unfortunately, the high loss that occurs in the electrical/optical and optical/electrical conversions still remains as a hardware limitation.

Despite those weaknesses, an optical beamformer firmly stands as a strong alternative for traditional electrical beamformer for UWB applications. For RFID readers, where precision, compactness, and weight are the decisive factors, the optical beamformer's strong points are even more emphasized. Being able to provide beam steering for multiple independent beams with wide RF bandwidth, the optical beamformer will undoubtedly elevate the system performance in searching range, scanning resolution, gain, and diversity.

REFERENCES

1. R. Das and P. Harrop, RFID Forecasts, Players & Opportunities 2008–2018 [Online]. Available at http://www.idtechex.com/products/en/view.asp? productcategoryid=151.
2. http://www.rfsaw.com/ (accessed 12 October 2008).
3. F. Bauchot, J. Y. Clement, G. Marmigere, and P. Secondo, Chipless RFID tag and method for communicating with the RFID tag, US Patent No. US20080084276, 2008.
4. X. Y. Jalaly and I. D. Robertson, Capacitively tuned microstrip resonators for RFID barcodes, in *Proceedings, 35th EUMC*, 2005, pp. 1161–1164.

5. J. McVay, A. Hoorfar, and N. Engheta, Theory and experiments on Peano and Hilbert curve RFID tags, *Proc. SPIE*, Vol. 6248, No. 1, 624808.1–624808.10, 2006.
6. S. Preradovic, I. Balbin, N. C. Karmakar, and G. Swiegers, A novel chipless RFID system based on planar multiresonators for barcode replacement, The 2008 IEEE International Conference on RFID, 2008, pp. 289–296.
7. S. Mukherjee, System for identifying radio-frequency identification devices, US Patent No. US20070046433, 2007.
8. L. Zheng, S. Rodriguez, L. Zhang, B. Shao, and L. Zheng, Design and implementation of a fully reconfigurable chipless RFID tag using inkjet printing technology, in *Proc. IEEE Int'l Symp. Circuit Systems, Seattle, USA*, 2008, pp. 1524–1527.
9. L. Zhang, R. S. Tenhunen, L. R. Zheng, An innovative fully printable RFID technology based on high speed time-domain reflections, HDP'06. Conf. 2006; 166–170.
10. S. Cataldo and N. C. Karmakar, Efficient rectenna design for emerging applications, in *Proceedings, 2008 Asia-Pacific Microwave Conference*, Hong Kong and Macau. 2008, CD-Rom.
11. N. C. Karmakar, Smart antennas for automatic radio frequency identification readers, in: *Handbook on Advancements in Smart Antenna Technologies for Wireless Networks*, Idea Group Inc., Chicago, pp. 449–473.
12. R. C. Hansen, *Microwave Scanning Antennas*, Academic Press, New York, 1964.
13. R. C. Hansen, Array feeds, in *Phased Array Antennas*, John Wiley & Sons, New York, 1998.
14. T. Peter, *Introduction to Radar Target Recognition*, Institution of Electrical Engineers, London, 2005.
15. N. A. Riza, *Selected Papers on Photonic Control Systems for Phased Array Antennas*, SPIE Milestone Series, SPIE, Bellingham, WA. 1997.
16. L. W. Maynard, M. Matt, and C. C. Raymond, Optically operated microwave phased-array antenna system, US Patent 3,878,520, 1975.
17. P. G. Sheehan and J. R. Forrest, The use of optical techniques beamforming in phased arrays, *Proc. SPIE*, Vol. 477, pp. 82–89, 1984.
18. W. J. Stewart, Optical phased array control, *Military Microwaves*, Vol. 00, pp. 287–291, 1984.
19. A. R. Nabeel, Acousto-optic architectures for multi-dimensional phased array antenna processing, *Proc. SPIE*, Vol. 1476, pp. 144–156, 1991.
20. I. D. Blanchflower and A. J. Seeds, Optical control of frequency and phase of GaAs MESFET oscillator, *Electron. Lett.*, Vol. 25, No. 5, pp. 359–360, 1989.
21. M. L. Arnold, Use of fiber optic frequency and phase determining elements in radar, in *Proceedings of the 33rd Annual Symposium on Frequency Control*, 1979, pp. 436–443.
22. C. Leo, Ultra-wideband microwave beamforming technique, *Microwave J.*, Vol. 28, No. 4, pp. 121–123, 126–131, 1985.
23. D. Dolfi, J. P. Huignard, and M. Baril, Optically controlled true time delays for phased array antenna, *Proc. SPIE*, Vol. 1102, pp. 152–161, 1989.
24. P. G. Akis and D. K. Davies, Hardware compressive 2-D fiber optic delay line architecture for time steering of phased-array antennas, *Appl. Optics.*, Vol. 29, No. 36, pp. 5353–5359, 1990.

25. M. Y. Frankel, P. J. Matthews, and R. D. Esman, Two-dimensional fiberoptic control of a true time-steered array transmitter, *IEEE Trans. Microwave Theory Tech.*, Vol. 44, No. 12, pp. 2696–2702, 1996.

26. H. Subbaraman, M. Y. Chen, and R. T. Chen, Photonic crystal fiber-based true-time-delay beamformer for multiple RF beam transmission and reception of an X-band phased-array antenna, *Lighwave Technol.*, Vol. 26, No. 15, pp. 2803–2809, 2008.

27. A. Molony, C. Edge, and I. Bennion, Fibre grating time delay element for phased array antennas, *Electron. Lett.*, Vol. 31, pp. 1485–1486, 1995.

28. J. L. Corral, J. Mart'i, S. Regidor, J. M. Fuster, R. Laming, and M. J. Cole, Continuously variable true time-delay optical feeder for phased-array antenna employing chirped fiber gratings, *IEEE Trans. Microwave Theory Tech.*, Vol. 45, No. 8, pp. 1531–1536, 1997.

29. D. B. Hunter, M. E. Parker, and J. L. Dexter, Demonstration of a continuously variable true-time delay beamformer using a multichannel chirped fiber grating, *IEEE Transactions on Microwave Theory and Techniques*, Vol. 54, No. 2, pp. 861–867, 2006.

30. B. Zhou, X. Zheng, X. Yu, H. Zhang, Y. Guo, and B. Zhou, Optical beamforming networks based on broadband optical source and chirped fiber grating, *IEEE Photonics Technol. Lett.*, Vol. 20, No. 9, pp. 733–735, 2008.

31. R. Kashyap, *Fiber Bragg Gratings*, Academic Press, New York, 1999.

32. P. Q. Thai, A. Alphones, D. R. Lim, and W. K. H. Vincent, Simplified optical dual beamformer employing multichannel chirped fiber grating and tunable optical delay lines, in *The 2008 IEEE Topical Meeting on Microwave Photonics*, jointly held with the *2008 Asia-Pacific Microwave Photonics Conference*, 2008.

33. R. Rotman, O. Raz, and M. Tur, Requirements for true time delay imaging systems with photonic components, in *IEEE International Symposium on Phased Array Systems and Technology*, 2003, pp. 193–198.

34. B. Zhou, X. Zheng, X. Yu, H. Zhang, Y. Guo, and B. Zhou, Impact of group delay ripples of chirped fiber grating on optical beamforming networks, *Optics Express*, Vol. 16, No. 4, pp. 2398–2404, 2008.

35. P. Q. Thai, A. Alphones, and D. R. Lim, Limitations by group delay ripple on optical beam-forming with chirped grating, *IEEE J. Lightwave Technol.*, Vol. 27, pp. 5619–5625, 2009.

36. R. A. Minasian, K. E. Alameh, and N. Fourikis, Wavelength-multiplexed photonic beam-former architecture for microwave phased arrays, *Microwave Optical Technol. Lett.*, Vol. 10, No. 2, pp. 84–88, 1995.

37. B. M. Min Jung and J. A. Yao, Two-dimensional optical true time-delay beamformer consisting of a fiber bragg grating prism and switch-based fiber-optic delay lines, *IEEE Photonics Technol. Lett.*, Vol. 21, No. 10, pp. 627–629, 2009.

38. P. Q. Thai, A. Alphones, and D. R. Lim, A novel simplified dual beam-former using multichannel chirped fiber grating and tunable optical delay lines, *J. Lightwave Technol.*, Vol. 26, No. 15, pp. 2629–2634, 2008.

39. B. Stephane, A. Mehdi, G. Karl, Q. Morgan, and M. Thomas, Optical Multibeamforming network based on WDM and dispersion fiber in receive mode, *IEEE Trans. Microwave Theory Tech.*, Vol. 4, No. 1, pp. 402–411, 2006.

40. S. Granieri, M. Jaeger, and A. Siahmakoun, Mulitple-beam fiber-optic beamformer with binary array of delay lines, *J. Lightwave Technol.*, Vol. 21, No. 12, pp. 3262–3272, 2003.

CHAPTER 10

ADAPTIVE ANTENNA ARRAYS IN RFID

MATTHEW TRINKLE and BEHNAM JAMALI
Department of Electrical and Electronic Engineering, The University of Adelaide, Adelaide, Australia

10.1 INTRODUCTION

RFID systems are used for automatic identification of objects in applications such as asset tracking, surveillance, and security. RFID systems consist of three main components: (a) a reader that sends an interrogation signal to the RF tag or transponder, (b) the RF tag itself, and (c) the middleware, which refers to the processing and hardware required to process the received signal, including the software protocols. To respond to the interrogating signal, the RF tag either needs to be powered up by the interrogation field so that it can reply with its unique coded message, or it backscatters a coded message. In low-frequency systems the magnetic coupling of the interrogation field is usually used to power up the tag, while in ultra-high-frequency (UHF) systems the interrogation signal is treated as a far-field electromagnetic wave and the tag responds by backscattering a coded message. There are several challenges in improving the reliability and range of the RFID systems. Firstly, enough energy needs to be transmitted to the tag for it to power up the on-board RF chip to respond to the reader; secondly, the reader needs to be sensitive enough to detect the transmitted or backscattered signal; thirdly, RF tags are often used indoors, which gives rise to a strong multipath environment causing signal fading in both the transmit and receive path. Finally, the polarization of the RF tag, transmitter, and reader should be matched for the best sensitivity.

One way to help overcome some of the above limitations is to use multiple antenna elements. By spatially distributing the transmit and receive antenna elements, the problem of signal fading can be reduced, because the areas of signal fading tend to be spatially localized. The key issue is how to distribute the transmit and receive signal over all the available antenna elements. The transmit and receive antennas can be considered in isolation or as a complete system. If we first consider only the

Handbook of Smart Antennas for RFID Systems, Edited by Nemai Chandra Karmakar

transmit antennas, then the transmit signal should ideally be distributed among the antennas to maximize the power at the tag location. This occurs if the signals add in-phase at the tag location. Similarly, if we consider only the receive antennas, then the signals from the tag should ideally be added in-phase to best enhance the signal-to-noise ratio (SNR) of the return signal. Finally, the overall system can be optimized by considering it as a multiple-input multiple-output system (MIMO). To optimally distribute the transmit and receive signals requires the phase and amplitude of each signal to be adjusted. This signal distribution and weighting is generally referred to as beamforming, and the amplitude and phase shift that are applied in each channel as the beamformer weights. These beamformer weights should ideally be adjusted in response to the environment, including the tag position, interference sources, and multipathing. Changing the weights based on the environment is generally termed adaptive beamforming, and a suitable adaptive algorithm needs to be employed to adjust the weights in response to the environment, usually based on data collected from the environment.

The amplitude and phase weighting can be done in either the analog or digital domain. Applying them in the analog domain requires variable phase shifters and variable gain attenuators/amplifiers to be implemented in each channel, and an RF summing or distribution network. An example of such a system is given in reference 1. Applying the weights in the digital domain simply requires a digital multiplication operation to implement the weighting function. However, a separate digital receiver including an ADC converter and RF front end is also required per antenna element. In the past, analog beamformers using phase shifters appear to have been the preferred method; however, as the cost of digital signal processing and ADC converters continues to drop, it is becoming more feasible to duplicate the receivers for each antenna element and do the array processing in the digital domain. Applying the phase shifts in the digital domain also has other advantages, such as being able to form many beams at the same time at minimal additional cost. All the beamformers with different weighting functions can use the same digital receivers and simply need to apply different weighting functions that can be implemented using digital multipliers.

The actual antenna elements themselves are also critical because the overall radiation pattern of the antenna array is the product of the radiation pattern of each element and the "array factor," which refers to the interference pattern generated by the superposition of the signals from each antenna.

The algorithm to control the weighting functions can simply consist of switching between antenna elements to more complex schemes that adjust both the gain and phase of each element. Simplified algorithms are particularly relevant to analog beamforming techniques where the phase and amplitude control is carried out with analog components and hence a simplification in the algorithm also results in a simplification of the overall hardware. A good overview of analog techniques is given in reference 2 and in Chapter 21 of reference 3. In the case of digital beamforming, optimal control algorithms can be implemented more simply, because all the calculations and weight updates occur in the digital domain.

This chapter considers adaptive beamforming techniques applied in the digital domain because these are the most flexible. As discussed above, beamforming techniques can be used to improve the power available to the tag by beamforming the transmitted signal, and they can also be used to improve the Signal-to-noise ratio (SNR) of the received signal by beamforming on the receive side. Recently, MIMO techniques have also been applied to RFID systems, which seek to optimze the overall performance of a multi-antenna transmit/receive system as well as incorporate multiple antennas on the tags [4]. Finally, antenna arrays can also be used to localize the position of the RF tag [5–7]. The adaptive antenna array is implemented as part of either the transmit or receive chain in the RFID reader.

This chapter will focus mainly on beamforming the received signal from the tag using relatively standard adaptive beamforming techniques. Experimental work has, however, suggested that it may be more beneficial to apply the beamforming on the transmit side, because the main factors limiting the range to the tag appear to be getting enough power to the tag for it to power up and respond. Once the tag does respond, the signal tends to be strong enough for a single antenna to receive it with sufficient SNR.

This chapter is organized as follows: Firstly, the main principle of an adaptive antenna array is introduced, and then several ways in which it may be applied to the RFID receive chain are summarized. Finally, the results of some data collected from an experimental system are discussed.

This section mainly considers RF tags in the 920-MHz band (902–928 MHz). Two main systems will be considered: The first is a chip-based RF tag based on a TDM (Time Division Multiplexing) principle where the reader first interrogates the tags and then sends a constant sine wave signal to energize the tag while it responds to the reader. A signal bandwidth of 500 kHz will be assumed which is only 0.05% of the carrier frequency. Frequency-domain tags will also be considered. These tags do not have a chip, but the data are encoded by the frequencies at which they respond to the reader. Because these tags do not have an RF chip that needs to be powered up before they respond, they can potentially operate at lower powers if the receiver is sensitive enough. As a result, the extra sensitivity obtained from a receive antenna array may be more useful for this class of tags (see Chapter 1).

10.2 ADAPTIVE ANTENNA ARRAY CONCEPT

Adaptive antenna arrays achieve an adjustable radiation pattern by combining the signals from multiple antenna elements with different phase and amplitude weightings. The final radiation pattern will depend on the choice of the weighting factors that allow nulls and beams to be steered in any desired direction. The weightings are usually calculated based on data captured from the environment in which the array is operating, and hence the radiation pattern adapts to the signal environment. A good overview of adaptive antenna arrays is given in reference 8. Adaptive antenna arrays usually adjust the weights to optimize some predefined optimality criterion, such as maximizing the SNR of a desired signal or minimizing the output power subject to having a unity gain constraint in the direction of the desired signal.

The adaptive antenna array is implemented as part of the receive chain in the RFID reader, as shown in Figure 10.1. It can be used to improve the SNR of the received signal as well as determine the direction of arrival of the RFID signal.

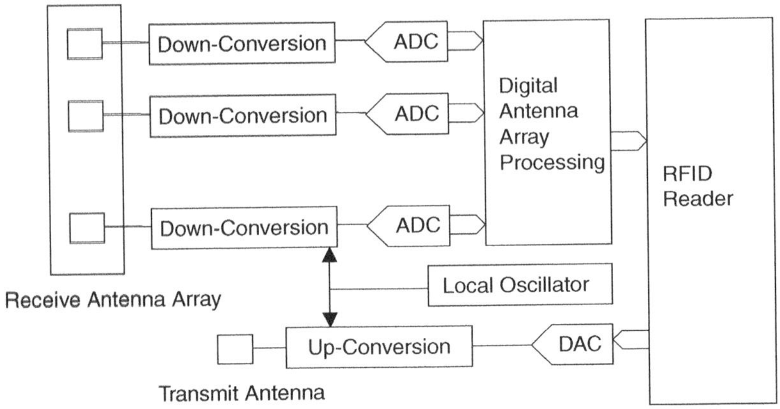

FIGURE 10.1. RFID reader with adaptive antenna array processing.

During the reading phase a sine wave will be transmitted to energize the tag. It is usually assumed that this transmitted sine wave will be used to demodulate the incoming signal, and the strong direct path component will thus be modulated to DC and removed by high-pass filtering in the down-conversion block prior to digitization by the ADC. A block diagram of a typical down-conversion chain is shown in Figure 10.2.

An alternative down-conversion block, using an IF (intermediate frequency) sampling ADC, is shown in Figure 10.3.

The IF sampling architecture would require a different local oscillator frequency from the transmitter. Also, a digital demodulator would need to be implemented after the ADC to convert the signal to baseband. The first bandpass filter removes the image frequency of the mixer, and the second bandpass filter is used for anti-aliasing; it can select any of the Nyquist bands of the ADC ranging from either $f_s/2$ to f_s, f_s to $3f_s/2$, $3f_s/2$ to $2f_s$, and so on, where f_s is the ADC sampling rate.

A possible implementation of the up-conversion block is shown in Figure 10.4.

An antenna array could also be implemented on the transmit side, and the transmit beam could be swept over the area to be covered. This may be a more useful system

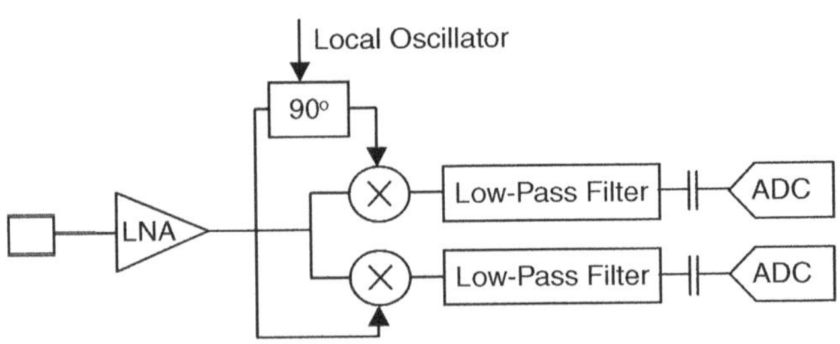

FIGURE 10.2. RFID down-conversion block.

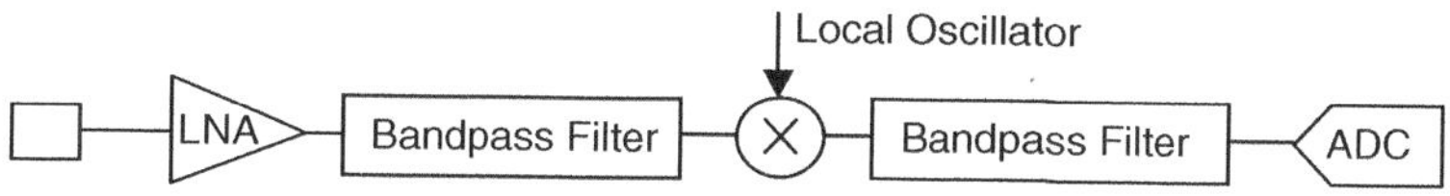

FIGURE 10.3. RFID down-conversion block.

in practice, because for a chip-based tag the main limitation in range to the tag appears to be due to insufficient power getting to the tag for it to power up and respond. The conventional beamforming techniques discussed in this chapter could be applied to this transmit antenna array. Finally, MIMO techniques are also currently being investigated that use multiple transmit and receive antennas, as well as multiple antennas on the actual tag [4]. These systems try to optimize the overall performance of the system rather than considering the receive and transmit arrays in isolation and should thus give better overall performance, at the expense of additional complexity.

Frequency-domain tags do not require a threshold power before they respond, and thus they may be more suitable to adaptive antenna array processing on the receiver side compared with chip-based tags.

10.2.1 Digital Antenna Array Processing

Antenna array processing algorithms combine the signals from all the antenna channels to either enhance the SNR of the desired signal via beamforming or determine the direction of arrival of the signal. First, signal enhancement via beamforming will be considered by introducing the narrowband beamforming structure followed by algorithms for determining the beamformer weights. Next algorithms for direction finding will be considered. Signal enhancement via beamforming is usually accomplished by either adding the desired signal in-phase, maximizing the SNR directly, or minimizing the output power of the array subject to a unity gain constraint in the direction of the desired signal [8]. A large number of direction-finding algorithms can be applied to antenna arrays; many of the more advanced ones are based on the Multiple Signal Classification (MUSIC) [9] algorithm.

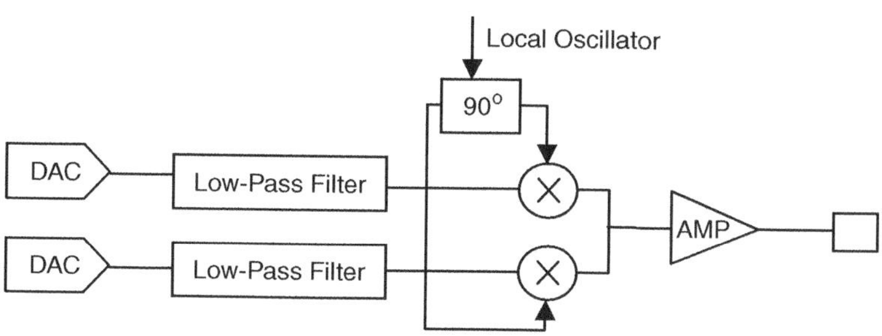

FIGURE 10.4. Up-conversion block.

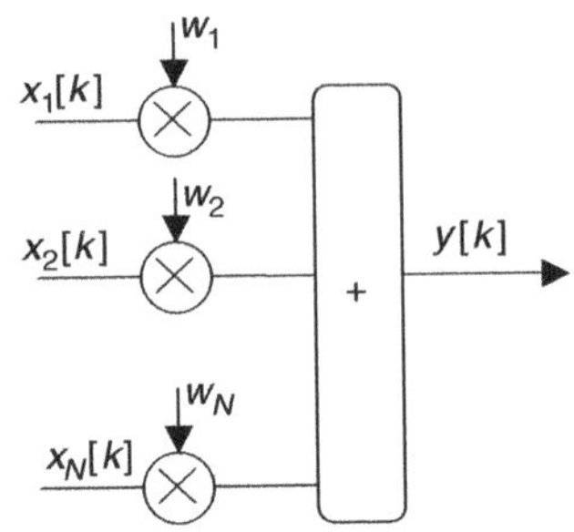

FIGURE 10.5. Beamformer architecture.

10.2.2 Beamforming Architecture and Mathematical Model

The basic beamformer architecture for an N element antenna array is shown in Figure 10.5.

The input channels, $x_n[k]$, are the inputs to the beamformer, consisting of complex I, Q data taken from each antenna channel after down-conversion with the local oscillator signal, and $y[k]$ is the beamformer output signal with improved SNR. The local oscillator signal is assumed to be the same sinewave signal used to energize the tag, and the DC offset produced by the direct path signal into the antenna array is assumed to be removed prior to digitization and beamforming.

The input signal to the beamformer is assumed to be the signal received at each antenna plus thermal noise. Thus,

$$x_n[k] = s[k - \tau_n]e^{-j2\pi\tau_n/\lambda} + n_n[k]$$

where $s[k]$ is the signal sent by the tag, λ is the wavelength of the carrier, and τ is the signal delay and is usually assumed to be zero at the first antenna. Hence τ_n corresponds to the relative signal delay between the nth antenna and the first antenna. $n_n[k]$ is complex random thermal noise with standard deviation σ. In general, the bandwidth is assumed to be significantly less than the carrier frequency leading to the narrowband assumption:

$$x_n[k] = s[k]e^{-j2\pi\tau_n/\lambda} + n_n[k]$$

This assumption is valid for the 900-MHz RFID system under consideration in this chapter, which has a fractional bandwidth of 0.05%.

Thus the signal in each antenna is assumed to be the same except for a different complex phase shift and independent random noise.

The output of the beamformer is obtained from the complex sum:

$$y[k] = \sum_{n=1}^{N} w_n^* x_n[k] = \mathbf{w}^H \mathbf{x}(k)$$

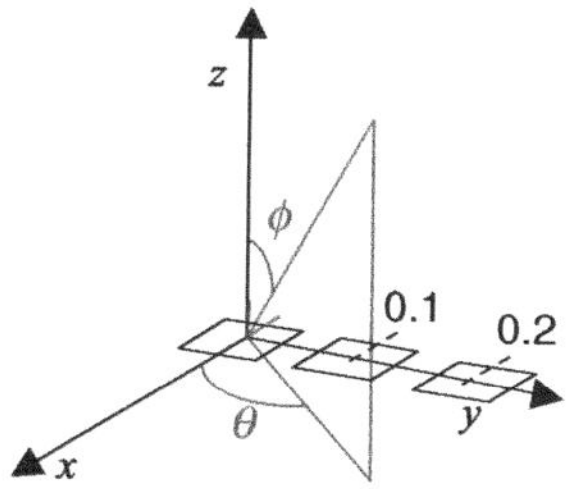

FIGURE 10.6. Assumed coordinate system.

where $\mathbf{w} = \begin{bmatrix} w_1 \\ \vdots \\ w_N \end{bmatrix}$ is a vector containing the complex weights of the beamformer and $\mathbf{x}(k) = \begin{bmatrix} x_1[k] \\ \vdots \\ x_N[k] \end{bmatrix}$ are the complex inputs to the beamformer.

The gain of the array in each direction depends on the geometry of the antennas as well as the choice of the complex weights. In general, to maximize the SNR of the desired signal, the complex weights should be chosen to add the desired signals in phase. This will maximize the array gain in the direction of the desired signal in the presence of thermal noise. Several methods for choosing the complex weights will be considered in the following sections.

Assuming that the antennas are placed at positions $u = \begin{bmatrix} x_1 & \cdots & x_N \\ y_1 & \cdots & y_N \\ z_1 & & z_N \end{bmatrix}$ and that the source is in the far field of the array, the expected phase shifts between antenna 1 and the remaining antennas are

$$\phi = k_s^T u$$

where $k_s = \frac{2\pi}{\lambda}\begin{bmatrix} \cos(\theta)\sin(\phi) \\ \sin(\theta)\sin(\phi) \\ \cos(\phi) \end{bmatrix}$ and θ and ϕ are the azimuth and elevation angles, as shown in Figure 10.6, which also shows a linear array in the y axis. The position vector for this array would be

$$u = \begin{bmatrix} 0 & 0 & 0 \\ 0 & 0.1 & 0.2 \\ 0 & 0 & 0 \end{bmatrix}$$

A common rule of thumb for the far field is that the distance d between the source and antenna should be given by $d > 2a^2/\lambda$; where a is the array aperture, which is given by $\lambda(N-1)/2$ for a linear antenna array spaced by half a wavelength. Hence $d > \lambda(N-1)^2/2$, or 6 m for a seven-element array at 902 MHz.

10.2.3 Conventional Beamforming

The conventional beamformer simply adds the desired signals on each channel in phase. Thus the complex weights are chosen to equal the expected phase shifts between the antenna elements, that is,

$$\mathbf{w} = \begin{bmatrix} w_1 \\ \vdots \\ w_N \end{bmatrix} = e^{i\mathbf{u}^T\mathbf{k}_s}$$

To calculate these phase shifts, the source will be assumed to be in the far field of the array, and thus the incoming wave is assumed to be a plane wave front across all antenna elements. The DOA of the tag also needs to be known along with the antenna array geometry. Because the tag DOA may not be known beforehand, it may be useful to run multiple beamformers in parallel, each looking in a different direction [10]. The look directions can be evenly spaced over the entire area that needs to be scanned. The number of beam directions will depend to the width of the beam. If the beam is pointing directly at the tag, conventional beamforming can provide an SNR enhancement up to a factor of N, where N are the number of antennas in the array. In general, to avoid a loss of more than 4 dB between beams, at least N beams will be required for an N-element antenna array [10]. The outputs of each beamformer can be monitored, and the one with highest power would be pointing at the tag. Note that only the signal processing needs to be replicated for each beamformer, not the antenna elements or RF down-conversion electronics or the ADCs.

The beam patterns of a seven-element linear antenna array spaced along the y axis at half-wavelength spacing is shown in Figure 10.7. The beam patterns of seven orthogonal conventional beamformers are shown. These seven beamformers should cover the entire angular space. Each orthogonal beamformer is chosen such that its response has a null in the look direction (maximum response) of all other beamformers. Note that at most 4-dB loss is incurred from the maximum response of any beamformer. Each beamformer would use the same antenna array and ADCs but a different weighting function, depending on the desired steering direction. The beamformers can all run in parallel. As the number of antenna elements increases, the beamwidth will become narrower and hence more beams will be required to cover the angular space. Note that the beamwidth is narrower at 90° to the array and gets broader toward the ends of the array. Thus the beams should be more dense at 90° to the array.

The beam patterns in Figure 10.7 are for an ideal array. In practice, the array will have phase and amplitude errors that will degrade the array pattern. Also, multipathing signals are a potential problem. The impact of multipath signals will depend on the degree to which they are correlated with the direct signal. Highly correlated multipath signals cannot be easily distinguished from the direct signal and will result in a curving of the wave front, resulting in further phase and amplitude errors that are also direction-dependent and hence difficult to calibrate. Multipath signals become correlated once the time–bandwidth product becomes less than unity. For a 500-kHz bandwidth signal, this would occur at a delay corresponding to 600 m. Because the actual delays are expected to be much less than this, the multipath components are expected to be highly correlated; and depending on the level of multipathing, these could result in significant phase errors as well as amplitude fading or enhancement.

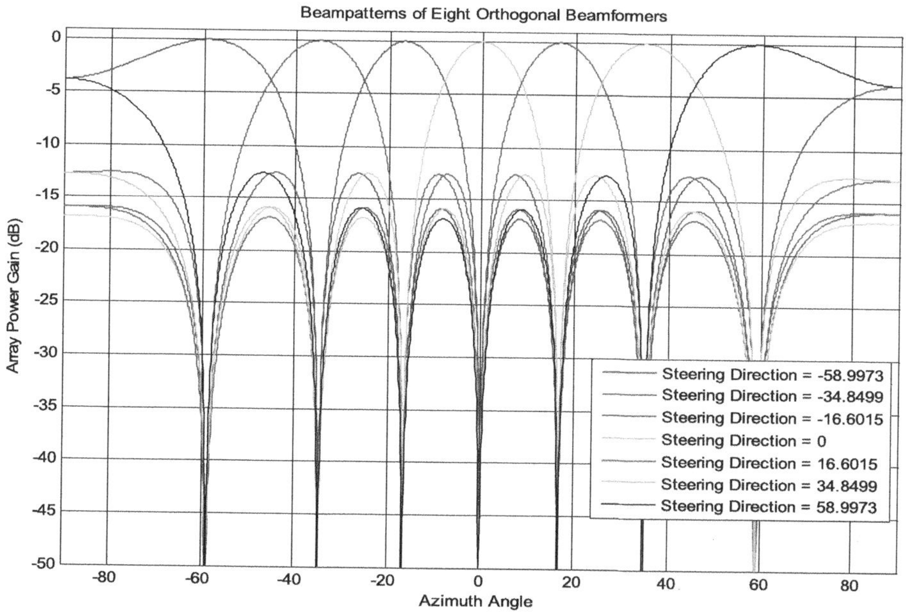

FIGURE 10.7. Parallel conventional beamformers.

10.2.4 Adaptive Beamforming

Adaptive beamforming can be used to automatically point the beamformer at the tag as well as remove interferences. In adaptive beamforming, the complex weights are calculated based on the input data $\mathbf{x}[k]$, and as a result the beam pattern is responsive to the signal environment. Thus these techniques also have the ability to adapt to phase errors in the array and phase errors introduced by signal multipathing.

A block diagram showing the basic operation and control flow of an adaptive beamformer is shown in Figure 10.8.

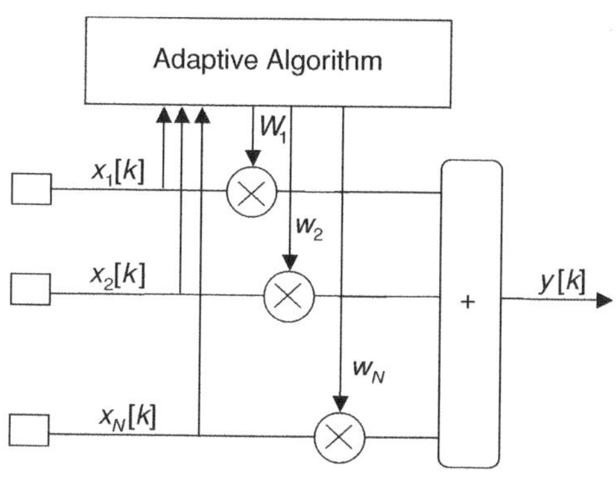

FIGURE 10.8. Adaptive beamformer architecture.

The adaptive algorithm accepts a block of data from the outputs of the antennas and uses them to calculated the beamformer weights that will best optimize the predefined optimality criterion. Two adaptive criterion appear to be particularly suited to RFID: the maximum SINR beamformer and the linearly constrained minimum variance (or LCMV) beamformer. A good overview of these techniques and beamforming in general can be found in reference 8.

The maximum SINR beamformer seeks to adjust the complex weights to maximize the SINR of the received signal directly. It requires no information apart from two data sets, the first containing only background noise and interference and the other containing background noise, interference, and the desired signal. It then uses these two data sets to adjust the complex weights to maximize the SINR in the output of the beamformer. Note that in RFID tags these two data sets can be obtained relatively easily because the tag will only transmit when interrogated, hence any data set that is captured before or after interrogation contains only noise and interference. It is suggested that the noise and interference data set be collected directly before or after interrogation so that the noise and interference statistics remain the same. This algorithm will automatically steer a beam at the tag as well as cancel interferences from other sources. Capturing the noise and interference only data may increase the overall reading time; however, depending on the amount of interference and how quickly the interference parameters are changing, the noise statistics may not need to be updated very frequently.

The algorithm automatically estimates the phase shifts of the desired signal across the array elements; as a result, any phase errors either due to the antenna elements or multipathing are automatically calibrated. Also, the algorithm should work in either the near or far field of the array, and the antenna locations do not need to be known, thereby giving more flexibility in the placement of antenna elements.

The beamformer weights that satisfy the maximum SINR criteria are given by the eigenvector corresponding to the maximum eigenvalue of the matrix: $R_{ni}^{-1} R_{nid}$, where R_{ni} is the noise plus interference covariance matrix captured while the tag was not transmitting, and R_{nid} is the covariance matrix while the tag was transmitting. The covariance matrix is defined as $R = E\{\mathbf{x}(k)\mathbf{x}(k)^H\}$. In practice the number of independent snapshots $\mathbf{x}(k)$ should be at least twice the number of antennas to accurately estimate the covariance matrix [11] to avoid a 3-dB loss in SNR. In practice a snapshot size of at least four times the number of antennas is thus recommended to make the loss in SNR due to parameter estimation negligible.

The LCMV beamformer [8] requires the direction of arrival of the desired signal to be known. It is similar to the conventional beamformer in that it steers a beam in the look direction but differs from the conventional beamformer in that it will also steer nulls at interferences coming from any direction other than the look direction. If there are no interferences, the LCMV beamformer is the same as the conventional beamformer. Hence its main advantage over the conventional beamformer is for interference mitigation. The LCMV beamformer calculates the complex weights by constraining the beam pattern to have unity gain in the look direction and then adjusting the complex weights to minimize the output power.

The beamformer weights that satisfy the LCMV criterion [8] are given by

$$\mathbf{w} = \frac{R^{-1}\mathbf{v}}{\mathbf{v}^H R^{-1}\mathbf{v}}$$

where the R matrix typically includes contributions from interference, noise, and the desired signal and $\mathbf{v}$ is the steering vector with the expected phase shifts of the desired signal across the array elements. These phase shifts can be calculated in the same manner as for the conventional beamformer, with the assumption that the source is in the far field of the array. This algorithm thus requires the phase shifts of the desired signal across the array elements to be known a priori, which is more difficult in the near field as the effect of range and DOA on the relative phase shifts become interdependent. The algorithm is also known to be quite sensitive to phase errors, or errors in the DOA of the desired signal, because the desired signal will also be canceled if it does not come exactly from the look direction. However, in RFID this problem can be avoided by using the data when the tag is not transmitting to calculate the complex weights. In this case, only interference and noise signals will be canceled by the minimization algorithms.

If the DOA to the tag is not known, multiple MVDR beamformers may need to be run in parallel to cover the entire scan area similar to the conventional beamformer system.

Overall, the maximum SINR beamformer appears to be the best choice because it does not require the DOA of the tag to be known and the antenna gain and phase errors are automatically compensated by maximizing the SINR directly. Also, there is no extra complexity if the tag is in the far or near field of the antenna array. Finally, the antenna positions do not need to be known by the algorithm, thereby allowing more freedom in how they are placed. However, if the system is to be used for direction finding, the other techniques may be more valuable and the antenna phase errors need to be estimated. When using multiple conventional or MVDR beamformers a rough direction estimate can be obtained directly from the steering direction of the beamformer with the maximum output power.

10.3 DOA ESTIMATION

Adaptive antenna arrays can also be used to estimate the DoA of the signal from the tag. Similarly to the beamforming algorithms discussed in the previous section, most commonly used DoA estimation algorithms assume that the wavefront is a plane wave. This means that the tag should be in the far field of the antenna array.

The conventional beamforming and LCMV algorithms can be used for DoA estimation, by scanning the beam over all possible angles and choosing the direction at which the maximum power is observed. However, more advanced DoA estimation algorithms such as MUSIC that are directly tailored to DoA estimation are usually applied [12]. The basic MUSIC algorithm looks for steering vectors, $\mathbf{v}$, that are orthogonal to the noise subspace. Assuming that there are M signals incident on the array, this is achieved by searching for the M peaks in the DoA spectrum:

$$\frac{1}{\mathbf{v}^H U U^H \mathbf{v}}$$

where $\mathbf{v}$ is the steering vector with the expected phase shifts of the desired signal across the array elements for the scanning angle, and U is an N by N–M matrix, containing the N–M eigenvectors corresponding to the N–M smallest eigenvalues of the receiver covariance matrix R. N is the number of antennas in the array.

DoA estimation algorithms such as MUSIC are sensitive to array phase errors. These can be introduced due to either the antenna patterns or signal multipathing; also, most DoA estimation algorithms assume that the signal is in the far field of the array. For a seven-element array, this is over 6 m at 920 MHz. In the near field, the effect of DoA and range on the relative phase shifts between antennas become interdependent, thereby significantly increasing the complexity of the problem. In reference 5 the problem of multipathing is addressed by taking measurements at different frequencies to enable the different paths to be resolved. The algorithm was applied to simulated data but does not appear to have been validated with real data. In references 6 and 7 a two-antenna system is proposed to simplify the near-field problem. In reference 6, experimental results are presented, showing reasonable DOA estimation performance. A standard deviation of 2.3° was observed in the DOA estimate before calibrating the phase response of the receive and transmit array, and 1.4° was observed after calibration. These results may not be directly relevant to a normal room with significant multipathing, because it appears from the photos in the paper that the experiment was carried out in a room with special anechoic material to remove reflections.

10.4 EXPERIMENTAL RESULTS

Experiments have been carried out to investigate the effectiveness of the maximum SINR algorithm for SNR enhancement. A four-element antenna array was set up together with a 920-MHz harmonic tag.

Harmonic tags, also referred to as frequency-domain tags, are classified as passive RFID tags [13]. A harmonic tag receives RF energy at one frequency and emits EM waves at higher frequencies that are harmonically related to the received frequency. For this experiment a harmonic tag was used because it has a longer read range than do micro-chip-based tags. Also, a frequency-domain tag operates at much lower power levels and responds at lower received power levels than does a micro-chip-based tag, and thus the extra sensitivity obtained from an antenna array is expected to be more useful. Figure 10.9 shows a block diagram of a typical harmonic tag that receives signals at 920 MHz and emits a strong second harmonic at 1.84 GHz.

A picture of the actual tag used in the experiment is shown in Figure 10.10.

The key idea behind frequency-domain tags is that the information is encoded in the frequencies at which they respond. The above tag is effectively only a single-bit

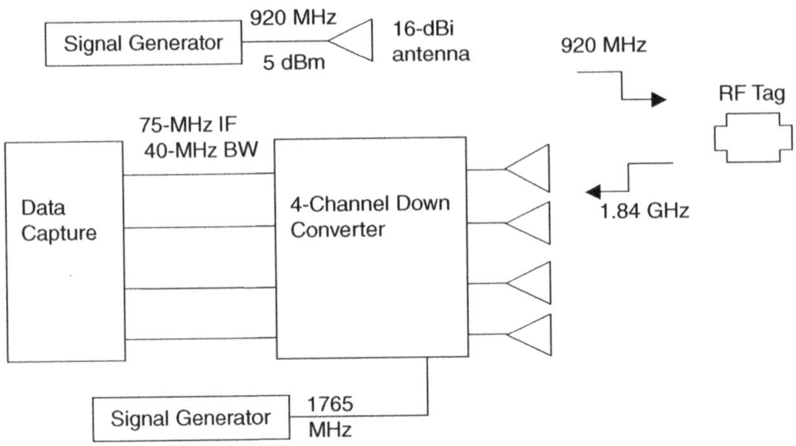

FIGURE 10.11. Experimental RFID system.

sampling, the signal is down-converted to baseband in the digital domain and filtered to a 0.5-MHz bandwidth. A 100,000-point data block is captured, corresponding to 0.2 ms.

The spectrum of the output data is shown in Figure 10.12. The response from the RFID tag is at 0 MHz, because the 1.84-GHz signal has been converted to baseband in the digital domain. The spectrum of each antenna channel is also shown in Figure 10.12, and the combined beamformed output is shown by a solid line.

The maximum SINR beamformer algorithm was used in this experiment. To calculate the beamformer weights, two covariance matrices are required: (a) the covariance matrix containing only noise and interference R_{ni} and (b) the covariance matrix including noise, interference, and the tag signal R_{nis}. In practice, R_{ni} would be obtained directly before the tag is energized, and R_{nis} after the tag is energized. In this experiment we did not have automatic control over the data capture process and transmit signal, and thus it was not possible to capture a data block directly before and after energizing the tag. To overcome this problem, a single data block was captured and the useful signal was removed from the first half of the data block by temporal filtering. This data block was then used to generate R_{ni}; and the second half of the data block, which still contained the useful signal, was used to generate R_{nis}.

As can be seen from Figure 10.12, the beamformer improves the sensitivity of the receiver by removing the interference (or noise) below the tag response frequency and by increasing the strength of the tag response.

A picture of the experimental setup is shown in Figure 10.13, showing the transmit antenna, receive array, down converter, and digital oscilloscope.

A similar experiment was carried out with a chip-based tag, which also operates at 920 MHz but responds with a binary signal also modulated at 920 MHz. In this experiment a different down-converter was used with an intermediate frequency of 22 MHz and a bandwidth of several megahertz. The signal is then converted to baseband in the digital domain, giving the only the baseband-modulated square-wave

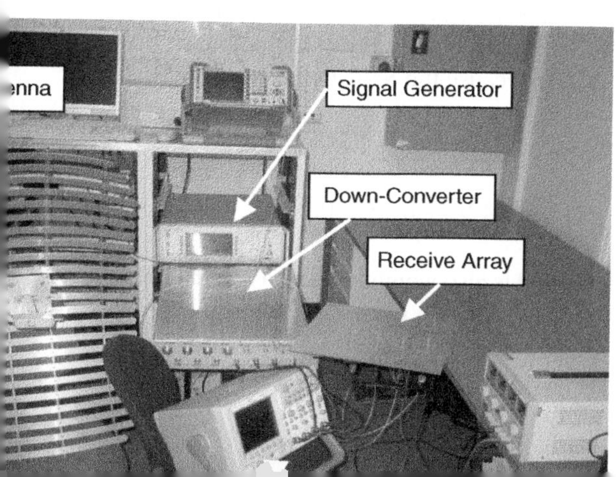

enna
Signal Generator
Down-Converter
Receive Array

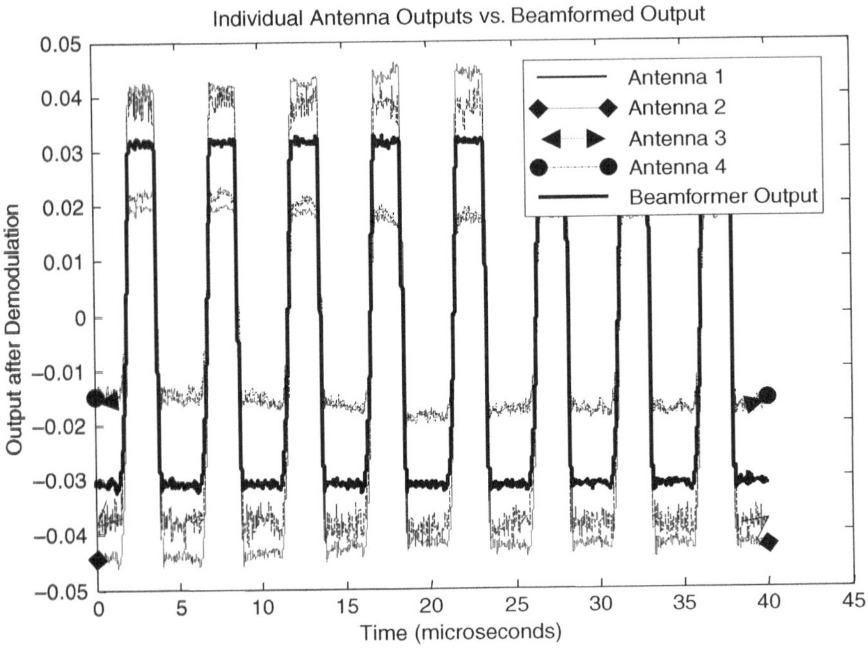

FIGURE 10.14. Output RFID signal.

signal transmitted by the tag. The baseband signal obtained from each antenna element is shown in Figure 10.14. These results show a relatively good SNR in each antenna channel; thus an antenna array is not really required to enhance the SNR further. The tag was moved further from the transmitter to try and reduce the SNR of the signal, but it was found that it could not be moved much before the tag did not get enough power to respond. As a result, the maximum SNR beamforming algorithm was applied to the relatively high SNR data to see if it could further improve the SNR.

To implement the maximum SNR beamforming algorithm, two data blocks were collected, one while the tag was transmitting and the other while it was turned off. Based on these two data blocks, the maximum SINR beamformer weights were calculated and applied to the antenna array. The results are shown in Figure 10.14.

From Figure 10.14 the maximum SINR beamformer is of little benefit in this scenario, and no significant improvement in SNR is observable at the output of the beamformer compared with antennas 2 and 3, because the SNR in these channels is already quite high.

The beamformer weights for the above scenario were as follows:

$$\mathbf{w} = [0.016,\ -0.24 - 0.32i,\ 0.18 - 0.42i,\ -0.13 + 0.049i]$$

Note that antenna 1 has the nosiest signal and is also weighted the least in the final beamformer solution. The noise in antenna 1 was actually mainly due to the dynamic range of the data acquisition hardware, but in practice a poor SNR may be observed on some of the antennas due to signal fading. The final beamformer output has the highest contribution from antennas 2 and 3, which also have the highest SNR.

It was found that using the maximum SINR beamformer did not increase the range to the RFID tag in the experimental scenario. The range was rather limited by the transmit power of the transmitting antenna. The tag had to be brought quite close to the transmitting antenna, which was co-located with the receive array, for the tag to power up. At this close range the reflected signal from the tag was quite strong and produced a good SNR in each of the antenna channels as shown in Figure 10.14. This is likely to be the case in most scenarios, and hence for chip-based tags it may be more beneficial to incorporate the beamforming algorithms into the transmit antenna, or consider a complete MIMO system as discussed in reference 4 to improve the range to the tag.

10.5 CONCLUSIONS

This chapter has considered the application of adaptive beamforming algorithms to the receive signal chain in RFID readers. An adaptive antenna array can potentially improve the robustness and range of an RFID reader by increasing the sensitivity of the receiver and by mitigating strong interferences. It can also be used to find the direction of arrival of the signal from the RFID tag.

The main focus of this chapter has been the use of adaptive antenna arrays to improve the SNR of the backscattered signal from the RFID tag. To do this, the basic theory of adaptive antenna arrays was introduced and standard adaptive beamforming algorithm that maximizes the SNR of the backscattered tag signal was applied to adjust the beamformer weights. This algorithm has been applied extensively in other areas and also appears to be well-suited to RFID. The algorithm requires two data blocks, one with the desired backscattered signal and one without. This can be readily obtained from an RFID system by turning the interrogator on or off. The beamforming algorithm is simple to implement and automatically adjusts for antenna phase and gain errors, interference, and multipathing to maximize the SNR of the received signal. In the absence of an interferene, the algorithm can improve the SNR of the tag by up to a factor of N, where N is the number of antenna elements in the array. But in the presence of an interference, the gain in SNR can be much more as a strong interference can be completely canceled by the beamformer.

Experimental results from two RFID systems were collected, one using a chip-based tag that produces a backscattered signal by modulating the incoming signal using a chip that needs to be powered from the interrogation signal, and a second system using a frequency-domain-based tag whose information is encoded in the frequencies at which it responds. This tag does not require a chip, but uses resonant structures to re-radiate the signal at specific frequencies. It was found that the receive adaptive antenna array provided limited advantage for the chip-based tag, because the tag had to be brought very close to the reader for it to get enough power to enable the chip. At this range the received signal was very strong and the additional sensitivity of the antenna array was not really required. For this type of system, beamforming at the transmitter appears to be more useful because the main limitation on range appears to be getting enough power to the tag to power up the RF chip. The second RFID system based on the frequency-domain tag does not need to power an RF chip before it will respond, and hence it appears to be more useful for this system to increase the sensitivity of the receiver and hence the range of the overall system by using an adaptive antenna array in the receive chain.

The use of adaptive antenna arrays for direction of arrival estimation were briefly reviewed, but no experiments were carried out in this area.

REFERENCES

1. P. Salonen and L. Sydanheimo, A 2.45 GHz digital beam-forming antenna for RFID reader, in *Proceedings, 55th Vehicular Technology Conference*, Vol. 4, 2002, pp. 1766–1770.
2. C. Sun, A. Hirata, T. Ohira, and N. C. Karmakar, Fast beamforming of electronically steerable parasitic array radiator antennas: Theory and experiment. *IEEE Trans. Antenna. Propag.*, Vol. 52, No. 7, 1819–1832, 2004.
3. C. Sun, J. Cheng, and T. Ohira, *Handbook on Advancements in Smart Antenna Technologies for Wireless Networks*, Information Science Reference, illustrated edition, August 5, 2008.
4. J. D. Griffin and G. D. Durgin, Reduced fading for RFID tags with multiple antennas, in *IEEE Antennas and Propagation Symposium* 2007, Honolulu, Hawaii, June 2007.
5. Y. Zhang, X. Li, and M. G. Amin, Array processing for RFID tag localization exploiting multi-frequency signals, in *SPIE Symposium on Defense, Security, and Sensing*, Orlando, FL, April 2009.
6. J. Wang, M. Amin, and Y. Zhang, Signal and array processing techniques for RFID readers, in *SPIE Symposium on Defense and Security*, Orlando, FL, April 2006.
7. Y. Zhang, M. G. Amin, and S. Kaushik, Localization and tracking of passive RFID tags based on direction estimation, *Int. J. Antennas Propag.*, Vol. 2007, Article ID 17426, doi:10.1155/2007/17426, 9 pages, December 2007.
8. B. D. Van Veen and K. M. Buckley, Beamforming: A versatile approach to spatial filtering, *IEEE ASSP Mag.*, Vol. 5, No. 5, pp. 4–24, April 1988.
9. H. L. Van Trees, Optimum array processing, Part IV of *Detection, Estimation, and Modulation Theory*, John Wiley & Sons, New York, 2002.
10. Y. Zheng, Adaptive antenna array processing for GPS receivers, The University of Adelaide Master's Thesis, July 2009.
11. I. S. Reed, J. D. Mallett, and L. E. Brennan. Rapid convergence rate in adaptive arrays, in *IEEE Trans. Aerospace Electron. Systems*, Vol. AES-10, No. 6, pp. 853–863, November 1974.
12. W. J. Krzysztofik and M. Fafara, Comparison & Examination the Methods of Direction-of-Arrival Estimation, in *IEEE Antennas & Propagation Society International Symposium*, Vol. 4a, 3–8 July 2005, pp. 154–157.
13. B. Jamali and B. Bates, A passive harmonic reradiator tag for animal tracking, in *Smart Structures, Devices, and Systems IV*, Vol. 7268, SPIE, Bellingham, WA, 2008.

CHAPTER 11

DESIGN OF PORTABLE SMART ANTENNA SYSTEM FOR RFID READER: A NEW APPROACH

JEFFREY S. FU
Department of Electronic Engineering, Chang Gung University, Taoyuan, Taiwan, Republic of China

WEIXIAN LIU
School of Electrical and Electronic Engineering, Nanyang Technological University, Singapore

NEMAI CHANDRA KARMAKAR
Department of Electrical and Computer Systems Engineering, Monash University, Clayton, Victoria, Australia

11.1 INTRODUCTION

Wireless communication has become an important part of our daily lives. Wireless local area network (WLAN) has become popular due to its emerging trend as a communication tool for networking in offices and homes. The heavy usage of wireless communication causes rapid electromagnetic (EM) radiation pollution in our environment. With the heavy use of the system, users' expectations on the performance of the wireless tools have increased. A poor design or noisy surroundings might cause the wireless devices to encounter interferences resulting in a poor communication system. Smart antenna systems, if integrated into such devices, will improve its signal-to-interference ratio (SIR) by using beamforming [1, 2] of signals in the desired direction while nulling the interference signals. These unique capabilities of smart antennas improve the SIR and the capacity of the system. However, implementation of the smart antenna system requires very expensive analog-to-digital converters (A/Ds), digital signal processing (DSP) chips and system, antenna modules, and sophisticated algorithm. These large amounts of hardware and software

Handbook of Smart Antennas for RFID Systems, Edited by Nemai Chandra Karmakar

designs hinder its application to RFID system. Hence, the idea of the work is to use general-purpose receivers and DSP chipsets to perform the task similar to an expensive smart antenna system, but with a low-cost solution. If the system is built with a readily available and low-cost general-purpose DSP chip system and if the system is compromised with a simple but robust direction of arrival (DoA) and digital beamforming (DBF) algorithm for fast operation, then the system can be made viable for the mass RFID market. The chapter has addressed the physical implementation of such a low-cost and portable smart antenna system that can be implemented in the RFID reader. The development has been addressed in the following areas of the smart antenna development:

1. A less computationally intensive Capon algorithm [3] is used for DoA as well as for DBF using a four-element uniform linear array (ULA) antenna made of low-cost patch antenna panels. The Capon algorithm is much less computationally intensive compared with the more comprehensive MUSIC [4], Least Mean Square (LMS), and ESPIRIT algorithm [5]. Thus a significant cost-saving measure is possible in the implementation of such algorithm in less costly DSP chips. Also, the operation with the simpler Capon algorithm will enhance the speed of the system.
2. Next, the proposed system is based on the robust theoretical concept and practical implementation of general-purpose DSP chipset. The selection and programming of the DSP chip will bring down the cost of the system significantly and make the smart-antenna-based RFID reader system a viable commercial product.
3. The practical design of the smart antenna systems for the RFID reader is comprised of the following modules: (i) an HP8648D signal generator with an antenna to emulate the RFID tag (which will be replaced by the low-cost active and passive tags), (ii) a receiver that is comprised of a four-element ULA of four moderately high gain patch antenna panels, and (iii) four-channel common block RF receiver modules that are connected to the antenna at the input ends and the general-purpose DSP interface at the output ends. The receiver is synchronized with the amplitude and phase shift with an external reference oscillator. The system constitutes a coherent reader with known reference signals (frequency is set to f_0; amplitude and phase shift are synchronized). The task of the receive is to convert the 2.4-GHz microwave signals from the transmitter (in this case the pseudo-RFID tag) to the baseband signal for the DSP block for further processing.
4. Since the DSP used is a general-purpose chipset with limited capability in order to reduce the cost of the system, a multiplexer is used to enhance the I/O capabilities of the DSP processor. The multiplexer interfaces the 80-bit digital signal to the 16 I/O ports of the DSP block. Immediately after receiving the I/O data, the DSP block converts the in-phase (I) and quadrature phase (Q) components of the received signal and computes the DoA, and DBF estimates the weight vectors for individual antenna elements in the four-element ULA.

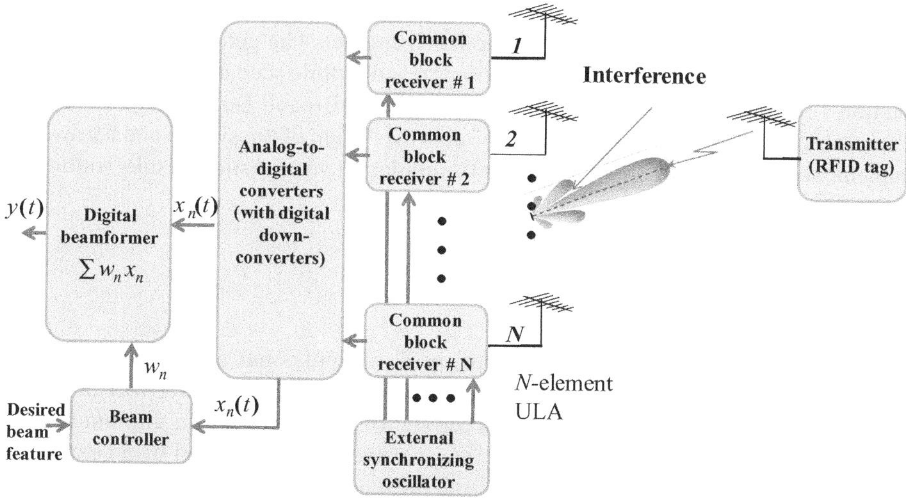

FIGURE 11.1. Proposed block diagram of the smart antenna for RFID reader.

The output of the DSP is also displayed via a light-emitting diode (LED) display unit, or it can be interfaced with a personal computer. A low-cost general-purpose Hitachi SH7615 DSP block is used as the low-cost solution of the smart antenna.

Figure 11.1 illustrates the proposed low-cost smart antenna system for the RFID reader. The practical implementation of the system successfully detects a desired tag at 30° angle of arrival (AoA) from the boresight direction of the four-element ULA. The motion tracking of the portable receiver as the handheld RFID reader is also investigated, and accuracy of about 10–20° within the DoA of the tag is detected. A compensation technique to improve the accuracy using the simple Capon algorithm is also proposed.

As stated in the introduction, the prime objective of the work is to make the smart antenna a commercially competitive product so that it can be mass-implemented in the RFID reader to improve the throughput and capacity of the RFID system. The proposed system is to be composed of cascaded general receivers that are being synchronized to an external reference oscillator. This approach makes upgrading of the system easy because more receivers can be added on to the reference oscillator easily.

In the development phase of the portable and low-cost smart antenna system, the immediate frequency (IF) signals are studied and analyzed first. Then simulation is performed on spectral-based DOA estimation algorithm such as Capon [3, 6]. The algorithm computes the baseband demodulated in-phase and quadrature (I/Q) signals from multiple incoherent sources. Simulations are carried out using LabView. The hardware setup of the receiver is implemented in the smart antenna laboratory. Finally the low-cost implementation issues and solution to improve the accuracy are envisaged.

The chapter is organized as follows: Section 11.1 introduces the proposed low-cost smart antenna system for the competitive RFID market. The general block diagram and practical implementation in both stationary and mobile scenarios are introduced. Section 11.2 presents the theoretical concept of the DBF and DoA estimation using Capon algorithm in brief. Section 11.3 presents the design of the system and hardware implementation. Section 11.4 presents both simulation and practical results followed by conclusion in Section 11.5.

11.2 THEORY

As mentioned above, the aim of this project is to implement a low-cost and less hardware-intensive smart antenna system for RFID applications that will be both cost-effective and portable. In order to keep the overhead costs to a minimum, the general-purpose DSP and cascaded common receivers, synchronized by an external reference oscillator are investigated. Since all the receivers shared a common external reference oscillator, it is assumed that the receivers are set for the same center frequency and bandwidth and that all phase shifts are identical.

The Capon algorithm [3, 6] is chosen as the basis of both DoA estimation and DBF implementations. It is a spectral-based beamforming algorithm, and it does not require any eigen decomposition, making the calculations less computation-intensive, at the expense of sacrificing some accuracy in estimation as compared to popular algorithms such as MUSIC, LMS, and ESPIRIT algorithms.

The objectives of the general-purpose DSP is to first determine a DoA estimation of the desired user based on the quadrature baseband I and Q components obtained from the receivers. Once the direction of the desired user is being recognized, the complex baseband signals are multiplied by the complex weights to apply the phase shift and amplitude scaling required for each antenna element.

Lastly, the results of each digital down-converter's (DDC) baseband output are multiplied by the complex weight for its antenna element; these are to be added together to produce a baseband signal with directional properties (DBF), at the same time nulling the potential interferences and thereby improving the SIR between the transmitter and receiver as illustrated in Figure 11.2a.

Assume that the demodulated in-phase (I) and quadrature (Q) signals in Eqs. (11.1) and (11.2) is corrupted with additive noise, $n(t)$. The expressions for the corrupted signal in our simulation are given by Eqs. (11.3) and (11.4). They will form four complex signal matrices:

$$I_K = A_K \cos(\omega t + \beta) \quad (11.1)$$

$$Q_K = A_K \cos(\omega t + \beta) \quad (11.2)$$

$$I_K = A_K \cos(\omega t + 2\pi dk \cos\varphi) + n(t) \quad (11.3)$$

$$Q_K = A_K \sin(\omega t + 2\pi dk \cos\varphi) + n(t) \quad (11.4)$$

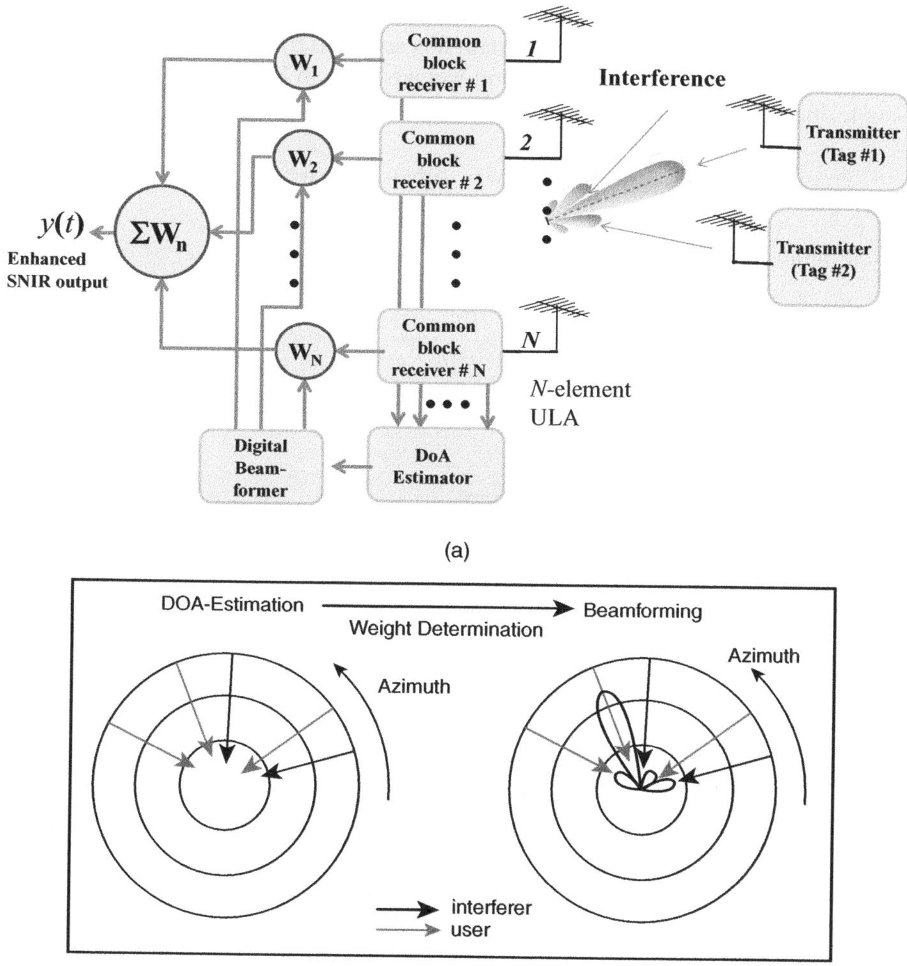

FIGURE 11.2. Adaptive beamforming for DoA system implementation for desired users and nulling interference. **(a)** Block diagram. **(b)** DoA and beamforming concept.

The computed complex signal matrix is the input to calculate the correlation matrix, which is able to describe how individual signals received by individual receivers are related. Hence the correlation matrix is crucial in determining the DoA of the desired signal. The complex signal matrix is comprised of k antenna elements, each with r samples of the time-domain signal. Equation (11.5) defines the power spectrum of the environment when the DSP does a DoA scan from 0° to 180°.

$$P_{\text{capon}}(\theta) = \frac{1}{e^H(\theta)R^{-1}e(\theta)} \tag{11.5}$$

where $e(\theta)$ is the steering vector and R is a special covariance matrix (the autocorrelation function) of the target. DoA estimation is achieved by locating the peak in the power spectrum. After that, beamforming weights, w_{opt} in Eq. (11.6), are computed using techniques such as Lagrange multipliers:

$$\mathbf{w}_{\text{opt}} = \arg \min_{\mathbf{w}} (\mathbf{w}^H \mathbf{R} \mathbf{w}) \quad \text{subject to} \quad \mathbf{w}^H e = 1$$

The optimum beamforming weight is

$$\omega_{\text{opt}} = \frac{R^{-1} e(\theta)}{e^H(\theta) R^{-1} e(\theta)} \tag{11.6}$$

The desired output is

$$y(t) = \sum_{k=0}^{K-1} \omega_k^H x_k(t) \tag{11.7}$$

It is expected that the beam formed in the desired direction suppresses the interference, and the combined overall signal $y(t)$ is able to achieve the optimum SIR within the azimuth pattern of the antenna array shown in Figure 11.2b.

11.3 SYSTEM DESIGN AND IMPLEMENTATION

Figure 11.3 shows the complete system. The emulated tag is comprised of a HP8648D signal generator connected to a transmitting antenna. The adaptive antenna is comprised of a four-element ULA of patch antenna elements, cascaded common receiver modules, a general-purpose DSP interface board (Hitachi SH7615), and an LED display. The sequence of the implementation of the proposed low-cost smart antenna is shown in Figure 11.3, and the important components are described as follows:

Emulated Tag: A probe-fed patch antenna placed at far field is hooked up to the signal generator to transmit a 2.4-GHz signal. This setup emulates the 2.4-GHz microwave tags (either active or passive).

RF Translator: The signal from the emulated tag is received by a four-element ULA, which is connected to four receivers that are being interleaved together (as shown in Figure 11.4) with cable assemblies from the antenna and to the beamforming DSP. An external reference oscillator is used to syncronize all four receivers.

DSP Board: The analog-to-digital converters (ADC) on-board the receivers are utilized to get the digital intermediate baseband frequency to be acquired by the DSP. However, the general-purpose DSP has only a limited number of

I/O ports. To minimize the use of I/Os for data acquisition, a DSP interface (Figure 11.5) is used to multiplex the digital signals and read by the DSP. An interface is designed as shown in Figure 11.5, which is also being used to syncronize the receiver block to the DSP and, at the same time, to regulate the power supply for the system. The DSP, upon acquiring the I/O data, computes the DoA estimation followed by implementing the DBF algorithm and, finally, displays the data via a LED display.

Portable Hardware Setup: Figure 11.6 shows the integrated adaptive beamforming system in a low-cost compact package. First, it is noted that we do not require the entire solution engine, but only the DSP. Second, there are available one-chip solutions in the market to replace the cascaded receivers. Third, the interface for the DSP might no longer be needed if the setup is to be incorporated into a portable system shown in Figure 11.6. In this work the setup in Figure 11.6 is only used for fast implementation and prototyping purposes. As can be seen, the cascaded common block receiver and the oscillator circuit are connected with flexible coaxial cables. The DSP interface and the common block receiver are connected via a ribbon cable followed by the connection with the SH7615 solution engine. The 15-MHz clock LED display is also visible. In the final phase of the handheld design the cable assemblies will be replaced with the PCB design. The hardware is set up as shown in Figure 11.6. Data acquisition of baseband signal is done by the DSP via the interface as shown in Figure 11.5. The interface multiplexed 80 digital signals to the 16-bit I/O port of the general-purpose Hitachi SH7615 DSP, enabling real-time data acquisition. Due to this heavy multiplexing of the baseband signals, the sampling rate of the baseband signals have to be compromised.

11.4 RESULTS

MATLAB Simulation. The main motivation of the simulation is to achieve the results as closely as possible to the real results achieved in the actual environment. A Matlab program is developed to estimate the DoA and separate the interference from the designed echoes. The flowchart of Capon beamforming is shown in Figure 11.7. As can be seen in the figure, the input data for the DBF are the operating frequency, the number of sensors (antenna elements) used, the interelement distance of the elements in the antenna array, DoA and the signal of interest (SoI) in degrees, signal-to-noise ratio (SNR) and interference-to-noise ratio (INR), the number of samples used, and the duration of samples. The echo (in this case the transmit signal) including the interference and the target is simulated first. The noise is created from the random function of Matlab. The echo is the summation of the SoI, the interference signal, and the noise. For comparison, a conventional beamforming (CBF) is performed along with the Capon beamforming. To calculate such beamforming, the required steps are: determination of the steering vector of SoI; computation of the CBF weight to input to the antenna elements; the output signal after CBF; and finally the output power spectrum. The Capon beamforming starting with the computation of space signal

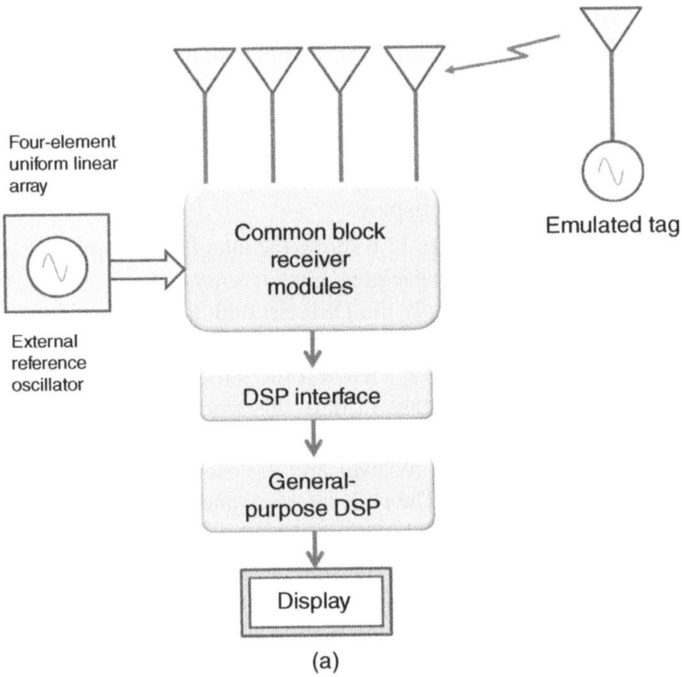

FIGURE 11.3. RFID adaptive antenna and tag system setup. **(a)** Block diagram. **(b)** Signal flow graph between blocks.

form the calculated echo, the covariance matrix, the Capon beamforming weight vector, the output signal, and, finally, the output spectrum.

Figure 11.8 shows the output signal (beam pattern) and the power spectrum obtained from the CBF computation. The antenna array has eight elements, and the interelement distance is 0.5λ. The DoA of the SoI is 0°, the DoA of interference is 20°, SNR = 0 dB, INR = 30 dB at each antenna element, and the sampling points are 256. As can be seen in the figure, CBF can direct the main beam toward SoI, which is 0°, but cannot position null toward the interference, which is positioned at 20°. Figure 11.8b shows the power spectrum calculated using CBF. As can be seen, the power spectrum peaks at both the signal and the interference. A Doppler frequency of −0.5 and 2 Hz are used to compute the power spectrum [7]. Doppler processing on each antenna element is used to separate the target and interference in frequency. An array snapshot on target Doppler cell is used to estimate target direction. Such processing reduces the effect of interference on the estimation of target direction.

Figure 11.9 shows the output signal (beam pattern) and the power spectrum obtained from the Capon beamforming computation. The antenna array has eight elements, and the interelement distance is 0.5λ. The DoA of the SoI is 0°, the DoA of interference is 20°, SNR = 0 dB; INR = 30 dB at each antenna element, and

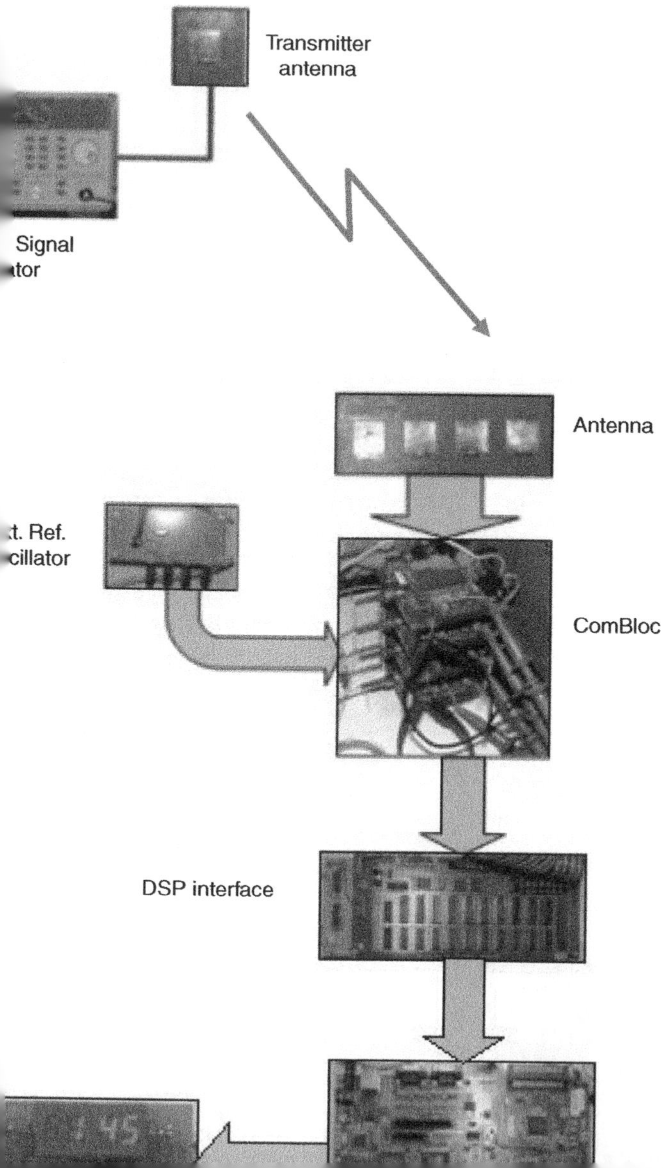

Transmitter antenna
Signal
Antenna
ComBloc
DSP interface

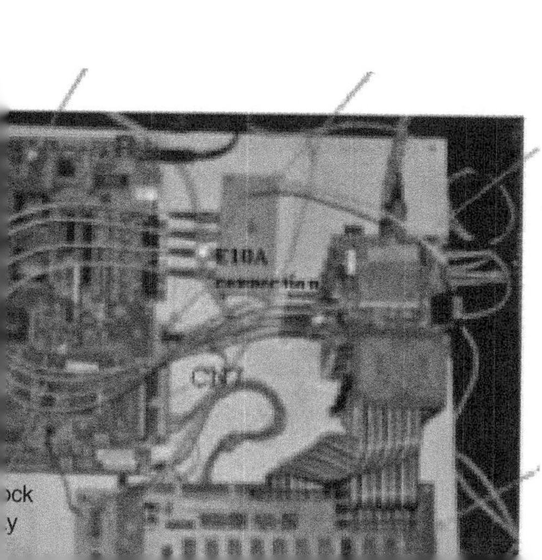
ock
y

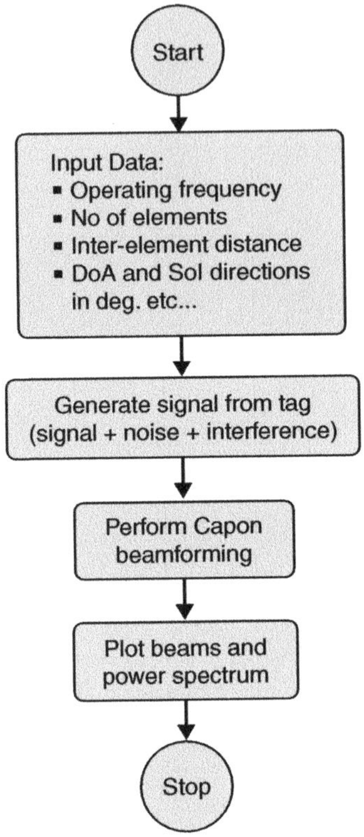

FIGURE 11.7. Flowchart of developing Capon algorithm for DBF.

used for synchronization. This inconsistent phase shift disrupted the estimate of DoA.

(ii) *Multipath Scattering*. Another factor to consider is the multipath scattering, which results in the presence of numerous signal multipath components to arrive at the receiver. This is due to reflections, diffractions, and signal scattering, caused by objects in the path between the transmitter and the receiver. Each signal component experiences a different path attenuation and phase rotation. These affect the received signals' amplitude, phase, and Doppler shift. As a result, the algorithm can only be achieved efficiently in an open environment or in a Faraday case where such interferences can be minimized.

Our results show that the DoA estimation can be done using common receiver blocks as long as there is synchronization between the receiver blocks. The results can be used for beamforming processing by applying the appropriate weights described in Eqs. (11.6) and (11.7).

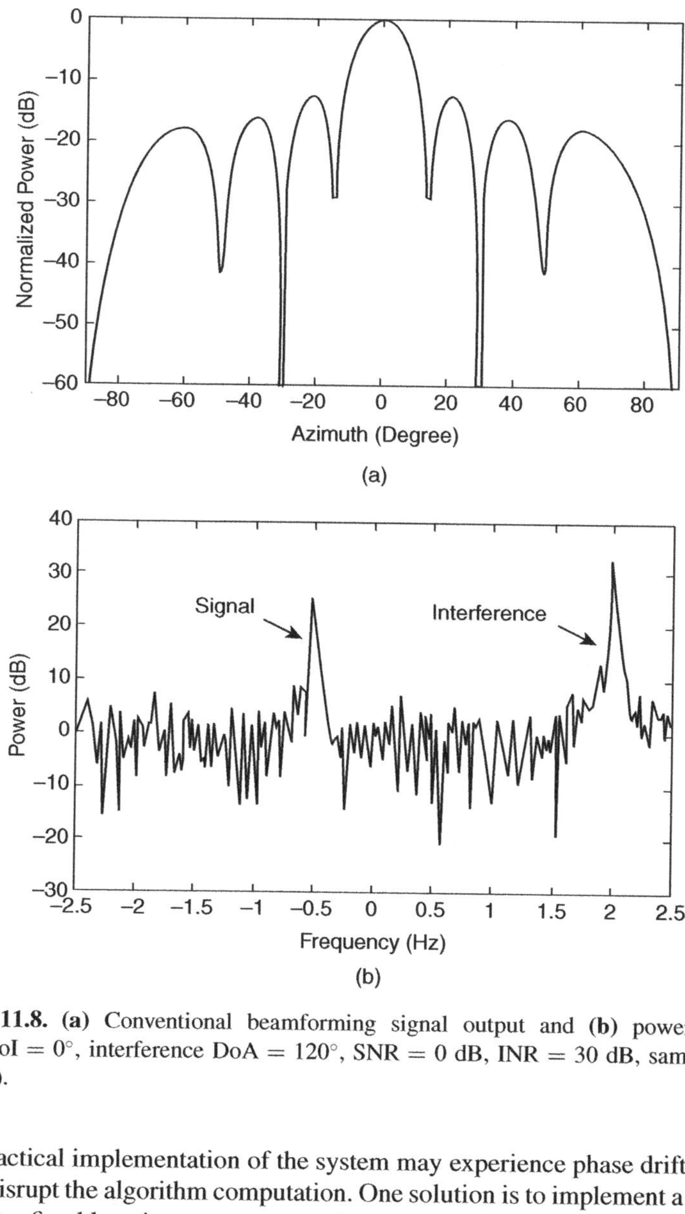

FIGURE 11.8. **(a)** Conventional beamforming signal output and **(b)** power spectrum (DoA of SoI = 0°, interference DoA = 120°, SNR = 0 dB, INR = 30 dB, sampling point 256 points).

The practical implementation of the system may experience phase drift problems that will disrupt the algorithm computation. One solution is to implement a fix source in space at a fixed location to act as a marker.

$$e(\theta) = \begin{bmatrix} 1 \\ e^{j2\pi dk\cos(\theta-\alpha)} \\ \vdots \\ e^{j2\pi d(K-1)\cos(\theta-\alpha)} \end{bmatrix} \tag{11.8}$$

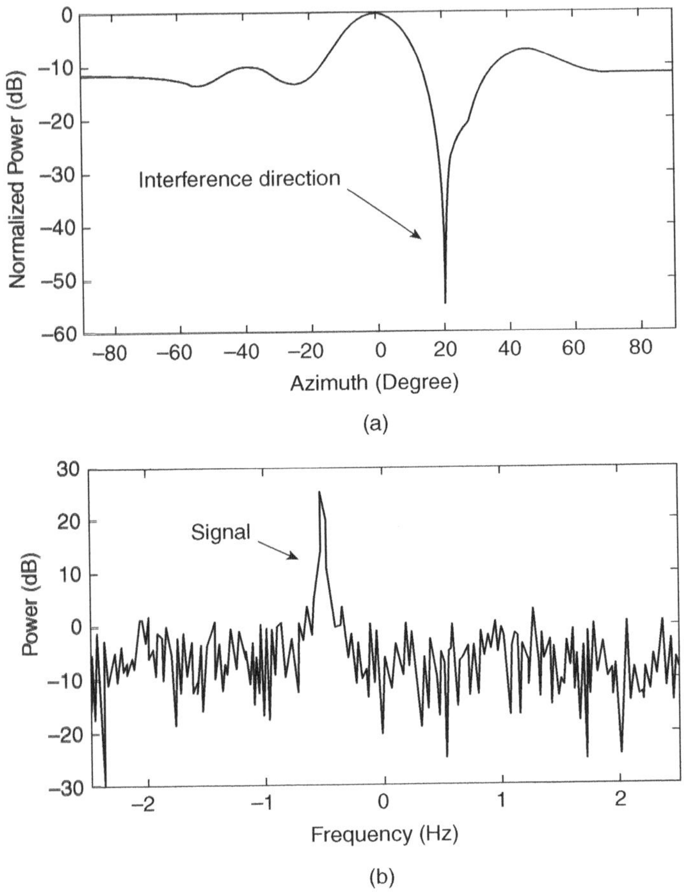

FIGURE 11.9. **(a)** Capon beamforming signal output and **(b)** power spectrum (DoA of SoI = 0°, interference DoA = 120°, SNR = 0 dB, INR = 30 dB, sampling point 256 points).

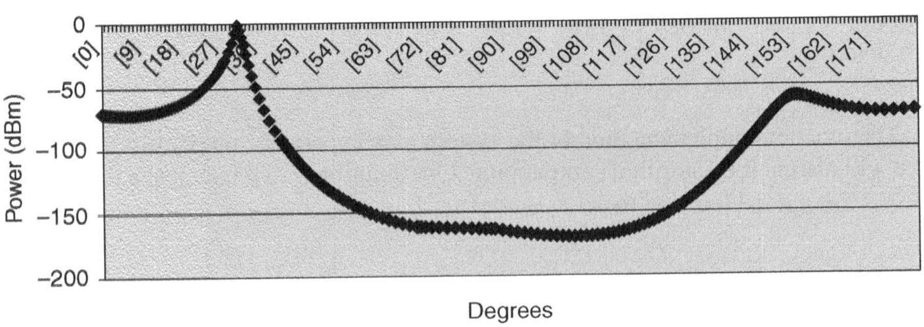

FIGURE 11.10. Simulated power spectrum to estimate DoA at 30° using four-element ULA.

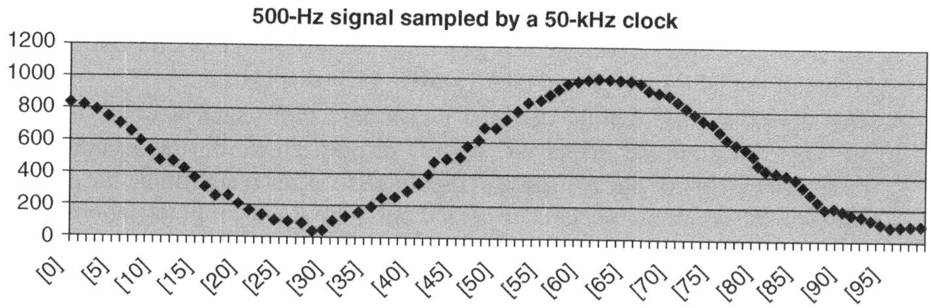

FIGURE 11.11. Sampled baseband signal by the DSP.

Then the correction factor, α, can be applied to the steering vector in Eq. (11.8) at a later stage to the signal processing portion to constantly correct the error presented by the phase drift.

11.5 CONCLUSION

In this chapter, we have described the hardware implementation of a portable smart antenna for the RFID reader to estimate DoA of RFID tags. Theory of the DoA and DBF based on the Capon algorithm, hardware setup, and simulation and experimental results have been presented. We also have discussed the limitation of the low-cost setup and implementation issues of improving the hardware to make a viable handheld or portable smart-antenna-based RFID reader. The following contributions have been made in the work:

1. A less computationally intensive Capon algorithm is implemented for DoA as well as DBF using a four-element ULA. The theoretical concept and practical implementation augment the portable system using a general-purpose and low-cost DSP chipset and cascaded common block receiver.
2. The practical design setup includes an emulated RFID tag, which is comprised of a HP8648D signal generator and a planar transmitting antenna of moderate gain. The frequency of operation is 2.4 GHz. The hardware can be replaced with a standard microwave tag in practical applications.
3. The receiver is comprised of (a) a four-element ULA of patch antenna elements and (b) a four-channel common block receiver connected to the antenna in the receive end and to a synchronized oscillator and a DSP interface in the output end. The reference external oscillator provides reference frequency and phase information similar to a coherent radar. The receiver converts the received RF signals for the transmitter to the baseband signal, which is to be processed by the DSP. A multiplexing unit is used to overcome the limited number of I/O ports for data acquisition in the DSP board (80 digital signals to 16 I/O ports in DSP).

4. The DSP board, upon receiving the I/O data, compiles the DoA estimation based on the Capon algorithm and generates the weight vectors for DBF. The output of the DSP is displayed on a 15-MHz LED display.
5. The algorithm successfully detected the desired tag, which was located at 30° DoA location and displayed on LED display. However, the accuracy is sacrificed by 10–20° due to the use of the simplistic Capon algorithm, the phase noises in the common block receiver, and noisy environment in the laboratory. A fixed tag at a known location for calibration may be used. The environmental issue can be overcome by testing the tag in scattering free open space or in a Faraday case.

ACKNOWLEDGMENT

The Matlab simulation results by Professor L. Yilong of Nanyang Technological University, Singapore are acknowledged.

REFERENCES

1. B. D. Van Veen and K. M. Buckley, Beamforming: A versatile approach to spatial filtering, *IEEE ASSP Mag.*, pp. 4–24, April 1988.
2. L. C. Godara, Application of antenna arrays to mobile communications, Part II: Beamforming and direction-of-arrival considerations, *Proc. IEEE*, Vol. 85, No. 8, pp. 1195–1245, August 1997.
3. J. Capon, High-resolution frequency-wave number spectrum analysis, *Proc. IEEE*, Vol. 57, pp. 1408–1418, 1969.
4. R. O. Schmidt, Multiple emitter location and signal parameter estimation, *IEEE Trans. Antennas Propag.*, Vol. AP-34, pp. 276–280, 1986.
5. H. Krim and M. Viberg, Two decades of array signal processing research, *IEEE Signal Processing Mag.*, pp. 67–94, July 1996.
6. P. Stoica, P. Händel, and T. Söderström, Study of Capon method for array signal processing, *Circuits Systems Signal Process*, Vol. 14, No. 6, pp. 749–770, 1995.
7. M. H. Li and Y. L. Lu, A refined genetic algorithm for accurate and reliable DOA estimation with a sensor array, *Wireless Personal Commun.*, Vol. 43, No. 2, pp. 533–547, 2007.

PART IV

DOA AND LOCALIZATION OF RFID TAGS USING SMART ANTENNAS

CHAPTER 12

DIRECTION OF ARRIVAL ESTIMATION BASED ON A SINGLE-PORT SMART ANTENNA FOR RFID APPLICATIONS

CHEN SUN and HIROSHI HARADA
National Institute of Information and Communications Technology (NICT), Japan

NEMAI CHANDRA KARMAKAR
Department of Electrical and Computer Systems Engineering, Monash University, Clayton, Victoria, Australia

12.1 INTRODUCTION

Radio-frequency identification (RFID) systems play important roles in service industries, purchasing and distribution logistics, industry, manufacturing companies, and material flow systems [1]. Over the last few years, there has been an increasing demand for positioning capability for RFID applications. For example, many department stores use RFID system to manage goods in stock. These goods and items bear their RFID tags. It would be convenient for the sales staff to find the wanted items if the RFID readers have the positioning ability.

The most efficient way could be using smart antennas [2–4] at the RFID reader. In this chapter, we introduce a DOA finding algorithm that can be used at the RFID reader with smart antennas for RFID positioning applications. Various DOA finding algorithms include conventional methods, linear prediction methods, eigenstructure methods, and estimation of signal parameters via rotational invariance techniques (ESPRIT) [5]. All these methods are based on the digital beamforming (DBF) antenna arrays. In DBF, signals, which are received by individual antenna elements, are down-converted to baseband signals. Then, they are digitized and fed into a digital signal processing (DSP) chip where the DOA estimation algorithm is executed. However, RF circuit branches connected to array elements, analog-to-digital converters (ADCs), and the baseband DSP chip consume a considerable amount of

Handbook of Smart Antennas for RFID Systems, Edited by Nemai Chandra Karmakar

DC power. Furthermore, each channel connected to the array sensor has the same structure, so the cost of fabrication increases with the number of array elements [6]. All these factors make DBF antenna arrays unsuitable for low-power and low-cost systems and thus hinder the mass implementation of smart antennas for positioning services.

To circumvent these problems of DBF arrays, we propose a novel DOA estimation technique based on a single RF port smart antenna. It is a parasitic array antenna [7], where one central element (connected to the sole RF port), and a number of surrounding parasitic elements form the array. Since the system has only one RF port and one consequent down-conversion circuitry, it obviously consumes much less power than a DBF array. Beam steering of parasitic array antennas, by either (a) selecting different sets of parasitic elements with the switching control circuitry in a manner similar to switched beam antennas (called "switched parasitic antennas" [7–11]) or (b) controlling the reactive loading at the parasitic elements to steer the beam continuously (called "reactively controlled directive arrays" [12] or "aerial beamforming (ABF)" [13–17], have been extensively studied. However, DOA estimation techniques are only confined to low-resolution estimation based upon signal powers received from different directions [13]. No such work has been done to explore the eigenstructure-based high-resolution technique.

The proposed direction finding technique is based on a reactively controlled directive array [12] structure and utilizes the MUSIC algorithm with periodic signals to explore the signal spatial information. RFID tags can be excited by an RFID reader to periodically to produce periodic signals for locating their positions. The beam pattern is shifted with a predefined angle. So, for the application of MUSIC, it is sufficient to know these angular shifts, without knowing exactly the response to one subset. This is similar to a uniform circular or linear antenna array in which we know the phase shift between consecutive antenna responses, without having to know those responses exactly. The MUSIC algorithm is carried out with the constructed signal vector over 1 period and the modified steering vector based on the beam pattern angle shift. Simulations show that the technique produces a high-resolution DOA estimation in an uncorrelated signal environment and is feasible to be implemented into practice.

Figure 12.1 shows such a scheme of DOA estimation of multiple tags using an array antenna connected to an array processor and DOA estimation functional unit followed by an RFID reader and a display. As shown in the figure, the reader sends command and control signals to the array processor to steer the beam in different angles. Note that it is not necessary that the beam is pointing to the direction of the RFID tag as done in spatial techniques. The MUSIC algorithm performs DOA estimation with rotated beam patterns and achieves 1° accuracy. Hence, the technique is very useful in department stores and warehouse scenarios when identification and positioning help to improve the logistic.

The chapter is organized as follows: Section 12.2 presents the configuration and the working principle of the reactively controlled directive array. The algorithm for DOA estimation is presented in Section 12.3. Results are produced in Section 12.4, followed by conclusions and recommendations in Section 12.5.

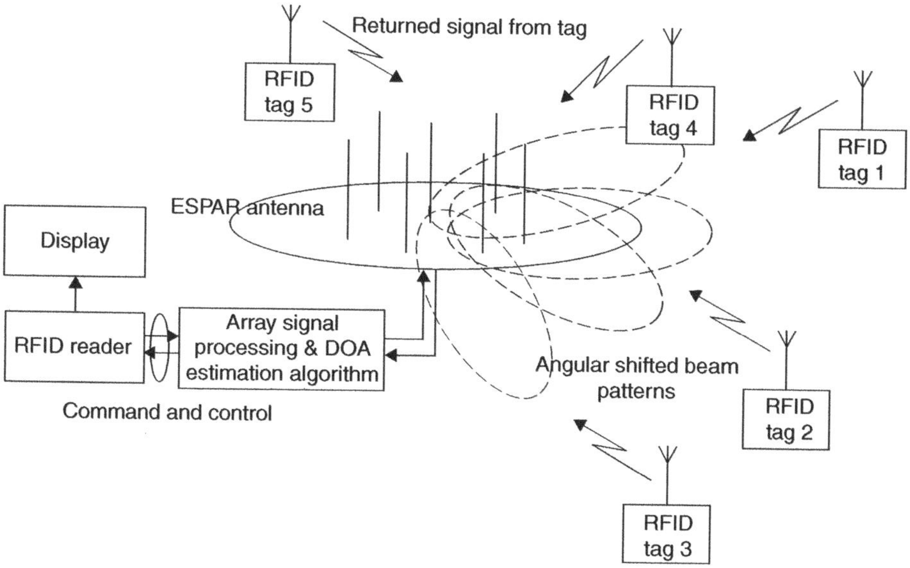

FIGURE 12.1. Scheme of DOA estimation of multiple tags using ESPAR antenna.

12.2 DESIGN OF THE ANTENNA

Figure 12.2 shows one typical structure of a reactively controlled directive array antenna, called electronically steerable parasitic array radiator (ESPAR) [6]. There, one active central element (monopole) is surrounded by parasitic elements on a circle of radius R on the circular grounded baseplate. The length of each monopole L and the radius R are one-quarter wavelength λ of the transmitting RF signal. The baseplate transforms monopoles with their images to dipoles with a length of $2L$. The central monopole is connected to an RF receiver, and each parasitic monopole is loaded with a tunable reactor.

The working principle of an ESPAR antenna is different from that for a DBF. The antenna generates a directional beam based on tuning load reactances $(x_1, x_2, \ldots, x_6)$ on the parasitic monopoles. Signals, which are received or transmitted from the central RF port, excite passive monopoles with substantial induced mutual currents on them. In the following discussion, we assume that the antenna works in transmitting mode. Vectors $\mathbf{i}$ and $\mathbf{v}$ represent the current and the voltage on the monopoles, respectively.

$$\mathbf{i} = [i_0\, i_1\, i_2\, i_3\, i_4\, i_5\, i_6]^T \tag{12.1}$$

$$\mathbf{v} = [v_0\, v_1\, v_2\, v_3\, v_4\, v_5\, v_6]^T \tag{12.2}$$

Superscript T represents the transpose. Mutual admittances are represented by the matrix $\mathbf{Y}$, with each entity y_{ij} denoting mutual admittance between two elements [15].

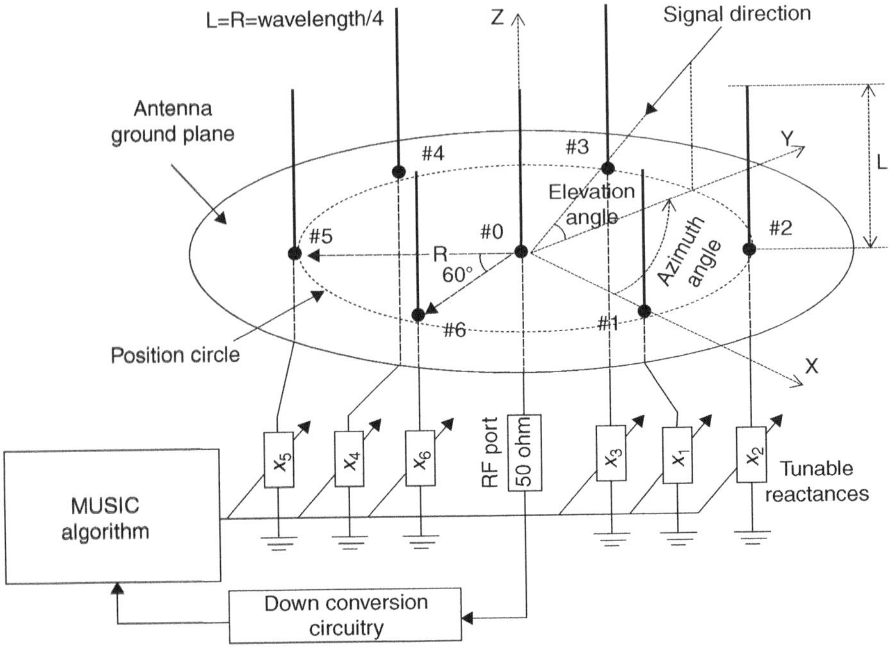

FIGURE 12.2. Configuration of a seven-element ESPAR antenna [6].

There, induced mutual currents are represented by mutual admittances as

$$\mathbf{i} = \mathbf{Y}\mathbf{v} = \begin{bmatrix} y_{00} & y_{01} & y_{02} & y_{03} & y_{04} & y_{05} & y_{06} \\ y_{10} & y_{11} & y_{12} & y_{13} & y_{14} & y_{15} & y_{16} \\ y_{20} & y_{21} & y_{22} & y_{23} & y_{24} & y_{25} & y_{26} \\ y_{30} & y_{31} & y_{32} & y_{33} & y_{34} & y_{35} & y_{36} \\ y_{40} & y_{41} & y_{42} & y_{43} & y_{44} & y_{45} & y_{46} \\ y_{50} & y_{51} & y_{52} & y_{53} & y_{54} & y_{55} & y_{56} \\ y_{60} & y_{61} & y_{62} & y_{63} & y_{64} & y_{65} & y_{66} \end{bmatrix} \times \begin{bmatrix} v_0 \\ v_1 \\ v_2 \\ v_3 \\ v_4 \\ v_5 \\ v_6 \end{bmatrix} \quad (12.3)$$

For the active central element, the voltages at monopoles are expressed as

$$v_0 = v_s - z_0 i_0 \quad (12.4)$$

where $z_0 = 50\ \Omega$ is the impedance at the RF port. For parasitic elements $l = 1, \ldots, 6$,

$$v_l = -jx_l i_l \quad (12.5)$$

where v_s represents the transmitted voltage signal source, with the amplitude and the phase, from the driving RF port at the central element #0. Represent Eqs. (12.4) and

(12.5) with a matrix form:

$$\mathbf{v} = [v_s \;\; 0 \;\; 0 \;\; 0 \;\; 0 \;\; 0 \;\; 0]^T - \mathbf{Xi} = v_s\mathbf{u}_1 - \mathbf{Xi} \tag{12.6}$$

and define

$$\mathbf{X} = \text{diag}[50, jx_1, \ldots, jx_6] \tag{12.7}$$

$$\mathbf{u}_1 = [1 \;\; 0 \;\; 0 \;\; 0 \;\; 0 \;\; 0 \;\; 0]^T \tag{12.8}$$

Combining Eqs. (12.3) and (12.6), we get

$$\mathbf{i} = \mathbf{Yv} = \mathbf{Y}(v_s\mathbf{u}_1 - \mathbf{Xi}) \tag{12.9}$$

After a simple mathematical manipulation, we obtain

$$\mathbf{i} = v_s\left(\mathbf{I}_{(7)} + \mathbf{YX}\right)^{-1}\mathbf{Yu}_1 = v_s\mathbf{E}_1 \tag{12.10}$$

where matrix $(\mathbf{I}_{(7)} + \mathbf{YX})^{-1}\mathbf{Yu}_1$ is represented as $\mathbf{E}_1$. $\mathbf{I}_{(7)}$ is a 7×7-dimensional identity matrix. The far-field radiation pattern is formed by the superposition of radiation patterns of all monopoles on the antenna ground plane [7, 18]. The far-field current signal, with its amplitude and the phase, in direction ϕ_k is

$$y_{\text{far}}(t) = \alpha(\phi_k)\mathbf{i} = v_s\alpha(\phi_k)\mathbf{E}_1 \tag{12.11}$$

Assume signals coming in the azimuth plane. The subscript k is the signal sources' index.

$$\alpha(\phi_k) = [1 \; e_1(\phi_k)\, e_2(\phi_k)\, e_3(\phi_k)\, e_4(\phi_k)\, e_5(\phi_k)\, e_6(\phi_k)] \tag{12.12}$$

$\alpha(\phi_k)$ is the steering vector corresponding to signal's direction and ϕ_k is given based on the ESPAR antenna geometry (cf. Figure 12.3).

$$e_i(\phi_k) = \exp\left\{j\frac{2\pi}{\lambda}R\cos\left(\phi_k - \frac{2\pi}{6}(i-1)\right)\right\}$$

$$= \exp\left\{j\frac{\pi}{2}\cos\left(\phi_k - \frac{\pi}{3}(i-1)\right)\right\} \quad \text{for } i = 1, \ldots, 6 \tag{12.13}$$

Setting the value of v_s in Eq. (12.11) as unity, we can get the array factor $\hat{\alpha}(\phi_k)$ of an ESPAR antenna as

$$\hat{\alpha}(\phi_k) = \alpha(\phi_k)\mathbf{E}_1 \tag{12.14}$$

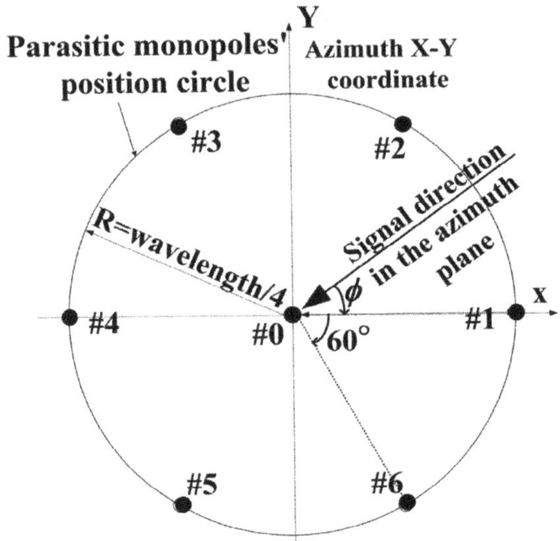

FIGURE 12.3. Geometry of a seven-element ESPAR antenna.

Thus, the far-field signal in Eq. (12.11) can be represented as

$$y_{\text{far}}(t) = \hat{\alpha}(\phi_k)v_s \tag{12.15}$$

According to the reciprocity theory for radiation patterns [19], in receiving mode, the received voltage signal $u(t)$ at the RF port is

$$u(t) = \hat{\alpha}(\phi_k)s_k(t) + n(t) \tag{12.16}$$

where $s_k(t)$ represents kth far-field incident current waves with the amplitude and the phase in the azimuth plane. $n(t)$ is the additive white Gaussian noise (AWGN) with the power of σ^2. When there are totally K impinging waves, the output signals from the RF port are represented as

$$u(t) = \sum_{k=1}^{K} \hat{\alpha}(\phi_k)s_k(t) + n(t) \tag{12.17}$$

Since the $\hat{\alpha}(\phi_k)$ is dependent on the reactance matrix $\mathbf{X}$, antenna beam patterns can be formed in desired directions with nulls in interferers' directions by adjusting reactance values. In practice, the variable reactance is realized by changing the bias voltage of Schottky [20] or varactor diodes loaded at parasitic monopoles.

Based on the previously explained theory, adaptive beamforming algorithms for the ESPAR antenna have been designed [15–17]. In our proposed DOA estimation technique, the ESPAR antenna is not used for adaptive beamforming, though it still has the ability to do so. Here, the antenna is employed to fulfill direction finding.

TABLE 12.1. Load Reactance in Each *Sampling Period* (#1 to #6) While the Antenna is Virtually Rotating Clockwise

Sampling Periods	Values of Load Reactances					
	#1	#2	#3	#4	#5	#6
#1	jX_1	jX_2	jX_3	jX_4	jX_5	jX_6
#2	jX_2	jX_3	jX_4	jX_5	jX_6	jX_1
#3	jX_3	jX_4	jX_5	jX_6	jX_1	jX_2
#4	jX_4	jX_5	jX_6	jX_1	jX_2	jX_3
#5	jX_5	jX_6	jX_1	jX_2	jX_3	jX_4
#6	jX_6	jX_1	jX_2	jX_3	jX_4	jX_5

Desired beam patterns are formed to explore the spatial information of the signal sources.

12.3 FORMING CORRELATION MATRIX FROM A SINGLE-PORT SMART ANTENNA

The MUSIC algorithm, proposed by Schmidt [21], is a relatively simple and high-resolution eigenstructure approach. It has been widely used as a model-based approach for finding the parameters of interest in direction finding or source location problems. There are many improved forms of the MUSIC algorithm, such as the Root–MUSIC Algorithm and the beamspace–MUSIC Algorithm [5]. All these algorithms require the snapshots of signals from antenna array elements to form the signal correlation matrix.

It is obvious that we can't apply the conventional MUSIC algorithm to an ESPAR antenna, because it needs groups of signal samples from linear or circular array elements to form a vector; whereas ESPAR antenna has only one RF port, forming signal sample vector is impossible at one time. This is the distinction between the MUSIC algorithm based on a single-port smart antenna and the conventional MUSIC algorithm based on a DBF array antennas as aforementioned. However, from the explanation in Section 12.4 we can see that received signals from an ESPAR antenna RF port contain signal sources' direction information. The signal correlation matrix can still be formed.

In operation, the antenna is virtually rotated by forming beams with changing the sequence of load reactances in six sampling periods (cf. Table 12.1). The sample signal vector $\mathbf{u}(t) = [u_1(t)\ u_2(t)\ u_3(t)\ u_4(t)\ u_5(t)\ u_6(t)]^T$ is constructed with signal samples from the RF output. Each entity $u_m(t)$ $(m = 1, \ldots, 6)$ is obtained in the sampling period #m consecutively. The maximum number of sampling periods $M = 6$ equals the number of the parasitic elements. For an ESPAR antenna with six parasitic elements, rotating $6 + 1$ times returns to the original status. Load reactance values of parasitic monopoles are the same for all the sampling periods. $\mathbf{E}_1$, defined in Eq. (12.10), remains constant as the antenna rotates virtually. However, signal directions with respect to antenna azimuth plane X–Y coordinate change. According to Eq. (12.17), the received signal samples at each sampling period #m are represented as

$$u_m(t) = \sum_{k=1}^{K} \hat{\alpha}\left(\phi_k - \frac{2\pi}{M}(m-1)\right)s_k(t) + n_m(t), \qquad m = 1, 2, \ldots, 6 \tag{12.18}$$

For simplicity we define

$$\hat{\alpha}_m(\phi_k) = \hat{\alpha}\left(\phi_k - \frac{2\pi}{M}(m-1)\right) \tag{12.19}$$

$\hat{\alpha}_m(\phi_k)$ contains the kth signal source direction information $\phi_k \cdot n_m(t)$ is the noise component at the mth sampling period. Suppose that each individual signal source transmits periodic signals, and the length of one signal period is the same as one sampling period. Received signal samples are the same for all sampling periods. Therefore, the received signal vector $\mathbf{u}(t)$ after six sampling periods is written in the following matrix multiplication equation form:

$$\mathbf{u}(t) = \begin{bmatrix} u_1(t) \\ u_2(t) \\ u_3(t) \\ u_4(t) \\ u_5(t) \\ u_6(t) \end{bmatrix} = \begin{bmatrix} \hat{\alpha}_1(\phi_1) & \hat{\alpha}_1(\phi_2) & \cdots & \hat{\alpha}_1(\phi_k) \\ \hat{\alpha}_2(\phi_1) & \hat{\alpha}_2(\phi_2) & \cdots & \hat{\alpha}_2(\phi_k) \\ \hat{\alpha}_3(\phi_1) & \hat{\alpha}_3(\phi_2) & \cdots & \hat{\alpha}_3(\phi_k) \\ \hat{\alpha}_4(\phi_1) & \hat{\alpha}_4(\phi_2) & \cdots & \hat{\alpha}_4(\phi_k) \\ \hat{\alpha}_5(\phi_1) & \hat{\alpha}_5(\phi_2) & \cdots & \hat{\alpha}_5(\phi_k) \\ \hat{\alpha}_6(\phi_1) & \hat{\alpha}_6(\phi_2) & \cdots & \hat{\alpha}_6(\phi_k) \end{bmatrix} \times \begin{bmatrix} s_1(t) \\ s_2(t) \\ \bullet \\ \bullet \\ \bullet \\ s_k(t) \end{bmatrix} + \begin{bmatrix} n_1(t) \\ n_2(t) \\ \bullet \\ \bullet \\ \bullet \\ n_6(t) \end{bmatrix} \tag{12.20}$$

Equation (12.20) has the same form as the output from an array antenna with six elements. In the first matrix of the right-hand side of the equation, we define each column as vector $\vec{\alpha}(\phi_k)$. It corresponds to the steering vector [22] for a six-element array antenna as in the array signal processing literature. But each entity is not a phase delay of a source signal at each antenna sensor, induced by the angle of arrival of the plane waves with respect to the array antenna in the azimuth plane. We call $\vec{\alpha}(\phi_k)$ as "ESPAR-direction vector" to differentiate it from the term "steering vector" in the array processing literature. The signal correlation matrix at the antenna output is $\mathbf{R}_{UU}$.

$$\mathbf{R}_{UU} = E(\mathbf{u}(t)\mathbf{u}^H(t)) = \frac{1}{Ns}\mathbf{U}\mathbf{U}^H \tag{12.21}$$

$E(\bullet)$ is the expected value operator. U without time argument represents the discrete samples with sample block length Ns. Superscript H denotes conjugate-transpose. Suppose all the noise components $n_m(t)\,(m = 1, \ldots, 6)$ at each sampling period #m are independent and identically distributed (i.i.d.) AWGN with the power of σ^2.

$\mathbf{R}_{UU}$ is represented as

$$\mathbf{R}_{UU} = [\vec{\alpha}(\phi_1)\,\vec{\alpha}(\phi_2)\ldots\vec{\alpha}(\phi_K)] \times \mathbf{R}_{SS} \times [\vec{\alpha}(\phi_1)\,\vec{\alpha}(\phi_2)\ldots\vec{\alpha}(\phi_K)]^H + \sigma^2\mathbf{I}_{(6)} \tag{12.22}$$

$\mathbf{I}_{(6)}$ is a 6×6-dimensional identity matrix. The source signal correlation matrix $\mathbf{R}_{SS}$ is

$$\mathbf{R}_{SS} = E\left([s_1(t)\,s_2(t)\cdots s_k(t)]^T \times [s_1(t)\,s_2(t)\cdots s_k(t)]^*\right) \tag{12.23}$$

where superscripts T and $*$ are transpose and complex-conjugate, respectively. Based on Eq. (12.21), MUSIC algorithm is executed in a similar procedure as the conventional MUSIC algorithm (cf. Figure 12.4). The ESPAR-direction vector is used to evaluate the MUSIC spectrum. The algorithm requires signals to be periodic. The same RFID tag produces the same signals. These RFID tags can be excited periodically to produce the desired periodic signals. Therefore, we call it MUSIC algorithm with periodic signals. Its procedure is given below:

1. Construct the signal sample vector given by $\mathbf{u}(t) = [u_1(t)\,u_2(t)\,u_3(t)\,u_4(t)\,u_5(t)\,u_6(t)]^T$, where $u_m(t)$, $m = 1, \ldots, 6$, is the signal sample when the antenna is at sampling period #m. Finally, form the signal correlation matrix $\mathbf{R}_{UU}$.
2. Eigendecompose the signal correlation matrix $\mathbf{R}_{UU}$, and form the noise subspace $\mathbf{E}_N$ with eigenvectors corresponding to the small eigenvalues.
3. Evaluate the MUSIC spectrum PMU versus the signal direction ϕ.

$$P_{MU} = \frac{1}{\left|\mathrm{E}_N^H\vec{\alpha}(\phi)\right|^2} \tag{12.24}$$

4. $\vec{\alpha}(\phi)$ is the ESPAR-direction vector corresponding to the azimuthal looking direction ϕ.
5. Finally, extract impinging waves' direction information.

We can see that besides the additional procedure of forming the signal sample vector, the proposed technique has the same procedure as that for the conventional MUSIC algorithm. An extra amount of time is required for saving data to form the signal sample vector.

12.4 SIMULATION

In our simulation, three directional binary phase shift keying (BPSK) signals with the equal power levels are randomly generated. Signal directions are set arbitrarily 120°, 150°, and 300° in the azimuth plane, respectively. During each of the six

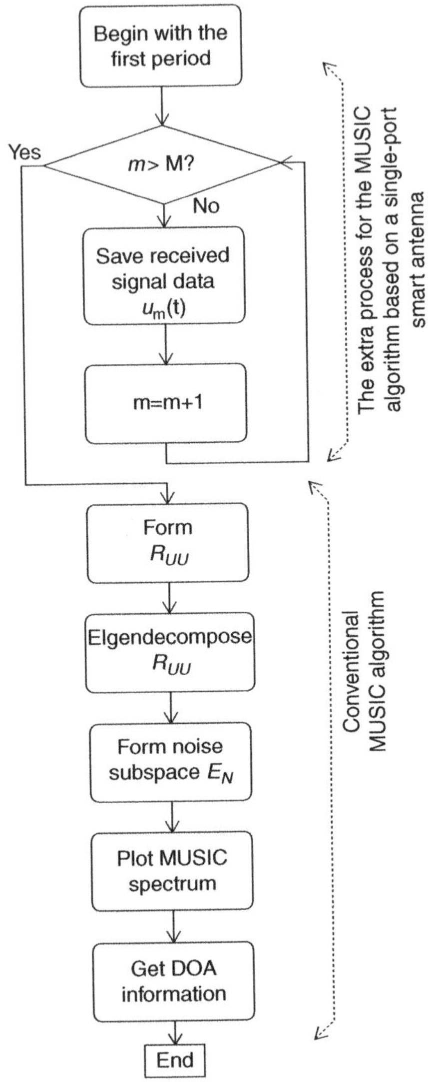

FIGURE 12.4. Flow chart of the MUSIC algorithm and the proposed MUSIC algorithm based on a single-port smart antenna.

antenna sampling periods, 16-bit signal samples are stored for further processing. This indicates that *Ns* is 16 in Eq. (12.21).

The monopoles on the ground plane of the ESPAR antenna are oriented vertically. Impinging signals are assumed to be co-polarized with the ESPAR antenna. Cross-polarization coupling typically produced by scatters and reflectors in the multipath propagation environment [23] is not considered.

TABLE 12.2. Setting of Six Parasitic Loads at Any Sampling Period[a]

Case	X_1 Ω	X_2 Ω	X_3 Ω	X_4 Ω	X_5 Ω	X_6 Ω	MUSIC Spectrum	DOA Information
1	30	30	30	30	30	30	Fig. 6	Not shown in the spectrum
2	0	30	0	30	0	30	Fig. 7	Not shown in the spectrum
3	0	30	0	30	30	30	Fig. 9	Shown in the spectrum
4	0	10	0	30	70	30	Fig. 9	Shown in the spectrum

[a]Once the setting is decided, these values do not change during the process of DOA estimation. The assignment of the loading changes as indicated in Table 12.1 to rotate the antenna.

12.4.1 Influence of Parasitic Load Reactance

We study four different reactance load settings to investigate their influence on the performance of DOA estimation. Table 12.2 summarizes the reactance values for these four cases. MUSIC spectra with different reactive load settings in uncorrelated signal environments with a signal-to-noise ratio (SNR) of 30 dB for each individual signal source are presented and explained as follows.

Case 1. Equal Load Reactance of 30 Ω. In Case 1, we set all the parasitic load reactances to 30 Ω. In this case, the antenna exhibits a nearly omnidirectional beam pattern as shown in Figure 12.5. The antenna cannot be virtually rotated in a way as described before. The MUSIC spectrum is plotted in Figure 12.6. From the figure we can observe that the information of signal direction cannot be extracted.

Case 2. Ordered But Unequal Load Reactances of 0 Ω and 30 Ω. In Case 2, we set the load to $x_1 = 0$ Ω, $x_2 = 30$ Ω, $x_3 = 0$ Ω, $x_4 = 30$ Ω, $x_5 = 0$ Ω, $x_6 = 30$ Ω. Two possible beam patterns (Case 2_1 & Case 2_2) for rotation with this load setting are shown in Figure 12.5. The MUSIC spectrum is plotted in Figure 12.7. It also cannot give the direction information. From the simulation studies in Cases 1 and 2, we conclude that we should not set the loads "ordered," which will result in insufficient beam patterns for rotation.

Case 3. Random Load Reactances of 0 Ω and 30 Ω. In Case 3, we arbitrarily set $x_1 = 0$ Ω, $x_2 = 30$ Ω, $x_3 = 0$ Ω, $x_4 = 30$ Ω, $x_5 = 30$ Ω, $x_6 = 30$ Ω. Six beam patterns (a) to (f) for antenna sampling period #1 to #6 are shown in Figure 12.8. Signals' DOA information is clearly indicated in the MUSIC spectrum in Figure 12.9.

Case 4. Random Load Reactances. Finally, we randomly set the reactance values to $x_1 = 0$ Ω, $x_2 = 10$ Ω, $x_3 = 0$ Ω, $x_4 = -30$ Ω, $x_5 = 70$ Ω, and $x_6 = 30$ Ω. The MUSIC spectrum is plotted in Figure 12.9. From the figure, we can see that if the reactance is not set ordered, they only slightly influence the MUSIC spectrum and not the performance. All the spectra reach the peak values at signal directions. The beam pattern for this case is shown in Figure 12.10. Note that the proposed technique

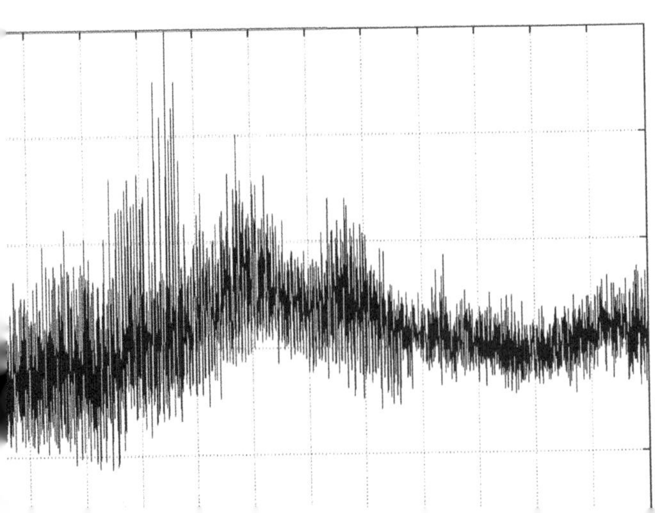

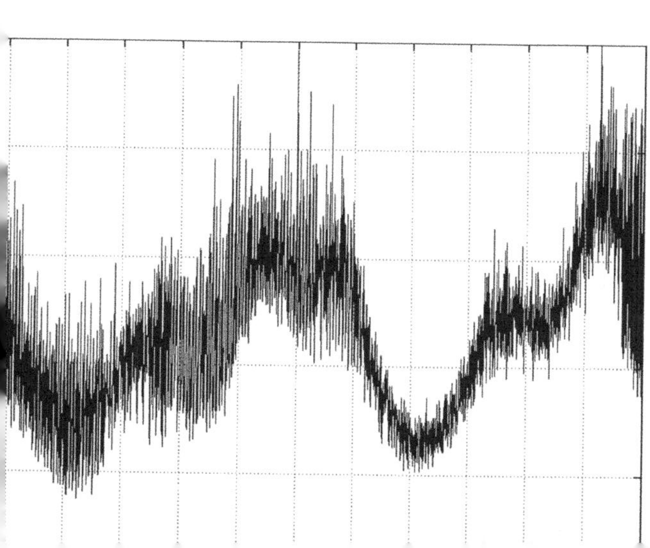

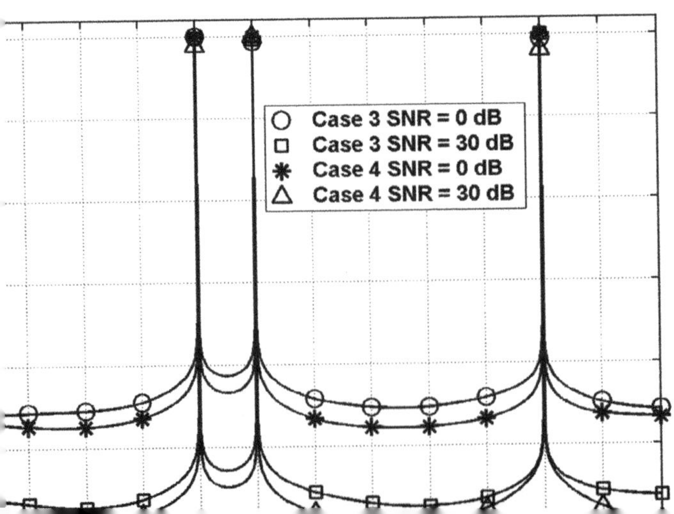
Case 3 SNR = 0 dB
Case 3 SNR = 30 dB
Case 4 SNR = 0 dB
Case 4 SNR = 30 dB

is used for DOA estimation and not for beamforming. So the beam pattern does not necessarily correspond to the directions of signals.

From the above investigation, we conclude that the manner of load settings greatly influences algorithm's DOAs estimation performance. When setting the reactive loads, we should avoid setting them ordered. Rather, we set their values arbitrarily to precisely estimate the DOA information. With a symmetric pattern, signals impinging from different directions may produce the same output at the RF port. Analogous to linear array antenna, image peaks appear. Rather, we can set the reactance values arbitrarily to precisely estimate the DOA information.

Forming directive beam patterns is not important to the performance. The technique is dependent on the angular shifting of the beam pattern. If only the pattern is not symmetric, the DOA can be estimated correctly. However, the load setting, which gives deep nulls in radiation patterns in any direction, should be avoided for the DOA estimation because signals coming from that region are received with low power levels, or they should even be nulled out. Consequently, the DOA information of signal in this null region cannot be resolved.

12.4.2 Performance in Different SNR Environments

We simulate MUSIC spectra in signal environments with different SNRs. Suppose that signal sources are totally uncorrelated and have equal power. The individual SNR for each signal source is equally set. MUSIC spectra with three different SNRs from 0 dB to 30 dB are shown in Figure 12.11. The figure shows that the performance of our proposed MUSIC algorithm degrades as SNR decreases, but three MUSIC spectra reach nearly the same peak values at signal directions.

12.4.3 Performance in a Multipath Propagation Environment

Our study assumes line-of-sight (LOS) propagations (Ricean channel). In a large store hose, we can assume that the LOS exists between the RFID reader and the RDIF tags. Ideally, signals arrive at the antenna without phase delays. The plotted MUSIC spectrum is shown in Figure 12.12 with a "×" marked solid line. In a real wireless communication environment, multipath propagation delays of the signals are expected [24]. Received signals are an ensemble of LOS components and multipath components from each of the far-field signal sources. Knowing that the LOS components are uncorrelated with the multipath components [25] and that LOS components have much higher signal power level, we can still reasonably represent received signals at the antennas as a summation of these LOS components from different directions with phase differences induced by the propagation delays. Signal directions are set as 120°, 150°, and 300° in the azimuth plane; The MUSIC spectrum for delayed signals is shown in Figure 12.12 with a "O" marked solid curve. The signal direction information can be extracted.

In the case where the RFID reader is mounted inside a room close to the ground to confine the signal coverage into a small local area, the RFID reader is located in a rich scattering environment. There are no LOS components. Signals received at the

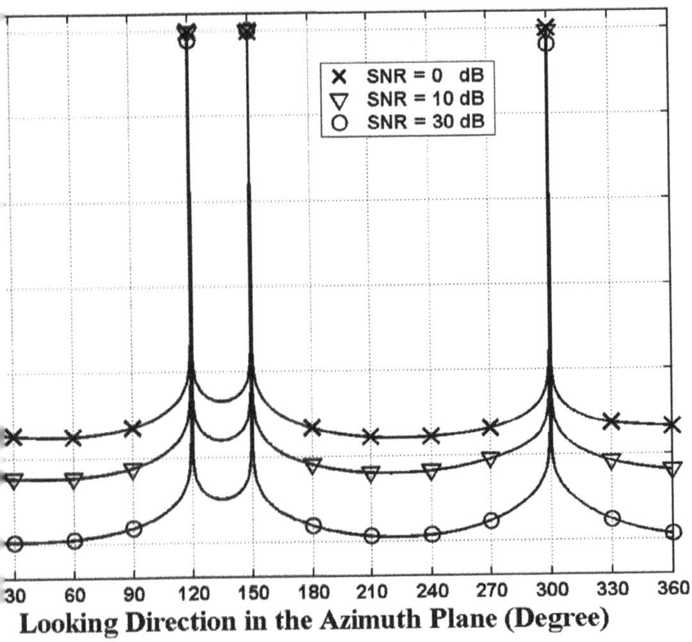

MUSIC spectra in different SNRs signal environments. Signal directions: 300° in the azimuth plane. Load setting is the same as in Case 4.

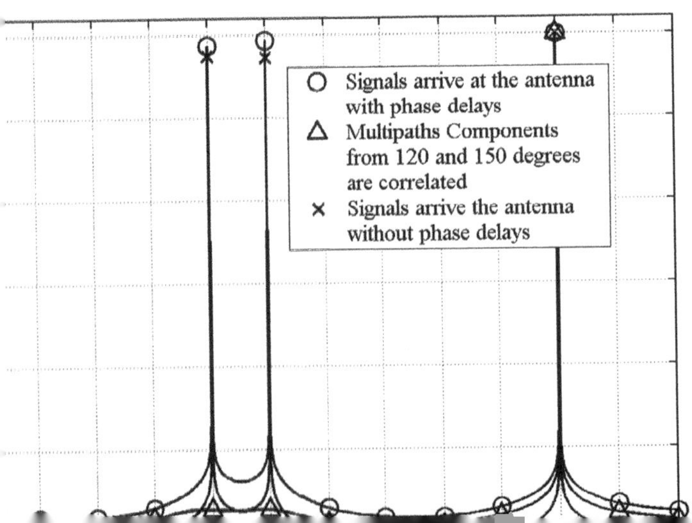

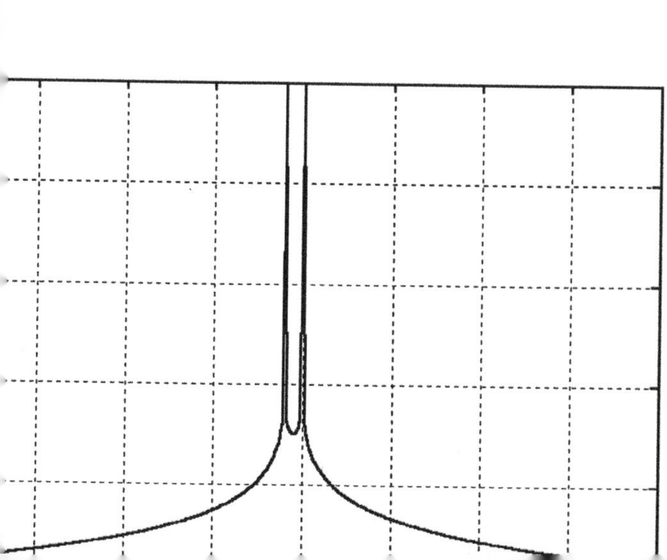

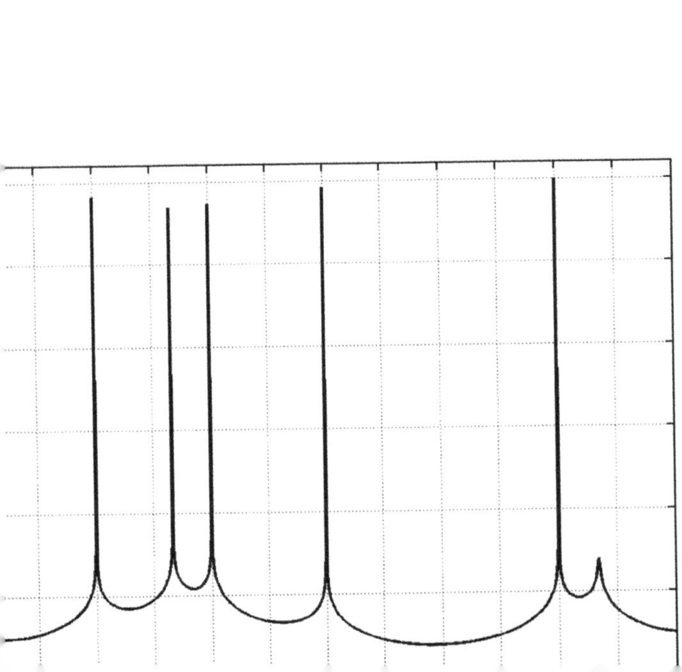

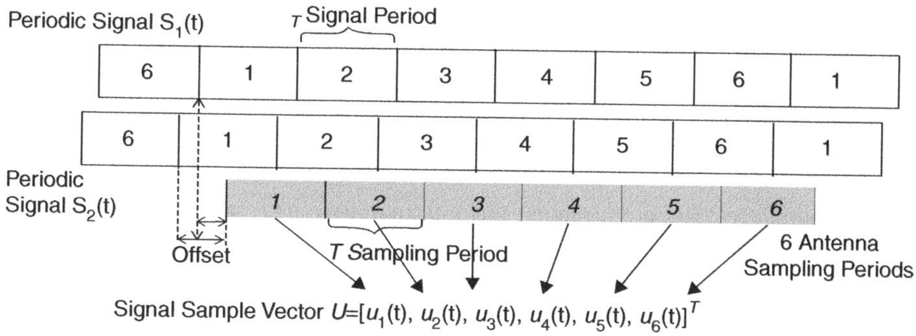

FIGURE 12.15. Periodic signal timeslots and antenna virtual sampling periods in the proposed DOA estimation technique. Each signal contains six slots. The antenna rotates six times. Time offsets are expected in a multipath environment.

six timeslots with a time length of 6.67 ms each. In each timeslot, 260 data bits are transmitted. The tuning of the reactance can be accomplished in nanoseconds [22]. These indicate that the DOA estimation is possible to be carried out within less than one TDMA timeslot.

12.5 CONCLUSION

Positioning can be achieved by using smart antennas at the RFID reader. In this chapter a novel approach for executing high-resolution MUSIC algorithm based on a single RF port smart antenna has been proposed. After presenting the configuration and the working principle of the antenna, the performance of the proposed technique for various aspects have been studied. The results have justified that the technique for a high-resolution DOA estimation of 1°, which is as good as a conventional MUSIC algorithm. The proposed technique for DOA estimation based on a signal RF port parasitic antenna has many advantages over that on a DBF. They are as follows:

1. Low power consumption and low complexity are obtained from using the single-port antenna. Therefore, this technique is highly applicable to commercial positioning services. This is the most outstanding benefit of our proposed DOA estimation technique.
2. Contrary to the DOA estimation based on a DBF, the mutual coupling is utilized to steer the beam [12]. Thus the modified MUSIC algorithm is free from negative influences of mutual coupling between elements. In a DBF this influence has to be compensated for the enhancement of system performance in a high-resolution DOA estimation as studied in [31–32].
3. High-quality performances occur in low SNR signal environments. The performances in an SNR = 0 dB environment are shown in Figures 12.9 and 12.11.

4. The proposed technique is very flexible in operation. When all loading reactance at parasitic monopoles are tuned adaptively to the signal environment, the antenna functions as a smart directional antenna serving to reduce the channel interferences and improve the channel capacity. At this time an adaptive beamforming algorithm needs to be executed at baseband DSP to control the reactive loadings to the parasitic monopoles.
5. Parasitic elements are arranged to form a circular array antenna. When used for DOA estimation, it gives the signal direction information covering 360° in the azimuth plane.

Simulation studies have also pointed out that further improvements are needed to enhance system performances to make it more applicable to practical RFID systems. The areas of improvements are as follows:

1. The proposed technique works in an uncorrelated signal environment. New methods should explore signal sources' spatial signatures in a correlated signal environment for positioning services.
2. The proposed antenna will not suffer from the negative influence from the mutual coupling between antenna elements. However, the calibration of the antenna aperture over DOA, frequency and temperature, weather environment, and fabrication error is, however, still unavoidable [1] and they influence the antenna performances.
3. The Doppler effect of the fast-moving RFID reader needs to be investigated. This will benefit the application of the antennas on moving terminals.

Finally, the state-of-the-art wireless positioning service is still a new area. New direction finding technologies that avail the realization of positioning services are under investigation. This chapter offers an economical approach to implement smart antennas into the RFID system, especially at the RFID reader to serve as an essential and fundamental technique of positioning services.

REFERENCES

1. K. Finkenzeller, *RFID Handbook: Fundamentals and Applications in Contactless Smart Cards and Identification*, 2nd edition, John Wiley & Sons, Hoboken, NJ, 2003.
2. K. J. Krizman, T. E. Biedka, and T. S. Rappaport, Wireless position location: Fundamentals, implementation strategies, and sources of error, in *IEEE Vehicular Technology Conference*, May 5–7, 1997, pp. 919–923.
3. G. V. Tsoulos, Smart antennas for mobile communication systems: benefits and challenges, *Elect. Commun. Eng. J.*, Vol. 11, No. 2, pp. 84–94, April 1999.
4. J. C. Liberti, Jr. and T. S. Rappaport, *Smart Antennas for Wireless Communications: IS-95 and Third Generation CDMA Application*, Prentice-Hall PTR, Upper Saddle River, NJ, 1999.

5. L. C. Godara, Application of antenna arrays to mobile communications. II. Beam-forming and direction-of-arrival considerations, *Proc. IEEE*, Vol. 85, No. 8, pp. 1195–1245, August 1997.
6. T. Ohira and J. Cheng, Analog smart antennas, *Adaptive Antenna Arrays*, Springer Verlag, Berlin, June 2004, pp. 184–204.
7. D. V. Thiel and S. Smith, *Switched Parasitic Antennas for Cellular Communications*, Artech House, Norwood, MA, 2002.
8. S. L. Preston, D. V. Thiel, T. A. Smith, S. G. O'Keefe, and J. W. Lu, Base-station tracking in mobile communications using a switched parasitic antenna, *IEEE Trans. Antennas Propag.*, Vol. 46, No. 6, pp. 841–844, June 1998.
9. R. Vaughan, Switched parasitic elements for antenna diversity, *IEEE Trans. Antennas Propag.*, Vol. 47, No. 2, pp. 399–405, Febuary 1999.
10. D. V. Thiel, S. G. O'Keefe, and J. W. Lu, Electronic beam steering in wire and patch antenna systems using switched parasitic elements, *IEEE Antenna Propag. Symp. Dig.*, Vol. 1, pp. 534–537, 1996.
11. S. L. Preston, D. V. Thiel, J. W. Lu, S. G. O'Keefe, and T. S. Bird, Electronic beam steering using switched parasitic patch elements, *Electron. Lett.*, Vol. 33, No. 1, pp. 7–8, January 2, 1997.
12. R. F. Harrington, Reactively controlled directive arrays, *IEEE Trans. Antennas Propag.*, Vol. AP-26, No. 3, pp. 390–395, March 1978.
13. T. Ohira and K. Gyoda, Hand-held microwave direction-of-arrival finder based on varactor-tuned analog aerial beamforming, in *Asia-Pacific Microwave Conference*, 2001, pp. 585–588.
14. K. Yang and T. Ohira, ESPAR antennas-based signal processing for DS-CDMA signal waveforms in ad hoc Network systems, in *IEEE 3rd Workshop on Signal Processing Advances for Wireless Communications*, 2001 (SPAWC '01), 2001, pp. 130–133.
15. J. Cheng, Y. Kamiya, and T. Ohira, Adaptive beamforming of ESPAR antenna based on steepest gradient algorithm, *IEICE Trans. Commun.*, Vol. E84-B, No. 7, pp. 1790–1800, 2001.
16. A. Komatsuzaki, S. Saito, K. Gyoda, and T. Ohira, Hamiltonian approach to reactance optimization in ESPAR antennas, in *Asia-Pacific Microwave Conferences, 2000*, Sydney, Australia, pp. 1514–1517.
17. C. Sun, A. Hirata, T. Ohira, and N. C. Karmakar, Fast beamforming of electronically steerable parasitic array radiator antennas: Theory and experiment, *IEEE Trans. Antennas and Propag.*, Vol. 52, No. 7, pp. 1819–1832, July 2004.
18. S. A. Leonov and A. I. Leonov, *Handbook of Computer Simulation in Radio Engineering, Communications, and Radar*, Artech House, Norwood, MA, 2001, pp. 175–176.
19. C. A. Balanis, *Antenna Theory Analysis and Design*, John Wiley & Sons, New York, 1997.
20. S. Y. Liao, *Microwave Solid-State Devices*, Prentice-Hall, Englewood Cliffs, NJ, 1985.
21. R. O. Schmidt, Multiple emitter location and signal parameter estimation, *IEEE Trans. Antennas Propag.*, Vol. AP-34, No. 3, pp. 276–280, March 1986.
22. D. H. Johnson and D. E. Dudgeon, *Array Signal Processing: Concepts and Techniques*, Prentice-Hall PTR. Englewood Cliffs, NJ, 1993.
23. D. C. Cox, R. R. Murray, H. W. Arnold, A. W. Norris, and M. F. Wazowicz, Cross-polarization coupling measured for 800 MHz radio transmission in and around houses

and large buildings, *IEEE Trans. Antennas Propag.*, Vol. AP-34, No. 1, pp. 83–87, January 1986.

24. W. C. Jakes, *Microwave Mobile Communications*, IEEE Press, New York, 1994.
25. C. Tepedelenlioglu, Modeling and Mitigation of Time-Selective and Frequency-Selective Fading in Single-Carrier and Multi-Carrier Communications, Ph.D thesis, University of Minnesota, May 2001.
26. J. Parsons, *The Mobile Radio Propagation Channel*, Halsted, New York, 1992.
27. A. Hirata, T. Aono, H. Yamada, and T. Ohira, Reactance-domain SSP MUSIC for an ESPAR antenna to estimate the DOAs of coherent waves, in *International Symposium on Wireless Personal Multimedia Communications*, WPMC2003, 3, Yokosuka, October 2003, pp. 242–246.
28. A. Hirata, E. Taillefer, T. Aono, H. Yamada, and T. Ohira, Correlation suppression performance for coherent signals in RD-SSP-MUSIC with a 7-element ESPAR antenna, in *European Conference on Wireless Technology, ECWT2004*, Amsterdam, October 2004.
29. E. Taillefer, E. Chu, and T. Ohira, ESPRIT algorithm for a seven-element regular-hexagonal shaped ESPAR, *European Conference on Wireless Technology*, ECWT2004, Amsterdam, October 2004.
30. R. O. Schmidt and R. E. Franks, Multiple source DF signal processing: An experimental system, *IEEE Trans. Antennas Propag.*, Vol. AP-34, No. 3, pp. 281–290, March 1986.
31. Y. Inoue, K. Mori, and H. Arai, DOA error estimation using 2.6 GHz DBF array antenna, in *Asia-Pacific Microwave Conference*, 2001, pp. 701–704.
32. K. R. Dandekar, H. Ling, and G. Xu, Effect of mutual coupling on direction finding in smart antenna applications, *Electron. Lett.*, Vol. 36, No. 22, pp. 1889–1891, October 2000.

CHAPTER 13

DOA GEO-LOCATION IN A REAL-TIME INDOOR WIFI SYSTEM UTILIZING SMART ANTENNAS

CHIN-HENG LIM
Temasek Laboratories, Nanyang Technological University, Singapore

BOON POH NG, MENG HWA ER, JONI POLILI LIE, and WENJIANG WANG
School of Electrical and Electronic Engineering, Nanyang Technical University, Singapore

13.1 INTRODUCTION

Indoor localization technologies have received considerable attention in recent years, and they are applicable in many fields. Applications include locating essential equipment in hospitals and specific items in warehouses by using radio-frequency identification (RFID) tags, tracking people with special needs, who are away from visual supervision, and navigating firefighters inside buildings. As with outdoor systems, they both suffer degradation in performance when multipath errors are present. Over the last couple of years, there has been growing interest in using Wireless Fidelity (WiFi) technology for localization within the RFID industry. As expected, this new RFID technology has also generated numerous commercial, military, and public safety applications in real-time location systems to monitor the position of assets and personnel [1–3]. WiFi, which is also known as 802.11 b/g, has become the preferred technology for wireless local area networking in both business and home environments [4–7]. Indoor localization technology finds important applications in commercial, public safety, and military applications. There is an increasing interest in accurate location finding techniques and location-based applications for indoor environments [8–10].

There are many kinds of location acquisition systems developed to get an object's location. The Global Positioning System (GPS) covers most of the earth's surface,

Handbook of Smart Antennas for RFID Systems, Edited by Nemai Chandra Karmakar

These robust techniques allow for the system to mitigate against the outliers caused by the multipath errors, therefore improving the location accuracy of the system.

Unlike other works, we consider the access point (AP) as the mobile station and consider the laptop as the base station, and this arrangement is also highly applicable in modern commercial applications. In some applications, the AP can also be replaced by an RFID tag or a personal digital assistant (PDA), which will provide more mobility. In this case, the use of the AP as a mobile target allows for the verification of our proposed robust methods at a lower cost.

The major contribution of this chapter is to report the indoor WiFi system setup used and the improvement that can be obtained by using robust methods for indoor localization systems. The utilization of these methods requires a software upgrade from what have been reported in reference 18, and simulation results show a better localization performance over conventional methods, which is critical for indoor localization applications. The limitations of our proposed system are the longer time taken to perform localization and the additional hardware required, which will be discussed in the following sections.

This chapter is organized as follows: The research methodology is stated in Section 13.2, and an overview of the IEEE 802.11 standard is provided in Section 13.3. Section 13.4 presents the indoor WiFi localization system setup. The proposed robust techniques are introduced in Section 13.5, and the experimental results are shown in Section 13.6. The main contributions and recommendation, which are further elaborated in Section 13.7, can be summarized as follows:

1. The system, based on the IEEE 802.11 standard, uses the existing wireless local area network infrastructure with minor changes. The localization is done by using the received signal strength information from employing smart antennas, and the system avoids the use of an off-line training phase, which is computationally intensive and requires a big database.
2. The use of robust techniques to mitigate the multipath effects and improve the localization accuracy is illustrated. Compared with conventional methods, the proposed robust techniques give a better localization performance and higher resolution accuracy, which are extremely beneficial for indoor localization systems.
3. The system can be further improved by using time-of-arrival information together with the received signal strength to improve the localization accuracy. This is the scope for future work.

13.2 RESEARCH METHODOLOGY

13.2.1 Smart Antennas System: An Overview

Smart antennas offer many ways to improve wireless system performance, such as providing enhanced coverage, improving system capacity, and reducing sensitivity to nonideal behaviors. Smart antennas use an array of low-gain antenna elements that are connected by a combining network [17]. Consider a uniform linear array of

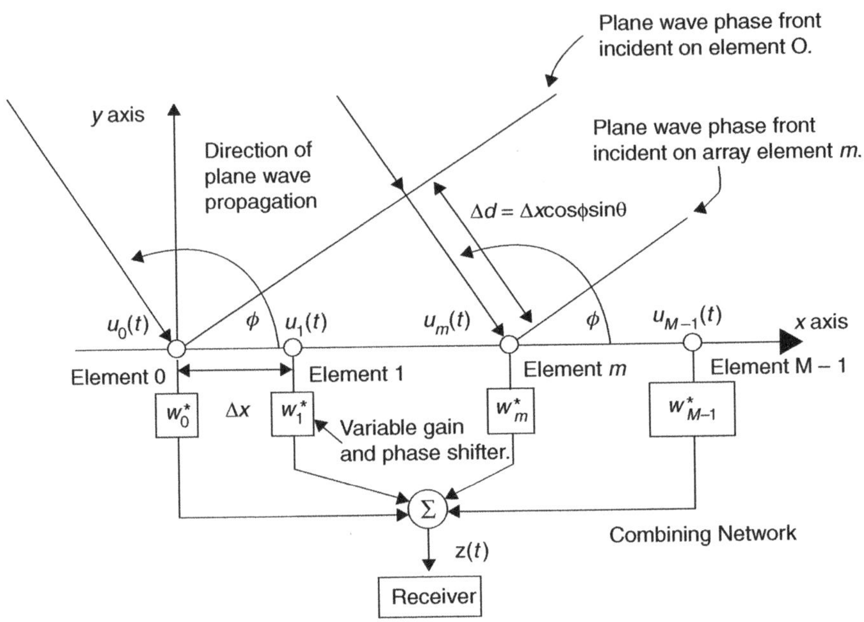

FIGURE 13.2. Uniform linear antenna array oriented along the x axis, receiving a plane wave from direction (θ, ϕ).

$M = 4$ omnidirectional elements, illustrated in Figure 13.2, oriented along the x axis with interelement spacing of Δd. For a plane wave incident on the array from the direction (θ, ϕ), the signal $z(t)$ at the array output is given by

$$z(t) = w^H u(t) = \sum_{m=0}^{M-1} w_m^* u_m(t) \tag{13.1}$$

For each of the smart antennas used in our proposed indoor localization system, the four antenna elements have been calibrated prior to the data collection process. As the scanning array uses a set of fixed steering weights, the lower side lobes of the array pattern will suppress the non-look-directional signals. This provides mitigation against multipath and/or reflected signals. Following this calibration process, the beamforming weights are subsequently obtained for each scan angle. These weights are pre-programmed into the client (introduced in Section 13.3), in order to determine the RSS information from each scan angle.

Figure 13.3 shows the client (laptop PC) equipped with the smart antenna. The smart antenna shown here is a prototype and a test bed. Although the use of this smart antenna gives us an improvement in performance over standard hardware for APs, there is a trade-off in terms of costs structure when the localization is to be done in a larger area. The operating size can be reduced, and work on this area is currently in progress. More details of the system's features and experimental setup are presented in Section 13.4 and Section 13.5, respectively.

In this case, a common method used to determine the position of a signal transmitter is by using triangulation: In a 2D coordinate system, if we know the position for the vertices of a triangle, we can calculate the position of the interior point (triangle incenter) [18]. Another method is to use the traditional least-squares method [8] since there are redundant measurements. However, we only consider the triangulation method in this work. Triangulation uses multiple APs to find a target based on the RSS at each AP, and its accuracy can be enhanced by using more APs within the localization system. Triangulation is more accurate than the closest neighbor method [9], which determines the target position as the AP with the greatest RSS. In the absence of big measurement errors, such as non-line-of-sight errors, the use of DOA information gives a better performance than the closest-neighbor method.

As shown in Figure 13.4, let the 2D coordinates of the three readers be defined as (X_{r1}, Y_{r1}), (X_{r2}, Y_{r2}), and (X_{r3}, Y_{r3}). Using the same coordinate axis, the equations of the three lines are given as

$$l_1: \quad y = -\tan\theta_1 \cdot x + X_{r1} \cdot \tan\theta_1 + Y_{r1} \tag{13.2}$$

$$l_2: \quad y = \tan\theta_2 \cdot x - X_{r2} \cdot \tan\theta_2 + Y_{r2} \tag{13.3}$$

$$l_3: \quad y = -\tan\theta_3 \cdot x + X_{r3} \cdot \tan\theta_3 + Y_{r3} \tag{13.4}$$

Given these three lines, the intercept points can then be calculated from any two of them. Considering l_1 and l_2, by subtracting Eqs. (13.2) from (13.3), the x-intercept point of B is given as

$$x = \frac{X_{r2} \cdot \tan\theta_2 + X_{r1} \cdot \tan\theta_1 - Y_{r2} + Y_{r1}}{\tan\theta_2 + \tan\theta_1} \tag{13.5}$$

Then, the y-intercept point of B can be solved using either Eq. (13.2) or Eq. (13.3). In this case, the three intercept points (A, B, and C) that are the peak points from the scan signal strength of a triangle can be calculated.

The lengths of the three legs of this triangle are labeled as a, b, and c respectively. Therefore, the origin of the triangle's incenter (taken to be the target's estimated location) will be

$$x_{\text{incenter}} = \frac{(b \cdot x_1 + c \cdot x_2 + a \cdot x_3)}{a + b + c} \tag{13.6}$$

$$y_{\text{incenter}} = \frac{(b \cdot y_1 + c \cdot y_2 + a \cdot y_3)}{a + b + c} \tag{13.7}$$

where (x_1, y_1) and (x_2, y_2) are the peaks A and B connected by leg a; (x_2, y_2) and (x_3, y_3) are connected by leg b; (x_3, y_3) and (x_1, y_1) are connected by leg c and

$$a = \sqrt{(x_2 - x_1)^2 + (y_2 - y_1)^2} \tag{13.8}$$

$$b = \sqrt{(x_3 - x_2)^2 + (y_3 - y_2)^2} \tag{13.9}$$

$$c = \sqrt{(x_1 - x_3)^2 + (y_1 - y_3)^2} \tag{13.10}$$

13.3 IEEE 802.11 STANDARD

The IEEE 802.11 Wireless LAN standard defines the media access control (MAC) and physical (PHY) layers for a LAN with wireless connectivity [4–7]. It addresses LAN where the connected devices communicate over the air to other devices that are within close proximity to each other. The IEEE 802.11 protocol is an extension of the IEEE 802.3 protocol for wired networks. The IEEE 802.3 protocol, which defines the Media Access Control (MAC) and physical layers for a wired network, is the most widely used standard for wired network that was developed out of the original work done on the Ethernet. Specifically, the 802.11 standard, similar in most respects to the IEEE 802.3 Ethernet standard, addresses the following points:

- Functions required for an 802.11 compliant device to operate either in a peer-to-peer fashion or integrated with an existing wired LAN.
- Operation of the 802.11 device within possibly overlapping 802.11 wireless LANs and the mobility of this device between multiple wireless LANs.
- MAC level access control and data delivery services to allow upper layers of the 802.11 network.
- Several physical layer signaling techniques and interfaces.
- Privacy and security of user data being transferred over the wireless media.

13.3.1 Mobility

Mobility of wireless stations is the most important feature of a wireless LAN (WLAN). A WLAN will not serve much purpose if stations are not able to move about freely from location to location either within a specific WLAN or between different WLAN "segments." For compatibility purposes, the 802.11 MAC must appear to the upper layers of the network as a "standard" 802 LAN. The 802.11 MAC layer is forced to handle station mobility in a fashion that is transparent to the upper layers of the 802 LAN stack. This forces functionality into the 802.11 MAC layer that is typically handled by upper layers.

13.3.2 802.11 WLAN Architecture

The 802.11 architecture, shown in Figure 13.5, comprises several components and services that interact to provide station mobility transparent to the higher layers of the network stack.

Wireless LAN Station. The station is the most basic component of the wireless network. It is a device that contains the functionality of the 802.11 protocol, that being MAC, PHY, and a connection to the wireless media. Typically, the 802.11 functions are implemented in the hardware and software of a network interface card (NIC). A station could be a laptop PC, RFID tag, or an AP. Stations may be mobile, portable, or stationary; and all stations support the 802.11 station services of authentication, de-authentication, privacy, and data delivery.

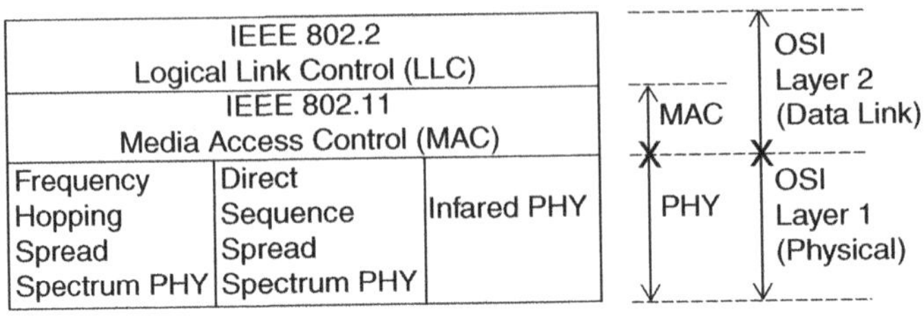

FIGURE 13.5. IEEE 802.11 architecture.

Basic Service Set (BSS). The Basic Service Set (BSS) is defined as the basic building block of an 802.11 wireless LAN. The BSS consists of a group of any number of stations.

13.3.3 802.11 Topologies

Independent Basic Service Set (IBSS). The most basic wireless LAN topology is a set of stations, which recognize each other and are connected via the wireless media in a peer-to-peer fashion. This form of network topology is referred to as an Independent Basic Service Set (IBSS) or an ad hoc network, which is shown in Figure 13.6.

In an IBSS, the mobile stations communicate directly with each other. Every mobile station may not be able to communicate with every other station due to the range limitations. There are no relay functions in an IBSS, therefore all stations need to be within range of each other and communicate directly.

Infrastructure Basic Service Set. An Infrastructure Basic Service Set, shown in Figure 13.7, is a BSS with a component called an AP. The AP provides a local relay function for the BSS. All stations in the BSS communicate with the AP and

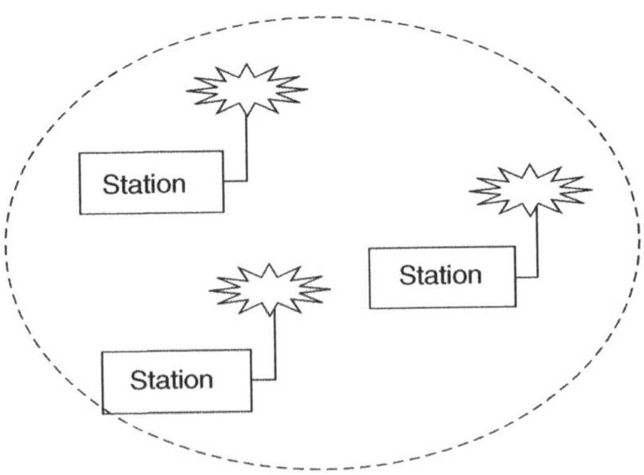

FIGURE 13.6. Independent Basic Service Set (IBSS).

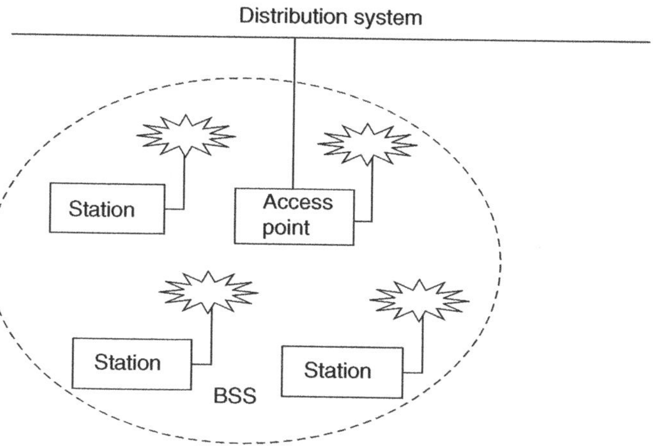

FIGURE 13.7. Infrastructure Basic Service Set (IBSS).

no longer communicate directly. All frames are relayed between stations by the AP. This local relay function effectively doubles the range of the IBSS. The AP may also provide connection to a distribution system.

The distribution system is the means by which an AP communicates with another AP to exchange frames for stations in their respective BSSs, forward frames to follow mobile stations as they move from one BSS to another, and exchange frames with a wired network.

13.3.4 802.11 Media Access Control

The 802.11 MAC layer provides functionality to allow reliable data delivery for the upper layers over the wireless PHY media. The data delivery itself is based on an asynchronous, best-effort, connectionless delivery of MAC layer data. There is no guarantee that the frames will be delivered successfully. The 802.11 MAC provides a controlled access method to the shared wireless media called Carrier-Sense Multiple Access with Collision Avoidance (CSMA/CA), which is similar to the collision detection access method deployed by 802.3 Ethernet LANs. This means that a station wishing to transmit must first sense the radio channel to determine if another station is transmitting. If the medium is not busy, the transmission may proceed.

Another function of the 802.11 MAC is to protect the data being delivered by providing security and privacy services. Security is provided by the authentication services and by Wireless Equivalent Privacy (WEP), which is an encryption service for data delivered on the WLAN.

13.3.5 802.11 Physical Layer (PHY)

The 802.11 physical layer (PHY) is the interface between the MAC and the wireless media where frames are transmitted and received. The PHY provides three functions.

First, the PHY provides an interface to exchange frames with the upper MAC layer for transmission and reception of data. Second, the PHY uses signal carrier and spread spectrum modulation to transmit data frames over the media. Third, the PHY provides a carrier sense indication back to the MAC to verify activity on the media. 802.11 provides three different PHY definitions: Both *Frequency Hopping Spread Spectrum* (FHSS) and *Direct Sequence Spread Spectrum* (DSSS) support 1- and 2-Mbit/s data rates.

Received Signal Strength Indicator (RSSI). The RSSI is a standard parameter found in most wireless devices, which has a value of 0 through RSSI Max. This parameter requires no additional hardware and is unlikely to impact local power consumption, sensor size, and thus costs. It is a measure of the PHY sublayer of the energy observed at the antenna used to receive the current physical layer convergence procedure (PLCP) protocol data unit (PPDU). The RSSI is measured between the beginning of the start frame delimiter (SFD) and the end of the PLCP header error check (HEC). RSSI is intended to be used in a relative manner and absolute accuracy of the RSSI reading is not specified.

In free space, the RSS varies as the inverse square of the distance r between the transmitter and the receiver. Let us denote this received power by $P_r(r)$. The received power $P_r(r)$ is related to the distance r through the Friis equation [8]

$$P_r(r) = \frac{P_t G_t G_r \lambda^2}{(4\pi)^2 r^2} \tag{13.11}$$

where P_t is the transmitted power in watts, G_t is the transmitter antenna gain, G_r is the receiver antenna gain, and λ is the wavelength of the transmitter signal in meters.

The propagation of a signal is affected by reflection, diffraction, and scattering. Of course, these effects are environment (indoors, outdoors, rain, buildings, etc.)-dependent. However, it is accepted on the basis of empirical evidence that it is reasonable to model the RSS $P_r(r)$ at any value of r at a particular location as a random variable with a distance-dependent mean value [19]. That is

$$P_r(r) = P_0(r_0) - 10 n_p \log_{10}\left(\frac{r}{r_0}\right) + X_\sigma \tag{13.12}$$

where $P_0(r_0)$ is a known reference power value in dB milliwatts (dBm) at a reference distance r_0 from the transmitter; n_p is the path loss exponent that measures the rate at which the RSS decreases with distance and the value of n_p depends on the specific propagation environment, and X_σ is a zero mean Gaussian distributed random variable with standard deviation σ that accounts for the random effect of shadowing.

From Eq. (13.12), we can conclude that given the RSS measurement P_{ij} between a transmitter g and a receiver h, a maximum likelihood estimate of the distance r_{ij}

between the transmitter and the receiver is

$$r_{ij} = r_0 \left(\frac{P_{ij}}{P_0 (r_0)} \right)^{1/n_p} \tag{13.13}$$

13.4 WIFI LOCALIZATION SYSTEM

13.4.1 Hardware Structure

In this section, we present the system setup and the data collection process. The prototype of this WiFi indoor localization system consists of the following:

1. The smart antenna used is a four-element uniform linear array; altogether, three smart antennas are used. The beamformer and scan direction are controlled by the laptop PC through the parallel port. The RSS information is connected to the laptop PC Personal Computer Memory Card International Association (PCMCIA) wireless network interface card (WNIC).
2. The laptop PC equipped with the smart antenna, henceforth referred to as the reader, is the client used to process the RSS data and send the results to the server PC. The WNIC equipped on the laptop PC is an Orinoco® 802.11b WaveLAN silver PCMCIA card and has a connector for external range extension.
3. The host PC is the server within the data processing network. It starts the data collection process of the readers. By using the angle and RSS information received from the three readers, the server then performs triangulation to estimate the mobile target's position.
4. The mobile target is a single LINKSYS® 802.11b wireless AP, which acts as the mobile transmitter. The data processing system is a client/server network, whereby all three PCs are connected within the same LAN. Figure 13.8 shows the structure of the WiFi indoor localization system.

At each reader, the antenna is deployed physically to a direction where the test points are covered equally to ensure that all the test points can be scanned by the antenna array. The software is designed to operate on the client/server mode function. Client-end software performs beamforming network control and network connection requests from each of the three readers. The server-end software in a separate PC acts as the center of information, which sends commands/requests to the clients or receives the real-time data from the clients. After each scan, the server performs the calculation and determines the localization results.

13.4.2 Software Structure

The system software is designed to operate on two parts: server and client. For the server part, the center of information sends commands and requests to the client PC

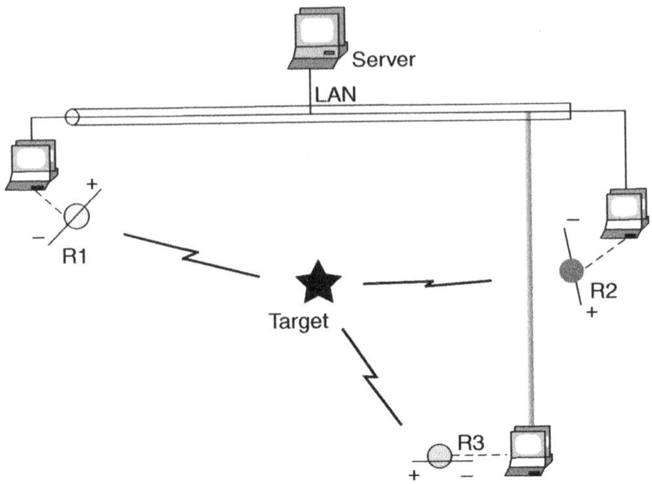

FIGURE 13.8. Structure of the WiFi indoor localization system.

to control the scanning procedure and receive the real-time data from the client PC, and then it performs calculation and refreshes the location result displays. For the client part, it performs the scanning procedure within a certain range and collects the RSS from the target. After the scanning procedure is done, the client PC sends the RSS and DOA results to the server PC through the LAN. The flow chart of the process is shown in Figure 13.9.

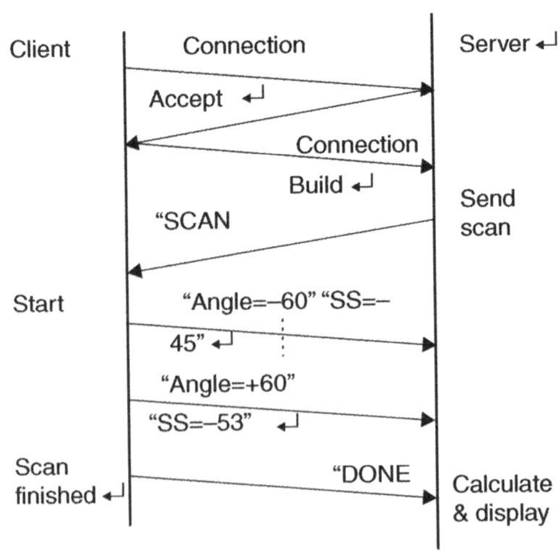

FIGURE 13.9. Flow chart of communication control between the server and client PC.

A brief description between the server and client PC communication is stated below:

- Both the server and client program start up independently. The server program waits for connection from the three client PCs.
- The client PC waits for the user to select which network device to be connected for scanning and tries to establish socket connection with the server PC.
- The server PC accepts connection from the client PC.
- Once all three readers are connected, the server PC sends the "SCAN" command to commence scanning.
- Client starts to scan from $-60°$ to $60°$ at a step size of $1°$. It gets the RSS from the WNIC and sends the results (angle and signal strength) to the server PC.
- Once scanning is completed, each client PC sends a "DONE" to the server PC to indicate the end of scanning.
- When all three readers indicate that the scanning is completed, the server PC starts to find the DOA of the mobile target for each reader. It uses the triangulation method to calculate the position of the mobile target and display the result on the server PC GUI.

13.5 ROBUST METHODS FOR MULTIPATH ERRORS MITIGATION

In early signal processing applications, the mean filter is the classic filter that is suited for random Gaussian noise. However, linear filters are not able to produce good results for the restoration of signals corrupted by nonlinear or non-Gaussian noise like impulsive noise [20]. Therefore, in many modern applications, nonlinear filters have been employed widely in removing non-Gaussian noise present in data. The most common and simplest type of these filters is the median filter, which shows very good performance for the removal of impulsive noise. Moreover, fast algorithms exist for implementing the median filter with reduced computational complexity. The median filter was introduced in reference 21 as a smoothing technique in time-series analysis. In reference 22, the median filter was used to smooth speech waveforms and the authors reported favorable results. They found that the median filter could preserve the discontinuities of a sufficient duration while eliminating roughness in the signal, whereas the linear filter was seen to be inadequate in that much information was lost due to noise.

To begin the scanning process, the server PC sends the "Scan" request to each reader. For each scan, each client PC is used to determine the RSS from a field of view of $[-60°, 60°]$. The scan step-size is $1°$and a total of three RSS measurements are obtained in each scan. The desired RSS value for each jth scan was determined by taking the mean of these three RSS measurements ($\bar{\mathbf{r}}_{i,j}$). Denote the three RSS measurements from the ith ($i = 1, 2, 3$) client PC as $\mathbf{r}_{i,j} = \lfloor r_{i,j,1} \quad r_{i,j,2} \quad r_{i,j,3} \rfloor$; then

$$\mathbf{r}_{i,j} = \frac{r_{i,j,1} + r_{i,j,2} + r_{i,j,3}}{3} \tag{13.14}$$

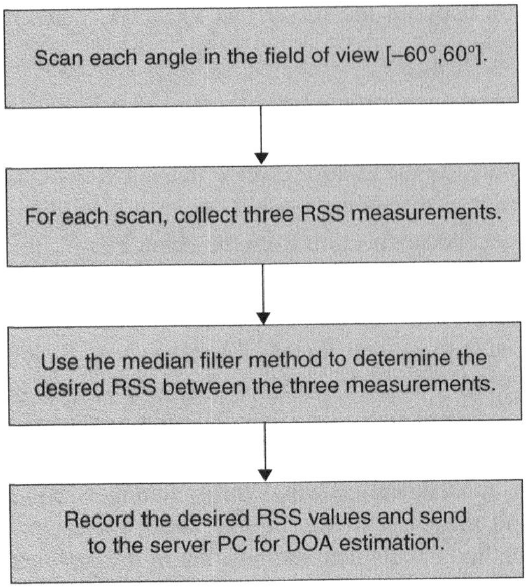

FIGURE 13.10. Flow chart detailing how the RSS measurements are obtained for each reader.

This is not very ideal, since outliers are prevalent in a multipath environment and they may result in a bias. Hence, we propose using the median filter to determine the desired RSS value ($r^m_{i,j}$). In this case, the three RSS measurements obtained are first sorted as $\mathbf{r}^S_{i,j} = [r^S_{i,j,1} \quad r^S_{i,j,2} \quad r^S_{i,j,3}]$, where $r^S_{i,j,1} \leq r^S_{i,j,2} \leq r^S_{i,j,3}$. Thereafter, the median (middle) measurement is taken to be the desired RSS value. Hence, $r^m_{i,j} = r^S_{i,j,2}$.

The desired RSS value from each jth scan are then recorded as $\mathbf{r}^m_i = [r^m_{i,1} \quad r^m_{i,2} \quad \cdots \quad r^m_{i,J}]$, where $J = 121$ is the total number of scan angles. Eventually all these RSS values ($\mathbf{r}^m_1, \mathbf{r}^m_2, \mathbf{r}^m_3$) and their respective scan angles are sent to the server PC for DOA estimation. Figure 13.10 shows the flow chart detailing the implementation of this robust technique on each client PC.

When all three readers have completed the scanning process, the server PC determines the DOA of the MT for each reader, by using the maximum RSS criterion. Thereafter, it computes the estimated MT's location by using the triangulation method. This DOA denotes the angle at each reader where the signal is arriving from the MT. In an ideal environment, the angle that has a direct line-of-sight from the receiver to the transmitter produces the maximum RSS. However, due to multipath errors, several angles may exist that corresponds to the same maximum RSS.

In reference 18, the method used was to sort the DOAs from these various RSS measurements and the median value is taken as the desired DOA. For example, let $\boldsymbol{\theta}^S_i = [\theta^S_{i,1} \quad \theta^S_{i,2} \quad \cdots \quad \theta^S_{i,L}]$ denote the sorted DOAs that correspond to the maximum RSS value for the ith reader and L can range from 1 to J. Then the desired DOA (θ^d_i)

is taken to be the median value of $\boldsymbol{\theta}_i^S$. This implementation is robust but fails when very extreme outliers are present, which results in erroneous DOA estimation.

Hence, we further propose using the trimmed-mean method at the server PC to determine the desired DOA. The trimmed mean [23] filter is a useful robust estimator against multipath errors [24] because it is less sensitive to outliers than the mean and will also give a reasonable estimate of central tendency or mean for almost all statistical models. It is therefore less susceptible to sampling fluctuation than the mean filter for extremely skewed distributions. The number of data values that are trimmed is controlled by the trimming parameter k, which is the number of measurements trimmed at both ends and assumes values between 0 and $N/2$ for each reader (where N is the number of the DOA measurements that have the maximum RSS value). Note that the trimmed-mean filter performs like a mean filter when k is close to 0. When k is close to $N/2$, the trimmed-mean filter behaves like a median filter.

By removing the extreme value at both ends (symmetric trimming) of the sorted DOAs [25], the desired DOA is taken to be the mean of the remaining DOAs after trimming. Let $\boldsymbol{\theta}_i^Z = [\theta_{i,1}^Z \quad \theta_{i,2}^Z \quad \cdots \quad \theta_{i,L'}^Z]$ denote the vector of DOAs after the trimming process, where the value of L' may vary for each reader; then the desired DOA is now calculated as

$$\theta_i^d = \frac{\theta_{i,1}^Z + \theta_{i,2}^Z + \cdots + \theta_{i,L'}^Z}{L'} \tag{13.15}$$

This trimmed-mean method is only used when there are four or more RSS measurements with the same maximum value, otherwise it is equivalent to the median filter in reference 18. Figure 13.11 shows the flowchart detailing how the trimmed-mean method is implemented on the server PC.

13.6 EXPERIMENT PROCEDURE

The test setup we chose was an (8 m × 9 m) area, which is located on the fourth floor of a six-story building. Figure 13.12 shows the layout of the laboratory where the indoor localization system is deployed, which consists of the positions of the three readers (R1, R2, R3) and 14 test points. *It should be noted that the test points are **not** meant for building any prior training database.*

The experiments were carried out when there was a substantial amount of human traffic occurring throughout the data collection process. We randomly picked 10 test points, and for each location point the RSS information was collected 15 times (150 results in total). The time taken for each scan, done concurrently for all three readers, is 30 seconds. The performance metric used in our study is the error distance in meters:

$$\text{Error distance} = \sqrt{(x - \hat{x})^2 + (y - \hat{y})^2} \tag{13.16}$$

where (x, y) and $(\hat{x}, \hat{y})$ denote the original and estimated positions of the mobile target, respectively.

Store room
R3
(-2,8)
R2
(3.5,7.5)
14
13
12
11
10
9
8
7
6
5
4
3
1
2
R1
(-1.5,2.5)

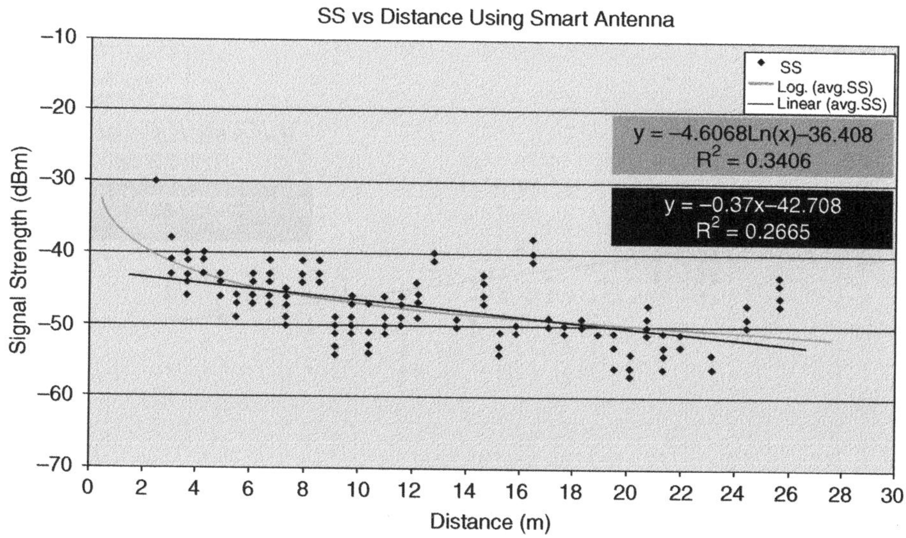

FIGURE 13.13. Trend-line analysis for raw SS using smart antenna.

13.6.1 Signal Strength Versus Distance

The first part of the experiment is to analyze the relationship between signal strength (SS) and distance for IEEE 802.11 WLAN and determine whether the mathematic model of signal propagation is reliable for indoor WLAN localization. Results are compared through the trend line with R^2 (the coefficient of determination, which is a useful measure for indicating the goodness of regression; high value of R^2 suggest that there is a good match between the estimated and the measured values of the signal strength).

The trend-line analyses are conducted with two types of data: raw SS and the averaged SS. We also compare the use of smart antenna and omnidirectional antenna. Based on the figures above, it is easy to conclude that:

1. The logarithm model has a better fit than the linear model, because the former R^2 is higher.
2. Smart antenna has a good performance in noise suppression, and raw data are apparently self-organized (Figure 13.13 versus Figure 13.14); this helps to improve the WLAN localization resolution for SS-based methods.
3. The model based on average SS is more reliable to fit the empirical data (Figure 13.13 versus Figure 13.15).

The analysis shows that the triangulation method will achieve a better result when smart antennas are used rather than the omnidirectional antenna.

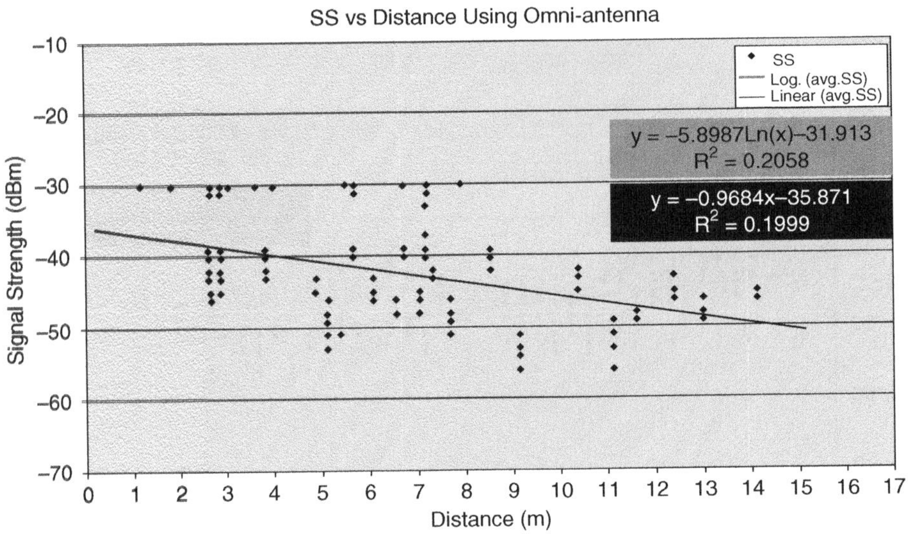

FIGURE 13.14. Trend-line analysis for raw SS using omnidirectional antenna.

13.6.2 Indoor Localization Performance

As discussed in Section 13.5, three RSS measurements are obtained for each scan angle of the client PC. Thereafter, either the mean filter or the median filter can be implemented to determine the estimated RSS value for that particular scan angle.

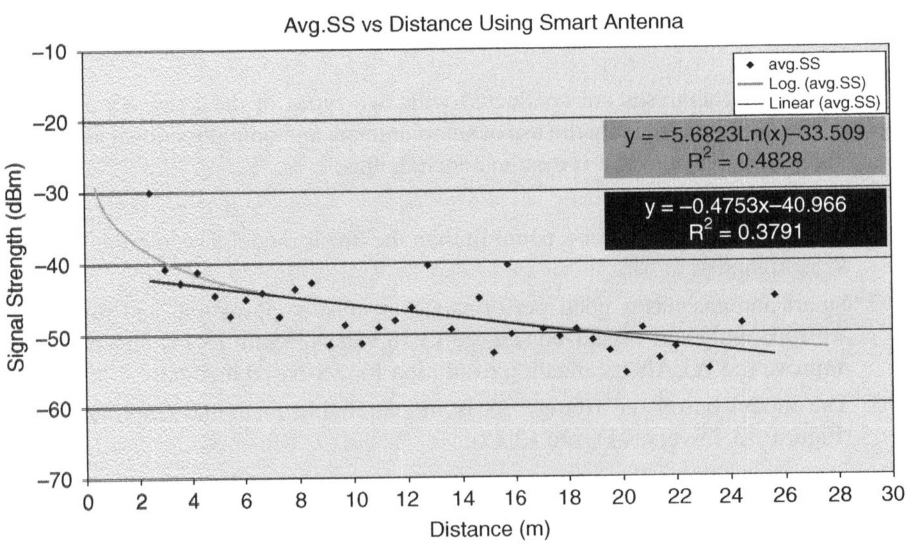

FIGURE 13.15. Trend-line analysis for averaged SS using smart antenna.

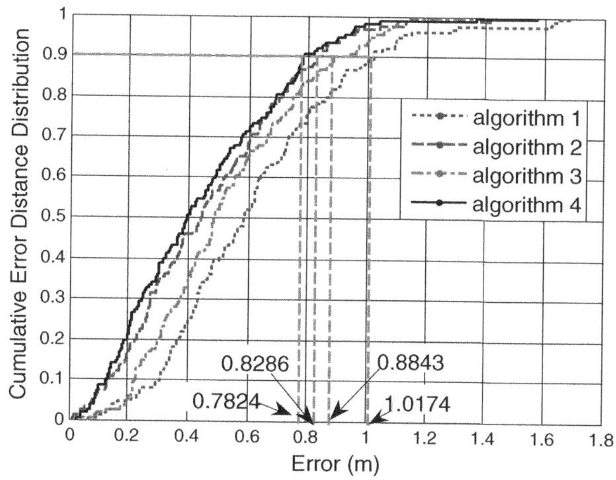

FIGURE 13.16. CDF plot of error distance for Algorithms 1–4.

After scanning through the entire field of view, the maximum RSS value and its corresponding scan angle will be sent to the server PC for localization. In the situation when more than one scan angle has the same maximum RSS value, the server PC can implement either the median filter or trimmed-mean filter to determine the resulting estimated DOA. The following experimental results will highlight the effects of implementing the afore-mentioned filters individually.

Figure 13.16 shows the empirical cumulative distribution function (CDF) of the error distance. The CDF plot is useful for examining the distribution of the errors and it indicates the estimation error for a certain confidence level.

Here we denote:

- *Algorithm 1:* Using the mean filter method at each reader and the median filter method on the server PC (implementation in reference 18).
- *Algorithm 2:* Using the median filter method at each reader and the median filter method on the server PC (to show the effects of implementing the robust median filter method at each reader).
- *Algorithm 3:* Using the mean filter method at each reader and the trimmed-mean method on the server PC (to show the effects of implementing the robust trimmed-mean method on the server PC).
- *Algorithm 4:* Using the median filter method at each reader and the trimmed-mean method on the server PC (to show the effects of implementing both robust techniques at the reader and server PC respectively).

From Figure 13.16, it can be seen that there is a significant improvement over the implementation (Algorithm 1) in reference 15 when we use the median filter at each

TABLE 13.1. Results Summary of Error Distance

Method	25th Percentile	50th Percentile	75th Percentile
RADAR [5]	1.92 m	2.94 m	4.69 m
RF fingerprinting [6]	0–0.6 m	0.25–1.0 m	1.2–1.5 m
DOA using smart antennas	0.21 m	0.39 m	0.66 m

reader to determine the desired RSS value (Algorithm 2). At the 90th percentile, there is an improvement of 0.19 m in resolution accuracy.

The benefits of using the trimmed-mean method at the server PC are shown (Algorithm 3). There is an improvement of 0.13 m at the 90th percentile. This improvement is not as great as implementing Algorithm 2, because occurrences of four or more DOAs having the same maximum RSS value are rare.

It can be seen that Algorithm 4 performs the best with an error of only 0.78 m at the 90th percentile. The results also indicate that the median filter and trimmed-mean methods provide mitigation against multipath effects.

The improvements shown here, over that reported in reference 18, result from the implementation of the robust techniques reported in Section 13.5. There is no need for any major hardware modification/upgrade or any highly complex software algorithms, because only a retrofit is required. The improvements, even by less than 0.5 m, are still very beneficial for indoor localization applications. Figure 13.15 shows that the errors occur at a distance less than 2 m, which is superior to existing techniques. Furthermore, the results show that robust techniques should be implemented as early in the process (reader) as possible, so as to minimize the possibility of magnifying any errors later on.

For comparison purposes, the accuracy information for the proposed method against two conventional methods is summarized in Table 13.1 for the 25th, 50th, and 75th percentiles, respectively. It can be observed that our proposed methodology compares favorably with those in the literature. In particular, at the 75th percentile, it gives an improvement of 4 m and 0.8 m over the RADAR and fingerprinting methods, respectively.

13.7 CONCLUSION

In this chapter the implementation of a real-time WiFi indoor location system that utilizes smart antennas has been discussed. The main contribution of the work is to report on the proposed WiFi system, incorporating various robust techniques, for more accurate localization. The system is based on the IEEE 802.11 standard, and it uses the existing wireless local area network infrastructure with minor changes. The localization is done by using the received signal strength information from employing smart antennas, which gives better localization performance over standard hardware used for WiFi APs. Our system avoids the use of an off-line training phase, which is computationally intensive and requires a big database.

In addition, the use of robust techniques to mitigate the multipath effects and improve the localization accuracy is also illustrated. The robust methods include the median filter method and the trimmed-mean method for the reader and server PC, respectively. Compared with conventional methods, the proposed robust techniques give a better localization performance and higher resolution accuracy, which are extremely beneficial for indoor localization systems, especially for the various critical applications mentioned here.

The system can be further improved by using time-of-arrival information together with the received signal strength to improve the localization accuracy. This is scope for future work.

REFERENCES

1. K. Pahlavan and A. Levesque, *Wireless Information Networks*, 2nd edition, John Wiley & Sons, Hoboken, NJ, 2005.
2. M. A. Assad, M. Heidari, and K. Pahlavan, Effects of channel modelling on performance evaluation of WiFi RFID localization using a laboratory testbed, in *IEEE Globecom Telecommunications Conference*, 2007, pp. 366–370.
3. R. Tesoriero, J. A. Gallud, M. Lozano, and V. M. R. Penichet, A location-aware systems using RFID and mobile devices for art museum, in *International Conference on Autonomic and Autonomous Systems*, 2008, pp. 76–81.
4. P. S. Henry, and H. Luo, WiFi: What's next? *IEEE Commun. Mag.*, Vol. 40, No. 12, pp. 66–72, 2002.
5. IEEE Standard 802.11 Association, Wireless LAN medium access control (MAC) and physical layer (PHY) specifications, 1999.
6. B. Crow, T. Widjaja, J. Kim, and P. Sakai, IEEE 802.11 wireless local area networks, *IEEE Commun. Mag.*, pp. 116–126, 1997.
7. T. Cooklev, *Wireless Communications Standards: A Study of IEEE 802.11, 802.15 and 802.16*, IEEE Press, New York, 2004.
8. K. Pahlavan, X. Li, and J. P. Makela, Indoor geolocation science and technology, *IEEE Commun. Mag.*, Vol. 40, No. 2, pp. 112–18, 2002.
9. P. Bahl, and V. Padmanabhan, RADAR: An in-building RD-based user location and tracking system, *IEEE INFOCOM*, Vol. 2, pp. 775–784, March 2002.
10. R. H. Jan, and Y. R. Lee, An indoor geolocation system for wireless LANs, *ICPPW*, pp. 29–34, October 2003.
11. T. S. Rappaport, *Wireless Communications: Principles and Practice*, Prentice-Hall, Englewood Cliffs, NJ, 1996.
12. T. S. Rappaport, J. Reed, and B. Woerner, Position location using wireless communications on highways of the future, *IEEE Commun. Mag.*, Vol. 34, No. 10, pp. 33–41, 1996.
13. A. Smailagic, and D. Kogan, Location sensing and privacy in a context aware computing environment, *IEEE Wireless Commun. Mag.*, pp. 10–17, 2002.
14. R. Casas, A. Marco, J. J. Guerrero, and J. Falco, Robust estimator for non-line-of-sight error mitigation in indoor localization, *EURASIP J. Appl. Signal Processing*, Vol. 2006, No. 17, pp. 156–163, 2006.

15. A. M. Ladd, K. E. Bekris, A. Rudys, L. E. Kavraki, D. S. Wallach, and G. Marceau, Robotics-based location sensing using wireless Ethernet, *MOBICOM*, pp. 00–00, September 2002.
16. J. Yin, Q. Yang, and L. Ni, Adaptive temporal radio maps for indoor location estimation, *IEEE PERCOM*, pp. 85–94, 2005.
17. J. S. Liberti and T. S. Rappaport, *Smart Antennas for Wireless Communications: IS-95 and Third Generation CDMA Applications*, Prentice-Hall Communications Engineering and Emerging Technologies Series, Englewood Cliffs, NJ, 1999.
18. C. H. Lim, Y. Wan, B. P. Ng, and C. M. See, A real-time indoor WiFi localization system utilizing smart antennas, *IEEE Trans. Consumer Electron.*, Vol. 53, No. 2, pp. 618–622, 2007.
19. R. Berhardt, Macroscopic diversity in frequency reuse radio systems, *IEEE J. Selected Areas Commun.*, Vol. 5, No. 5, pp. 862–870, 1987.
20. R. Oten and R. J. P. de Figueiredo, Adaptive alpha-trimmed mean filters based on asymptotic variance minimization, in *Proceedings of the IEEE International Symposium on Circuits and Systems*, Vol. 3, 1999, pp. 45–48.
21. J. W. Tukey, Nonlinear (non-superposable) methods for smoothing data, *Congress Records, EASCON*, 1974, p. 673.
22. L. R. Rabiner, M. R. Sambur, and C. E. Schmidt, Applications of a nonlinear smoothing algorithm to speech processing, *IEEE Trans. Acoustic Speech Signal Processing*, Vol. ASSP-23, pp. 552–557, 1975.
23. R. A. Maronna, *Robust Statistics: Theory and Methods*, John Wiley & Sons, Hoboken, NJ, 2006.
24. G. W. Webb, I. V. Minin, and O. V. Minin, New technique to combat multipath fading in wireless networks, in *Proceedings of SPIE, Wireless Sensing and Processing*, Vol. 6248, 2006, p. 24.
25. R. Oten and R. J. P. de Figueiredo, Adaptive alpha-trimmed mean filters under deviations from assumed noise model, *IEEE Trans. Image Processing*, Vol. 13, No. 5, pp. 627–639, 2004.

CHAPTER 14

DIRECTION-OF-ARRIVAL (DOA) ESTIMATION OF IMPULSE RADIO UWB RFID TAGS

JONI POLILI LIE, BOON POH NG, and CHONG MENG SAMSON SEE
School of Electrical and Electronic Engineering, Nanyang Technological University, Singapore

CHIN-HENG LIM
Temasek Laboratories, Nanyang Technological University, Singapore

14.1 INTRODUCTION

Next-generation radio-frequency identification (RFID) technology demands a high-accuracy localization capability. Because impulse radio (IR) ultra-wideband (UWB) has the potential to provide centimeter-level location accuracy, the RFID industry has been developing a strong interest in IR-UWB. There exist two ways of locating the RFID tags: ranging- and directional-based localization. In the context of IR-UWB, while ranging-based approach enjoys high popularity, direction-of-arrival (DOA)-based approach receives less attention. This is due to the additional cost incurred for the use of antenna arrays and the increased complexity of the problem when a realistic dense multipath propagation is considered. The RFID industry has seen a growing interest in using impulse radio ultra-wideband (IR-UWB) as the wireless technology for the RFID [1–5]. Not only does the IR-UWB promise high-accuracy location estimation, but also the RFID tag based on IR-UWB requires lesser transmission power and can be built with simpler architecture [2]. These two traits make IR-UWB real-time location system (RTLS) different from sinusoidal-transmission-based RFID [6].

The indication of the promising future of IR-UWB RTLS can be seen from the increasing number of new product releases and the amount of venture funding received by the key industry players who develop the product [7–10]. Besides the strong industry engagement, the numerous research papers discussing IR-UWB ranging reflect also the strong interest from the research community [11–16].

Handbook of Smart Antennas for RFID Systems, Edited by Nemai Chandra Karmakar

The location sensing techniques used in RTLS can be grouped into two categories based on the information required: ranging and direction finding [17]. The first utilizes received signal strength, time-of-arrival (TOA), or time-difference-of-arrival to estimate the range between the transponder and the reader [18]. The latter utilizes direction-of-arrival (DOA) to estimate the bearing of the transponder with respect to the reader [19]. It is also possible to utilize both the TOA and DOA for better accuracy [20]. In the context of IR-UWB, both techniques are required to process the received analog or subsampled signal because the use of Nyquist-rate sampling is not possible with the currently available sampling technology.

The IR-UWB DOA estimation technique faces several additional challenges [21]. First, the well-known formulation of the high-resolution narrowband DOA estimation is no longer applicable because of the ultrawide-bandwidth characteristics of IR-UWB. Second, the broadband DOA estimation technique cannot be used because the Nyquist-rate-sampled signal is not available. And lastly, in multipath propagation, only the DOA of the direct path propagation is meaningful for the location estimation.

In this chapter, we consider the DOA estimation techniques for IR-UWB using low-complexity receiver architecture and without Nyquist-rate sampling. Section 14.2 discusses the system model and preliminaries with regards to IR UWB transmission and reception in realistic environment. Section 14.3 presents the first proposed technique, which uses a channelization receiver structure without very high Nyquist-rate sampling. Section 14.4 describes the second proposed technique, which utilizes a simple time-domain processing without ADC. The conclusion is given in Section 14.5.

14.2 SYSTEM MODEL AND PRELIMINARIES

Consider that an IR-UWB RFID tag is transmitting a single IR pulse periodically. Let the received UWB signal at UWB antenna from a multipath channel be

$$r(t) = \sum_{j=-\infty}^{\infty} w_{\mathrm{mp}}(t - jT_f - \tau_{\mathrm{toa}}) + \eta(t) \tag{14.1}$$

where $w_{\mathrm{mp}}(t)$ denotes the one-period received waveform capturing the effects of the multipath channel and is given by

$$w_{\mathrm{mp}}(t) = \sqrt{E_s} \sum_{l=1}^{L} a_l w_l(t - \tau_l) \tag{14.2}$$

where E_s is the pulse energy, w_l is the pulse waveform of the lth multipath component with unit energy, and (a_l, τ_1) are the multipath fading coefficients and delays, respectively. Note that the multipath delays have been ordered such that $\tau_1 < \tau_2 < \cdots < \tau_L$. Also, the first-arriving multipath delay is nil ($\tau_1 = 0$) because the delay due to the first-arriving multipath delay has been accounted for in the term τ_{toa}, the TOA of the received signal. The frame index and interval are denoted by j and T_f, respectively. $\eta(t)$ is the zero-mean additive white Gaussian noise (AWGN) with power spectral density $N_0/2$, and $N_0/2$ also denotes the variance of the noise.

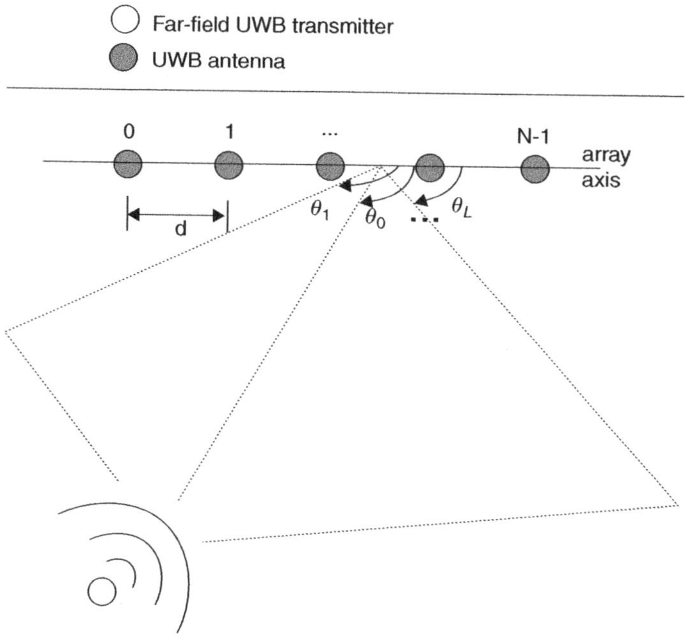

FIGURE 14.1. An illustration of the multipath propagation environment with L number of multipaths. The transmitter is a far-field transmitter and the receiver is an array.

The transmission is arranged such that the frame duration T_f is sufficiently large so that the last-arriving multipath component does not overlap with the first-arriving multipath component of the subsequent pulse.* That is,

$$T_f > \tau_L + \tau_{\text{toa}} \tag{14.3}$$

Each of the multipath component carries unique directional information denoted as $\{\theta_l\}_{l=1}^{L}$, measured with respect to the array axis (see Figure 14.1). If the propagation environment is an ideal multipath-free propagation, the array perceives only one impinging direction as shown in Figure 14.2.

To capture the directional information, an array of UWB sensors is used at the receiving end. Let (x_n, y_n) denote the position of the nth sensor in the two-dimensional space. The differential delay due to the lth multipath component can be expressed as

$$\tau_{l,n} = \frac{x_n \sin(\theta_l) + y_n \cos(\theta_l)}{c} \tag{14.4}$$

where c is the speed of radio wave propagation. A simple illustration of the differential delay is given in Figure 14.3, assuming a uniform linear array (ULA) receiver.

*In a realistic dense multipath propagation, the delay due to the last-arriving multipath component is 300–400 ns [25].

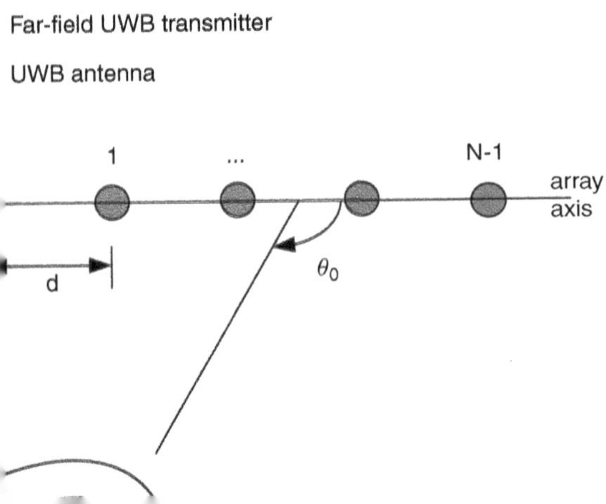

Far-field UWB transmitter
UWB antenna
1
...
N-1
array axis
d
θ_0

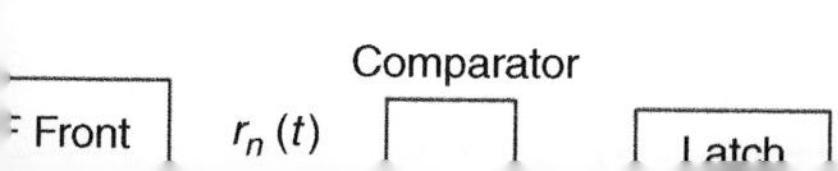
Comparator
Front
$r_n (t)$

Let $\breve{r}_n(t)$ denote the output of the LTD at the nth antenna of the array

$$\breve{r}_n(t) = p\left(t - \tau_p^{(n)}\right) \tag{14.7}$$

where $p(t)$ denotes the rectangular pulse waveform with rise time at reference 0 and pulse width T_{latch}. The parameter $\tau_p^{(n)}$ is expected to be $(\tau_{\text{toa}} + \tau_{1,n})$ ideally. Because of the possibility of detecting the noise, it is not always the case. The error from this pre-processing will contribute to the error in DOA estimation, as will be discussed in Section 14.3.4.

14.3.2 Algorithm Development

Consider the addition of the LTD's output from all the elements of the array. The one-period expression of this addition can be written as

$$y(t) = \sum_{n=o}^{N-1} b_n w_1(t - \tau_{\text{toa}} - \tau_{1,n}) + \eta(t, n) \tag{14.8}$$

where b_n is the attenuation level at the nth antenna and we have assumed the LTD at each element of the array to be ideal. Transforming (14.8) into frequency domain, we arrive at

$$\frac{Y(\omega)}{R_0(\omega)} = \sum_{n=0}^{N-1} b_n \exp(-i\omega\tau_{1,n}) \tag{14.9}$$

where $R_0(\omega) = W_1(\omega)e^{-i\omega\tau_{\text{toa}}}$ is the spectrum of the received signal at the array phase center, b_n is real and unknown, and $i = \sqrt{-1}$. Since the frequency range of UWB signal is wide in nature, the spectrum of the array output can be channelized into a series of discrete frequency channels $\{\omega_1, \ldots, \omega_k\}$. This results in a matrix representation of Eq. (14.9) as

$$\boldsymbol{x} = \boldsymbol{A}(\theta_1)\boldsymbol{b} \tag{14.10}$$

where $= \left[\frac{Y(\omega_1)}{R_0(\omega_1)} \quad \ldots \quad \frac{Y(\omega_M)}{R_0(\omega_M)}\right]^T$; $\boldsymbol{A}(\theta_1) = [\boldsymbol{a}_0 \quad \ldots \quad \boldsymbol{a}_{N-1}]^T$, where $\boldsymbol{a}_n = [e^{-i\omega_1\tau_n(\theta_1)} \quad \ldots \quad e^{-i\omega_M\tau_n(\theta_1)}]^T$; and $\boldsymbol{b}_n = [b_1 \quad \ldots \quad b_{N-1}]^T$. Note that $[.]^T$ denotes matrix transpose. The least-squares solution for $\hat{\boldsymbol{b}}$ is straightforward if the matrix $\boldsymbol{A}(\theta_1)$ is known; that is,

$$\hat{\boldsymbol{b}} = [\boldsymbol{A}(\theta_1)\boldsymbol{A}^H(\theta_1)]^{-1}\boldsymbol{A}^H(\theta_1)x \tag{14.11}$$

And the minimum least-squares error expression can be obtained as

$$J_{\min}(\theta_1) = \boldsymbol{x}^H\boldsymbol{x} - \boldsymbol{x}^H\boldsymbol{A}(\theta_1)[\boldsymbol{A}(\theta_1)\boldsymbol{A}^H(\theta_1)]^{-1}\boldsymbol{A}^H(\theta_1)\boldsymbol{x} \tag{14.12}$$

Since θ_1 is unknown, but we know that it is within a finite range of values, the estimate can be found by scanning for the direction that gives minimum least-squares error

$$\hat{\theta}_1 = \arg\min_{\theta_s}\{J_{\min}(\theta_s)\} \tag{14.13}$$

Note that the cost function expression in (14.12) is positive definite.

From the linear model, we can deduce the minimum M required so that the linear system of equations gives a solution. The requirement depends on the number of unknowns in the system. In this case, the unknowns are $\tau_{1,n}$ and b_n. Since $\tau_{1,n}$ is a function of θ_1, the number of unknowns is $N+1$, which comprises $\{b_0, \ldots, b_{N-1}\}$ and θ_1. Therefore, the channelization structure designed is required to fulfill the following inequality: $M \geq N+1$. In the case when $M > N+1$, the linear system is an overdetermined system and the least-squares-error-based scanning is able to solve for θ_1 as expressed in Eq. (14.13).

14.3.3 Receiver Structure

To get around the requirement for Nyquist-rate sampling, we propose a channelization-based receiver structure. With this structure, the observation matrix $\boldsymbol{x}$ in Eq. (14.10) can be obtained with much lower sampling rate.

Channelization is known as the process of subdividing a wideband spectrum into multiple sub-bands [23, 24]. In our context, we are interested to extract multiple line spectra (as many as $N+1$) from the wide bandwidth of UWB spectrum. To do so, the spectrum can be shifted down to near baseband using an analog complex mixer and the desired line spectrum can be extracted using an analog low-pass filter. The line spectrum $Y(\omega_m)$ can be computed from the output of the analog low-pass filter. The same procedure can be used to compute $R_0(\omega_m)$ during the calibration stage, whereby the transmitter is placed at 0° from the receiver [28]. Overall, the proposed receiver structure is shown in Figure 14.5. The structure comprises an LTD at each element, an adder, and a channelization structure.

14.3.4 Numerical Results

In the following simulations, a second-derivative Gaussian pulse is considered to be the received waveform. The mathematical expression of Gaussian second-derivative pulse can be found in [23] and is given by

$$w(t) = \frac{1}{\sqrt{2\pi}\sigma^3}\left(1 - \frac{t^2}{\sigma^2}\right)\exp\left(-\frac{t^2}{2\sigma^2}\right) \tag{14.14}$$

The parameter σ is defined such that the duration* of the Gaussian pulse, $T_p \approx 1$ ns. As suggested in [23], σ is set such that 99.99% of the energy of the pulse is contained

*Even though the duration of the Gaussian pulse and all of its derivatives are infinite, the simulation truncates the pulse so that the duration is T_p.

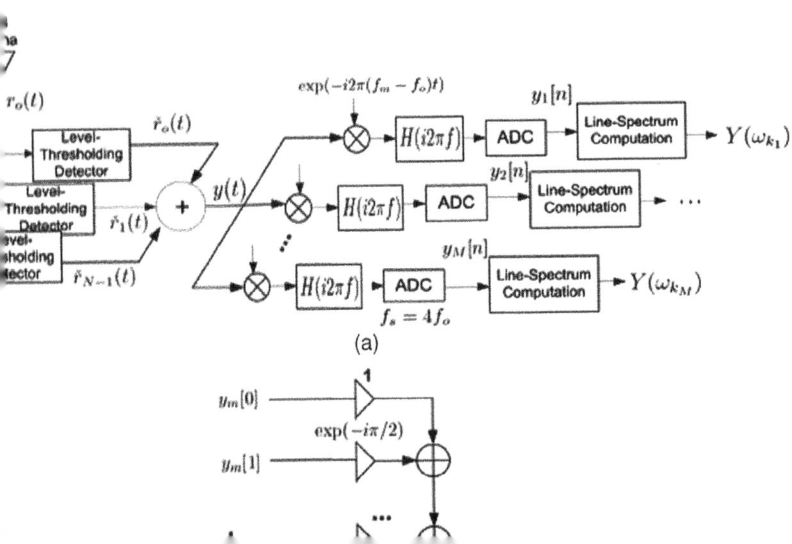

$r_o(t)$
Level-Thresholding Detector
Level-Thresholding Detector
evel-sholding ector
$\hat{r}_o(t)$
$\hat{r}_1(t)$
$\hat{r}_{N-1}(t)$
+
$y(t)$
$\exp(-i2\pi(f_m - f_o)t)$
$H(i2\pi f)$
ADC
$y_1[n]$
$y_2[n]$
$y_M[n]$
Line-Spectrum Computation
$Y(\omega_{k_1})$
$Y(\omega_{k_M})$
$f_s = 4f_o$
(a)
$y_m[0]$
$y_m[1]$
1
$\exp(-i\pi/2)$

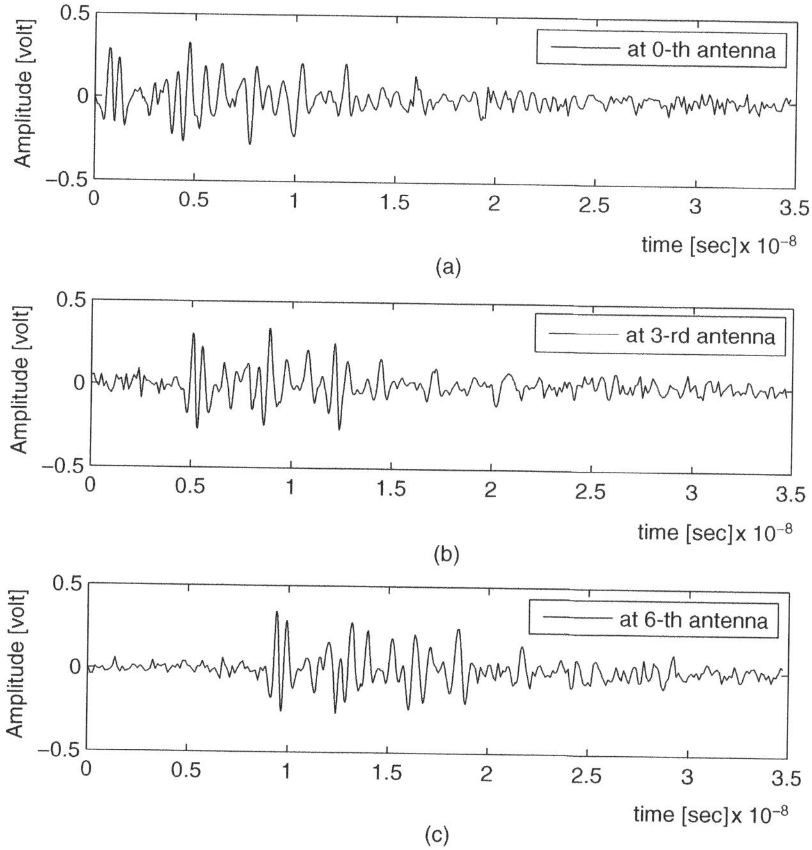

FIGURE 14.6. A realization of a received UWB pulse at the (**top**) 0th, (**middle**) 3rd, and (**bottom**) 6th antenna that forms the array for the multipath case.

As shown in Figure 14.5, the received signal is first passed through the LTD. The threshold level for the LTD is $\gamma = 0.3$ for all antennae. The detection may introduce errors due to missed and false-alarm detection. This nonideality will cause the inaccuracies in the DOA estimation as will be shown later.

The outputs of the LTD are passed to the adder and then to the channelization structure. The frequency f_m in the channelization structure is arbitrarily chosen. As many as 10 frequencies are used for the channelization structure. These frequencies are chosen within the 1- to 4-GHz frequency range because the received power spectral density (PSD) is the highest at this range. The line spectra of the sum-received signals are extracted from the output of the channelization architecture.

The estimated DOA is obtained by finding the scanning direction that maximizes the inverse of $J_{\min(\theta_s)}$ as defined in Eq. (14.12). Both the extracted line spectra and the knowledge of the array geometry are required to calculate this function.

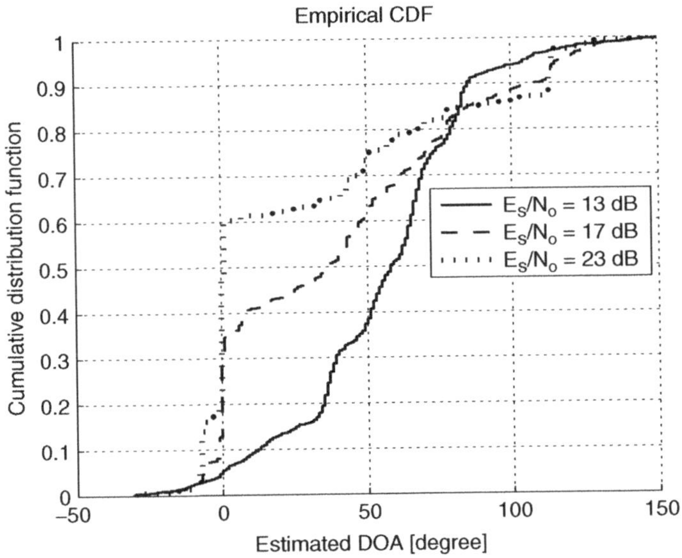

FIGURE 14.7. Cumulative distribution function (cdf) plots of the estimation error for various E_s/N_0 in multipath environment.

From 1000 realizations of the DOA estimation, the cumulative distribution function (cdf) curve of the estimation error is plotted. We vary the signal-to-noise ratio by varying the parameter E_s/N_0. Figure 14.7 shows the cdf plot for various E_s/N_0. As E_s/N_0 increases, the number of the correctly estimated DOA also increases. However, the distribution shows the presence of outliers even when E_s/N_0 is as high as 23 dB. This will then degrade the estimation's bias and standard deviation.

To improve on the bias and standard deviation, we propose to collect more estimation values and take the mode* of those values as the final estimate while keeping the system under consideration to be stationary when collecting these values. For example, if 30 estimation values are collected, the system is required to be stationery for $30T_f$ duration. For $T_f = 400$ ns, the required duration is only 12 μs. Therefore, within 12 μs, the system under consideration is assumed to be stationary. This is indeed a very practical assumption.

The question now is, How many values need to be collected? To answer this, we define a confidence level metric (CL) that measures the probability level that the estimation error is confined within a predefined interval. For example, we use the limit $(-5, +5)$ degrees for the CL. We simulate the case when 5, 10, 20, and 30 estimation values are collected for the mode calculation and assess the CL, bias and standard deviation of the proposed estimation as compared to the case when only *one* estimation value is considered. The simulation still considers 1000 realizations for each scenario where E_s/N_0 is fixed at 23 dB. The resulting bias, standard deviation, and CL are listed in Table 14.1. When 30 estimation values are used, the CL achieved

*Here, the mode refers to statistical mode, the largest occurrence values.

TABLE 14.1. The Performance of the Proposed Estimator Using the Mode as the Final Estimate

Number of estimates collected:	1	5	10	20	30
Bias (degrees)	27.0710	2.1140	1.1100	0.1350	0.1010
Standard deviation (degrees):	42.3325	20.7547	12.7311	5.1598	3.6794
CL (%):	46	72	88.5	97.8	99

can be as high as 99%. This result proves that the proposed DF method using digital channelization architecture is effective.

14.4 ANALOG-DIFFERENTIATION-BASED DOA ESTIMATION

14.4.1 Background and Motivation

The approach proposed in Section 14.3 is based on frequency-domain processing. The DOA is estimated by extracting the line spectra of the sum of received signals at the array. In this section, we present an approach based on time-domain processing. Unlike the approach explained in Section 14.3, this approach involves only simple analog time-domain processing by using an analog circuitry. Therefore, the receiver's complexity is lower than the previous approach.

Consider the ULA antenna array and the array output $y(t)$ as the sum of all the elements of the array after the LTD. If the latch circuitry is designed such that the pulse width T_{latch} (of the rectangular pulse $p(t)$) is large enough, the array output will form a staircase-shape waveform as illustrated in Figure 14.8. Then DOA estimation can be achieved by finding the slope of the staircase-shape waveform. The following paragraph explains how the slope can be formed.

Let the number of elements in the array be $N = 2$, for simplicity, and the first element is placed at the array phase center. The rectangular pulse at the output of the first element appears earlier as compared to the second element by $\tau_{1,1}$ seconds. This delay is termed the inter-element propagation delay. If T_{latch} is set to be equal to $\tau_{1,1}$, the resulting sum of the two rectangular pulses is another rectangular pulse with $2T_{\text{latch}}$ pulse width. When $T_{\text{latch}} < \tau_{1,1}$, the resulting sum is two rectangular pulses separated by $\tau_{1,1} - T_{\text{latch}}$ interval. Only when $T_{\text{latch}} \geq \tau_{1,1}$, then the resulting sum will form a staircase-shape waveform. Figure 14.8 helps to illustrate the three cases.

Recall that $\tau_{1,1}$ is a function of DOA θ_1: $\theta_1 = \frac{d}{c}\cos\theta_1$, where d denotes the interelement distance and c is the propagation speed. Therefore, the maximum possible value of $\tau_{1,1}$ is achieved when $\theta_1 = 0°$. Hence, in general, to form a staircase-shape waveform, the pulse width has to satisfy the following inequality: $T_{\text{latch}} > \frac{d}{c}(N-1)$.

As $N \to \infty$, the resulting staircase-shape waveform can be seen approximately as a pair of upward and downward slopes. Both slopes reflect the source DOA. In fact, the slopes can be expressed as

$$y(t) \approx f(x) = \begin{cases} \dfrac{1}{\tau_{1,1}}t, & \text{upward slope} \\ -\dfrac{1}{\tau_{1,1}}t, & \text{downward slope} \end{cases} \qquad (14.15)$$

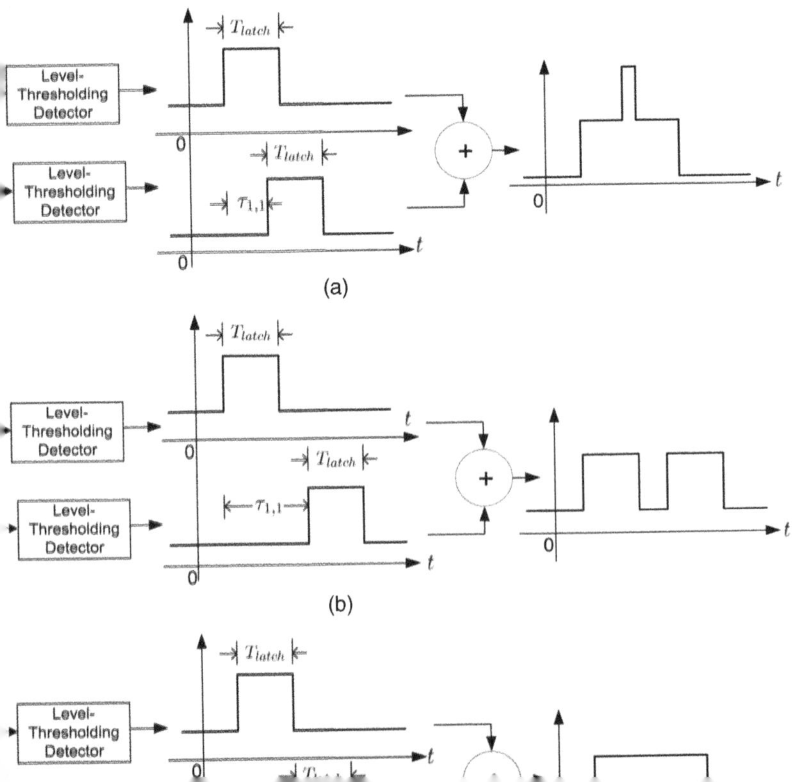
Level-Thresholding Detector
Level-Thresholding Detector
T_{latch}
T_{latch}
$\tau_{1,1}$
0
0
0
t
t
+
(a)
Level-Thresholding Detector
Level-Thresholding Detector
T_{latch}
T_{latch}
$\tau_{1,1}$
0
0
0
t
t
t
+
(b)
Level-Thresholding Detector
T_{latch}
0
t

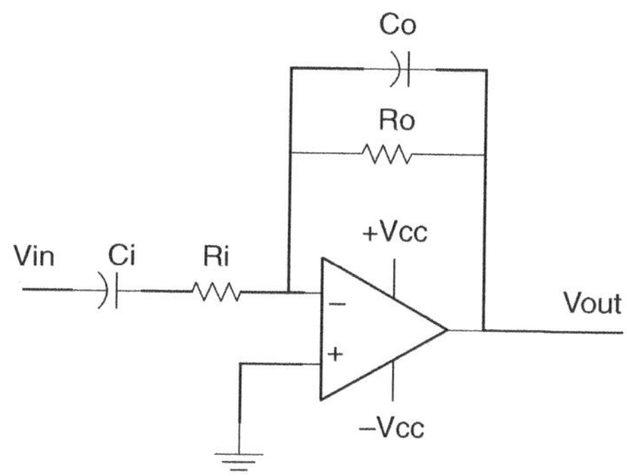

FIGURE 14.9. A typical op-amp differentiator circuitry.

To realize the differentiation process, it is proposed to use an analog differentiator to avoid high frequency sampling. In particular, we use an op-amp to design an analog differentiator. Figure 14.9 shows the circuitry of a general op-amp differentiator. Its transfer function is given by

$$H(i2\pi f) = \frac{-i2\pi f C_i R_o}{(1 + sC_o R_o)(1 + sC_i R_i)} \tag{14.17}$$

To determine the value of the passive components (resistors and capacitors) that are suitable for the design, the operating frequency range of the differentiator has to be defined. The frequency range is chosen in such a way that all the possible wave periods of the staircase-shape waveform are within the frequency range. The wave period is in fact the pulse repetition interval (PRI) of the transmitted pulse.

Let f_L denote the low-frequency limit and f_H denote the high-frequency limit of the operating frequency of the op-amp used, the choice of the passive components is governed by the following equations*: $2\pi f_L = \frac{1}{C_i R_i}$ and $2\pi f_H = \frac{1}{C_o R_o}$. The corresponding impulse response can then be derived and written as

$$h(t) = ae^{-\frac{1}{C_O R_O}t} + \beta e^{-\frac{1}{C_i R_i}t} \tag{14.18}$$

In addition to the differentiator, a peak-and-hold detector is implemented to track-and-hold the peak of the output. This is because the slope of the staircase-shape waveform is observed only within an extremely short duration, that is, $(N-1)\tau_{1,1}$ nanoseconds. Overall, the proposed receiver structure is shown in Figure 14.10.

14.4.3 Numerical Results

To help illustrate the operational theory of the proposed approach, a numerical realization of the system is utilized. The simulation considers a Gaussian second-derivative

*This is true when both poles are assumed to be the dominant factors in the differentiator's frequency response.

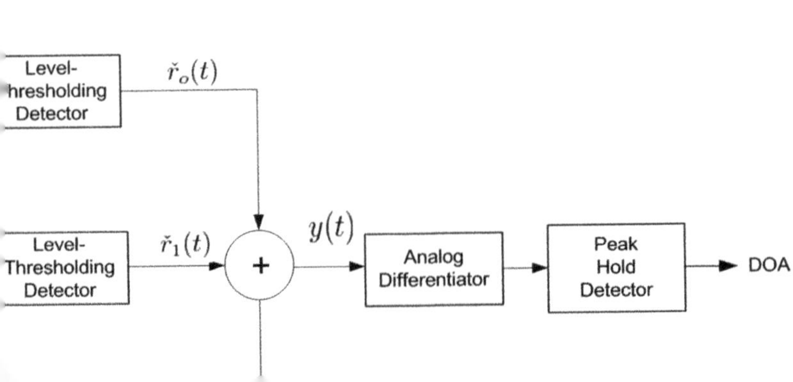

Level-
hresholding
Detector
$\check{r}_o(t)$
Level-
Thresholding
Detector
$\check{r}_1(t)$
+
$y(t)$
Analog
Differentiator
Peak
Hold
Detector
DOA

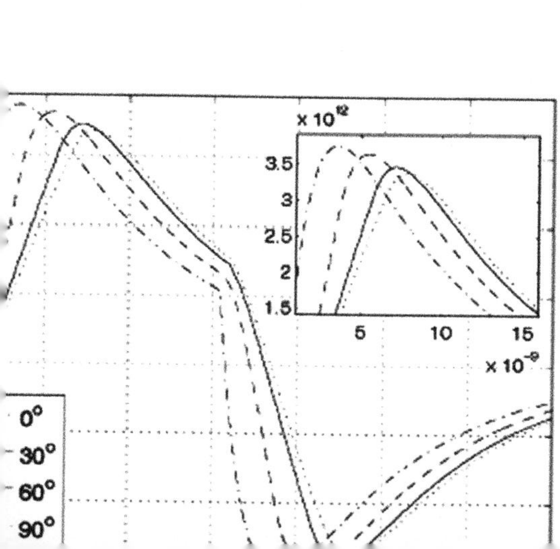

x 10^12
3.5
3
2.5
2
1.5
5
10
15
x 10^-9
0°
30°
60°
90°

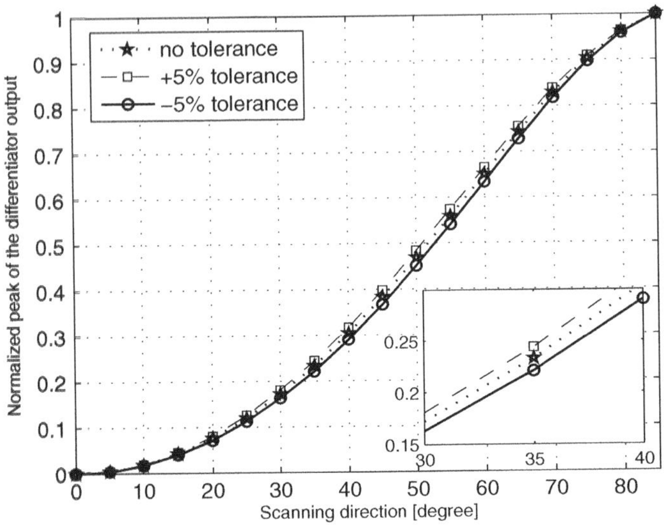

FIGURE 14.13. Normalized peak of the differentiator output as a function of the scanning direction for the case when the passive components in the analog differentiator's circuit are subjected to ±5% tolerance and for an ideal case.

The peaks indicate the slopes of the outputs. Before reaching the peak, the output experiences a transition time, which is dependent on the duration of observing the slope of the staircase-shape waveform. Notice that this duration is smaller than 6 ns. For different values of θ_1, the curves reach different peak's values. The highest slope is obtained when $\theta_1 = 90^\circ$.

Prior to estimation, the proposed structure has to be calibrated by measuring the differentiator's peaks for a few known directions. The calibration will produce a lookup table that maps the differentiator's peak and its corresponding direction. Given this lookup table, one is able to estimate the unknown DOA from the observed differentiator's peak using interpolation.

14.4.3.1 Effect of the Tolerance from the Passive Components. We first investigate the effect of the tolerance from the passive components (resistors and capacitors) on the performance of the estimator. When the passive components in the analog differentiator's circuit are subjected to ±5% tolerance, the impulse response of the differentiator will also deviate as well. The deviation in the impulse response will then result in a different differentiator's outputs and peak values. To see how far the peak of the differentiator's output will deviate, the following simulations consider the case when the passive components are subjected to the tolerance and that of an ideal case. Figure 14.13 shows the resulting normalized peak of the differentiator's output as a function of the scanning direction. From the figure it can be concluded that no significant deviation is observed when the passive components are subjected to ±5% tolerance.

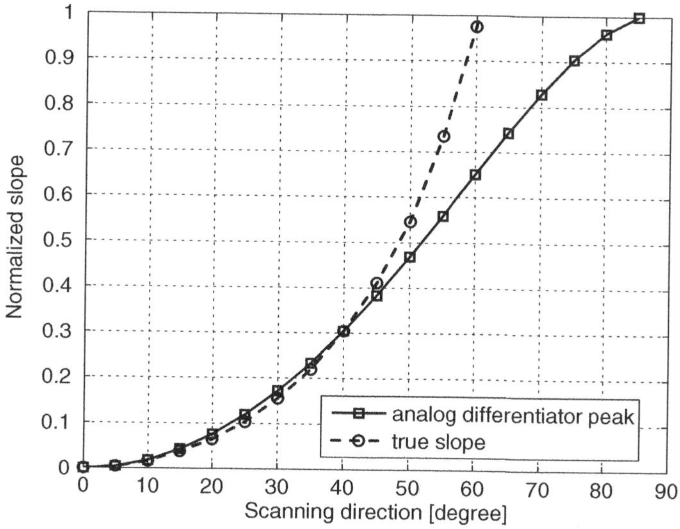

FIGURE 14.14. Plots of the slope as a function of scanning direction. The comparison shows that the peak of the differentiator starts to deviate from the true slope at higher scanning direction.

14.4.3.2 Limitations of the Approach. Generally speaking, a ULA antenna array is unable to distinguish the signal wavefront impinging from the front and back of the array. In addition to such an ambiguity, the DOA estimation based on slope calculation suffers from ambiguity in differentiating the signal wavefront impinging from the left and right sides of the array. To be precise, the resulting staircase-shape waveform is the same when the signal is impinging from θ and $(180^\circ - \theta)$, where θ is measured with respect to the array axis.

Besides the ambiguity limitations, there is also a limitation on the accuracy due to the nonideality of the differentiator on the proposed approach. Ideally, the slope of the staircase-shape waveform is inversely proportional to the interelement propagation delay $\tau_{1,1}$. Here, it is shown that the peak of the differentiator's output does not match exactly with the inverse of the interelement propagation delay.

Consider the realization of the proposed system for propagation's direction from 0° to 90° with a 5° interval step-size. The peak of the differentiator's output is recorded and normalized. This peak should be equal to the slope or inversely proportional to the interelement propagation delay $\tau_{1,1}(\theta_1)$. To compare the result with the ideal case, Figure 14.14 is plotted. The normalization is taken with respect to the highest peak, obtained when the signal propagates from the array broadside (90°). The ideal case (true slope) is plotted as the inverse of $\tau_{1,1}(\theta_1)$. As shown, the peak of the differentiator's output and the true slope are not an exact match. Although it shows a good match for directions below 45°, it starts to deviate when the direction increases further. The reason is due to the incapability of the differentiator to respond to very short-duration changes.

TABLE 14.3. The Performance of the Proposed Estimator Using the Mode as the Final Estimate

Number of estimates collected	1	5	10	20	30
Bias (degrees)	4.5090	10.1730	4.5350	1.2820	0.2860
Standard deviation (degrees)	25.5328	11.7613	8.5631	5.4000	4.0527
CL (%):	41.8	45.4	65.9	82.4	89.4

14.4.3.3 Improving the Confidence Level. Similar to the channelization-based approach, the differentiation-based approach also suffers from the nonideality of the LTD. To improve the estimation, we apply the same technique to the differentiation-based approach (see Section 14.3.4). That is, we collect more estimation values and take the mode of those values as the final estimate while keeping the system under consideration to be stationary when collecting these values. Therefore, the next simulation considers collecting 5, 10, 20, and 30 estimation values where E_s/N_0 is fixed at 27 dB. The improvement can be observed from the increase in the confidence level (defined previously in Section 14.3.4), when more estimation values are used, as listed in Table 14.3.

Compared with the channelization-based approach, the differentiation-based approach performs worse under the same propagation condition. In other words, the differentiation-based approach requires a higher SNR and more estimates to achieve the same CL, as compared to that of the channelization based approach. For example, the channelization-based approach is able to achieve 99% CL when using 30 initial estimates, while the differentiation-based approach is only capable of achieving 89.4%.

Although it performs worse than the channelization approach, the differentiation-based approach provides an alternative means of having a lower complexity system. With its hybrid digital–analog* structure, almost no computational resource is required to implement such system.

14.4.4 Receiver Architecture Comparison

Compared with the digital channelization receiver proposed in Section 14.2 (see Figure 14.5), the proposed structure based on the analog differentiation requires less components. The channelization based receiver requires complex mixers, BPFs and ADCs while the differentiation based receiver requires only a differentiator and a peak hold detector.

To estimate the DOA, the channelization-based receiver requires a digital processing unit to perform the complex additions and multiplications. The processing unit is also required to have some memory to store some complex values temporarily. On the other hand, the differentiation-based receiver requires only a lookup table in

*Front-end systems (from antenna to level threshold detector) are an analog subsystem. Latch circuit produces digital output in the form of an rectangular pulse. The subsystem after latches is another analog subsystem.

TABLE 14.4. Comparison of the Receiver's Complexity Between the Channelization-Based DF Receiver and the Differentiation Counterpart, where N Represents the Number of Antennae and M Represents the Number of Channels for the Channelization Architecture

Hardware	Channelization	Differentiation	Quantity Required
Level threshold detector	Yes	Yes	N
Mixer	Yes	No	2M
BPF	Yes	No	2M
ADC	Yes	No	2M
Computing power	Yes	No	Complex + and ×
Memory	Yes	No	Complex values
Differentiator	No	Yes	1
Peak-hold detector	No	Yes	1
Lookup table	No	Yes	1

order to associate the direction with its respective peak of the differentiator's output. No computing power is required for this process.

In summary, the differentiation-based DF receiver requires less hardware than the channelization counterpart. Table 14.4 shows the receiver architecture comparison of the two DF receivers. Compared with a DF receiver using conventional Nyquist rate sampling, both approaches are more practical because they do not require an extremely high sampling rate and tremendous computational resources to process a huge amount of data.

14.5 CONCLUSIONS

In this chapter, we introduce a technique for estimating the DOA of UWB RFID tags using low-complexity receiver architecture. In particular, we consider multiple transmissions of low-duty cycle and noncontinuous impulse-shaped electromagnetic wave from different directions. After propagating through a multipath channel, the proposed receiver aims to estimate the DOA of the transmitters. Through the study presented in this chapter, the conclusions are highlighted as follows:

- The suppression of incoming multipath is achieved using a simple level-thresholding method, which exploits the multipath diversity of IR UWB.
- With the multipath suppressed, we proposed two solutions, namely channelization-based and differentiation-based estimation techniques. The channelization approach is derived based on a simple linear model formulation of the observed line spectra of the sum of received signals, while the differentiation approach is formulated based on the observation of staircase-shape waveform whose slope indicates the DOA.
- Numerical experiments show the effectiveness of both solutions, where the estimation bias and standard deviation achieved are less than 0.3° and 5°, respectively.

According to the results presented in this chapter, some recommendations on estimating the DOA of IR-UWB RFID tags is made as follows:

- Channelization-based approach requires the implementation of some RF components that may increase the implementation cost. Nevertheless, the accuracy achieved is better than the differentiation-based approach.
- Differentiation-based approach is preferred over the channelization-based one if the estimation accuracy can be traded-off with implementation cost and system complexity.
- A localization system based on the DOA estimation discussed in this chapter may comprise the DOA-enabled receivers based on both approaches.
- Further research should focus on the extension to various array geometries, where some array geometries are preferred over others for practical reason.

REFERENCES

1. K. A. Finkenzeller, *RFID Handbook: Fundamentals and Applications in Contactless Smart Cards and Identification*, John Wiley & Sons, Hoboken, NJ, 2003.
2. Z. Zhuo, M. Baghaei-Nejad, H. Tenhunen, and L.-R. Zheng, An efficient passive RFID system for ubiquitous identification and sensing using impulse UWB radio, *Elektrotech. Informationstech.*, Vol. 124, No. 11, pp. 397–403, 2007.
3. R. Want, Enabling ubiquitous sensing with RFID, *Computer*, Vol. 37, No. 4, pp. 84–96, 2004.
4. W. Hirt, Ultra-wideband radio technology: Overview and future research, *Computer Commun.*, Vol. 26, No. 1, pp. 46–52, 2003.
5. S. Hu, C. L. Law, W. Dou, and H. Chen, Detection Range Enhancement of UWB RFID Systems, in *Identification 2007 IEEE International Workshop on, Anti-counterfeiting, Security*, 2007, pp. 431–434.
6. S. Gezici, Z. Tian, G. B. Giannakis, H. Kobayashi, A. F. Molisch, H. V. Poor, and Z. Sahinoglu, Localization via ultra-wideband radios: A look at positioning aspects of future sensor networks, *IEEE Signal Processing Mag.*, Vol. 22, No. 4, pp. 70–84, 2005.
7. Multispectral Solutions, Inc. website, http://www.multispectral.com/.
8. Time Domain Corporation website, http://www.timedomain.com/.
9. Parco Wireless, Parco Wireless Completes UWB RFID at Washington Hospital Center, webpage: http://embedded-system.net/parco-wireless-completes-uwb-rfid-at-washington-hospital-center.html.
10. P. Steggles and S. Gschwind, Ubisense-a smart space platform, Technical Report Document, May 2005.
11. J. P. Lie, C. M. See, and B. P. Ng, UWB ranging with high robustness against dominant jammer and multipath, *IEEE Microwave Wireless Components Lett.*, Vol. 15, No. 12, pp. 907–909, 2005.
12. C. Falsi, D. Dardari, L. Mucchi, and M. Z. Win, Time of arrival estimation for UWB localizers in realistic environments, *Eurasip J. Appl. Signal Processing*, Vol. 2006, pp. 152, 2006.

13. R. J. Fontana, E. Richley, and J. Barney, Commercialization of an ultra wideband precision asset location system, in *Proceedings, IEEE Conference on Ultra Wideband Systems and Technologies*, 2003, pp. 369–373.
14. Guvenc and Z. Sahinoglu, Threshold-based TOA estimation for impulse radio UWB systems, in *ICU 2005: 2005 IEEE International Conference on Ultra-Wideband, Conference Proceedings*, Vol. 2005, 2005, pp. 420–425.
15. Guvenc and Z. Sahinoglu, TOA estimation with different IR-UWB transceiver types, in *ICU 2005: 2005 IEEE International Conference on Ultra-Wideband, Conference Proceedings*, Vol. 2005, 2005, pp. 426–431.
16. S. J. Ingram, D. Harmer, and M. Quinlan, Ultra wide band indoor positioning systems and their use in emergencies, in *Record—IEEE PLANS, Position Location and Navigation Symposium*, 2004, pp. 706–715.
17. K. Pahlavan, L. Xinrong, and J. P. Makela, Indoor geolocation science and technology, *IEEE Commun. Mag.*, Vol. 40, No. 2, pp. 112–118, 2002.
18. L. M. Ni, Y. Liu, Y. C. Lau, and A. P. Patil, LANDMARC: Indoor location sensing using active RFID, in *Proceedings of the First IEEE International Conference on Pervasive Computing and Communications, 2003 (PerCom 2003)*, 2003, pp. 407–415.
19. Y. Zhang, M. G. Amin, and K. Shashank, Localization and tracking of passive RFID tags based on direction estimation, *Int. J. Antennas Propagation*, Vol. 2007, pp. 1–9, 2007.
20. M. Navarro and M. Najar, TOA and DOA estimation for positioning and tracking in IR-UWB, Ultra-Wideband, 2007, in *IEEE International Conference on ICUWB 2007*, 2007, pp. 574–579.
21. J. P. Lie, C. M. S. See, and B. P. Ng, Direction finding receiver for UWB impulse radio signal in multipath environment, *Int. J. Commun. Syst.*, accepted 2010, doi: 10.1002/dac.1123.
22. M.-G. D. Benedetto and G. Giancola, *Understanding Ultra Wide Band Radio Fundamentals*, Prentice-Hall, Upper Saddle River, NJ, 2004.
23. L. Feng and W. Namgoong, An oversampled channelized UWB receiver with transmitted reference modulation, *IEEE Trans. Wireless Commun.*, Vol. 5, No. 6, pp. 1497–1505, 2006.
24. W. Namgoong, A channelized digital ultrawideband receiver, *IEEE Trans. Wireless Commun.*, Vol. 2, No. 3, 2003.
25. J. Foerster, *Channel Modeling Sub-committee Final Report*, IEEE 802.15.SG3a, 2002.
26. Q. Spencer, M. Rice, B. Jeffs, and M. Jensen, Statistical model for angle of arrival in indoor multipath propagation, in *IEEE Vehicular Technology Conference*, Vol. 3, 1997, pp. 1415–1419.
27. J. P. Lie, B. P. Ng, and C. M. S. See, Multiple UWB emitters DoA estimation employing time hopping spread spectrum, *Prog. Electromagn. Res.*, Vol. 78, pp. 83–101, 2008.
28. J. P. Lie, C. M. See, and B. P. Ng, Ultra wideband direction finding using digital channelization receiver architecture, *IEEE Commun. Lett.*, Vol. 10, No. 2, pp. 85–87, 2006.
29. J. P. Lie, B. P. Ng, and C. M. See, Hybrid digital–analog technique for UWB direction finding, *IEEE Commun. Lett.*, Vol. 10, No. 2, pp. 79–81, 2006.

CHAPTER 15

ENABLING LOCALIZATION SERVICES IN SINGLE AND MULTIHOP WIRELESS NETWORKS

VASILEIOS LAKAFOSIS, RUSHI VYAS, and MANOS M. TENTZERIS
Department of Electrical and Computer Engineering, Georgia Institute of Technology, Atlanta, Georgia, USA

15.1 INTRODUCTION

Localization is the process of determining the physical positions of devices with a specific degree of accuracy in indoor or outdoor environments. The physical location of a user, device, or mote within an area covered by a wireless network can prove to be a very useful or even indispensable functionality in many applications. First of all, high correlation between data captured—for instance, environmental— and locality may be required for the data to be meaningful. Device tracking, involving location and bearing, is another type of application that makes use of different localization techniques. Location awareness is also the basic component of a special category of routing protocols, namely, geographic aware ones, where the traffic is relayed to or from a particular area of the sensing field. Finally, context-aware applications can make smarter decisions in terms of user interface or behavior when knowledge of the physical location of the nodes is available.

This chapter reviews the most representative and reliable localization techniques, most of which can be easily deployed in existing networks, as well as presents how these have been expanded to provide localization solutions suitable for environments where mostly multihop connectivity can be established with anchor nodes; nodes whose exact location coordinates are known a priori.

It should be noted that most of these techniques are based on methods used since ancient and medieval times by Thales from Miletus, Claudius Ptolemaus, Leon Battista Alberti, and others [1] and are still used mostly in navigation systems.

Handbook of Smart Antennas for RFID Systems, Edited by Nemai Chandra Karmakar

15.2 LOCALIZATION TECHNIQUES

15.2.1 Single-Hop Environments

Before starting to describe the most prominent localization techniques, it should be noted that these techniques can either be deployed directly in single-hop wireless networks or serve as the basis for techniques used in multihop network environments, as described in the next section.

15.2.1.1 Tri- and Multi-Lateration. Lateration is the approach in which distances from three (trilateration) or more (multilateration) anchor nodes are used to estimate the node location, relative to the anchors or to absolute coordinates if the anchor positions are given in absolute coordinates. The minimum number of anchors required for lateration on a plane is three (noncolinear), whereas in a three-dimensional (3D) space, four are required. The most important methods of lateration are described in the following subsections.

Received Signal Strength Indicator (RSSI). Many proposed solutions rely on the received signal strength indicator (RSSI) returned by the transceiver after the reception of a packet from another node. Initially, knowledge of the effective isotropically radiated power (EIRP), which takes into account the transmission power, the antenna gain, and the cable losses, is assumed. Plugging this value along with the RSSI into the Friis equation for a specific path-loss coefficient and model, which can be very sophisticated with dependency on the environment, the distance from the emitter can be estimated. The required computation can be done in a localized or a centralized fashion, depending on whether the RSSI value used is of a packet transmitted by an anchor node or by the node whose location is under estimation, respectively.

An example of the centralized case is presented in [2], an application used mostly by car auction dealers for quickly locating a vehicle in huge parking lots and tracking its trajectory over the course of its stay in their premises for cost optimization reasons. Specifically, a paper-based, batteryless, solar-powered tag hung from the front mirror of a car transmits time-stamped, identical, unique identification packets in regular intervals. In Figure 15.1, such packets are shown to be captured from five different, very low-cost anchor nodes and relayed to a central location along with an RSSI value each.

The two main characteristics that have rendered this method attractive are that neither additional hardware nor additional communication overhead is required. However, there are quite a few parameters that can degrade the accuracy of the RSSI approach. First, incorrect estimations are introduced when the RSSI is extracted from packets that have followed an indirect path due to multipath fading [3], regardless of whether they have been emitted from an anchor node in line-of-sight with the receiver or not. Second, the fast-fading effect, as well as the dynamic nature of the environment, can result in serious oscillations in the RSSI measurements over time. Contrary to the previous problem, this effect can be alleviated using statistical techniques [4] in conjunction with repeated measurements. The authors used the studentized residuals

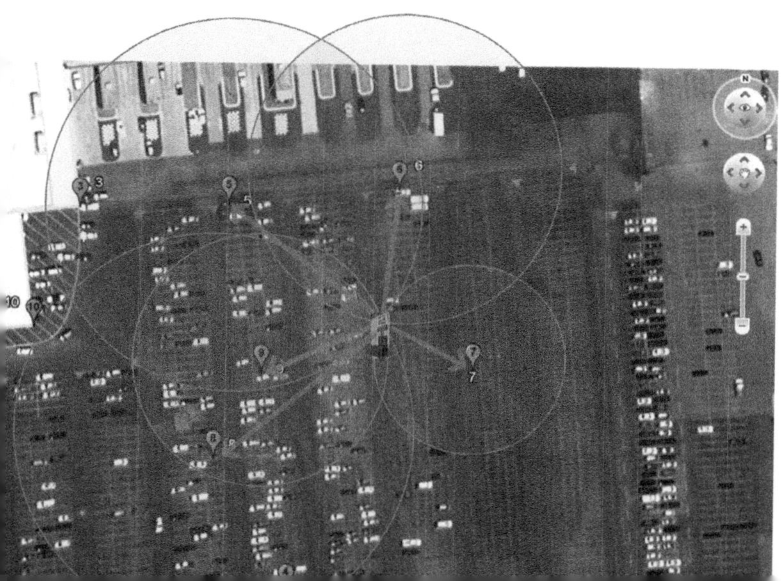
3
5
6
10
7
N

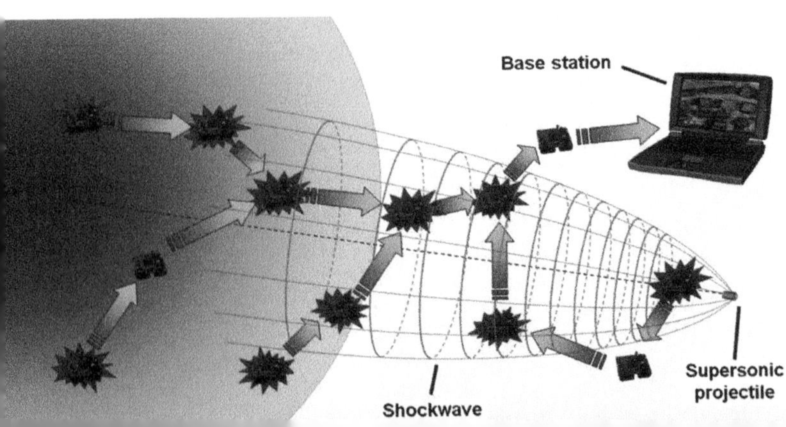
Base station
Supersonic projectile
Shockwave

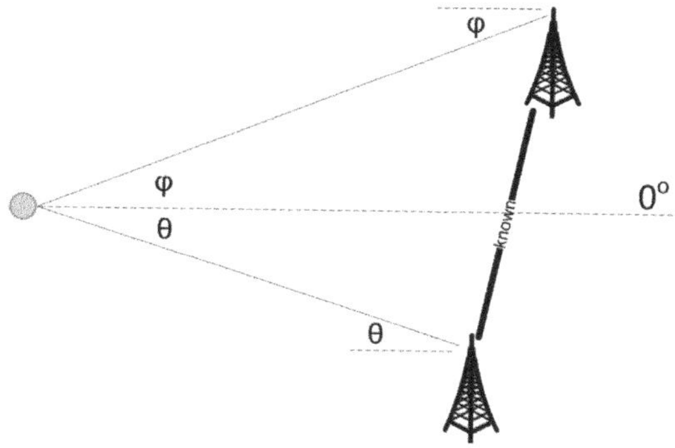

FIGURE 15.3. Location estimation on a plane based on angular information (φ, θ) from two reference points.

centralized management or control and requires no explicit coordination between anchor nodes. This system allows end devices to compute their physical location locally on their own and provide this information to any user application, thus guaranteeing user privacy.

15.2.1.2 Angulation. Angular information extracted in reference to multiple anchor nodes, whose positions are well known, can be used to estimate the location of a node. An example of a two-dimensional (2D) location estimation system is given in Figure 15.3. Here, since the length of one side of the triangle drawn and two angles are known, the position of the third vertex can be unambiguously found.

Nasipuri and Li [10] propose an angle-of-arrival (AoA) estimation technique according to which at least three fixed anchor nodes continuously transmit a unique RF signal on a narrow directional beam that is rotated at a constant angular speed, known to all nodes. This is depicted in Figure 15.4. A node equipped with a low-power transceiver translates the time difference of arrivals (TDoAs) of the different beacon signals to angular values and, eventually, evaluates its angular bearings and location with respect to the beacon nodes using trigonometry. According to the simulation results reported, the maximum error is within 2 m in a 75×75 m area; the performance does not depend on the absolute dimensions of the network area, but narrow beamwidths of 15° or less are assumed.

An interesting alternative approach is the lighthouse laser-based location system [11]. Here, lighthouse base stations use broad horizontal beams that rotate at a constant speed. A node equipped with a photodetector measures the start and end time of the beam passing by and uses this sweep time along with the known complete rotation time to estimate its position at high precision. Indicatively, using an early 2D prototype of the system, node locations could be estimated with an average accuracy of about 2% and an average standard deviation of about 0.7% of the node's distance to the base station.

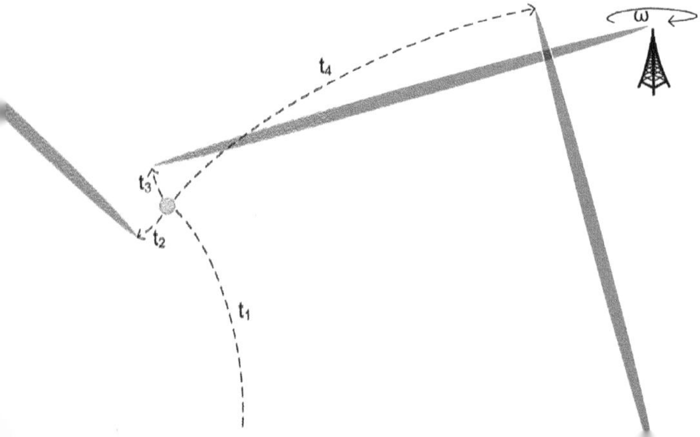

ω
t4
t3
t2
t1

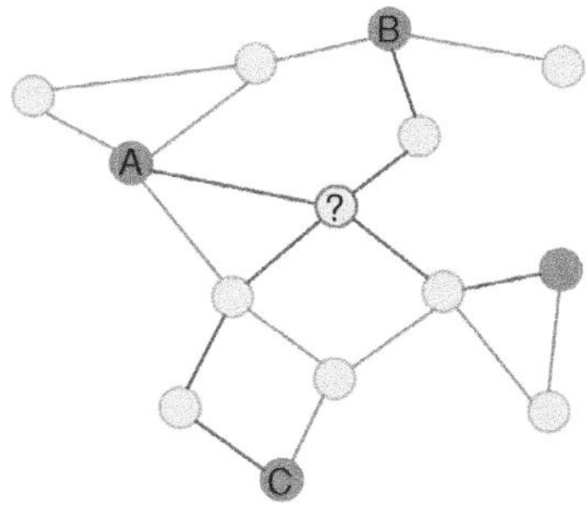

FIGURE 15.5. Single and multihop connectivity of a node, whose location is unknown, to three anchor nodes.

emitted from points in a virtual dense mesh, covering the whole parking lot right after the initial deployment. This capturing phase is recommended to be carried out for different vehicle occupation conditions of the lot (empty, full, etc.) so that multipath and other RF effects are accounted for as accurately as possible, thus resulting in different field fingerprinting profiles. Another application making use of a similar approach in indoor environments is the RADAR system [5], where the received signal strength values from multiple anchors are compared with premeasured, stored ones.

15.2.2 Localization in Multihop Environments

In multihop wireless networks, often a node whose location is to be determined, does not have direct connectivity to at least three anchor nodes. A representative example is shown in Figure 15.5, where the aforementioned node (denoted with a question mark in the center of the topology) is within the single-hop neighborhood of only one anchor node (A) and has multihop connectivity to two others (B and C).

15.2.2.1 Proximity. This is the simplest of all the lateration methods, since it exploits the inherent finite topology of the wireless transmissions without the need for any numerical calculation. Bulusu et al. present the application of this connectivity-metric method in outdoor environments, where anchors at fixed reference points having overlapping coverage regions periodically transmit beacon signals [16]. A node infers its location from the intersection of the coverage areas, assuming that the anchors are arranged in a mesh, as in Figure 15.6. An idealized radio model, under which the transmissions of fixed power from an anchor can be received within a circular area of predetermined radii, is also assumed. The ratio of the number of beacon signals successfully received to the total number of signals transmitted is defined in [16] as a connectivity metric for a specific pair of nodes. The accuracy of localization is dependent on the number of anchors and their relative distance and should not be expected to be high since the actual coverage range is usually not a perfect circle.

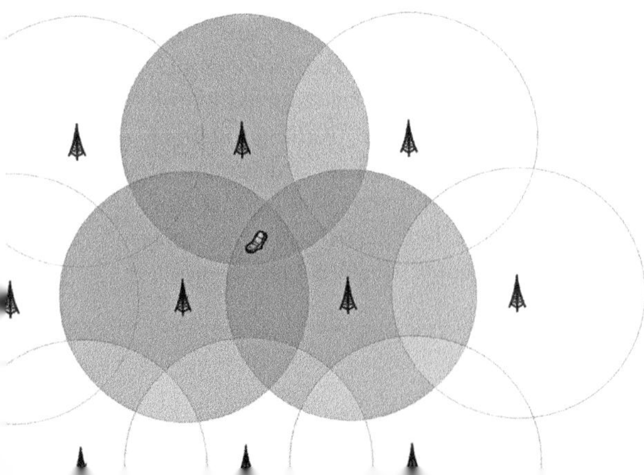

algorithm, even when the spatial density of the anchor nodes is small, by obtaining not only one but a number of local overlapping maps of individual groups of nodes in the entire field and stitching them together. An example of merging two such local maps based on their common nodes is shown in Figure 15.7.

A very similar approach to this optimized concept of [18] is proposed by Ji and Zha [20], who demonstrate through simulation that their multidimensional scaling-based distributed sensor positioning method can accurately estimate the sensors' positions in a network with complex terrain and anisotropic topology, where the nodes are not spatially uniformly located.

15.2.2.3 Distributed Localization. The first approach to be presented in this category is developed by Niculescu and Nath [21], who consider using distance vector-like range estimations exchanged between nodes by multihop communication. In the range-free distance–vector–hop technique, all anchor nodes start independently flooding the network with their location coordinates through multihop packet broadcasts. The hop count field in these messages, which accounts for the number of hops that the latter have traversed from their sources, is updated as they hop from node to node. This value allows each node to maintain a shortest path table to every anchor. Every anchor estimates the average single-hop distance based on the hop count and the known locations of the other anchors and propagates it into the network. A nonanchor node can then use this estimated hop distance, as well as the hop count to other anchors, to perform multilateration. If, instead of hop count values, measured distances between neighboring nodes can also be propagated, then these can be used similarly, resulting in the distance–vector–distance technique. In this case, Euclidean distances from anchor nodes a number of hops away can be estimated, thus increasing the overall accuracy. Angulation information has also been successfully used in the same framework by Niculescu and Nath [22].

In [6], Savvides et al. define atomic, iterative, and collaborative multilateration. As described previously, in the atomic multilateration, a node can estimate its location being within range of at least three beacons. In the iterative multilateration, as soon as connectivity to at least three anchors is established and a node estimates its location, it becomes a beacon for other nodes. This process can be repeated until all nodes with eventually three or more beacons estimate their positions. But even after these two multilateration processes have been completed, a node may still have less than three neighboring beacon nodes, in which case collaborative multilateration should be applied. The ad hoc localization system (AhLOS) proposed in [6] uses iterative multilateration and reveals that this type of multilateration can be problematic in regions where anchor densities are low. Additionally, error propagation becomes an issue in large networks.

The collaborative multilateration, extensively presented in [23], addresses the above two issues. This multihop operation enables nodes found a number of hops away from anchor nodes to collaborate with each other and estimate their locations. The operation takes place in four phases. In the first phase, the nodes whose coordinates can be uniquely determined self-organize into collaborative subtrees. During the second phase, each non-anchor node uses simple geometric relationships to estimate

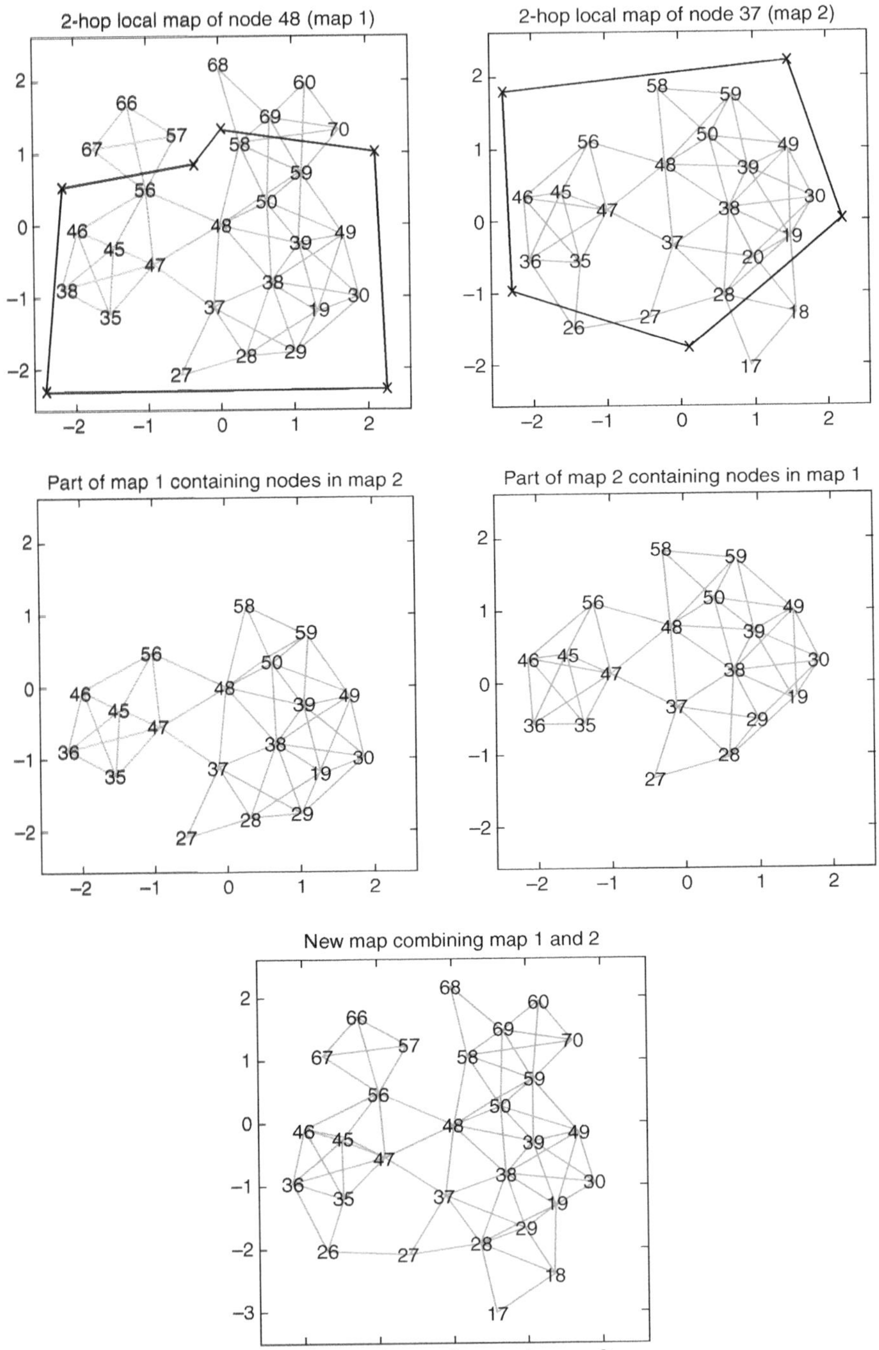

FIGURE 15.7. Improved multidimensional scaling-based localization merging two local overlapping range-free connectivity maps based on their common nodes. (From [19], used with permission of the IEEE.)

its location based on known anchor locations and measured distances obtained with a distance–vector-like algorithm, similar to the one described previously. Iterative least-squares trilateration is applied in the next phase on the initial location estimates to refine them. Finally, all the new location information is used to further refine the location of each node that does not belong to a collaborative subtree.

A similar two-phased approach is presented by Savarese et al. [24]. In the start-up phase, the distance vector-like Hop-TERRAIN algorithm is run to overcome the sparse anchor node problem and obtain rough location estimates. In the refinement phase, the accuracy of the initial location estimates is increased iteratively using the measured distances between neighboring nodes by means of a least-squares algorithm. In contrast to [23], where the refinement process might not converge, the convergence in this work is achieved in almost all cases due to the addition of confidence weights to the position estimates. These confidence weights are close to one for anchors and are lower for nodes with less faith in their position estimates.

An extensive quantitative comparison of the approaches in [21, 23, 24] is presented in [25]. The main conclusion is that no single algorithm performs best; rather, the preference to any of these is application-dependent.

15.3 A REMOTE TRACKING APPLICATION

A novel remote localization system utilizing renewable solar energy to power and trigger communication between end devices and a wireless network is presented. The novelty of the system is the unique node design that allows for battery-less operation using ambient solar energy. Hardware and software design considerations involved in making the localization system operational in spite of the limitations of solar power are outlined. A remote localization system with very good accuracy utilizing this tag communicating with a WSN is showcased as an end application with tremendous potential.

This chapter presents a fully operational localization system that tracks the position of a remotely placed end device through lateration. This service has been found to be very helpful in large parking lot environments; not only does this localization service offered help the user (for example, a car auction dealer) to be able to directly find any vehicle fast and accurately, but it also enables the user to exploit the location data for optimization and cost reduction of its everyday operation.

The goal of the work presented in this chapter is the provision of location information of a solar powered battery-less RFID tag placed on the dashboard of a vehicle in large-scale parking lots to customers. The localization technique deployed is the RSSI (received signal strength indicator)-based lateration, described in section 15.2.1.1.

The reference points were wireless sensor network (WSN) nodes made up of Crossbow's MICA2 wireless nodes. The distance of the end device from the anchor nodes was estimated from the strength of the wireless signal transmitted from the end device as measured by the RSSI on the TI CC1000 transceiver [3] that the MICA2 motes relied upon. RSSI measurements from the anchor node were pooled together on a central computer where they were selectively parsed for computations that would estimate the position of the end device.

The novelty of the system was that the tag not only used ambient light to power itself, but also used it as a triggering mechanism for initiating communication between itself and the WSN nodes, thereby allowing for a completely stand-alone, batteryless asynchronous communication link from itself to the WSN used in the lateration. Section 15.3 describes the power scavenging hardware design of the tag; Section 15.4 describes considerations taken to establish a robust wireless communication link between the tag and WSN given the power limitations; Section 15.5 covers the wireless front end, namely the amplifier characterization and antenna design of the tag; Section 15.6 describes the localization techniques used; the final results of the localization estimation using a commercial mapping software are covered in Section 15.7.

A combination of different technologies—namely, photovoltaics, basic analog circuit design, embedded systems, and RF and microwave/antenna design—are utilized to establish an asynchronous wireless link between the solar-powered tag and a commercial wireless (WSN) network mote using a simplified protocol in the absence of a regulated battery supply. The design utilizes supercapacitors, which are much cleaner to dispose environmentally and have much higher recharge lifetimes compared to batteries. Specifically, a paper-based microcontroller-enabled wireless sensor prototype has been developed for the first time for the UHF frequency band (centered at 904.2 MHz).

Battery technology is considered the Holy Grail for a wide range of industries such as automobile, cellular communications, and so on. The economic and environmental costs of developing and discarding them are still high. While some applications absolutely require a regulated power source such as a battery, many applications through a slightly more involved co-design of different aspects in their hardware and software can be remotely powered using ambient renewable energy forms. Wireless networks present such an application.

15.3.1 Hardware

The hardware system for the tag that was to be remotely tracked by the WSN was designed to meet the following criteria for successful lateration: constant RF power output during each transmission, long range, omnidirectionality, and wireless connectivity. The fundamental problem with using a finite-sized solar cell array was its scarce output power. The palm-sized solar cell array that was stacked in a parallel configuration was capable of generating a maximum of 15 mW. The most power hungry portion of the tag was the wireless front end, which consumed a peak power of 48 mW in transmit mode. A comparison of the two previous numbers showed that, while it was not possible to continuously power the tag with the solar cell array in a relatively short period of time, enough solar energy could be harnessed from the environment to supply the tag with just enough power for communication for a short period of time. A system level diagram of the tag is shown in Figure 15.8.

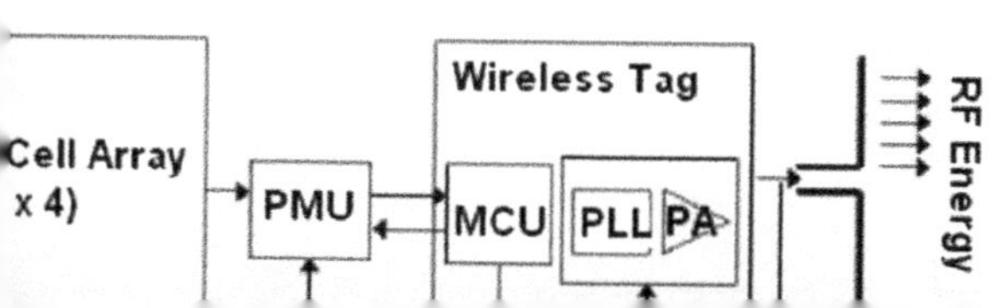
Cell Array
x 4)
PMU
Wireless Tag
MCU
PLL
PA
RF Energy

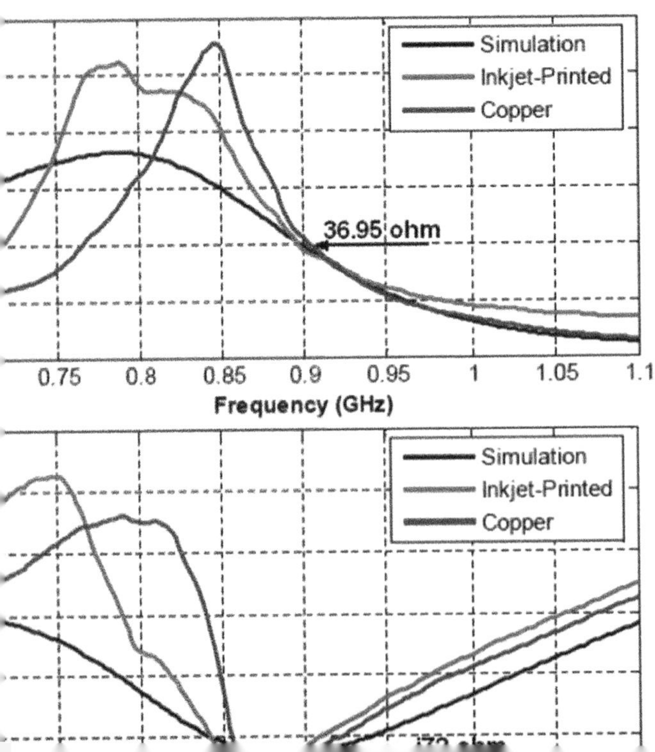

Simulation
Inkjet-Printed
Copper
36.95 ohm
0.75
0.8
0.85
0.9
0.95
1
1.05
1.1
Frequency (GHz)
Simulation
Inkjet-Printed
Copper

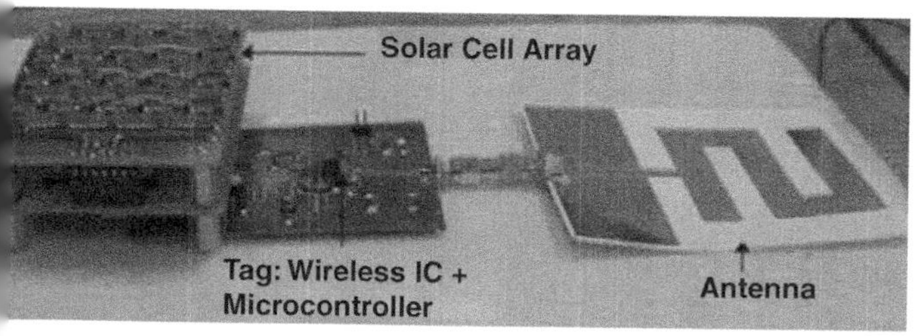

FIGURE 15.10. Solar-powered tag prototype.

· Cell Array. The solar cells were arranged in an array of 5 by 4, in a palm-sized circuit board. Each solar cell unit was a stack-up h open-circuit voltage and short-circuit current characteristics with tensity shown in Figure 15.14. Measurements in Atlanta, GA, USA ity to vary from 4000 lux on a cloudy day to 60,000 lux on a clear intensity measurements were performed using a LUX meter. From

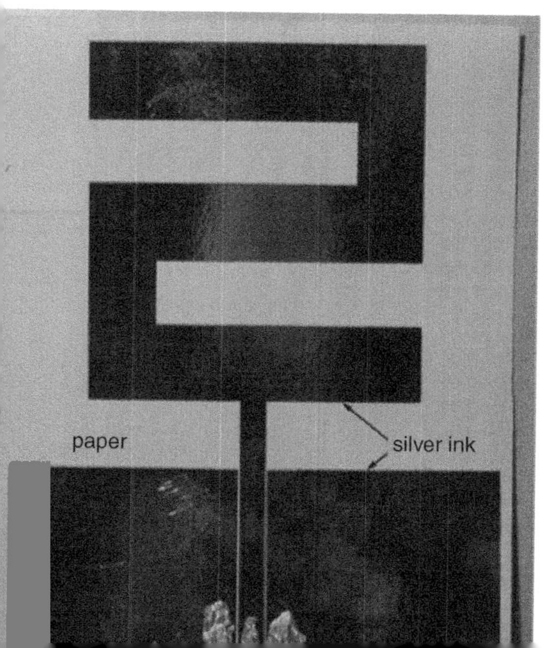

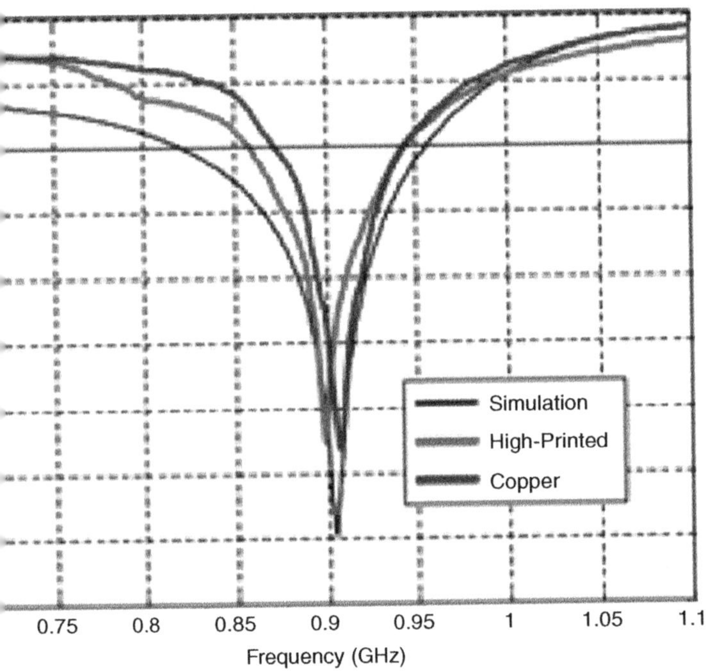

E 15.12. Simulated and measured tag antenna return loss (S_{11}) [32].

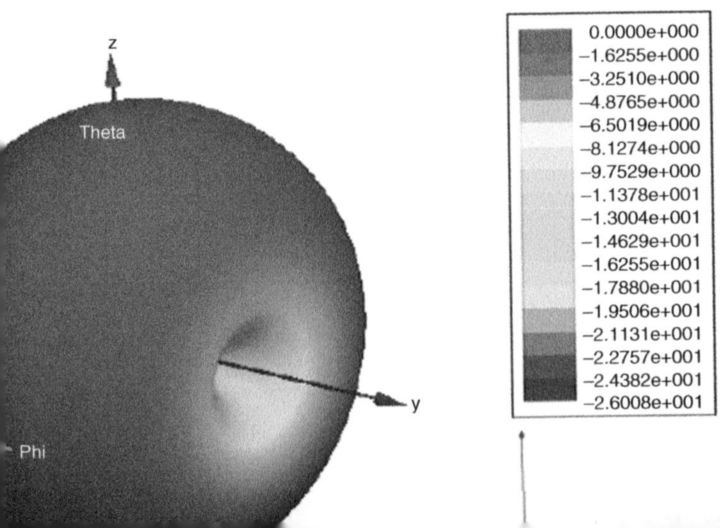

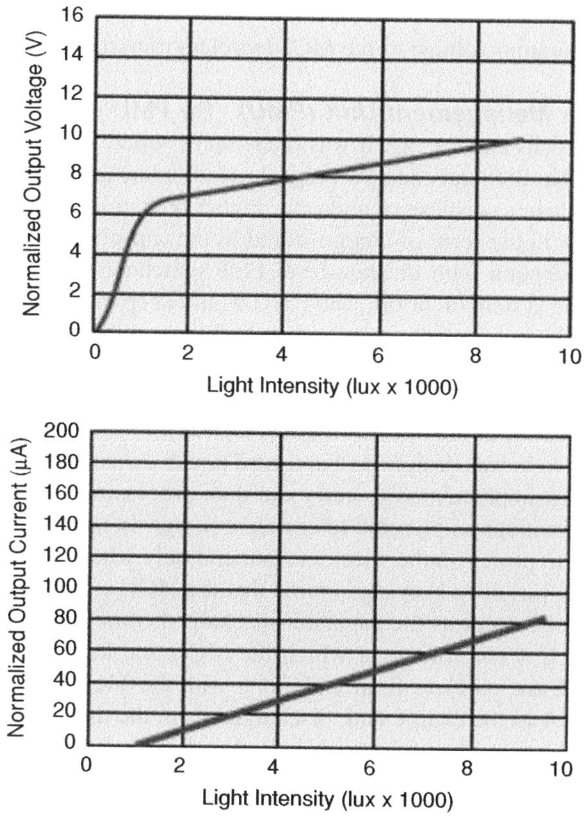

FIGURE 15.14. Characteristics of single solar cell used in array [31].

Figure 15.14, the open-circuit voltage node and short-circuit current of each solar cell of the PV array would vary from 7 to 10 V and 0 to 80 uA based on external light conditions. The parallel configuration of the solar cells would give a maximum short-circuit output current of 1.6 mA while ensuring the same output voltage. Solar cells with a higher stackup that generate a higher output voltage were used based on the choice of the capacitor used in the power management unit.

In the absence of batteries, the solar energy was to be collected in a supercapacitor (charge tank) for use by the tag [28]. A higher solar cell output voltage across the capacitor would provide a faster charge-up time for different light conditions, which is important for more frequent communication by the tag as will be shown later. The drawback of using a supercapacitor along with an ambient light variable power source, such as a solar cell, is its lack of a stable regulated voltage output when different components within the tag are powered on. A stable regulated power output is important to ensure that the RF power radiated from the tag is consistent during each transmit to the WSN to ensure accurate lateration. To rectify this, a novel PMU was devised that regulated the solar cell/supercapacitor output to between 3 and 5.5 V, which is the operating voltage of the MCU/wireless transceiver.

15.3.1.3 Power Management Unit (PMU).

The PMU served as the interface between the solar cells and the tag. It was made up of supercapacitors connected in parallel that served as the solar energy storage device. Supercapacitors are much more environmentally clean to dispose of and offer higher recharging lifetimes compared to batteries. Power in the form of charge stored in the capacitors was monitored by the MCU in conjunction with discrete-level FET switches that were also used to switch power to the system on or off. The PMU would keep the tag in either "off" or "sleep" mode until enough of the solar energy had been collected across the charge tank capacitors. Once the capacitor voltage had reached a certain user set voltage threshold (V_{TH}), the PMU would trigger the tag on. Once "on," the MCU of the tag would take over control of the operation of the tag [28].

The MCU firmware was designed to optimize power consumption of the tag, and the MCU code was implemented to carry out data processing of the wireless data bits in as few instructions as possible to conserve power. In addition, the MCU was also programmed to power on the wireless front end only when the tag was ready to transmit data. This was done keeping in mind that the MCU consumed only between 540 and 480 μA of current as the capacitor discharged from a threshold voltage of 3.8 to 2.2 V, which is the voltage at which the MCU and hence the tag shuts off. On the other hand, the wireless front end along with the MCU consumed between 15.65 and 15.37mA as the charge tank discharged from the threshold voltage of 3.8 to 2.7 V, which was the voltage at which only the wireless front end unit of the tag turned off. Upon completion of communication, the MCU would disable the wireless front end and put itself in "sleep" mode consuming microAmps of current, thereby allowing the already discharged charge tank capacitor to replenish itself using solar energy and repeat the process once it got to the threshold voltage [28].

Setting a fixed voltage threshold (V_{TH}) with the PMU ensures an identical PA output and hence an identical RF radiated power output independent of the external light conditions. The maximum available transmit time is determined by the value of the amount of solar energy harnessed or conversely the charge tank capacitance. A quick way of estimating the total transmit time (T_{XMIT}) available for a given value of charge tank capacitance is given by Eq. (15.1) [28]:

$$T_{XMIT} = -\ln\left(\frac{V_{OFF}}{V_{TH}}\right) \cdot R \cdot C_{TANK} \tag{15.1}$$

where V_{OFF} is the turn-off voltage of the wireless front end, V_{TH} is the threshold voltage set in the PMU, R is the mean load resistance of the tag operating between V_{TH} and V_{OFF}, and C_{TANK} is the value of the charge tank capacitor.

For the tag prototype developed, the threshold voltage (V_{TH}) by the PMU was set to 3.8 V and the tag's turn-off voltage was measured to be 2.7 V when transmitting using frequency shift keying (FSK) modulation around a center frequency of 904.29 MHz. The mean load resistance of the tag was measured to be 194 Ω. Using Eq. (15.1), the total transmit time was determined to be 42.23 ms for the charge tank capacitor

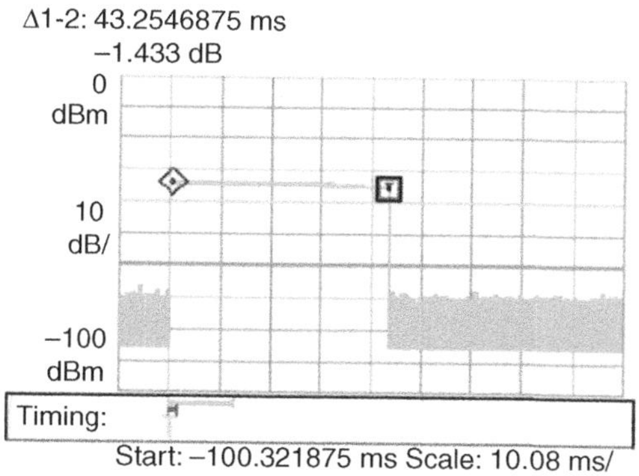

FIGURE 15.15. Power versus time wireless transmission profile of the tag's transmission captured by a real-time spectrum analyzer. (From [2], used with permission of the IEEE.)

of 637 μF used in the prototype. The approximation in Eq. (15.1) does not take into account the power consumed in the initial processing done by the MCU, which is less than 4% of the total power consumed and the power supplied by the solar cell as the charge tank capacitor discharges. The total wireless transmit time available for a charge tank capacitance of 637 μF and a PMU threshold of 3.8 V in the tag is shown in the wireless signal captured by a Tektronics RSA-3408A Real-Time Spectrum Analyzer (RTSA) connected to an AN-400 RFID reader antenna in the form of power versus time in Figure 15.15. The measured transmit time was observed to be 43.25 ms, very close to the theoretically predicted value. The charge tank capacitance along with external light conditions would also set the time interval between adjacent transmits. Light intensity measurements carried out on the tag using a charge tank of 637 μF show a linearly proportionate relationship between light intensity and the rate of wireless transmissions as can be seen from Table 15.1. The light intensities were replicated in the lab with the help of halogen bulbs and may not represent the entire spectrum present in sunlight, therefore presenting the worst-case scenario [28].

15.3.2 Wireless Communication with a WSN

The data packets were transmitted by the tag to the WSN nodes within its range using FSK modulation at a center frequency of 904.4 MHz. The data packets consisted of the following fields: Preamble, Sync, Addr, Type, Group, Length, Data, and CRC.

TABLE 15.1. Localization Estimation Performance Results

Light Intensity Exposed on End Node	Time Interval between Consecutive Wireless Transmissions
10.5 kLux	4.4 s
20 kLux	2.7 s
40 kLux	1.8 s
70 kLux	1.4 s

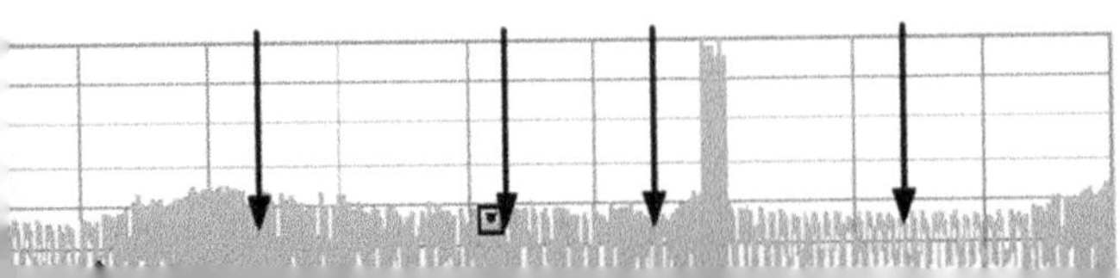

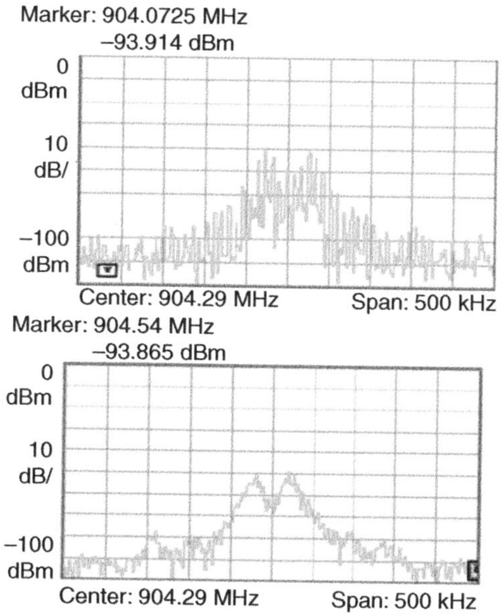

FIGURE 15.17. Calibrated tag transmitter and MICA2 transceiver FSK modulation profile.

was 3.5 m and the height of the tag was around 1.5 m, the value of the radius of the cross section of the ellipsoidal of the first Fresnel zone at the middle of the distance, typically 75 m, was approximately 2.8 m and the same value just 0.5 m away from the mote side was around 0.3 m.

As mentioned above, the localization technique was based on the RSSI (Received Signal Strength Indicator) lateration. After the successful reception from each WSN node of the beacon message transmitted by the tag, significant information (namely RSSI, unique tower ID, and time stamp) is appended and forwarded to a central location. All these message entries are imported into an Sqlite database. The first type of process performed is data validity checks, according to which a number of filters are applied to reject invalid entries. Then the RSSI, after it is filtered as explained later, is used for the distance calculation from a particular WSN node with the use of a free space or a two-ray tracing radio propagation loss model. The time-stamp information availability enables the use of a timer function, which loads data from specific time intervals back in the past relative to the time the location estimate was initiated. Finally, the trilateration is performed as a database procedure calling [29]; and, in turn, functions are called to display the returned WGS 84 latitude–longitude coordinates on Google Earth [30].

15.3.4 Localization Results

In order to verify the feasibility of this technology for remote tracking applications, actual measurements of the location estimate error were taken at different positions of the tag in the field covered by a number of fixed motes. These 24 different predetermined positions are shown in Figure 15.18 with capital letters; the numbers

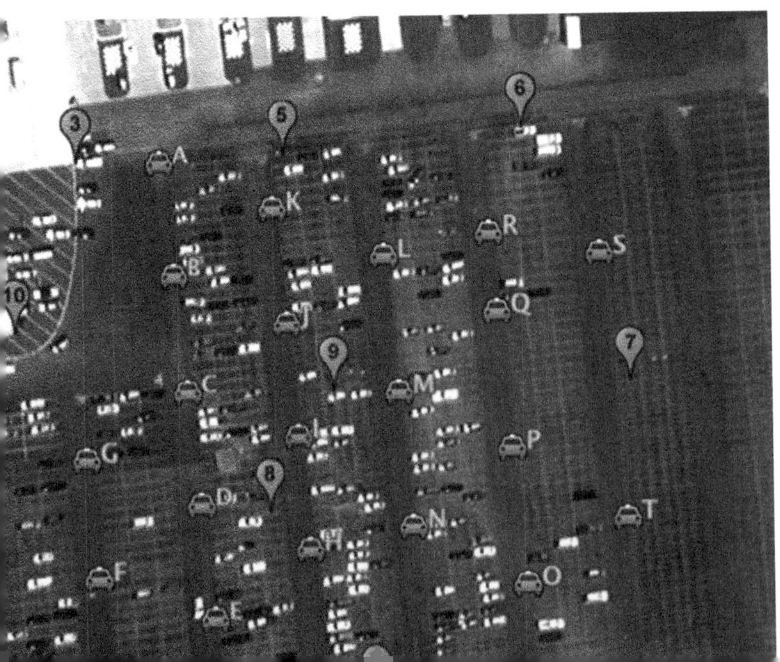
3
5
6
A
K
R
L
S
B
10
Q
J
9
7
C
M
I
G
P
8
D
T
N
H
F
O
E

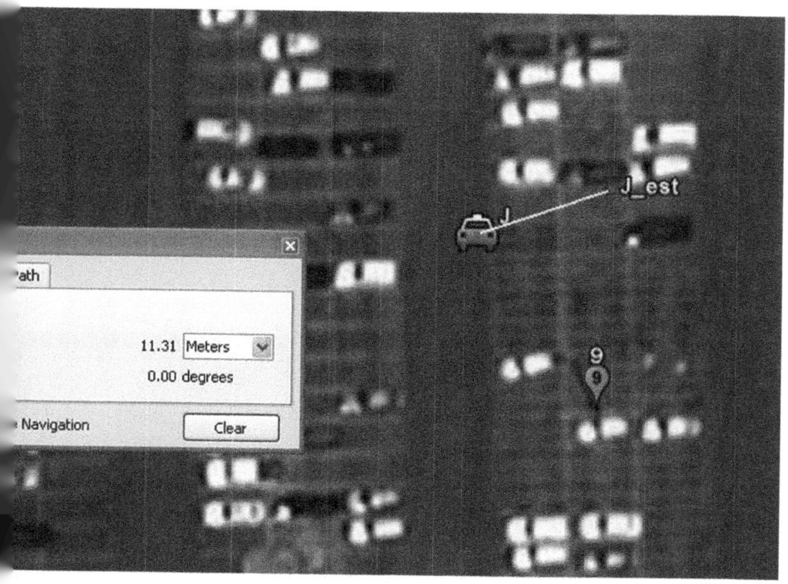

FIGURE 15.19. An example of a localization estimation error.

gh all these proposed solutions are addressing localization needs in dif- cal environments with exclusively single-hop or additionally multihop to reference nodes and have been either hardware-implemented or just nulated, an attempt to summarize their characteristics in a somewhat fashion is shown in Table 15.2.

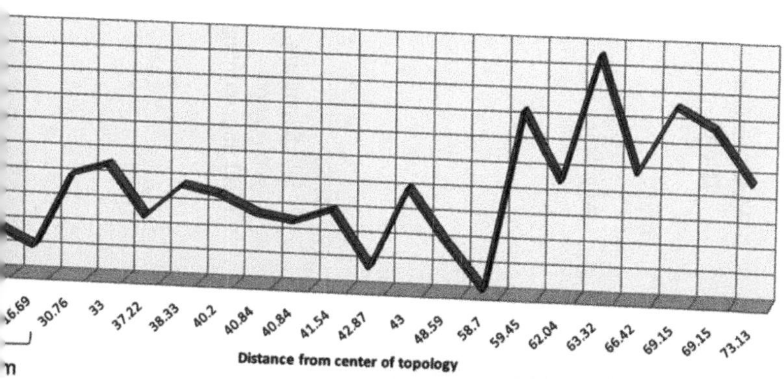

n

TABLE 15.2. Localization Techniques

Technique	Solution	Accuracy		Scale	Cost	Implemented in Real Scenario?
Multilateration—RSSI	Batteryless, solar-powered wireless tag [2]	4 ft	Outdoor		*Inexpensive*	Yes
Multilateration—ToA	Countersniper system [7] (only muzzleblast fusion results)	Avg. error 1.3 m < 1 m for 46%, and <2 m for 84%	Outdoor	56 motes in the central area of the McKenna village	*Inexpensive*	Yes
Multilateration—TDoA	Active office [4]	<14 cm for 95%	Indoor	16 celling recevers, over a volume of some 75 m^3	—	Yes
	Cricket [9]	4 × 4 feet granularity	Indoor	2 beacons over an area of 36 ft^2	$10/component	Yes
Angulation	Nasipuri—directionality [8] (for ideal propagation characteristics)	<2m	—	75 × 75 m	N/A	Simulation only
	Lighthouse [9]	Overall mean relative offset of the mean locations from ground truth locations is 2.2%	Indoor	5 × 5 m		Yes
Radio interferometric geolocation	Radio interferometric geolcation [12]	<6 cm	Indoor	3 anchor nodes over an area of 18 × 18 m	low	Yes
Field fingerprinting	RADAR [15]	<3 m	Indoor	3 base stations over an area of 43.5 × 22.5 m	—	Yes

Multihop localization—proximity	Bulusu—GPS—less [16]	Avg. error 1.83 m	Outdoor	4 reference points over an area of 10 × 10 m	low	Yes
	Convex position estimation [17]	Variable over a wide range depending on simulation scenario	—	200 nodes over an area of 10 × 10 m	N/A	Simulation only
	Shang—connectivity [18]	Variable over a wide range depending on simulation scenario	—	>79 nodes in all 3 different scenarios	N/A	Simulation only
Multihop localization—multidimensional scaling	Improved MDS-based localization [19]	Improved compared to ones from [18]	—	>79 nodes in all 3 different scenarios	N/A	Simulation only
	Ji—MDS [20]	0–50% of average radio range	—	400 nodes over an area of 100 × 100 m	N/A	Simulation only
Distributed multihop localization	APS [21]	Variable over a wide range depending on simulation scenario	—	100 nodes	N/A	Simulation only
	APS with AoA [22]	Variable over a wide range depending on simulation scenario	—	1000 nodes	N/A	Simulation only
	Savvides [6]	<2 cm	Indoor	9 Medusa nodes	low	Yes
	n-Hop multilateration [23]	Avg. error 2.77 cm	—	Varying from 10 to 100	N/A	Simulation only
	Savarese [24]	Less than 33% for 5% range measurement error and 5% anchor population		400 nodes over an area of 200 × 200 units	N/A	Simulation only

The parameter that eventually plays the major role in choosing a particular localization technique is the ease of deployment in a new or an existing wireless network infrastructure. This involves the existence of a real-world implementation of a technique with low hardware complexity, low or no packet transmission overhead, low cost, and minor limitations. As for the hardware complexity, for instance, solutions based on the RSSI multilateration require only minor software module additions, whereas the lighthouse solution [11] requires the installation of quite complex anchor nodes with rotating mirrors and laser diodes.

Hopefully, this work can serve as an initial small step and reference for researchers in their effort to choose the one localization technique that is most suitable for the requirements and special characteristics of their own application.

The first-ever operational remote tracking system using a battery-less, solar-power scavenging unique node has been demonstrated. It is the first time reported that the ambient solar energy not only provides the necessary operation power of the tag being remotely located, but also functions as the trigger of the data communication. A successful solar-triggered wireless link between the tag and WSN was successfully established providing excellent localization accuracy in a sustainable way.

ACKNOWLEDGMENT

This work was supported by the National Science Foundation (NSF), the New Energy and Industrial Technology Development Organization (NEDO)-Japan, and the Interconnect Focus Center (IFC) of Semiconductor Research Corporation (SRC).

REFERENCES

1. J. B. Harley and D. Woodward, *The History of Cartography: Cartography in Prehistoric, Ancient, and Medieval Europe and the Mediterranean*, Humana Press, Totowa, NJ, 1987.
2. R. Vyas, V. Lakafosis, N. Chaisilwattana, Z. Konstas, and M. M. Tentzeris, Design and characterization of a novel battery-less, solar powered wireless tag for enhanced-range remote tracking applications, in *Proceedings, European Microwave Conference*, Rome, September 2009.
3. N. Bulusu, D. Estrin, L. Girod, and J. Heidemann, Scalable coordination for wireless sensor networks: Self-configuring localization systems, in *Proceedings, 6th International Symposium on Communication Theory and Application*, Ambleside, Lake District, U.K., 2001, pp. 1–6.
4. A. Ward, A. Jones, and A. Hopper, A new location technique for the active office, *IEEE Pers. Commun.*, Vol. 4, No. 5, pp. 42–47, 1997.
5. K. Whitehouse and D. Culler, Calibration as parameter estimation in sensor networks, in *Proceedings, 1st ACM International Workshop Sensor Networks and Applications (WSNA)*, Atlanta, GA, 2002, pp. 59–67.

6. A. Savvides, C.-C. Han, and M. Srivastava, Dynamic fine-grained localization in ad-hoc networks of sensors, in *Proceedings, 7th Annual International Conference on Mobile Computing Networking*, ACM Press, Rome, Italy, 2001, pp. 166–179.
7. A. Ledeczi, A. Nadas, P. Volgyesi, G. Balogh, B. Kusy, J. Sallai, G. Pap, S. Dora, K. Molnar, M. Maroti, and G. Simon, Countersniper system for urban warfare, *Proc. ACM Trans. Sensor Netw.*, Vol. 1, No. 1, pp. 153–177, 2005.
8. P. Volgyesi, G. Balogh, A. Nadas, C. Nash, and A. Ledeczi, Shooter localization and weapon classification with soldier-wearable networked sensors, in *Proceedings, MobiSys 2007*, San Juan, PR, June 11–14 2007, pp. 113–126.
9. N. B. Priyantha, A. Chakraborty, and H. Balakrishnan, The cricket location-support system, in *Proceedings, 6th International Conference on Mobile Computing and Networking (ACM Mobicom)*, Boston, MA, 2000, pp. 32–43.
10. A. Nasipuri, and K. Li, A directionality based location discovery scheme for wireless sensor networks, in *Proceedings, 1st ACM International Workshop Sensor Networks and Application (WSNA)*, Atlanta, GA, 2002, pp. 105–111.
11. K. Romer, The lighthouse location system for smart dust, in *Proceedings, ACM/USENIX International Conference on Mobile Systems, Applications, and Services (MobiSys)*, San Francisco, CA, 2003, pp. 15–30.
12. M. Maroti, B. Kusy, G. Balogh, P. Volgyesi, A. Nadas, K. Molnar, S. Dora, and A. Ledeczi, Radio interferometric geolocation, in *Proceedings, ACM 3rd Conference on Embedded Networked Sensor Systems (SenSys)*, 2005, pp. 1–12.
13. CC1000 single chip very low power RF transceiver [Online]. Available: http://focus.ti.com/lit/ds/symlink/cc1000.pdf. 2009.
14. B. Kusy, J. Sallai, G. Balogh, A. Ledeczi, V. Protopopescu, J. Tolliver, F. DeNap, and M. Parang, Radio interferometric tracking of mobile wireless nodes, in *Proceedings, MobiSys 2007*, San Juan, PR, June 11–14 2007, pp. 139–151.
15. P. Bahl and V. N. Padmanabhan, RADAR: an in-building RF-based user location and tracking system, in *Proc. IEEE INFOCOM*, Tel Aviv, Israel, 2000, pp. 775–784.
16. N. Bulusu, J. Heidemann, and D. Estrin, GPS-less low cost outdoor localization for very small devices, *IEEE Pers. Commun. Mag.*, Vol. 7, No. 5, pp. 28–34, 2000.
17. L. Doherty, L.El Ghaoui, and K. S. J. Pister, Convex position estimation in wireless sensor networks, in *Proceedings, IEEE INFOCOM*, Anchorage, AK, 2001, pp. 1655–1663.
18. Y. Shang, W. Ruml, Y. Zhang, and M. Fromherz, Localization from connectivity in sensor networks, *IEEE Trans. Parallel Distributed Syst.*, Vol. 15, No. 11, pp. 961–974, 2004.
19. Y. Shang and W. Ruml, Improved MDS-based localization, in *Proceedings, IEEE INFOCOM*, Vol. 4, 2004, pp. 2640–2651.
20. X. Ji and H. Zha, Sensor positioning in wireless ad-hoc sensor networks using multidimensional scaling, in *Proceedings, IEEE INFOCOM*, Vol. 4, 7–11 March 2004, pp. 2652–2661.
21. D. Niculescu and B. Nath, Ad hoc positioning system (APS), in *Proceedings, IEEE GlobeCom*, San Antonio, AZ, 2001, Vol. 5, 25–29 November 2001, pp. 2926–2931.
22. D. Niculescu and B. Nath, Ad hoc positioning system (APS) using AOA, in *Proceedings, IEEE INFOCOM*, San Francisco, CA, 2003, Vol. 3, 30 March–3 April 2003, pp. 1734–1743.
23. A. Savvides, H. Park, and M. B. Srivastava, The *n*-hop multilateration primitive for node localization problems, *Mobile Netw. Appl.*, Vol. 8, No. 4, pp. 443–451, 2003.

24. C. Savarese, J. Rabay, and K. Langendoen, Robust positioning algorithms for distributed ad-hoc wireless sensor networks, in *Proceedings, Annual USENIX Tech. Conference*, Monterey, CA, 2002, pp. 317–328.
25. K. Langendoen and N. Reijers, Distributed localization in wireless sensor networks: A quantitative comparison, *Comput. Netw.* (Special Issue on Wireless Sensor Networks), Vol. 43, No. 4, pp. 499–518, 2003.
26. MICA2 Datasheet, Crossbow [Online]. Available: http://www.xbow.com/Products/Product_pdf_files/Wireless_pdf/MICA2_Datasheet.pdf.
27. C. Balanis, *Antenna Theory*, Wiley, New York, 1997.
28. L. Barclay, *Propagation of Radiowaves*, 2nd edition, IEE, Van Nuys, CA, 2003, pp. 123—135.
29. GTAthena Localization, *Ecological Software Solutions LLC*, Version 1.11, Hegymagas, Hungary.
30. [Online]. Available at http://earth.google.com.
31. Clare, CPC 1832 Solar Cell, Data Sheet, 2008.
32. Konstas, Zissis, Design and Fabrication of a Novel Conductive Inkjet Printed Antenna on Paper Substrate for RFID Applications, Diploma thesis, October 2008.

PART V

MULTI-ANTENNA RFID TAGS

CHAPTER 16

MULTI-ANTENNA BACKSCATTERED CHIPLESS RFID DESIGN

ISAAC BALBIN and NEMAI CHANDRA KARMAKAR
Department of Electrical and Computer Systems Engineering, Monash University, Clayton, Victoria, Australia 3800

16.1 INTRODUCTION

The adoption of RFID technology has been flagged in various commercial industry sectors including farming, retail, manufacturing, and others; however, there has not been universal adoption. This is due to many factors, including the tag cost, system reliability, and privacy concerns, but many of these issues could be addressed once the technology is matured to the point where it is ready for large-scale implementations. Since RFID has such a broad definition, systems can be applied in a variety of different forms, depending upon the specific requirements of the client [1]. System parameters such as reading range, read accuracy, tag size, and tag cost will determine which form of RFID is most suitable for a particular application. For low-cost solutions the most appropriate system is chipless RFID.

Chipless tags have their identification code written once only. This means that their functionality is limited; however, generally since there is no requirement for surface-mounted components on-board, we can mass-produce the transponders at a cost low enough to be competitive with barcodes. They generally operate by encoding an identification number into the time domain or frequency domain, creating digital bits that we refer to as *time-signature* and *frequency-signature*. If the transponder is in the interrogation zone defined by the reader unit, then the reader will receive a response that is backscattered from the transponder containing its identification information. The only successful commercialization of chipless RFID technology has been achieved by RFSAW®, whose tags are designed using surface acoustic wave (SAW) technology. These tags encode their identification code in the time domain since surface waves propagate around three orders of magnitude slower than

Handbook of Smart Antennas for RFID Systems, Edited by Nemai Chandra Karmakar

signals propagating in free space. The transponders are rugged and have been used for asset tracking on the International Space Station [2] but must always be made rigid and bulky, even if the antenna portion is flexible, making it unsuitable for some applications.

The other set of chipless RFID systems that can be found in the published literature encode their identification number using frequency signatures. The main difference between the different systems is the way in which they generate the *frequency signatures* and the way in which the transmission and reception signals are separated at the reader terminals. The encoding principles for such systems tend to be similar, in that the transponder generates amplitude variations in a swept frequency band [3–6]. However, the number of information bits that can be stored on the transponder is limited by the higher-order modes of the lowest frequency resonances. These systems have also encountered orientation sensitivity issues arising from the distortion in the radiation pattern of the transponder's antenna elements at multiple frequencies.

This has led research in RFID technology to incorporate transponders with multiple antennas. Griffin and Durgin [7] have performed a simulated study of the merits of implementing multiple antennas on both the reader and tag for passive RFID systems and found that significant improvements in performance can be achieved. It was found that UHF RFID systems operate in a pinhole channel that experiences small-scale fading, even when a line-of-sight (LOS) is available and unperturbed. Through the use of multiple antennas it was found that the fading can be significantly reduced. The study investigated the benefits of multiple antennas for active and semiactive RFID systems; however, the principal conclusions of the report can be applied to chipless RFID systems to increase the data contained on a transponder.

The classical *RF barcode* consists of a series of resonant half-wave antennas with variable load impedance [3]. Jalaly and Robertson [3] present a chipless RFID tag where the resonances are created by using an N-element dipole array where there are different gap distances (which can be expressed as a capacitance or an inductance) between the elements to shift the resonant frequency. They have designed a transponder that only operates in frequency bands designated as industrial, scientific, and medical (ISM) use that contains 11 bits of data. However, due to the effect of the nonresonant scatterers, the dipole resonances would have a far-field radiation pattern that is directional and is linearly polarized, limiting its practicality. The number of bits in the measured system indicate that it is not possible to generate a tag using such a method that can contain 180 bits of data.

McVay et al. [4] reported that they designed a system using the same conceptual basis, but with a fractal geometry replacing the straight line dipole arm. The Hilbert and Peano curve geometries that are used have an excellent spatial efficiency due to their space-filling natures that allow them to be much smaller than their equivalent dipoles. They have demonstrated a 5-bit RFID tag operating within a bandwidth of 300 MHz. In both of these systems there is a 1-to-1 correspondence between the existence of each antenna and its associated frequency signature. A third system has been developed with this concern in mind with an attempt to overcome the problem. Preradovic et al. [9] designed a transponder that uses two wideband orthogonally polarized monopole antennas coupled with a series of stop-band resonators. The resonant stop-band filter that is used is a spiral that is gap-coupled to the interconnecting microstrip line. A strip of metal "shorting" the spiral, which in effect shifts the resonant frequency outside of the operating range, allows the spiral resonators to be turned "on" or "off," which is an extremely useful and cost-effective feature of the transponder. The resonators that are "off" will allow signals in their resonant frequency band to pass through both antennas (and thus change their polarization to an orthogonal state) and return to the reader. In this case we have a 1-to-1 correspondence between each spiral resonator and each resonance, which allows much greater spatial efficiency since the transponder contains only two antennas.

In this study, we will investigate the benefits of encoding the transponder's identification code in the phase of the backscattered signal. It has been shown that the backscattered phase is a much more stable and predictable feature when compared with the backscattered magnitude. By making use of the backscattered phase, we are able to inject much more information in each frequency signature than the simple on/off systems that have been developed thus far. However, when we try to precisely control the backscattered signal's phase, it requires that the complexity of the design of both the reader unit and the transponders will increase significantly. It has been experimentally verified that such a transponder will be able to generate a predictable phase shift in the backscattered signal, which means more data are squeezed into the available portion of the frequency spectrum [8]. A measured 30° phase shift was achievable on a transponder containing three resonant signatures in a preliminary study. This chapter will present the design process and associated theory required to design a chipless RFID transponder encoded in the phase domain.

This chapter is organized as follows: We initially give a brief description of the chipless RFID systems that make use of spectral signatures, followed by a description of how we use the same spectral space, but insert more data by making use of the phase component of the backscattered signal along with a brief description of the radar theory concepts we are exploiting. Section 16.3 investigates the radiation properties of a single radiating element and then gives an analysis of the radar properties, particularly in regard to the backscattered signal. Section 16.4 investigates the radiation and scattering properties of an array of radiating elements and the associated benefits that this introduces. In Section 16.5 we investigate the performance of a transponder that consists of multiple arrays of radiating elements, resonant at different resonant frequencies with performance verified by measured results. The conclusions are given in Section 16.6.

16.2 SYSTEM DESCRIPTION

16.2.1 Existing Technology

A brief description of the system operation will now be presented. All RFID systems contain three basic components, as seen in Figure 16.1. The tags (or transponders) are attached to the item of interest and contain a unique identification code. In some systems there is additional information that is stored on the tag. The reader is designed

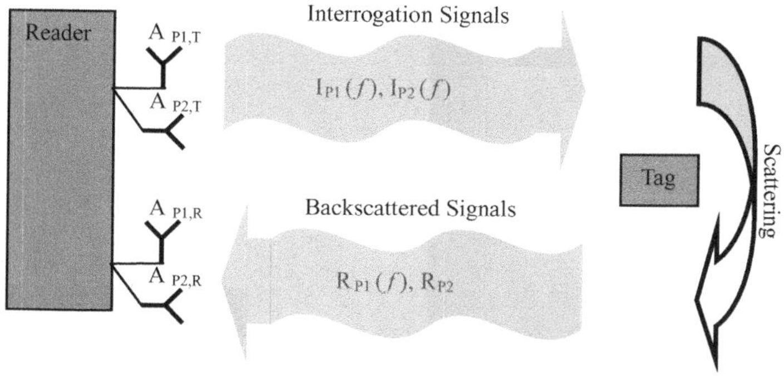

FIGURE 16.2. Interrogation and backscattered signals in a phase-encoded chipless RFID system.

the reader with a single dual-polarized antenna or two separate orthogonally polarized antennas (A_{P1} and A_{P2} for polarization states P1 and P2, respectively) as shown in Figure 16.2. The system can also function using a monostatic setup where the same antennas are used for transmission and reception or in a bistatic setup (Figure 16.2) where separate antennas are used for transmission ($A_{P1,T}$ and $A_{P2,T}$) and reception ($A_{P1,R}$ and $A_{P2,R}$). The reader generates the interrogation signal ($I_{P1}(f)$ and $I_{P2}(f)$) at a frequency f(GHz), and then it divides the power equally between $A_{P1,T}$ and $A_{P2,T}$. There are now two signals that propagate through free space toward the tag. The signals reach the transponder and are scattered according to its radar cross section (RCS) characteristics. The tag is designed so that the RCS characteristics differ in the orthogonal polarization states P1 and P2. The phase shift is specified by the designer so that the return signals ($R_{P1}(f)$ and $R_{P2}(f)$) that propagate back toward the reader have a phase difference of, $\Delta\varphi_N$.

The signals are always orthogonally polarized and will travel through the same propagation path, so the path loss should be almost the same. The difference between the return signals is only due to differences in the scattering properties of the transponder for the two orthogonal polarization states. The phase difference is then converted into a set of discrete bins called *phase signatures* that are used to obtain digital data at each resonant frequency. The concept is illustrated in Figure 16.3, which shows a set of seven frequency signatures. Each frequency signature will contain a discrete phase difference between 0° and 180°. In the presented example we are able to resolve each phase signature to the nearest 10°, creating 18 possible combinations for each frequency signature instead of the standard two combinations (binary).

At a number of distinct frequencies $f_1 - f_N$, we generate a phase difference, φ_N defined in Eq. (16.1):

$$\varphi_N = \angle R_{P1}(f_N) - \angle R_{P2}(f_N), \qquad \text{where } 0 \leq |\varphi_N| \leq 180 \qquad (16.1)$$

If we are able to resolve the phase of the backscattered signal with a tolerance of $\pm T/2$ degrees, then we can create $180/T$ unique combinations for data use. This means we

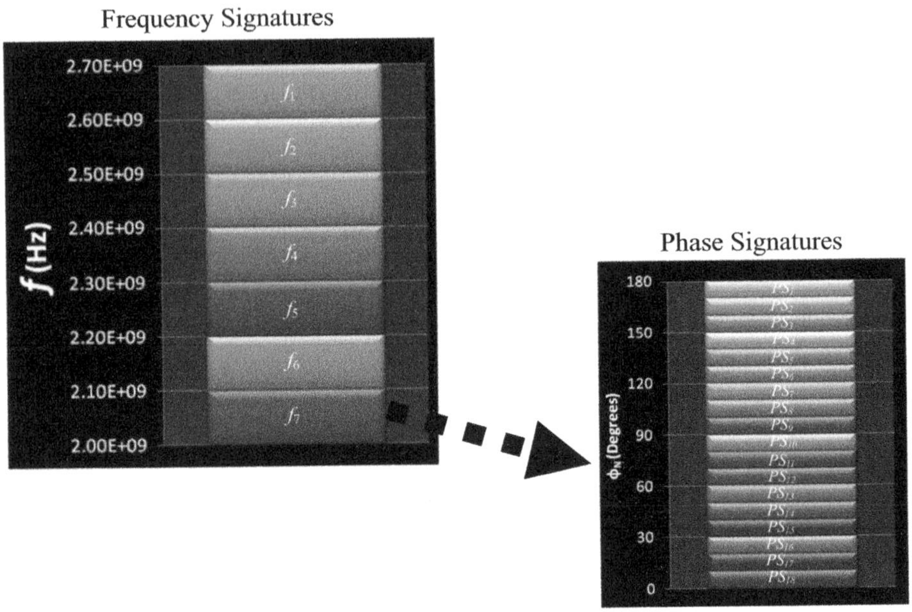

FIGURE 16.3. Frequency and phase signatures.

can calculate the total number of bits of information in the transponder, which is presented in Eq. (16.2):

$$\text{Bits} = \log_2 \left(\frac{180}{T} \right) * N \tag{16.2}$$

We will illustrate the concept with a simple example.

If we refer to Figure 16.3, we can see seven frequency signatures between 2GHz and 2.7 GHz, meaning that $N = 7$. Each of the frequency signatures can be broken up into phase signatures with a tolerance, $T/2 = 5$ degrees, so we have $T = 10$ degrees. This means that the number of bits can be calculated as

$$\text{Bits} = \log_2 \frac{180}{10} * 7 = 29$$

So for a 700-MHz section of the EM spectrum which previously held 7 bits of data, it now contains 29 bits.

It is easy to conclude that with such a scheme it will be easier to design transponders for applications that require a large number of data bits. The backscattered phase difference is generated by precisely controlling the backscattered RCS of the transponder in two orthogonal polarizations. The next section presents a short description of the radar theory required to describe the system.

16.2.3 Radar Theory

It is appropriate to model our system as a short-range radar system so that appropriate theoretical considerations can be made. If a single antenna is used to send and receive signals in each polarization state, then we have a monostatic situation, the radar range equation of which is shown in Eq. (16.3):

$$\frac{P_R}{P_T} = \frac{G^2\lambda^2\sigma}{(4\pi)^3 r^4} \tag{16.3}$$

where P_R is the received power (W), P_R is the transmitted power (W), G is the reader antenna gain, r is the distance between the tag and the reader (m), and σ is the RCS of the transponder. It is clear from Eq. (16.3) that if the transponder is in the interrogation zone of the reader, we can vary the RCS with respect to frequency and create a frequency spectrum where distinct changes in the phase and amplitude should be observable. In radar theory, if the target being scanned is an antenna, there are additional design equations regarding the RCS. It consists of two components called the *antenna mode* and the *structural mode* scattering. The structural mode scattering is an unavoidable portion of the RCS that occurs due to the structure itself, and it exists for all possible radar targets. In general we can assume that this scattering component will not exhibit a phase difference between its orthogonally polarized components. The antenna mode is a function of the radiation characteristics of the antenna itself and can be designed using standard antenna theory. This gives us the ability as designers to be able to design the radiation characteristics of the antenna to give us a satisfactory scattering performance.

The two scattering component parameters defined with respect to the total electric field scattered from an antenna are shown in Eq. (16.4) [10]:

$$E^s(Z_L) = E^s(Z_A^*) - \frac{I_m^*}{I_t}\Gamma^* E^t \tag{16.4}$$

where $E^s(Z_L)$ is the electric field scattered by the antenna when it is loaded with an impedance of Z_L, $E^s(Z_A^*)$ is the electric field scattered by the antenna when it is conjugate-matched, Z_A^* is the conjugate antenna input impedance, I_m^* is the current induced when the antenna is in transmitting mode and is conjugate-matched, I_t is the current induced when the antenna is in transmitting mode, Γ^* is the conjugate-matched reflection coefficient, and E^t is the time varying electric field.

The first term in Eq. (16.4) represents the antenna mode, and the second term the structural mode scattering. It implies that if we change the loading impedance of the antenna, then we will also change the antenna mode scattering component of the RCS. In the proposed system, we will therefore implement a dual polarized microstrip patch antenna that is loaded with open-circuit stubs.

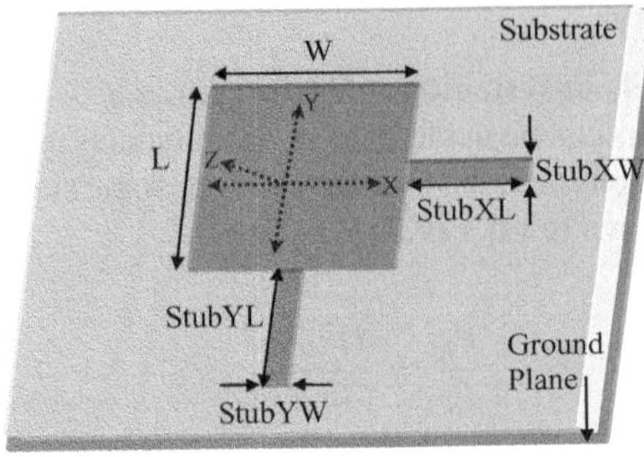

FIGURE 16.4. Layout of SLMPA with dimensions (mm).

16.3 STUB-LOADED MICROSTRIP PATCH ANTENNA (SLMPA)

To illustrate the concept, we will present the development of a system that encodes data with phase signatures, using two orthogonal linear polarization states; this is the simplest case. We use a series of dual-polarized microstrip patch antenna reflector elements to generate the phase difference in the orthogonally polarized backscattered signals. The phase of the backscattered signal is controlled by altering the antenna mode scattering of the transponder. This is achieved by loading the microstrip patch antennas with open-circuited microstrip loading stubs in each polarization state. The ground plane of the antenna reflector element makes the transponder more insensitive to the contents of the tagged item.

16.3.1 Antenna Mode

The layout of the SLMPA is shown in Figure 16.4. The element is square so that it is dual-resonant ($L = W$)—in other words, resonant at the same frequency in both polarizations. We use an edge-feed for attaching the loading stubs so as not to disturb the radiation properties of the antenna. The open-ended microstrip stub is a good choice to load the antenna since it is the cheapest and simplest termination to fabricate. The input impedance of the stub can be calculated using Eq. (16.5) [11]:

$$Z_{\text{IN}} = -jZ_0 \cot(\beta L) \tag{16.5}$$

where Z_0 is the characteristic impedance of the loading stub, $\beta = 2\pi/\lambda$ is the wavenumber, and L is the length of the loading stub. As an example, we have designed an antenna to operate in the ultra-wideband (UWB) frequency band, resonating at 6.12 GHz on a 1.5-mm-thick FR4 ($e_r = 4.9$) substrate. We set $L = W = 10$ mm, StubXL = StubYL = 1.5 mm, and StubXW = StubYW = 0.2 mm. To simulate the operation of the circuit, we use the full-wave EM solver software CST Microwave Studio™ (MS). We thus place simulation ports with an input impedance equal to that of free space at the end of the loading stubs, Port1 at the end of StubY and Port 2

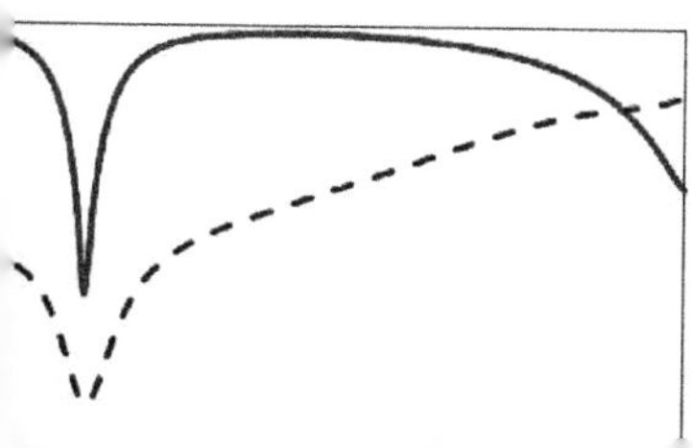

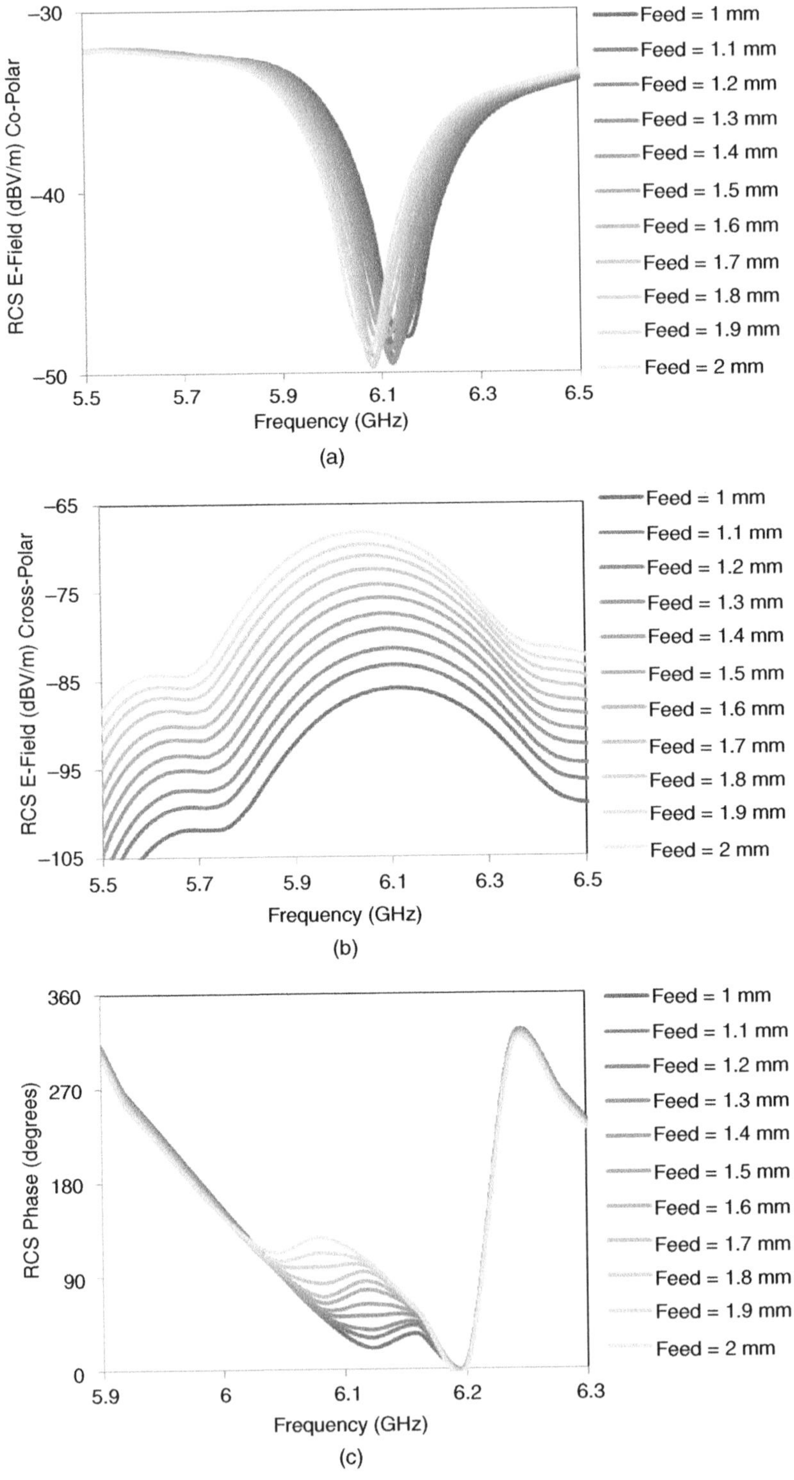

FIGURE 16.6. RCS E-field co-polar **(a)** cross-polar **(b)**, and co-polar phase **(c)** of SLMPA on 1.5 mm FR4 ($e_r = 4.9$) with dimensions $L = W = 10$ mm, StubXL = StubYL = 1.5 mm, StubXW = StubYW = 0.2 mm.

The resonant peak changes by 80 MHz for a 1-mm change in the loading stub length; however, this is not significant since the bandwidth of the resonance is 200 MHz and the design frequency of 6.12 GHz remains in the operating band at all times. The cross-polar plot shows that the axial ratio is never poorer than 15 dB, which is important so that the interrogation signal in one polarization does not influence the backscattered signal in the orthogonal polarization.

When we design our transponder, we will design multiple SLMPAs to generate multiple frequency signatures and maximize the amount of data that is stored on the tag. It is therefore desirable that the phase shift in the nonresonant portions of the spectrum be 0, as exhibited in the figure. With a larger transponder real estate, the structural mode scattering component will increase significantly due to the increased number of patch antenna reflectors. It is therefore necessary to modify the radiating circuits that will operate at each frequency signature.

16.4 SLMPA ARRAY

If a single element is used for each operating frequency, then regardless of the layout we choose there will always be extra grating lobes created, along with a tilting or splitting of the main radiated beam due to the effect of the nonresonant antennas. Moreover, the structural mode scattering from a tag large enough to accommodate multiple SLPMAs will be significant compared with the antenna mode scattering of a single antenna. To increase the gain and introduce an element of symmetry into the design, we use an array of elements at each frequency signature. This section will present the design process for a two-by-two array of SMLPAs.

When we place multiple elements together in an array, there are a number of factors that will affect the overall radiation characteristics of the structure. These include the layout of the array (linear, planar, etc.), the distance between the elements, and the excitation amplitude and phase of each element. The transponder will be illuminated with a plane wave with uniform magnitude and phase, and so the excitation amplitude and phase are equal.

The array consists of four identical SLMPAs arranged as shown in Figure 16.7. The elements are arranged symmetrically along a square of side length S. The interelement spacing is described by the parameter S, and the element orientation is described by the translation parameter D, which exists from $-0.5{*}S$ to $0.5{*}S$. The translation described by the parameter D for each element is in the direction indicated in the figure. The transponder ground plane should be square so that the structural mode scattering in each orthogonal polarization is equal, and its edge length is labeled as L.

16.4.1 Array Theory

When we have an array of identical radiating elements we can describe the total radiated field as a product of the single element radiation pattern and the *array factor* (AF) as shown in Eq. (16.6).

$$\mathbf{E}_{\text{total}} = \mathbf{E}_{\text{single element}} * \text{AF} \tag{16.6}$$

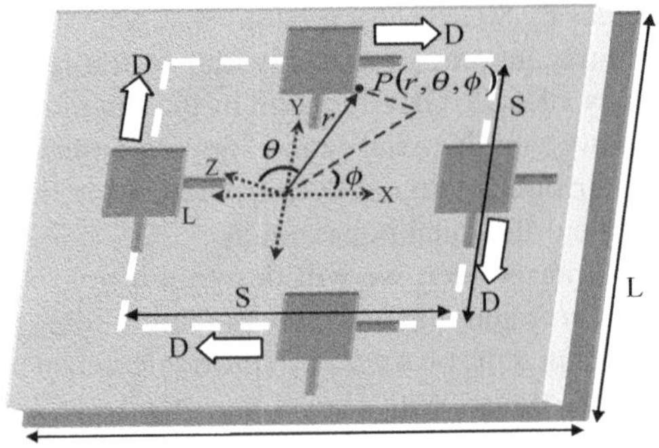

FIGURE 16.7. NSEW layout of two-by-two array of SLMPA with array parameters S, D, and L (mm).

For an $N \times M$ element planar array where the x separation is denoted by d_x, the y separation is denoted by d_y and each element is excited by amplitude I_{mn} and phase β, the array factor is given in reference 10.

$$AF = \sum_{n=1}^{N} I_{1n} \left[\sum_{m=1}^{M} I_{m1} e^{j(m-1)(kd_x \sin\theta \cos\phi + \beta_x)} \right] e^{j(n-1)\left(kd_y \sin\theta \sin\phi + \beta_y\right)} \quad (16.7)$$

For the 2×2 array layout described in Figure 16.7 the array factor expands to

$$\begin{aligned} AF &= e^{jk\frac{S}{2}\sin\theta\sin\phi} e^{jkD\sin\theta\cos\phi} + e^{-jk\frac{S}{2}\sin\theta\sin\phi} e^{-jkD\sin\theta\cos\phi} \\ &\quad + e^{jk\frac{S}{2}\sin\theta\cos\phi} e^{-jkD\sin\theta\sin\phi} + e^{-jk\frac{S}{2}\sin\theta\cos\phi} e^{jkD\sin\theta\sin\phi} \end{aligned} \quad (16.8)$$

There are two special cases that exist which we shall call the NSEW case and the Corners case. The NSEW case exists when $D = 0$, and the Corners case exists when $D = \pm 0.5 * S$. In the Corners case, Eq. (16.8) becomes

$$\begin{aligned} AF &= e^{jk\frac{S}{2}\sin\theta\sin\phi} e^{jk\frac{S}{2}\sin\theta\cos\phi} + e^{-jk\frac{S}{2}\sin\theta\sin\phi} e^{-jk\frac{S}{2}\sin\theta\cos\phi} \\ &\quad + e^{jk\frac{S}{2}\sin\theta\cos\phi} e^{-jk\frac{S}{2}\sin\theta\sin\phi} + e^{-jk\frac{S}{2}\sin\theta\cos\phi} e^{jk\frac{S}{2}\sin\theta\sin\phi} \end{aligned} \quad (16.9)$$

Equation (16.8) is the same array factor as for a standard 2×2 planar array with an interelement separation of S. The normalized form can be expressed as

$$AF_n = \frac{1}{4} \frac{\sin(kS\sin\theta\sin\phi)}{\sin\left(k\frac{S}{2}\sin\theta\sin\phi\right)} \frac{\sin(kS\sin\theta\cos\phi)}{\sin\left(k\frac{S}{2}\sin\theta\cos\phi\right)} \quad (16.10)$$

According to reference 10, we can calculate the angle, θ, where the main and grating lobes of the array factor will exist which is given by

$$\theta = \sin^{-1}\left(\frac{\pm m\lambda}{S\cos\phi}\right) = \sin^{-1}\left(\frac{\pm n\lambda}{S\sin\phi}\right), \qquad m, n = 0, 1, 2 \ldots \quad (16.11)$$

For the NSEW case, Eq. (16.8) reduces to

$$AF = e^{jk\frac{S}{2}\sin\theta\sin\phi} + e^{-jk\frac{S}{2}\sin\theta\sin\phi} + e^{jk\frac{S}{2}\sin\theta\cos\phi} + e^{-jk\frac{S}{2}\sin\theta\cos\phi} \quad (16.12)$$

This can then be normalized to

$$AF_n = \cos\left(k\frac{S}{4}\sin\theta(\sin\phi + \cos\phi)\right) * \cos\left(k\frac{S}{4}\sin\theta(\sin\phi - \cos\phi)\right) \quad (16.13)$$

Using the same procedure as before, we find that the maximum of the array factor will occur when

$$\begin{aligned} k\frac{S}{4}\sin\theta\,(\sin\phi + \cos\phi) &= m\pi, \qquad m = 0, \pm 1, \pm 2, \ldots \\ k\frac{S}{4}\sin\theta\,(\sin\phi - \cos\phi) &= n\pi, \qquad n = 0, \pm 1, \pm 2, \ldots \end{aligned} \quad (16.14)$$

which means that the main and grating lobes of the array will occur at

$$\theta = \sin^{-1}\left(\frac{\pm 2m\lambda}{S(\sin\phi + \cos\phi)}\right) = \sin^{-1}\left(\frac{\pm 2n\lambda}{S(\sin\phi - \cos\phi)}\right), \qquad m, n = 0, 1, 2, \ldots \quad (16.15)$$

Equations (16.11) and (16.15) clearly indicate that the larger the separation between the elements defined by the parameter S, the smaller the angle θ where the first grating lobe will occur. In other words, we expect that the closer the elements are to each other than the better the sidelobe performance will be. Moreover, we expect that the NSEW orientation should have its optimum size with larger values of S since there is a factor of 2 in Eq. (16.15) that does not exist in Eq. (16.11).

16.4.2 Antenna Mode

We will now simulate the performance of the 2×2 SLMPA array using CST MS. We will begin by investigating the radiation characteristics of the structure since they are directly related to the antenna mode scattering component. The transponder will be illuminated by a uniform plane wave, and so the excitation for each port is uniform with no phase offset. To make the result comparable with Figure 16.4, we will only deliver 25% of the power used previously to each antenna element.

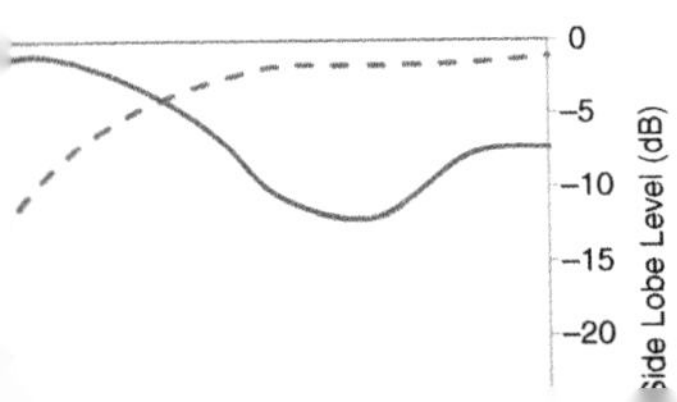

0
−5
−10
−15
−20
Side Lobe Level (dB)

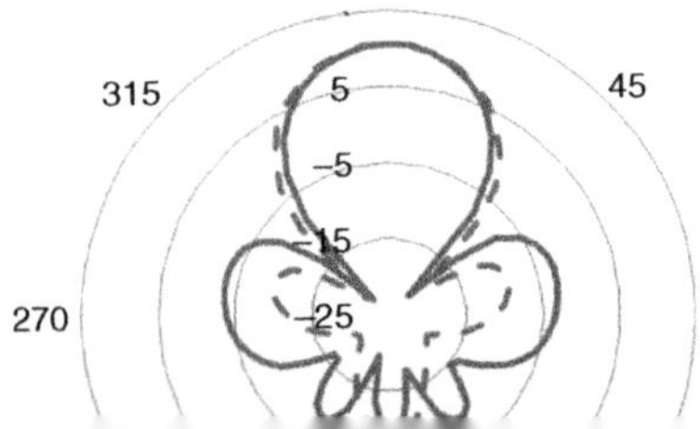

315
45
270
5
−5
−15
−25

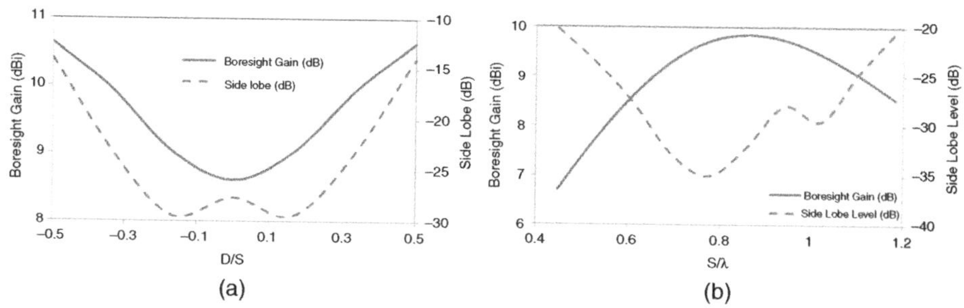

FIGURE 16.9. Boresight gain and side-lobe levels of SLMPA 2 × 2 array on 1.5-mm FR4 ($\varepsilon_r = 4.9$) with dimensions $L = 10$ mm, StubXL = StubYL = 1.5 mm, StubXW = StubYW = 0.2 mm, $S = 30$ mm, D vary **(a)** and $D = 0$, S vary **(b)**.

of the structure including the substrate and the ground plane. If we use a structure that is too large, then the antenna mode scattering component will be too small compared with the structural mode scattering component and no longer be observable.

Figure 16.10a shows the RCS of an optimally placed 2 × 2 array of SLMPAs ($L = 10$ mm) when the ground size is varied. It is clear that as the edge length L, is increased, the structural mode scattering increases significantly and the resonance is no longer observable. When the ground is too small, the radiation of the elements is disturbed at the edges. To determine the optimal size of the ground plane, we take the ratio of the total RCS and the structural mode component of the RCS at resonance (Figure 16.10b). This ratio is largest when L is 64 mm since this is the smallest size where the radiation of the antenna elements is undisturbed.

Now that we have optimized the size of the structure we vary the length of the loading stub and investigate the change in the scattering characteristics (Figure 16.11). Once again there is a slight change in the resonant frequency of 80 MHz within the resonant bandwidth of the optimally matched structure. The co-polar phase shows a shift of 189° for a change of 1 mm in the loading stub length.

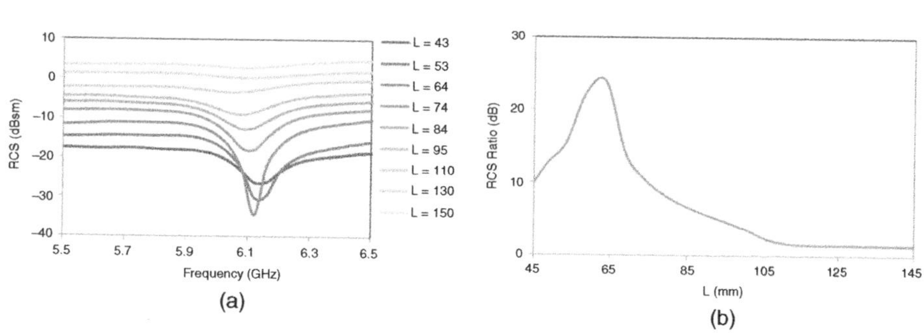

FIGURE 16.10. RCS **(a)** and RCS ratio **(b)** of SLMPA 2 × 2 array on 1.5-mm FR4 ($e_r = 4.9$) with dimensions $L = 10$ mm, StubXL = StubYL = 1.5 mm, StubXW = StubYW = 0.2 mm, $S = 30$ mm, $D = 0$ mm.

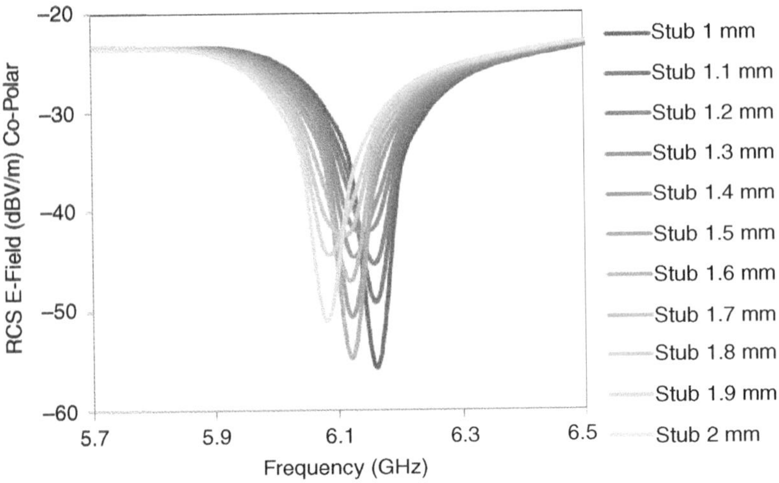

(a)

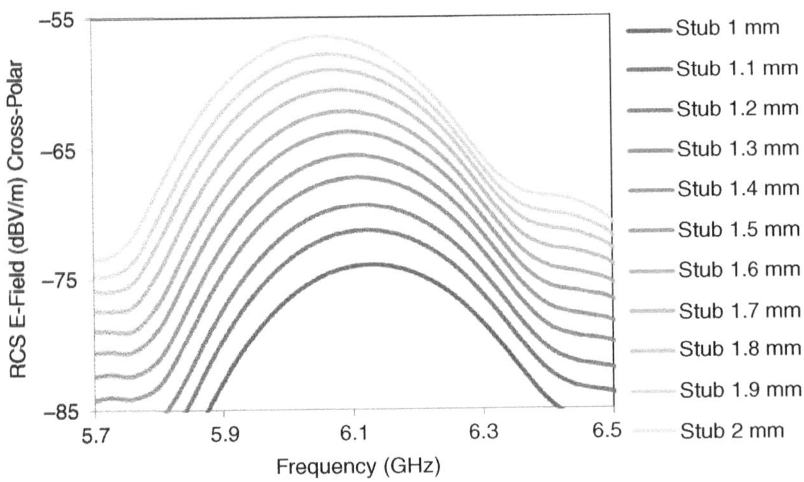

(b)

FIGURE 16.11. RCS *E*-field co-polar **(a)**, cross-polar **(b)**, and co-polar phase **(c)** of SLMPA 2 × 2 array on 1.5-mm FR4 ($e_r = 4.9$) with dimensions $L = 10$ mm, StubXL = StubYL = 1.5 mm, StubXW = StubYW = 0.2 mm, $S = 30$ mm, $D = 0$ mm, $L = 64$ mm and backscattered phase comparison with single element **(d)**.

It is clear that we have improved the range of the phase difference in the orthogonally polarized backscattered signals realized by changing the length of the loading stub by 1 mm from 82° to 189°. The significant effect of the structural mode scattering on the RCS is verified, and design techniques implemented to minimize its magnitude will improve the performance of the structure. We now have a theoretical basis for designing a transponder, and so the next section will present the novel design.

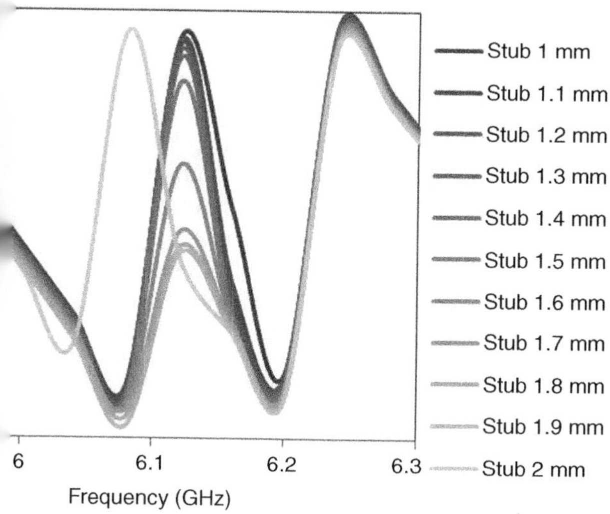
Stub 1 mm
Stub 1.1 mm
Stub 1.2 mm
Stub 1.3 mm
Stub 1.4 mm
Stub 1.5 mm
Stub 1.6 mm
Stub 1.7 mm
Stub 1.8 mm
Stub 1.9 mm
Stub 2 mm
6
6.1
6.2
6.3
Frequency (GHz)

(c)

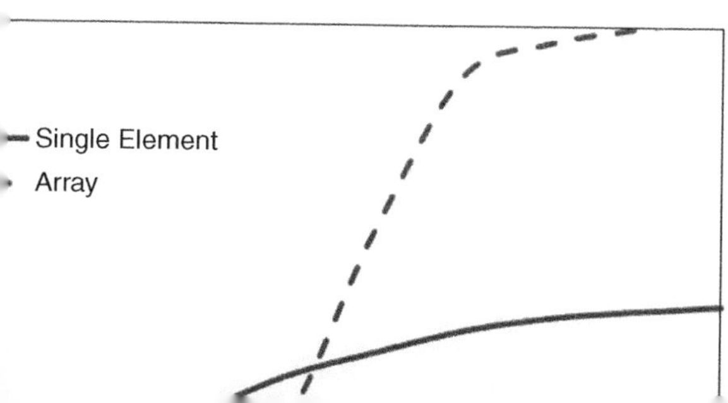
Single Element
Array

TABLE 16.1. Fabricated SLMPA Element Parameters

Patch Width (mm)	Resonant Frequency (GHz)	Stub Length (mm)	Stub Width (mm)
9	7.24	1.1	0.2
10	6.47	1.5	0.2
11	5.91	1.6	0.2
12	5.45	1.75	0.2
13	5.06	2	0.2
14	4.7	2.3	0.2

we prototype the design on an inexpensive low-performance PCB (FR4, $e_r = 4.9$) substrate material with a thickness of 1.5 mm and a loss tangent of 0.025.

16.5.1 Six-Signature Chipless RFID Transponder

The design of a six-signature chipless RFID transponder begins with designing an optimally matched SLMPA at six resonant frequencies that do not interfere. This is done for six SLMPAs, and their parameters are listed in Table 16.1. Now we can begin the task of arranging the two-by-two arrays of SMLPAs on the transponder. In Section 16.4 two cases for the array orientation were investigated, the NSEW case and the Corners case. It was found that the optimal spacing distance (S) for the Corners case is larger than for the NSEW case. Therefore we should place SLMPA arrays with shorter resonant wavelengths at a smaller S, and with D closer to 0. The concept will be illustrated with the help of the fabricated transponder shown in Figure 16.12. In the Figure 16.12 the six sets of 2×2 SLMPA arrays (9, 10, 11, 12, 13, 14 mm) are labeled with a number that indicates their patch width so that the layout of the arrays can be discussed. The highest-frequency (smallest-size) SLMPAs are placed on the inner layer that has space for two sets of 2×2 SLMPA arrays (9, 10 mm). In the next layer we can fit four sets of 2×2 SLMPA arrays with the highest-frequency array (11 mm) in the NSEW orientation, and we can fit the lowest-frequency array (14 mm) in the Corners orientation. The intermediate-frequency arrays (12, 13 mm) are placed in between with $|D|$ closer to 0 for the higher-frequency array. A bend is introduced on the loading stubs to allow the elements to be placed closer together while minimizing mutual coupling effects.

16.5.2 Simulation Results

The primary purpose for the transponder is to create a difference in the backscattered phase, so we use our simulation software (CST MS) to investigate the effect of altering the stub lengths for each 2×2 array. Since we are using a single polarization, we only need to vary the length of the loading stubs aligned with the polarization of the electric field vector of the excitation wave. To clearly illustrate the operation of the circuit, we will only vary the stubs for single patch width (9,10,11,12,13,14 mm) at a time. The results are presented in Figure 16.13, with shaded areas indicating the patch length associated with the resonance; a summary is presented in Table 16.2.

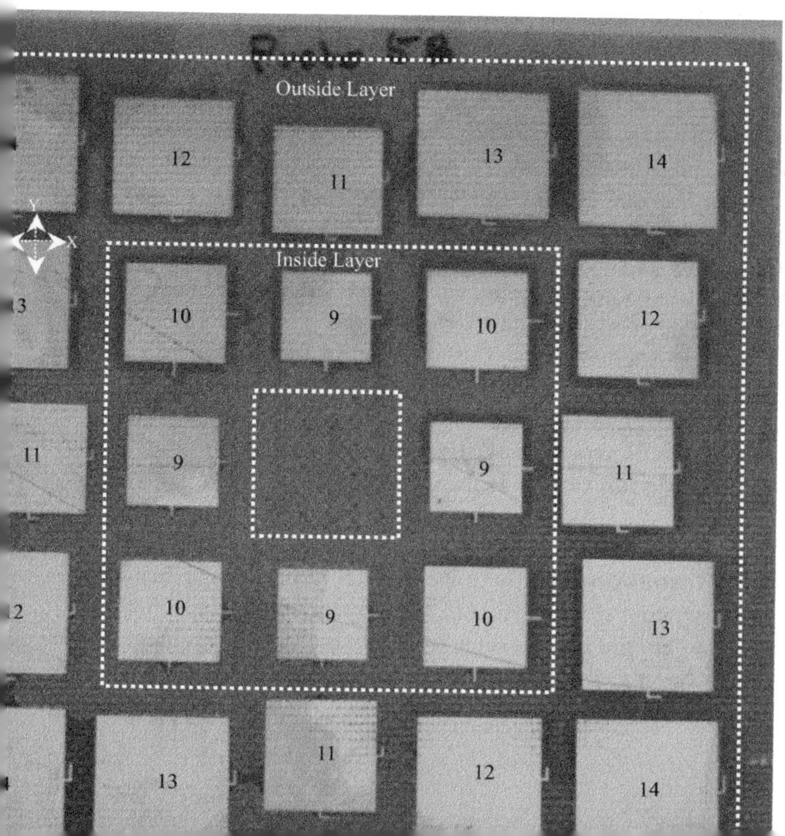
Outside Layer
Inside Layer
Y
X
12
11
13
14
10
9
10
12
11
9
9
11
10
9
10
13
13
11
12
14

The RCS gain (Figure 16.13a) is for a transponder with optimally matched loading stub lengths. It shows six resonant peaks consistent with the six sets of 2 × 2 SLMPA arrays. We condense the frequency band to observe the performance of each resonant peak as we vary the appropriate loading stub, since the spectrum in the rest of the band does not show noticeable change (Figure 16.13b–m). The co-polar E-field amplitude of the backscattered signal shows a slight shift in the resonant frequency, while the design frequency (which corresponds to the resonant frequency of the median loading stub length in each case) remains in the operating band of the resonance. The co-polar E-field phase shows a distinct phase shift at the design frequency while remaining unchanged in the nonresonant bands. These results are consistent with the results from Section 16.4.3. We will now fabricate and measure a single transponder and reader antennas to observe the resonant frequency signatures.

16.5.3 Experimental Results

The proposed phase-encoded chipless RFID system is shown in Figure 16.2. The experimental setup is designed based on the operating principles as described in Section 16.2.2. In this section the experimental results that verify the predicted operation of the transponder is shown.

16.5.3.1 Reader Antenna. To be able to extract the information from the transponder, we need a reader antenna that is operational over the entire band and also linearly polarized. The antenna that is chosen to be used for a reading of our transponder is a co-planar waveguide (CPW)-fed circular monopole antenna shown in Figure 16.14a. The antenna is chosen due to its wideband impedance bandwidth properties and almost omnidirectional radiation pattern across this band as well as for ease of design and fabrication. The antenna was designed and simulated using CST MW and was then fabricated on a Taconic TLX-0 ($e_r = 2.45$) substrate with a thickness of 0.787 mm. A photograph of the fabricated circuit is shown in Figure 16.14 along with the measure return loss and radiation pattern over the band. The return loss shows a 10-dB bandwidth from 4.49 to 8 GHz and beyond (Figure 16.14b). However, the return loss at 5.5 GHz is significantly better than at 4 and 7 GHz, so we expect to see varied performance throughout the band. At the higher frequencies we approach the second radiating mode and thus expect to see some splitting in the radiation pattern as shown in Figure 16.15. The radiation pattern shows good omnidirectional performance at both 4.5 and 6 GHz in the H-plane (Figure 16.15). The E-plane measurements show directionality and show a null at the top and bottom apex of the antenna. There is 13-dB suppression at 4 GHz and more than 40-dB suppression at 6 GHz, which is a useful feature of the antenna for our experimental setup.

16.5.3.2 Chipless RFID Transponder. The experimental setup used to read the prototype is a bistatic radar setup (Figure 16.16a). We have used bistatic setup rather than monostatic setup to simplify the testing by removing the requirement for additional active signal processing components. The experiment was carried out

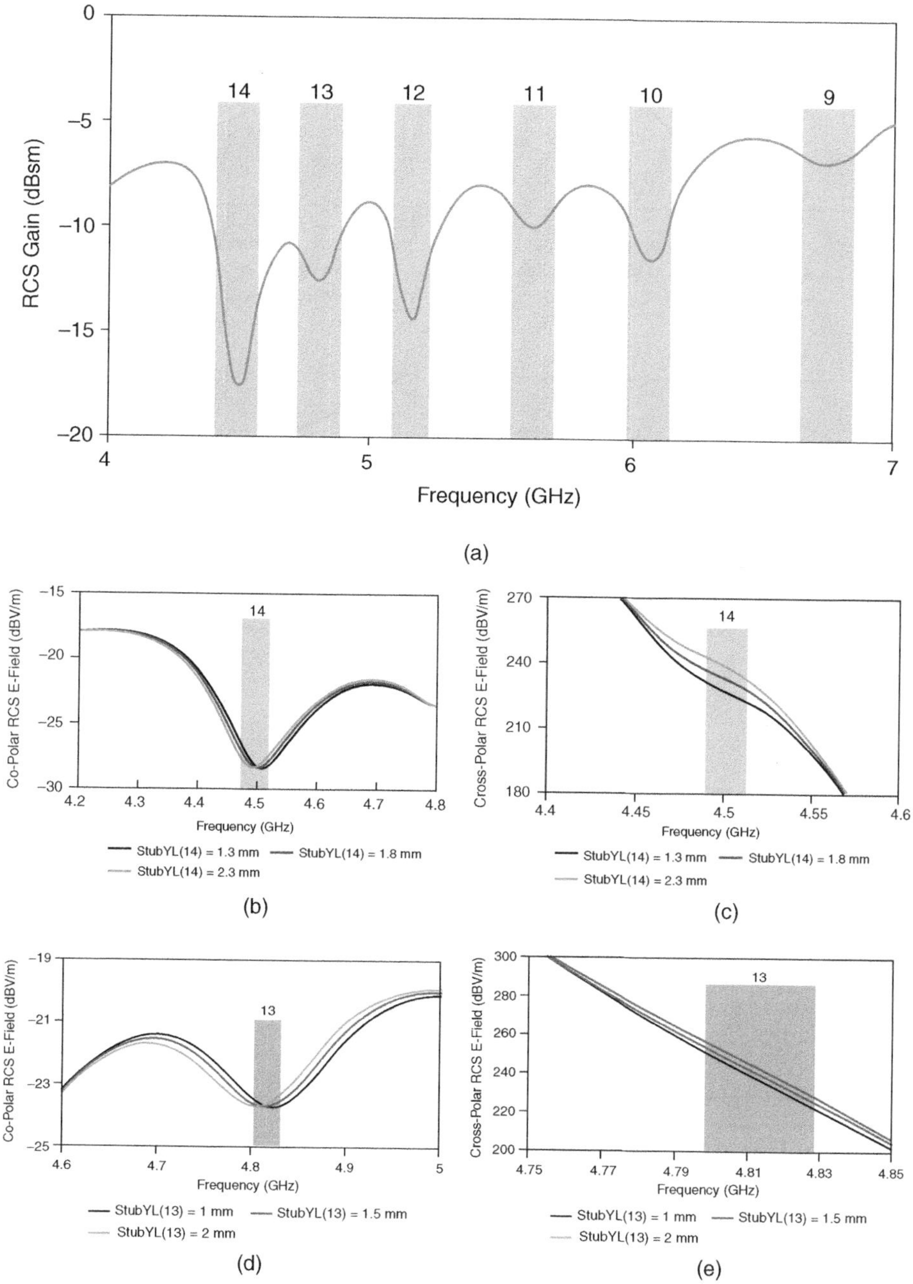

FIGURE 16.13. Simulated RCS of novel chipless RFID transponder (Figure 16.12) illuminated by linearly polarized plane wave **(a)** with varying loading stub length StubYL(14) **(b, c)**, StubYL(13) **(d, e)**, StubYL(12) **(f, g)** , StubYL(11) **(h, i)**, StubYL(10) **(j, k)**, StubYL(9) **(l, m)**.

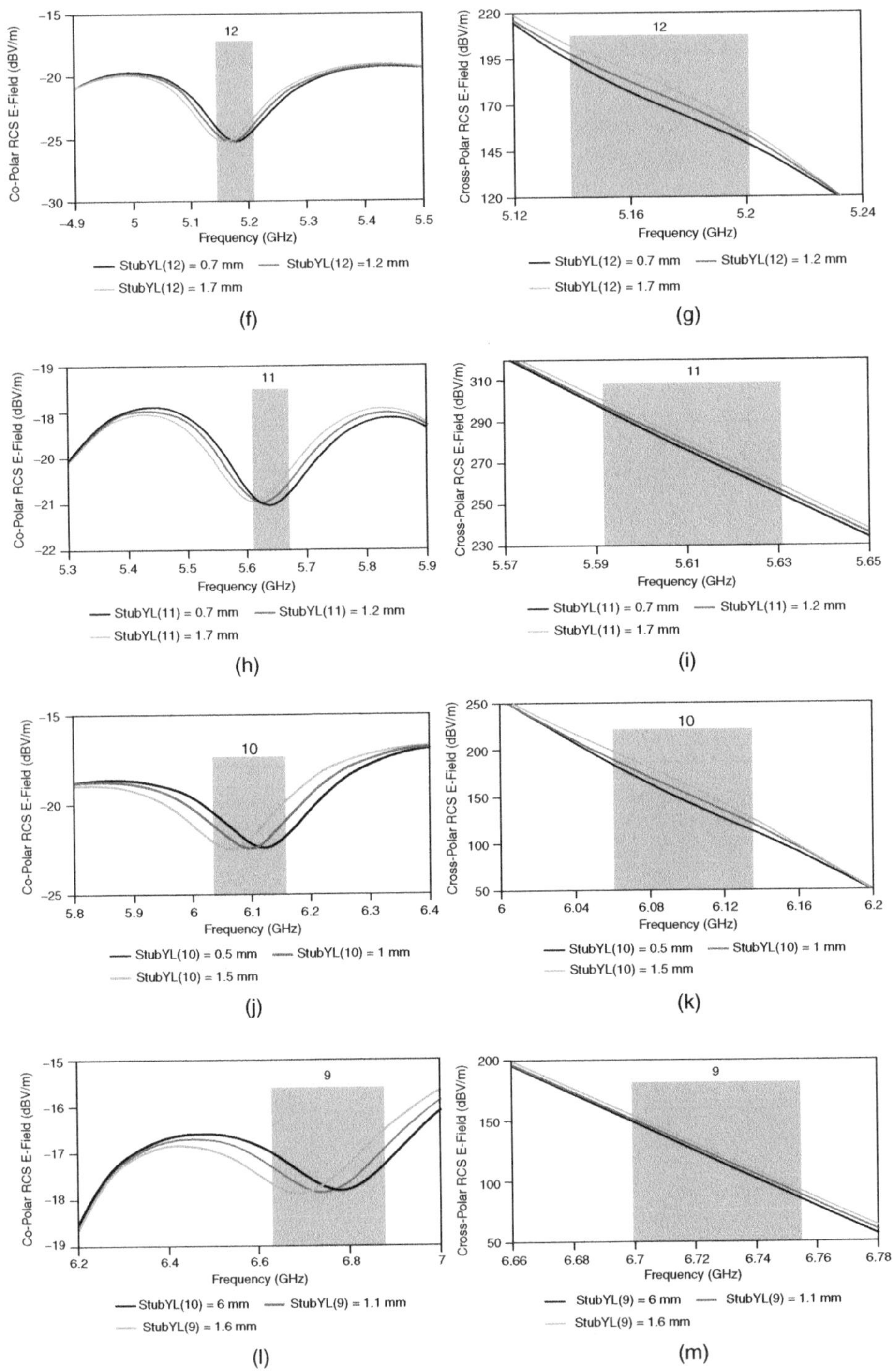

FIGURE 16.13. (continued)

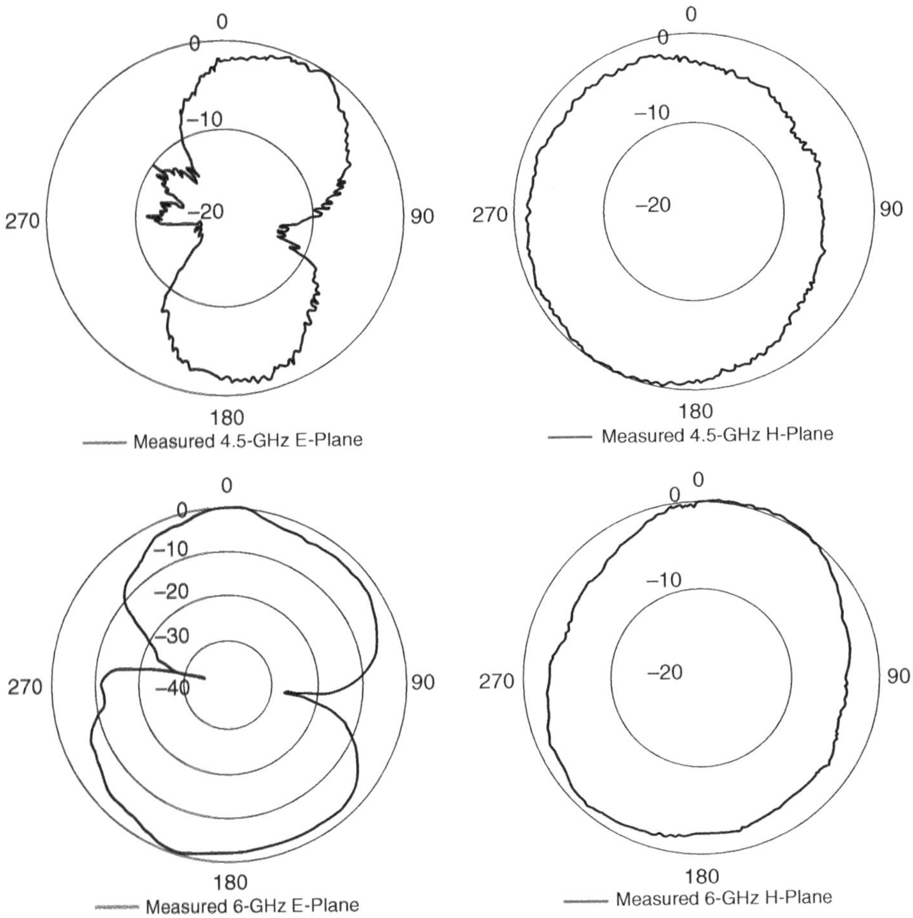

FIGURE 16.15. 2D *E*-plane and *H*-plane radiation pattern of CPW-fed UWB monopole antenna at 4.5 and 6 GHz.

The RCS magnitude for when the reader antennas are polarization-matched with the transponder's *Y* axis, along with the RCS phase difference between the two polarization states, is shown in Figures 16.16c–n. At each frequency signature we can observe a very slight change in the resonant frequency and an easily detectable change in the phase that shows similar characteristics to the simulated results of Figure 16.12. The results are summarized in Table 16.3. There is an average phase shift of 41° in each frequency signature for this prototype design, which is an excellent result. We have now shown that we are able to generate frequency signatures with a backscattered phase difference that can be controlled by adjusting the length of the loading stubs of an array of SLMPA antenna elements on a transponder that contains six such arrays.

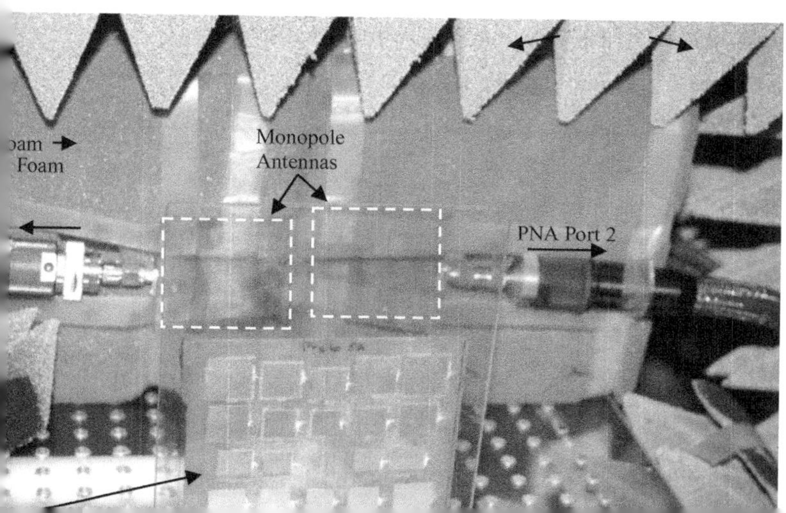

oam
Foam
Monopole
Antennas
PNA Port 2

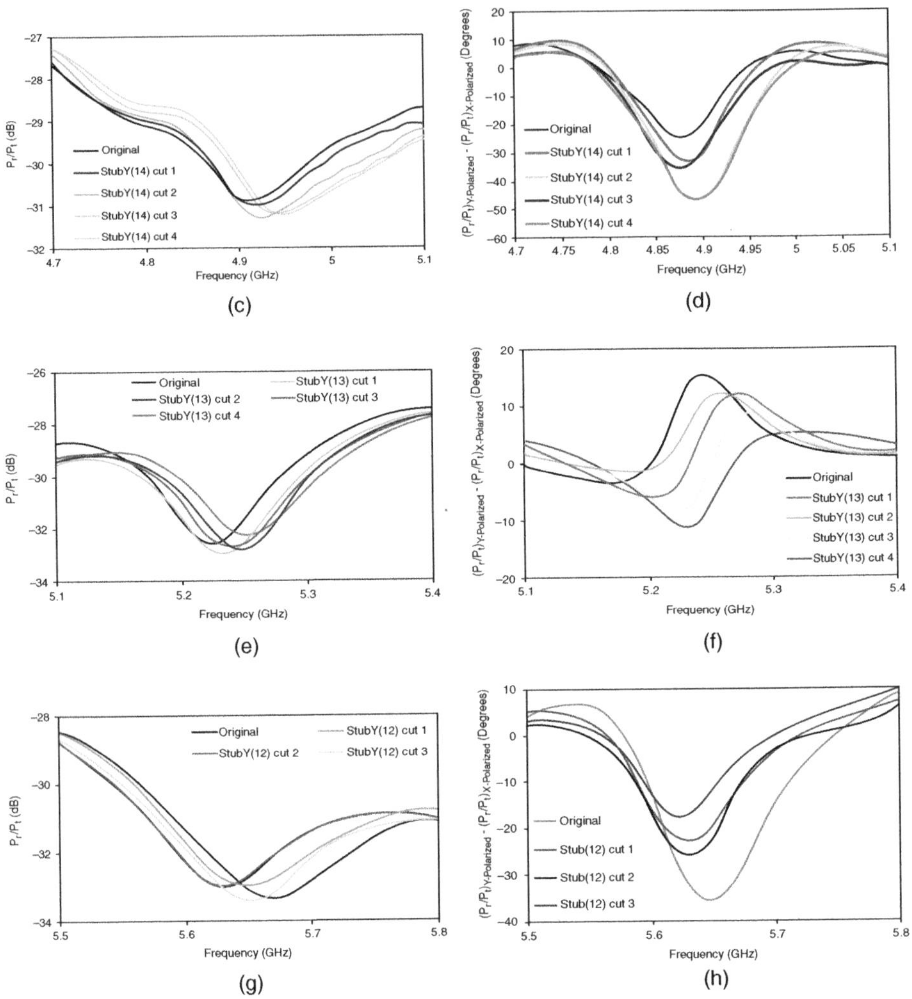

FIGURE 16.16. (continued)

TABLE 16.3. Six-Signature Chipless RFID Transponder Measurement Summary

Patch Width (mm)	Resonant Frequency (GHz)	StubYL Initial (mm)	StubYL Final (mm)	Phase Shift (Degrees)
9	6.987	0.6	0	15.75
10	6.591	1	0	162.73
11	6.118	1.2	0	14.21
12	5.621	1.2	0	11.29
13	5.244	1.5	0	24.65
14	4.878	1.8	0	19.66

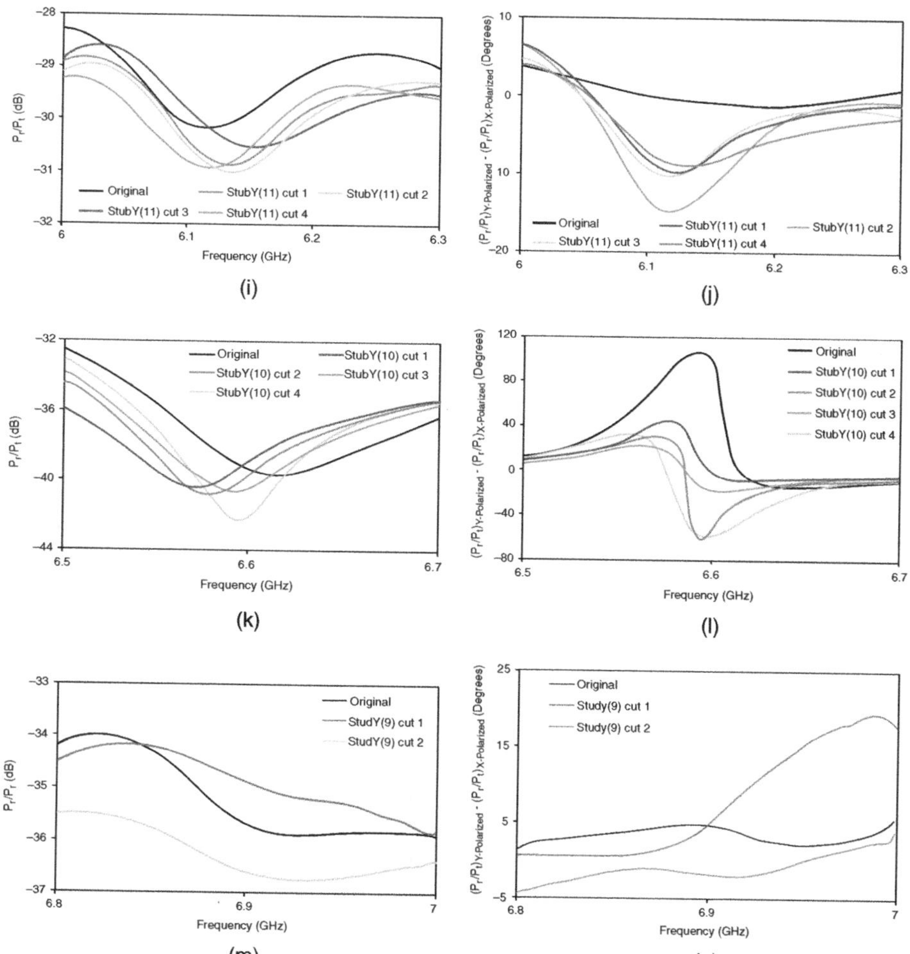

FIGURE 16.16. (continued)

The total size of the transponder is 9 cm × 9 cm, which is too large for many applications. However, the operating frequency of the design can be adjusted to higher frequencies (40 GHz, for example), which allows the transponder to easily fit onto a credit card size.

16.6 CONCLUSION

This chapter has presented a novel chipless RFID transponder that operates based on generation of a precise phase difference between the backscattered signal's phase in two orthogonal polarization states. This is the only chipless RFID system that uses phase-encoding to increase the data stored in each frequency signature rather than each signature simply being a binary on/off switch. The encoding is controlled by using a reactive stub to load a square, dual-resonant microstrip patch antenna. To enhance the RCS performance of the proposed chipless RFID transponder, a 2×2 array of antenna elements is designed and simulated. The array structure has a significant enhancement in the magnitude and phase of the backscattered signal. A final design with six 2×2 arrays of different dimensions has been produced which was measured to generate an average backscattered phase shift of 41°. If 32 phase combinations are possible, then the bit size has increased from 6 to 30.

In this chapter a theoretical formulation was developed to describe the array factor of a 2×2 SLMPA array based on a size parameter and an orientation parameter. It was shown that two extreme cases exist, the Corners case and the NSEW case. Simulated results are presented that show how the effect of the antenna mode scattering component on the total RCS has been optimized using a combination of optimized loading reactance's, spacing distances, and orientations. Guidelines have also been developed using established array theory to describe how the optimized spacing and orientation parameters are designed.

The chipless RFID transponder that is developed can be applied in a number of different applications such as robot identification procedures, sensors in automated industrial systems, or sensors for any large-scale, low-cost applications that require a high number of data bits such as transport ticketing and library cataloging.

REFERENCES

1. P. Harrop, The price–sensitivity curve for RFID, IDTechEx, Aug 23, 2006. Available at http://www.idtechex.com/printelecreview/en/articles/00000488.asp.
2. P. Brown, P. Hartmann, A. Schellhase, A. Powers, T. Brown, C. Hartmann, and D. Gaines, Asset tracking on the international space station using global SAW tag RFID technology, in *Proceedings, IEEE Ultrasonics Symposium*, pp. 72–75, October, 2007, IEEE Ultrasonics Symposium Proceedings, IUS.
3. I. Jalaly and I. D. Robertson, Capacitively tuned microstrip resonators for RFID barcodes, in *Proceedings, 35th EUMC, 2005*, Paris, France, 4–6 October 2005, pp. 1161–1164.
4. J. McVay, A. Hoorfar, and N. Engheta, Theory and experiments on Peano and Hilbert curve RFID tags, *Proc. SPIE*, Vol. 6248, No. 1, pp. 624808-1–10, 2006.
5. S. Preradovic, I. Balbin, N. C. Karmakar, and G. Swiegers, A novel chipless RFID system based on planar multiresonators for barcode replacement, in *Proceedings, 2008 IEEE International Conference on RFID*, Las Vegas, NV, 16–17 April, 2008, pp. 289–96.
6. S. Preradovic, I. Balbin, S. M. Roy, N. C. Karmakar, and G. Sweigers, Radio frequency transponder, Australian Provisional Patent Application (Ref: P30228AUP1), 14 April 2008.
7. J. D. Griffin and G. D. Durgin, Reduced fading for RFID tags with multiple antenna, in *IEEE AP International Symposium*, 2007, pp. 1201–1204.
8. I. Balbin and N. C. Karmakar, Phase-encoded chipless RFID transponder for large-scale low-cost application, *IEEE Microwave Wireless Compon. Lett.*, Vol. 19, No. 8, pp. 509–511, August 2009.

9. S. Preradovic, I. Balbin, N. C. Karmakar, and G. Swiegers, Multiresonator-based chipless RFID system for low-cost item tracking, *IEEE Trans. Microwave Theory Tech.*, Issue 5, part 2, pp. 1411–1419, May 2009.
10. C. A. Balanis, *Antenna Theory: Analysis and Design*, 2nd edition, John Wiley & Sons, Hoboken, NJ, 2005.
11. D. M. Pozar, *Microwave Engineering*, 3rd edition, John Wiley & Sons, Hoboken, NJ, 2005.

CHAPTER 17

LINK BUDGETS FOR BACKSCATTER RADIO SYSTEMS

JOSHUA D. GRIFFIN
Disney Research, Pittsburgh, Pennsylvania

GREGORY D. DURGIN
School of Electrical and Computer Engineering, Georgia Institute of Technology, Atlanta, Georgia

17.1 INTRODUCTION

Backscatter radio is a term that refers to systems that communicate wirelessly by scattering electromagnetic waves. Such systems have found application in many radio-frequency identification (RFID) and passive sensor applications. In a typical backscatter radio system, a small, low-power transponder, or *radio-frequency (RF) tag*, communicates with an interrogator, or reader, by modulating the waves scattered from its antenna. While this communication method allows the transponder to communicate while consuming very little power, system design is challenging because of the complex electromagnetic interactions between the reader, RF tag, and the radio channel. These interactions affect backscatter radio performance by limiting the amount of power received by the RF tag and/or the amount of modulated power backscattered to the reader. If either of these powers drop below a certain threshold, RF tag operation will cease and/or the reader will be unable to reliably detect the backscattered signal. Therefore, it is extremely important for the design and operation of modulated backscatter systems that these received powers be predicted. This chapter, which has been adapted from Griffin and Durgin [1] and Griffin [2], describes two radio link budgets that account for all of the major propagation mechanisms affecting backscatter radio performance. Each link budget parameter is explained in a tutorial fashion, and the link budgets are demonstrated through two practical examples: a conventional RFID portal operating at 915 MHz and a multi-antenna backscatter radio system operating at 5790 MHz.

Handbook of Smart Antennas for RFID Systems, Edited by Nemai Chandra Karmakar

17.1.1 Backscatter Radio Link Budgets

The first link budget, the *power-up link budget*, describes the one-way power flow from the transmitter portion of the reader to the RF tag. Specifically, the power P_t is the amount of RF power coupled into the radio-frequency integrated circuit (RFIC), discounting any loss factors internal to the chip, and is given by the following linear-scale link budget [1]

$$P_t = \frac{P_T G_T G_t \lambda^2 X \tau}{(4\pi r)^2 \Theta B F_p} \tag{17.1}$$

where

P_t is the power coupled into the RFIC (W)
P_T is the power transmitted by the reader (W)
G_T is the load-matched, free-space gain of the TX antenna
G_t is the load-matched, free-space gain of the tag antenna
λ is the carrier frequency wavelength (m)
X is the polarization mismatch
τ is the power transmission coefficient
r is the reader-to-tag separation distance (m)
Θ is the RF tag antenna on-object gain penalty
B is the excess blockage loss
F_p is the power-up link, small-scale multipath fading loss

Much like a conventional radio link budget, Eq. (17.1) depends on the separation distance r, antenna gains G_T and G_t, and transmit power P_T. Additionally, the unitless loss terms B, τ, Θ, F_p, and X that appear in Eq. (17.1) are described in Section 17.2. The power-up link budget does not apply to semi-passive tags—sometimes called battery-assisted passive (BAP) tags—that use a small battery or other onboard power source to provide DC power for tag operation. Instead, it applies only to *passive* RF tags that operate their circuitry using power rectified from the incident wave transmitted by the reader. In both passive and semi-passive tags, no power is used to amplify and transmit signals; instead, the tag sends information to the reader receiver by modulating the waves scattered from its antenna—that is, *modulated backscatter*. The amount of modulated backscatter power received by the reader receiver, P_R, is governed by the *backscatter link budget* [1]

$$P_R = \frac{P_T G_R G_T G_t^2 \lambda^4 X_f X_b M}{(4\pi)^4 r_f^2 r_b^2 \Theta^2 B_f B_b F} \tag{17.2}$$

where

r_f is the reader-to-tag link, or *forward link*, separation distance (m)
r_b is the tag-to-reader link, or *backscatter link*, separation distance (m)

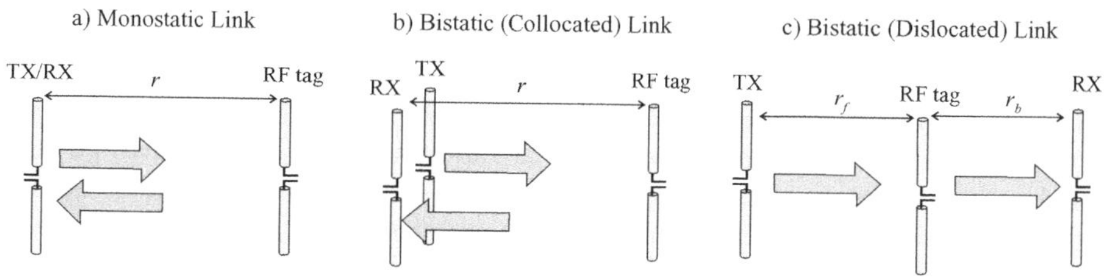

FIGURE 17.1. The three main antenna configurations in the backscatter link: **(a)** monostatic, **(b)** bistatic collocated, and **(c)** bistatic dislocated. (Reprinted from Griffin and Durgin [1], copyright 2008 IEEE.)

X_f is the forward-link polarization mismatch
X_b is the reverse-link polarization mismatch
B_f is the forward-link blockage loss
B_b is the backscatter-link blockage loss
M is the modulation factor
F is the backscatter-link, small-scale multipath fading loss

All other terms are as defined in Eq. (17.1). The backscatter link budget [Eq. (17.2)] will take different forms depending upon the antenna configuration of the RF tag reader, shown in Figure 17.1. In the *monostatic* configuration, the reader uses a single antenna to transmit and receive. In this case $r_f = r_b$, $X_f = X_b$, $B_f = B_b$, and F is equal to the monostatic fade margin, F_2. In the *bistatic* configuration, where separate antennas are used at the reader to transmit and receive, two forms of Eq. (17.2) are possible. If the reader antennas are closely spaced, then the *bistatic collocated* fade margin, F_α, is used; however, if the antennas are widely separated, the *bistatic dislocated* fade margin, F_β, is used. A complete discussion of the fade margins is provided in Section 17.2.4, and the link budget terms determined by the reader antenna configuration are summarized in Table 17.1.

17.2 CHANNEL IMPAIRMENTS

This section discusses each term of the power-up and backscatter link budgets in detail.

TABLE 17.1. Backscatter Link Parameters that Are Determined by the Reader Antenna Configuration: Monostatic, Bistatic Collocated, or Bistatic Dislocated

Monostatic	Bistatic Collocated	Bistatic Dislocated
$r_f = r_b$ $X_f = X_b$ $B_f = B_b$ $F = F_2$	$F = F_\alpha$ $B_f = B_b$	$F = F_\beta$

17.2.1 Polarization Mismatch

The power that an antenna receives is maximized when the polarization of the incident wave is matched to that of the antenna. When this is not the case, the power lost relative to the maximum power received with a polarization match is given by two power polarization mismatch factors X_f and X_b for the forward and backscatter links, respectively. In a backscatter system, polarization mismatch is of great importance because the orientation of the RF tag is often arbitrary; therefore, the power coupled into the RF tag antenna will change depending on the orientation of the antenna with respect to the polarization of the incident wave. One way to make sure that the RF tag receives power regardless of its orientation is to transmit a circularly polarized wave from the reader transmitter. Another method is to use two linearly polarized antennas on the RF tag that are oriented at 45° with respect to each other [1].

17.2.2 Power Transmission Coefficient

Impedance is the ratio of voltage to current that a given device can accept at its input terminals. When a voltage (or current) wave traveling in a device with impedance Z_A is fed into a device with impedance Z_L, assuming that $Z_A \neq Z_L$, the voltage (or current) wave must increase or decrease to conform to the voltage-to-current ratio specified by Z_L. The change in voltage (or current) wave results in a reflected wave that carries energy in the opposite direction of the original wave. Consequently, not all of the power carried by the original wave is transferred to the second device. This result is general and pertains to any situation in which transmission line theory can be applied, including an RF tag antenna attached to the RF port of a tag integrated circuit. Given an antenna impedance Z_A and a load impedance Z_L, the normalized amount of power delivered to the load, the *power transmission coefficient* τ [3,4], is

$$\tau = \frac{4\mathrm{Re}\{Z_A\}\mathrm{Re}\{Z_L\}}{\mathrm{Re}\{Z_A + Z_L\}^2 + \mathrm{Im}\{Z_A + Z_L\}^2}, \qquad 0 \leq \tau \leq 1 \tag{17.3}$$

where $\mathrm{Re}\{\cdot\}$ and $\mathrm{Im}\{\cdot\}$ denote the real and imaginary parts, respectively. When $Z_A = Z_L^*$, where $(\cdot)^*$ is the complex conjugate operator, maximum power transfer occurs and $\tau = 1$. RF tags are particularly susceptible to the impedance mismatch losses described by τ. RF tag impedance detuning caused by object attachment [4–6], close object proximity [7], RF tag antenna deformation [8, 9], and variations in received power that change the input impedance of the RFIC [5, 10] will all reduce τ. For example, Figure 17.2 shows that the simulated input impedance of a half-wave folded dipole drops from its design value of $300 + j100\,\Omega$ to a virtual short circuit as it is brought close to a conductive plane.

17.2.3 RF Tag Antenna Modulation Factor

An RF tag typically modulates power backscattered from its antenna by switching the load impedance between two states, shown in Figure 17.3. The amount of modulated

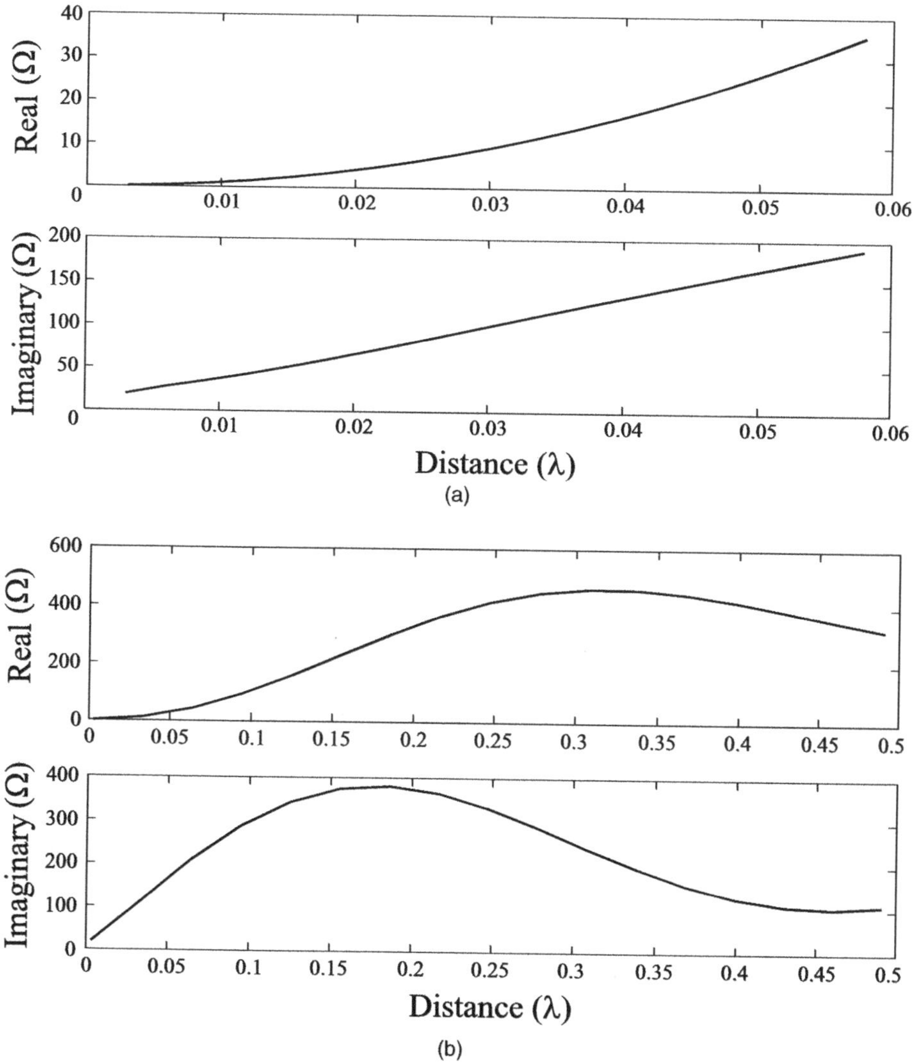

FIGURE 17.2. Folded dipole antenna impedance (simulated using NEC) as a function of the electrical distance from a perfectly conducting half-plane at 915 MHz [11] for electrical spacings of **(a)** 0.005λ to 0.06λ and **(b)** 0.005λ to 0.5λ. (Reprinted from Griffin and Durgin [1], copyright 2008 IEEE.)

backscatter power is described by the modulation factor M, which Prothro and Durgin [11] define as

$$M = \frac{1}{4} |\Gamma_A - \Gamma_B|^2 \tag{17.4}$$

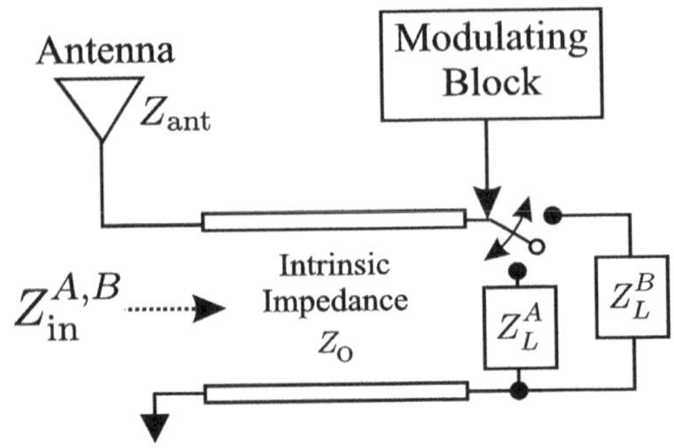

FIGURE 17.3. A simplified block diagram of an RF tag. The input impedance of the tag's RFIC $Z_{\text{in}}^{A,B}$ is changed between states A and B by switching the load between Z_L^A and Z_L^B. (Reprinted from Griffin and Durgin [1], copyright 2008 IEEE.)

where the reflection coefficient is defined as [12]

$$\Gamma_{A,B} = \frac{Z_{\text{in}}^{A,B} - Z_{\text{ant}}^{*}}{Z_{\text{in}}^{A,B} + Z_{\text{ant}}} \tag{17.5}$$

and $Z_{\text{in}}^{A,B}$ is the input impedance of the RF port of the RFIC, Z_{ant} is the input impedance of the antenna, and $(\cdot)^*$ is the complex conjugate operator. Equation (17.4) shows that the modulated backscatter power depends on the *difference* between the reflection coefficients of each state and not on the absolute reflected power from the RF tag. This equation assumes that the DC component of the received backscattered signal is removed, the backscatter signal has a 50% duty cycle, and an equal number of tag impedance states (data bits) are received. If these assumptions are not valid, then the definition of the modulation factor will differ from that in Eq. (17.4) [13].

The choice of the modulation factor presents a trade-off in design parameters [13]. If the reflection coefficients are switched between an open and short circuit ($M = 1$), then all of the incident power will be backscattered and the power-up circuitry of the tag's RFIC will be starved of available power. Instead, many RF tags use amplitude-shift-keying (ASK) modulation and switch the reflection coefficient between a matched load and a short, $M = 0.25$, to balance the power backscattered and absorbed by the RFIC [14]. Another method is to use phase shift keying (PSK) and simply modulate the reactive component of the chip impedance [15]. This allows constant power to be supplied to the RFIC regardless of the modulation state.

The power backscattered from an RF tag can also be characterized using the tag's effective *radar cross section* (RCS). Over the years, much research on the RCS of loaded antennas has been done with the goal of making "stealthy" antennas—antennas with zero RCS (nonscattering)—for military applications. For modulated backscatter applications, however, the goal is to maximize an antenna's RCS while still absorbing enough power to turn on the tag's RFIC. The theory of loaded antenna RCS for modulated backscatter applications has been presented by Nikitin and Rao [16], Penttilä et al. [17], and Fuschini et al. [18], among others. In general, the RCS σ_{RCS} can be written in terms of a load-dependent *antenna component* and a load-independent *structural component* [19],

$$\sigma_{\text{RCS}} = \frac{\lambda^2}{4\pi} G_t^2 |A_s - \Gamma_{A,B}|^2 \tag{17.6}$$

where A_s is a complex-valued term that represents the structural component and $\Gamma_{A,B}$, defined in Eq. (17.5), represents the antenna component. Since the backscattered signal is proportional to the difference between modulation states, the differential RCS is [12]

$$\Delta\sigma_{\text{RCS}} = \frac{\lambda^2 G_t^2}{4\pi} |\Gamma_A - \Gamma_B|^2. \tag{17.7}$$

Note that the differential RCS is dependent on only the antenna component, not on the structural component of the RCS.

17.2.4 Multipath Fading Loss

Waves propagating in a homogeneous medium exhibit well-defined wavefronts (i.e., contours of constant phase) that travel away from their source in time. In such a medium, the amplitude of the wave at any given observation point is constant, discounting path loss attenuation. For wave propagation in the real world, however, the homogeneous medium assumption does not hold because waves are commonly diffracted and reflected. The result is that, at any given position in space, the total field will be the linear combination of the original incident wave and several reflected and diffracted waves—*multipath waves*—that have different amplitudes, phases, and angles of arrival. As these waves combine constructively and destructively, the RF tag's received power will vary as a function of its position, a phenomena known as *multipath fading*.

A common method to analyze multipath fading is to model the variation of the total field using stochastics.* In this approach, the behavior of the total field is modeled as a random variable whose value is determined by a prescribed probability distribution. The type of distribution used is determined by the link budget in question and the propagation characteristics of the channel. Once an appropriate probability distribution has been chosen, a safety factor, or *fade margin*, is included in the link budgets to ensure that the backscatter radio system can operate with a certain

*It is worth noting that, although multipath fading can be modeled using stochastics, the fields *are* deterministic and governed by Maxwell's equations. It is the complexity and sheer number of field/obstacle interactions that occur in a typical channel that make stochastic process theory a useful analysis tool.

outage probability. For any channel, the fade margin in the logarithmic scale can be calculated as [1]

$$\text{Fade margin} = 10\log_{10}\left[\frac{\left(F_R^{-1}\,(\text{Outage Probability})\right)^2}{P_{\text{av}}}\right] \tag{17.8}$$

where F_R is the cumulative distribution function (CDF) of the received signal envelope (with units of $\sqrt{\text{watts}}$) and P_{av} is the average power in the channel. The following sections describe the appropriate distributions for several backscatter channel configurations discussed in this chapter.

17.2.4.1 The Power-Up Fade Margin, F_p. The two probability distributions most commonly chosen to model fading in the power-up link budget Eq. (17.1) are the Rayleigh and Rician distributions. A Rician distribution is used when a strong line-of-sight (LOS) path exists between the reader and the RF tag; otherwise, the Rayleigh distribution is used [20]. The level of fading is described by the Rician K factor—the ratio of the power in the LOS signal to the power in the nonspecular, scattered signal. Reported Rician K factors for backscatter channels are $-\infty$ dB and 2.8 dB [21], although higher values are certainly possible. These numbers represent the K factors of the individual reader-to-tag and tag-to-reader portions of the link, where it has been assumed that each half of the link has the same K factor.

17.2.4.2 The Monostatic Backscatter Link Fade Margin, F_2. Fading in the monostatic backscatter link, although caused by the same physical mechanisms, is described by radically different distributions than those of the power-up radio link. The reason for this difference is that the fading in the reader-to-tag link is multiplied by the fading in the tag-to-reader link. The fading of the backscattered signal can be modeled using a product-Rayleigh (for the non-line-of-sight (NLOS) case) or a product-Rician (for the LOS case) distribution. Since a single antenna is used to transmit and receive in a monostatic reader, the forward and backscatter channels are identical. Intuitively, this means that the forward and backscatter channels may fade simultaneously, resulting in the deepest fading of all the backscatter links [22].

17.2.4.3 The Bistatic Collocated Backscatter Link Fade Margin, F_α. The bistatic collocated link differs from the monostatic link in that two separate, but closely spaced, reader antennas are used to transmit and receive. In terms of fading, this means that the reader-to-tag and tag-to-reader links will be less correlated. Hence, when compared to the monostatic backscatter link, fading in the bistatic collocated link will always be less severe because the likelihood that both halves of the link will fade simultaneously is less. However, the bistatic collocated link will have more severe fading than the bistatic dislocated case; therefore, $F_2 \leq F_\alpha \leq F_\beta$.

Although fading in this channel configuration is less severe than the monostatic case, in practice, this configuration can suffer from severe self-interference—that is, the interference caused by the strong, unmodulated carrier that is received directly from the transmitter when the transmitter and receiver antennas are closely spaced.

TABLE 17.2. Small-Scale Multipath Fade Margins for One-Way (F_p), Monostatic (F_2), and Bistatic Dislocated (F_β) RF Tag Links for RF Tags that Use a Single Antenna[a]

Outage Probability	$K = -\infty$ dB			$K = 0$ dB			$K = 3$ dB			$K = 10$ dB		
	F_p	F_2	F_β	F_p	F_2	F_β	F_p	F_2	F_β	F_p	F_2	F_β
0.5	2	6	4	1	5	3	1	3	2	0	1	1
0.1	10	22	15	9	20	14	7	16	11	3	7	5
0.05	13	29	19	12	26	17	10	21	15	4	9	6
0.01	20	43	28	19	40	26	16	34	22	6	13	9
0.005	23	49	32	22	46	29	19	40	26	7	15	10
0.001	30	63	40	29	60	37	26	54	33	9	20	13

[a]Fade margins are reported in decibels.

Source: Reprinted from Griffin and Durgin [1], Copyright 2008 IEEE.

When practical, the better configuration is the bistatic dislocated configuration, discussed in the next section.

17.2.4.4 The Bistatic Dislocated Backscatter Link Fade Margin, F_β.

In the bistatic dislocated backscatter link, the reader transmitter and receiver antennas are separated by a very large electrical distance. This large separation causes the fading in the reader-to-tag and tag-to-reader links to be statistically uncorrelated. The exact separation distance required varies depending upon the spatial coherence of the channel [20]; however, previous work [23] achieved uncorrelated links with a spacing of 6.5λ in an NLOS channel, although it is possible that this could be achieved with a smaller separation distance. Therefore, the fading in this link will be the least severe of all the backscatter links, but still worse than the one-way link in terms of deep-fade frequency.

17.2.5 Calculated Fade Margins

Table 17.2 shows fade margins for several different channels and levels of fading. The left column gives the outage probability defined as $\Pr[P_R \leq (P_{av}/FM)]$ — that is, the probability that the received power P_R has dropped below the average power P_{av} by an amount equal to the fade margin FM. Therefore, for the power-up link with $K = 0$ dB, the required fade margin F_p to guarantee that signal fades render the system inoperable with only a 10% probability is 9 dB. In other words, an additional 9 dB of power must be transmitted to limit system failures to 10%.

17.2.6 On-Object Antenna Gain Penalties

Though an RF tag antenna may perform well when separated by several wavelengths from conductive and dielectric materials, operation may cease completely when it is brought close or attached to an object. Aside from altering the input impedance

TABLE 17.3. On-Object Gain Penalties for Various Materials Measured at 915 MHz in the Decibel Scale

Cardboard Sheet	Acrylic Slab	Pine Plywood	De-ionized Water	Ethylene Glycol	Ground Beef	Aluminum Slab
0.9	1.1	4.7	5.8	7.6	10.2	10.4

Source: Reprinted from Griffin and Durgin [1], Copyright 2008 IEEE.

(discussed in Section 17.2.2), object attachment reduces the antenna's radiation efficiency and distorts the antenna pattern, limiting both the backscattered power for communication and, for a passive RF tag, the power available for RFIC operation. In Eqs. (17.1) and (17.2), the on-object gain penalty Θ accounts for these losses and is defined in the linear scale as the ratio of the load-matched, free-space gain of the RF tag antenna to the gain of the RF tag antenna when attached to an object (both gains are in the linear scale) [24]

$$\Theta = \frac{G_t}{G_{\text{on-object}}}. \tag{17.9}$$

In Eq. (17.9), the power used to calculate G_t and $G_{\text{on-object}}$ should be averaged over the half-space facing away from the object so that Θ is a useful "rule of thumb" for design engineers regardless of the angle-of-arrival of waves at the RF tag. Unfortunately, calculating $G_{\text{on-object}}$ is difficult analytically and is complicated by the Θ's dependence on object material properties, object geometry, frequency, and antenna type. Therefore, the best method to determine $G_{\text{on-object}}$ is through careful simulation or measurement.

Measurements for several representative cases, shown in Table 17.3 , have shown that Θ can range from 0.9 dB on cardboard to 10.4 dB on an aluminum slab [24]. Initially, these gain penalties may not seem significant; however, recall that backscatter links decrease as $1/\Theta^2$ (in the linear scale). Therefore, an aluminum slab may result in over a 20-dB loss in backscattered power. Furthermore, the power-up link [Eq. (17.1)] budget decreases as $1/\Theta$ (in the linear scale); though these losses are not as severe as those seen in the backscatter link, RF tags are generally more sensitive to changes in the power-up link.

17.2.7 Path Blockages

When an object intrudes on or completely blocks the LOS path between the reader and the RF tag, the power received by the RF tag or backscattered to the reader will decrease. In the wireless literature, the large-scale power fades caused by blockages are often modeled with a log-normal distribution [25]

$$f_B(b) = \frac{1}{\sigma_B\sqrt{2\pi}} \exp\left(-\frac{(b-\mu_B)^2}{2\sigma_B^2}\right) \tag{17.10}$$

TABLE 17.4. Summary of the RF Tag System Parameters and Propagation Environment Described in Section 17.3

Operating frequency:	915 MHz
Reader antenna configuration:	Monostatic and bistatic dislocated operation with 5% desired outage probability
Reader antennas:	Right-hand circularly-polarized patch antennas with 7-dBi gain
Reader receiver sensitivity:	−80 dBm
RF tag antenna:	Linearly polarized folded dipole with free-space terminal impedance of $300 + j100\ \Omega$ and 2-dBi antenna gain
RF tag modulation:	ASK modulation switching between a matched and short-circuit load
RF tag sensitivity:	−13 dBm
Channel:	A LOS is present, giving $K = 3$ dB
RF tag object attachment:	Cardboard and aluminum objects

where b is the index of the distribution, μ_B is the average value of B, and σ_B is the standard deviation of B, all in the decibel scale. No study of path blockages has been reported for backscatter radio because many commercial, passive backscatter systems depend on a LOS channel for proper operation. Therefore, it is usually assumed that $B = 1$ (in the linear scale). However, as more backscatter systems operate in NLOS channels — which is currently possible for semi-passive tags and will become more common for passive tags as their power consumption is reduced — path blockage statistics will become more significant in link budget calculations.

17.3 RFID PORTAL EXAMPLE

In this section, the link budgets described in Section 17.1 are demonstrated through an example of an RFID portal operating at 915 MHz. This section defines the RF tag system parameters, describes the propagation environment, and then plots the link budgets as a function of the reader-to-tag separation distance for two different object attachments. A summary of the system parameters and propagation environment is given in Table 17.4.

17.3.1 RF Tag Reader

In this system, tagged objects pass through a passageway, or portal, to which the reader's antennas are fixed. The reader has the ability to operate with a single transmitter and receiver antenna (monostatic mode) or with two separate, widely spaced transmitter and receiver antennas (the dislocated bistatic mode). In either mode, the transmitter and receiver antennas are right-hand circularly-polarized patch antennas that resonate at 915 MHz with a gain of 7 dBi. The sensitivity of the reader is −80 dBm [5, 26].

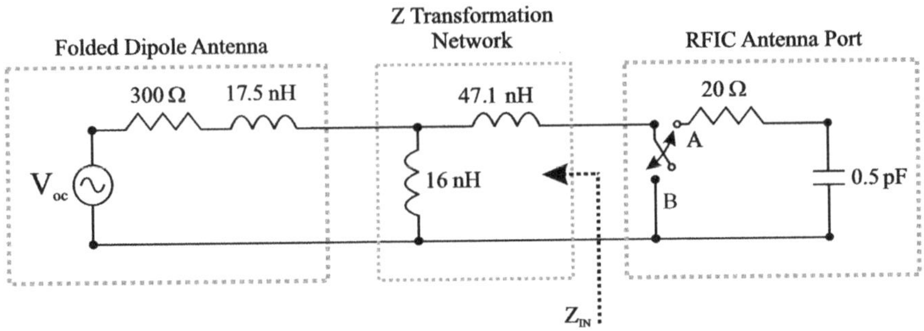

FIGURE 17.4. The equivalent circuit of the folded dipole RF tag antenna, impedance transformation network, and RFIC antenna port used in the 915-MHz portal example. (Reprinted from Griffin and Durgin [1], copyright 2008 IEEE.)

17.3.2 RF Tag

Figure 17.4 shows the equivalent circuit of the RF tag antenna, impedance transformation network, and antenna port of the tag RFIC. The RF tag uses a single folded dipole that is linearly polarized with a free-space, load-matched gain of 2.1 dBi and a free-space terminal impedance of approximately $300 + j100\ \Omega$ [11]. The input impedance at the antenna port of most tag RFICs can be modeled as a series resistor of a few ohms (often less than 30 Ω) in series with a capacitor that is a fraction of a picofarad [13]. For this example, we will assume that the tag RFIC uses ASK modulation and switches its impedance between two states, $Z_A = 20 - j350\ \Omega$ and $Z_B = 2 - j0.1\ \Omega$ (an approximate short circuit). The nonzero input impedance of Z_B reflects the fact that field-effect transistors (FETs) cannot provide a true short circuit, but are modeled as a small shunt resistance in parallel with a small capacitance whose values depend upon the transistor implementation and biasing. For efficient transfer of power from the antenna to the RFIC, a matching network, shown in Figure 17.4, is used to transform the antenna impedance from $300 + j100\ \Omega$ to $Z_{IN} = 20 + j350\ \Omega$. The network creates a conjugate impedance match with the RFIC in impedance state A. In most RF tags, lumped element impedance transformation networks are not used; instead, the necessary inductance or capacitance is incorporated directly in the structure of the planar antenna. In this example, the matching network will be treated separately from the antenna structure for illustrative purposes. As mentioned previously, the power required for a tag to power-up varies widely by design; in this example, we will assume that −13 dBm is required at 915 MHz [5, 27].

17.3.3 The Propagation Environment

The RF tag system operates in a cluttered environment that experiences small-scale fading caused by multipath propagation. As the RF tag passes in front of the reader's antennas, a LOS path exists, resulting in a Rician K factor of 3 dB. Furthermore, it is assumed that no blockages impair the LOS path and a 5% outage probability is

TABLE 17.5. 915-MHz RF Tag Portal Example Parameters in the Linear Scale

Material	P_T (mW)	$G_{T,R}$	G_t	$X_{f,b}$	F_p	F_2	F_β	λ(m)	M	τ	Θ	B
Cardboard	800	5	1.6	0.5	10	126	32	0.33	0.25	1	1.2	1
Aluminum	800	5	1.6	0.5	10	126	32	0.33	3.5×10^{-5}	6.2×10^{-3}	11	1

Source: Reprinted from Griffin and Durgin [1], Copyright 2008 IEEE.

desired. For illustration of the extreme object attachment scenarios, we assume that the tagged objects passing through the portal are made of cardboard and aluminum.

17.3.4 Link Budget Calculations

The information above combined with Tables 17.2 and 17.3 provide all of the parameters necessary to calculate the terms in the link budgets. A brief description of each is provided below and is summarized in Table 17.5.

Transmit Power, P_T: In the United States, the Federal Communications Commission (FCC) limits the power that an RF tag system can transmit in the 902 to 928-MHz frequency band to 4 W of equivalent isotropic radiated power (EIRP). EIRP power is simply the product of the transmitted power and transmitter antenna gain ($P_{\mathrm{EIRP}} = G_T P_T$) in the linear scale. Since the reader's antennas have a gain of 7 dBi (5 in the linear scale), the maximum continuous-wave (CW) power transmitted is limited to 29 dBm (or 800 mW).

Modulation Factor, M, and Power Transmission Coefficient, τ: The RF tag modulates backscatter by switching its input impedance between two states. When the RF tag is attached to a cardboard object, its impedance changes little from its free-space value: $Z_{\mathrm{IN}} = 20 + j350\ \Omega$ as seen at the output of the impedance transformation network. Therefore, using Z_{IN}, Z_A, and Z_B in Eq. (17.4), the modulation factor on cardboard is found to be $M = 0.25$. Using Eq. (17.3), the power transmission coefficient on cardboard is $\tau = 1$.

As the RF tag is brought close to an aluminum object, the impedance of the folded dipole antenna drops rapidly, as shown in Figure 17.2. For a very small separation distance of 0.005λ, which is a good approximation for an RF tag attached to a metallic object, the terminal impedance of the antenna drops to approximately $0.5 + j25\ \Omega$. The corresponding impedance seen at the output terminals of the impedance transformation network is $Z_{\mathrm{IN}} = 0.31 + j290\ \Omega$. This assumes that the values of the impedance transformation network are not altered by the close metal proximity—an assumption that may not be valid for a matching network incorporated into the antenna structure. Using Z_{IN}, Z_A, and Z_B in Eqs. (17.4) and (17.3) shows the modulation factor and power transmission coefficient to be $M = 3.5 \times 10^{-5}$ and $\tau = 6.2 \times 10^{-3}$, respectively (both in the linear scale).

Gain Penalty, Θ: From Table 17.3, the gain penalty for cardboard and metal are $\Theta = 0.9$ dB (1.2 in the linear scale) and $\Theta = 10.4$ dB (11 in the linear scale), respectively.

Fade Margins: To maintain a 5% outage probability in a multipath environment with $K = 3$ dB, the required fade margins, from Table 17.2, are $F_p = 10$ dB (10 in the linear scale) for the power-up link, $F_2 = 21$ dB (126 in the linear scale) for the reader in monostatic mode, and $F_\beta = 15$ dB (32 in the linear scale) for the reader in bistatic dislocated mode.

Polarization Mismatch, X: Since the reader's antennas are circularly polarized and the RF tag antenna is linearly polarized, a 3-dB polarization mismatch will result on both the reader-to-tag and tag-to-reader links, $X_f = X_b = 0.5$ in the linear scale.

17.3.5 Discussion

Figure 17.5a shows plots of the power-up link, described by Eq. (17.1), for three different cases. In the first power-up link budget, X, τ, Θ, B, and F_p are all equal to 1 (in the linear scale). This is the Friis free-space link budget [3]—a link budget that assumes free-space power attenuation, no impedance or polarization mismatches, no blockages, and no multipath fading. The second case is the power-up link for an RF tag attached to cardboard, and the third is for an RF tag attached to an aluminum object (link budget parameters are given in Table 17.5). Note the extreme optimism of the Friis free-space link budget. Compared to realistic scenarios, an RF tag system designed using this link budget may overestimate its range by several meters. Even though cardboard itself does little to affect the tag's operation, fading and polarization mismatches reduce the range of the RF tag to about 2 m. The third case reflects the severe effects of metal attachment which prevent the RF tag from turning on.

Similarly, Figure 17.5b shows the plots of the backscatter links described by Eq. (17.2) for an RF tag attached to cardboard and aluminum. A brief comparison of Figure 17.5a and Figure 17.5b shows that the range of the RF tag is limited by the power-up link [Eq. (17.1)], not the backscattered power. The effect of the reader antenna configuration is also seen; the monostatic link is several decibels worse than the bistatic, dislocated case regardless of material attachment.

17.4 HIGH-FREQUENCY MULTIPLE ANTENNA BACKSCATTER RADIO SYSTEMS

As was seen in Section 17.3, even a small change in the link budget parameters can have a significant effect on the range of an RF tag. Therefore, any method to increase antenna gain, minimize material effects, and/or reduce fading is welcome news to the backscatter tag designer. One way to realize improvements is to use multiple antennas at both the reader and RF tag in order to reduce multipath fading. Using multiple antennas is easiest to do at frequencies higher than those used in conventional

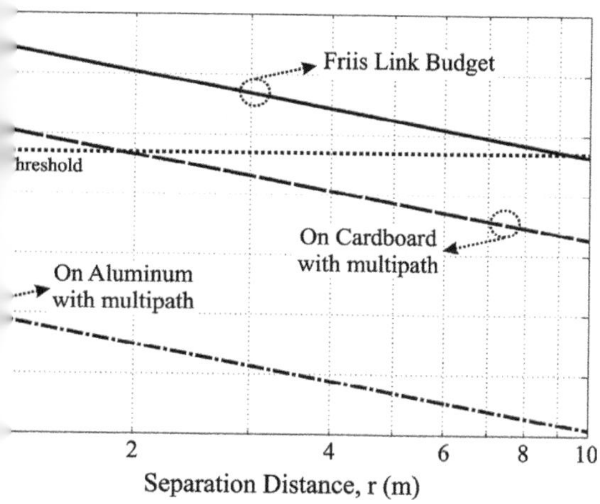

Friis Link Budget
hreshold
On Cardboard
with multipath
On Aluminum
with multipath
2
4
6
8
10
Separation Distance, r (m)

(a)

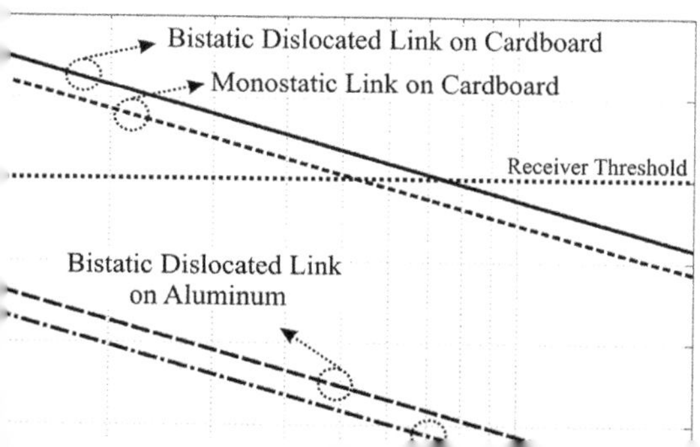

Bistatic Dislocated Link on Cardboard
Monostatic Link on Cardboard
Receiver Threshold
Bistatic Dislocated Link
on Aluminum

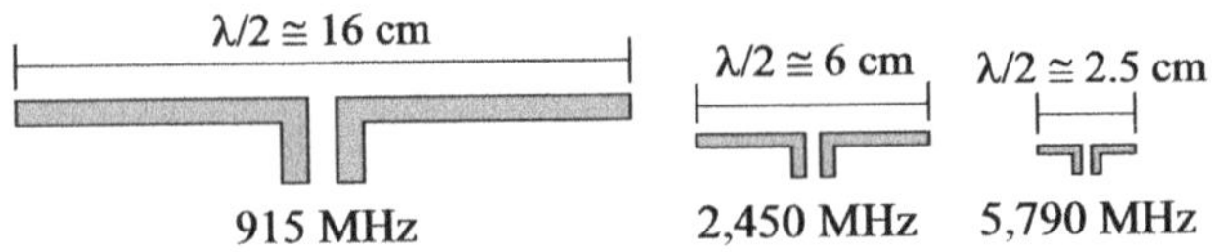

FIGURE 17.6. Half-wave dipoles for 915 MHz, 2450 MHz, and 5790 MHz drawn proportionally (not actual size) to show the relative decrease in size with increasing frequency. (Reprinted from Griffin and Durgin [1], copyright 2008 IEEE.)

Multiple antennas can be used at either the reader transmitter, the reader receiver, or the RF tag. When multiple antennas are used at the reader, the phase of the signal transmitted or received from each antenna is adjusted to cancel the phase shift caused by the channel. When this is done, the signals combine coherently—that is, with the same phases—and result in a *diversity gain* that is simply defined as the signal-to-noise-plus-interference ratio (SINR) gain that results from the decreased multipath fading. If multiple antennas are used at the reader transmitter, the amplitudes and phases of the signals are adjusted before they are transmitted [28]. If multiple antennas are used at the reader receiver, the received signal amplitudes and phases from each antenna are adjusted before they are combined into a single signal. This type of combining scheme is known as *gain combining*. Although more than one gain combining algorithm exists, the optimal algorithm is maximal ratio combining (MRC) because it maximizes the SINR of the combined signal [20]. Another technique to derive a diversity gain from multiple antennas is *switch combining* in which the receiver simply selects the antenna with best SINR. Unlike gain combining algorithms, which require knowledge of the channel estimated from a transmitted training sequence, switch combining algorithms only require an estimate of the SINR at the antenna terminals. Although switch combining does not perform as well as gain combining algorithms, it is much less expensive to implement.

If multiple antennas are used to modulate the signal backscattered from the RF tag, no channel knowledge or diversity combining techniques are required. Instead, the signals modulated by each RF tag antenna combine at the reader receiver's antennas in such a way that multipath fades are reduced. Intuitively, the probability that the signal scattered by each RF tag antenna fades at the same time is small; therefore, when more than one RF tag antenna is used to modulate the backscattered signal, fading on the overall signal is reduced. This reduction in fading is known as a *pinhole diversity gain* and is greatest in NLOS backscatter channels [23, 29]. Using multiple RF tag antennas has other benefits as well. Mi et al. [30] have shown that multiple, closely spaced antennas can be used to increase power received by the RF tag. Multiple RF tag antennas may also allow diversity schemes to be employed in the RF tag to mitigate fading in the power-up link budget [31].

In addition to gains realized using multiple antennas, the use of high frequency may itself provide several benefits. First, fade mitigation and direction finding will be aided by the very compact antennas arrays that are available for readers in the the 5725–5850 MHz band. At these frequencies, it may be possible to place antenna arrays on mobile and handheld readers.

Second, greater reader and tag antenna gains are possible in the 5725–5850 MHz band. For a fixed antenna aperture, Eq. (17.11) dictates that the gain of the RF tag antenna will increase with frequency [3]:

$$G = \frac{4\pi}{\lambda^2} A_{\text{eff}} \tag{17.11}$$

where G is the antenna gain in the linear scale, λ is the wavelength (m), and A_{eff} is the effective area of the antenna (m^2).

Third, operating in the 5725–5850 MHz band may improve an RF tag's on-object performance compared to lower frequencies by increasing the electrical separation between the tag and the object. For example, Figure 17.2 shows that the real part of the antenna impedance at an electrical separation distance of 0.01λ is $\approx 2\ \Omega$ at 915 MHz. If the physical separation distance is unchanged and the frequency is increased to 5790 MHz, the electrical distance now becomes 0.06λ with a corresponding real impedance of $\approx 40\ \Omega$. The result is a larger power transmission coefficient τ and modulation factor M. The increased object immunity at high frequencies is based upon the fact that the physical size of an antenna's near-field becomes smaller as the frequency is increased. As the near-field region becomes smaller, nearby objects will intrude less into the near-field and, as a result, have a smaller effect on its fields. The NEC simulations, shown in Figure 17.2, indicate that this is the case for the impedance of a folded-dipole antenna as its distance from a perfectly conducting half-plane is increased; however, the antenna's behavior on dielectric objects has not been investigated.

Fourth, more bandwidth is available in the 5725–5850 MHz ISM band than in the other two ISM bands used for RF tags. This additional bandwidth may enable high data rate and spread-spectrum backscatter communication [32, 33].

17.4.1 Link Budget Example

As an example of the gains available using multiple antennas and higher frequencies, consider the same RF tag system described in Section 17.3 redesigned to operate at 5790 MHz with multiple antennas at both the reader and the RF tag. The resulting changes in the link budget parameters are described in the following paragraphs and are summarized in Table 17.7a and 17.7b.

17.4.1.1 Increased Antenna Gain: If the effective area of the antennas remains constant, Eq. (17.11) dictates that an additional 16.4 dB of antenna gain is available as λ decreases from 33 cm at 915 MHz to 5 cm at 5790 MHz. Therefore, at 5790 MHz, the reader antenna gains are now 23.4 dBi (219 in the linear scale) and the RF tag antenna gain is now 18.5 dBi (71 in the linear scale).

17.4.1.2 Multiple Tag Antennas and MRC: As with the 915-MHz link budget example in Section 17.3, the additional power that must be transmitted to overcome multipath fading when using multiple antennas is given by a fade margin. When using multiple antennas, the fade margin is determined by the desired outage probability,

TABLE 17.6. Fade Margins (in dB) for Bistatic, Dislocated Backscatter Radio Systems Using Multiple Antennas in a Rician Fading Backscatter Channel with $K = 3$ dB

	No MRC		MRC	
Outago Probability	$1 \times 1 \times 1$	$1 \times 2 \times 1$	$1 \times 1 \times 2$	$1 \times 2 \times 2$
0.5	2.2	1.3	1.5	1.0
0.1	11	8.1	8.7	6.2
0.05	15	11	12	8.1
0.01	22	17	18	12
0.005	26	20	21	14
0.001	33	27	28	19

the fading characteristics of the channel, the combining algorithm used at the reader, the number of reader transmitter or reader receiver antennas, and the number of RF tag antennas. Table 17.6 shows the fade margins F_β for several bistatic, dislocated RF tag systems that use multiple antennas. The fade margins were calculated from Monte Carlo simulations for a backscatter channel in which the K factor of the forward and backscatter links was 3 dB. The first two columns show the fade margins for a backscatter radio system using one and two RF tag antennas with no diversity combining at the reader.* The third and fourth columns show F_β for RF tags with one and two antennas and using MRC from two reader receiver antennas. Note that when multiple RF tag antennas are used in conjunction with diversity combining at the reader receiver, pinhole diversity gains and conventional diversity gains combine for greater benefits.

For this example, the power-up fade margin F_p remains unchanged from the 915-MHz example.

17.4.1.3 Increased Object Immunity. From the example in Section 17.3, the RF tag antenna attached to a metal object is at a distance of 0.005λ from the metal surface at 915 MHz. If the RF tag now operates at 5790 MHz and its physical distance from the metal surface is unchanged, the electrical separation distance becomes 0.033λ. At this distance, Figure 17.2 shows that the antenna's terminal impedance is now $10 + j100\ \Omega$, which corresponds to $Z_{\mathrm{IN}} = 2.3 + j319\ \Omega$ at the output terminals of the impedance transformation network. From Eqs. (17.3) and (17.4), the power transmission coefficient and modulation factor for metal attachment are now $\tau = 0.13$ and $M = 4.4 \times 10^{-3}$ (both in the linear scale), respectively. Because of cardboard's weak effect on the antenna impedance, τ and M for cardboard attachment do not change from the 915-MHz example.

To complete this example, the transmitted power must be decreased to 12.6 dBm (or 18.3 mW) to meet FCC power limitations with the increased reader antenna gains.

*In this chapter, the number of antennas used in a backscatter radio system is given by $M \times L \times N$, where M is the number of reader transmitter antennas, L is the number of RF tag antennas, and N is the number of reader receiver antennas.

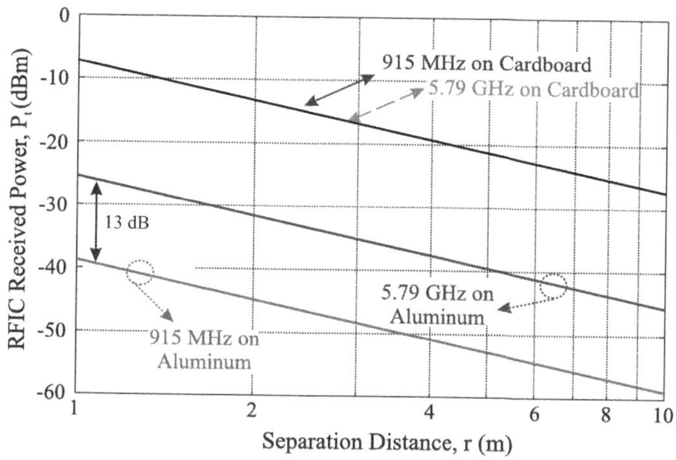

FIGURE 17.7. The power-up links plotted as a function of reader-tag separation distance, r. In this figure, $P_T = 18.3$ mW, $G_{T,R} = 219$, $G_t = 62$, and $X_{f,b} = 0.5$ all in the linear scale. A complete summary of the system and channel parameter is given in Table 17.7a and 17.7b. (Reprinted from Griffin and Durgin [1], copyright 2008 IEEE.)

Since no measurements of gain penalties at 5790 MHz are available, 915-MHz gain penalty values from Table 17.3 are used in this example.

17.4.2 Discussion

Power-Up Link. Figure 17.7 compares the power-up link budgets for a $1 \times 1 \times 1$ backscatter radio system operating at 915 MHz, from Section 17.3, and at 5790 MHz, whose parameters are given in Table 17.7a and 17.7b. For cardboard attachment, the links are almost identical; the power sacrificed to higher path loss at 5790 MHz is

TABLE 17.7a. 5790-MHz RF Tag Portal Example Parameters in the Linear Scale

<table>
<tr><th>Material</th><th>P_T (mW)</th><th>$G_{T,R}$</th><th>G_t</th><th>$X_{f,b}$</th><th>F_p</th><th>λ (m)</th><th>M</th></tr>
<tr><td>Cardboard</td><td rowspan="2">18.3</td><td rowspan="2">219</td><td rowspan="2">71</td><td rowspan="2">0.5</td><td rowspan="2">10</td><td rowspan="2">0.05</td><td>0.25</td></tr>
<tr><td>Aluminum</td><td>4.4×10^{-3}</td></tr>
</table>

Source: Expanded from Griffin and Durgin [1], Copyright 2008 IEEE.

TABLE 17.7b. 5790-MHz RF Tag Portal Example Parameters in the Linear Scale

<table>
<tr><th rowspan="2">Material</th><th rowspan="2">τ</th><th rowspan="2">Θ</th><th rowspan="2">B</th><th colspan="4">F_β</th></tr>
<tr><th>$1 \times 1 \times 1$</th><th>$1 \times 2 \times 1$</th><th>$1 \times 1 \times 2$ MRC</th><th>$1 \times 2 \times 2$ MRC</th></tr>
<tr><td>Cardboard</td><td>1</td><td>1.2</td><td rowspan="2">1</td><td rowspan="2">32</td><td rowspan="2">13</td><td rowspan="2">16</td><td rowspan="2">6.5</td></tr>
<tr><td>Aluminum</td><td>0.13</td><td>11</td></tr>
</table>

Source: Expanded from Griffin and Durgin [1], Copyright 2008 IEEE.

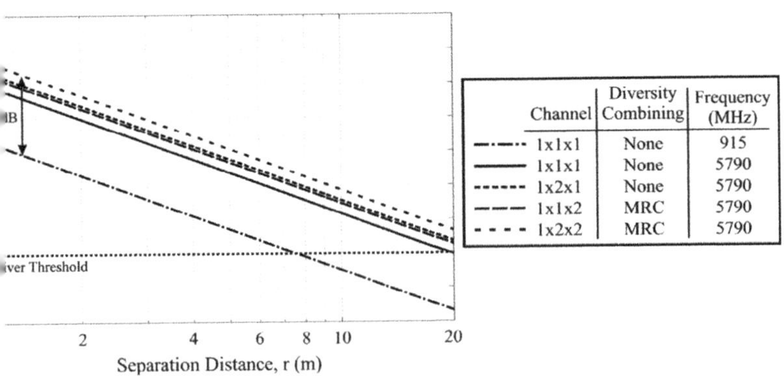

(a) Cardboard attachment

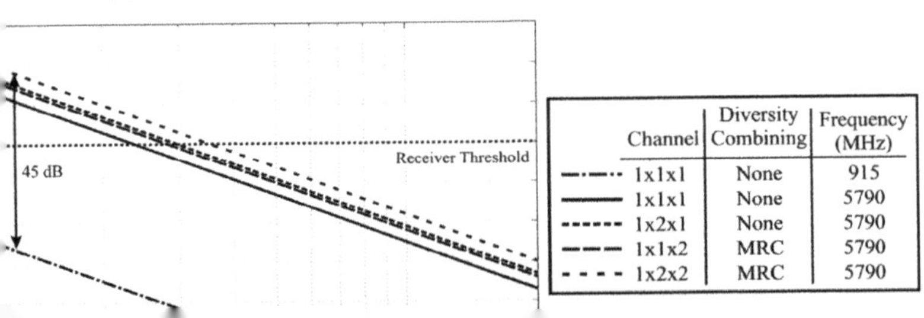

MRC at 5790 MHz and a 1 × 1 × 1 system at 915 MHz. While these gains still do not allow the RF tag to operate while attached to the aluminum slab, they may do so for other, less-extreme material attachments and result in significant range and reliability improvements.

These examples show that using backscatter radio systems with multiple antennas at high frequencies can improve their performance significantly; however, a few qualifications should be kept in mind.

First, for passive tags, the range benefits shown in Figure 17.7 will be limited in some commonly used tag RFIC technologies by parasitic capacitance that increases power consumption at high frequencies. However, with careful circuit layout and use of technologies other than conventional silicon, it should be possible to adequately reduce the power consumption of high-frequency tag RFICs at reasonable manufacturing costs. Such high-frequency, low-power RFICs will have operating ranges comparable to their low-frequency counterparts, but with the added advantage of reduced fading and increased object immunity. An example of a high-frequency tag RFIC with low power consumption was reported by Curty et al. [14] to consume 2.7 μW (−25 dBm) at 2.45 GHz. This RFIC was fabricated using a 0.5-μm silicon-on-sapphire technology.

Second, semi-passive tags will immediately benefit from the range gains evident in Figures 17.8a and 17.8b since their performance is not constrained by the more stringent power-up link, as is the case with passive tags. Though the range of passive tags will likely be limited by the power-up link, they will still benefit from reduced fading on the backscatter link and increased object immunity.

Third, it should be noted that while the RF tag antenna gain at 5790 MHz will increase the RF tag range, it will also increase its directivity. The result will be an RF tag that is more sensitive to orientation than that at 915 MHz. If the RF tag's orientation is fixed or its range is very small, the increased directivity will not be an issue; however, for other applications, it may pose a serious, but not insurmountable, problem. In such cases, multiple reader antennas can be used to power the RF tag from different spatial positions. Furthermore, as tag RFIC technology advances and the required turn-on power is decreased, the RF tag antenna gain and, as a result, directivity can be reduced without sacrificing tag range.

17.5 CONCLUSION

Successful backscatter radio design requires more than just accurate link budget calculations; it requires an understanding of the propagation mechanisms that affect both the power-up and backscatter links. This chapter has provided these link budgets along with a detailed description of the modulation factor, on-object gain penalties, path blockage losses, polarization mismatch losses, impedance mismatch losses, and small-scale multipath fading losses. A realistic 915-MHz backscatter radio example was presented to demonstrate use of the link budgets and has shown that object attachment and multipath fading can significantly impact RF tag range. Furthermore, it has been shown that backscatter tag systems using compact reader antenna arrays and multiple RF tag antennas available in the 5725–5850 MHz frequency band will experience a number of propagation benefits—increased antenna gain, reduced small-scale multipath fading, and increased object immunity—over lower-frequency systems using only single antennas. Though some obstacles must be overcome, such high-frequency, multi-antenna systems may provide reliable communication for compact backscatter RF tags.

REFERENCES

1. J. D. Griffin and G. D. Durgin, Complete link budgets for backscatter radio and RFID systems, *IEEE Antennas Propag. Mag.*, Vol. 51, No. 2, pp. 11–25, April 2009.
2. J. D. Griffin, *High-frequency modulated-backscatter communication using multiple antennas*, Ph.D. dissertation, The Georgia Institute of Technology, 2009 [Online]. Available at http://www.propagation.gatech.edu/Archive/PG TR 090120 JDG/PG TR 090120 JDG.pdf. Accessed May 7, 2009.
3. W. L. Stutzman and G. A. Thiele, *Antenna Theory and Design*, 2nd edition, John Wiley & Sons, Hoboken, NJ, 1998.
4. K. V. S. Rao, P. V. Nikitin, and S. F. Lam, Antenna design for UHF RFID tags: A review and a practical application, *IEEE Trans. Antennas Propag.*, Vol. 53, No. 12, pp. 3870–3876, 2005.
5. P. Nikitin and K. V. S. Rao, Performance limitations of passive UHF RFID systems, in *Proceedings of IEEE Antenna and Propagation Society International Symposium*, Albuquerque, New Mexico, 2006, pp. 1011–1014.
6. K. V. S. Rao, P. V. Nikitin, and S. F. Lam, Impedance matching concepts in RFID transponder design, in *IEEE Workshop on Automatic Identification Advanced Technologies*, 2005, pp. 39–42.
7. D. M. Dobkin and S. M. Weigand, Environmental effects on RFID tag antennas, in *IEEE MTT-S International Microwave Symposium Digest*, 2005, pp. 135–138.
8. Y. Tikhov and J. H. Won, Impedance-matching arrangement for microwave transponder operating over plurality of bent installations of antenna, *Electron. Lett.*, Vol. 40, No. 10, pp. 574–575, 2004.
9. J. Siden, P. Jonsson, T. Olsson, and G. Wang, Performance degradation of RFID system due to the distortion in RFID tag antenna, in *Conference Proceedings of Microwave and Telecommunication Technology*, Weber Co., Sevastopol, Crimea, Ukraine, 2001, pp. 371–373.
10. G. De Vita and G. Iannaccone, Design criteria for the RF section of UHF and microwave passive RFID transponders, *IEEE Trans. Microwave Theory Tech.*, Vol. 53, No. 9, pp. 2978–2990, 2005.
11. J. T. Prothro and G. D. Durgin, Improved Performance of a Radio Frequency Identification Tag Antenna on a Metal Ground Plane, Master's thesis, The Georgia Institute of Technology, 2007. [Online]. Available: http://www.propagation.gatech.edu/Archive/PG_TR_070515_JTP/PG_TR_070515_JTP.pdf, last accessed on March 29, 2009.
12. P. V. Nikitin, K. V. S. Rao, and R. D. Martinez, Differential RCS of RFID tag, *Electron. Lett.*, Vol. 43, No. 8, pp. 431–432, 2007.

13. D. M. Dobkin, *The RF in RFID: Passive UHF RFID in Practice*, Newnes, Burlington, MA, 2008.
14. J. P. Curty, N. Joehl, C. Dehollain, and M. J. Declercq, Remotely Powered Addressable UHF RFID Integrated System, *IEEE J. Solid-State Circuits*, Vol. 40, No. 11, pp. 2193–2202, 2005.
15. U. Karthaus and M. Fischer, Fully integrated passive UHF RFID transponder IC with 16.7-μW minimum RF input power, *IEEE J. Solid-State Circuits*, Vol. 38, No. 10, pp. 1602–1608, 2003.
16. P. V. Nikitin and K. V. S. Rao, Theory and measurement of backscattering from RFID tags, *IEEE Antennas Propag. Mag.*, Vol. 48, No. 6, pp. 212–218, 2006.
17. K. Penttilä, M. Keskilammi, L. Sydänheimo, and M. Kivikoski, Radar cross-section analysis for passive RFID systems, *IEE Proc. Microwaves, Antennas Propag.*, Vol. 153, No. 1, pp. 103–109, 2006.
18. F. Fuschini, C. Piersanti, F. Paolazzi, and G. Falciasecca, Analytical approach to the backscattering from UHF RFID transponder, *IEEE Antennas Wireless Propag. Lett.*, Vol. 7, pp. 33–35, 2008.
19. R. B. Green, The General Theory of Antenna Scattering, Ph.D. dissertation, Ohio State University, 1963.
20. G. D. Durgin, *Space-Time Wireless Channels*, Prentice-Hall, Upper Saddle River, NJ, 2003.
21. D. Kim, M. A. Ingram, and W. W. Smith, Jr., Measurements of small-scale fading and path loss for long range RF tags, *IEEE Trans. Antennas Propag.*, Vol. 51, No. 8, pp. 1740–1749, 2003.
22. J. D. Griffin and G. D. Durgin, Link Envelope correlation in the backscatter channel, *IEEE Commun. Lett.*, Vol. 11, No. 9, pp. 735–737, 2007.
23. J. D. Griffin and G. D. Durgin, Multipath Fading Measurements for Multi-Antenna Backscatter RFID at 5.8 GHz, in 2009 *International IEEE Conference on RFID*, Orlando, FL, 2009, pp. 322–329.
24. J. D. Griffin, G. D. Durgin, A. Haldi, and B. Kippelen, RF tag antenna performance on various materials using radio link budgets, *IEEE Antennas Wireless Propag. Lett.*, Vol. 5, pp. 247–250, 2006.
25. J. D. Parsons, *The Mobile Radio Propagation Channel*, 2nd edition, John Wiley & Sons, West Sussex, England, 2000.
26. S. Chiu, I. Kipnis, M. Loyer, J. Rapp, D. Westberg, J. Johansson, and P. Johansson, A 900 MHz UHF RFID reader transceiver IC, *IEEE J. Solid-State Circuits*, Vol. 42, No. 12, pp. 2822–2833, 2007.
27. D. M. Dobkin and S. M. Weigand, UHF RFID and tag antenna scattering, Part I: Experimental results, *Microwave J., Euro-Global Edition*, Vol. 49, No. 5, pp. 170–190, 2006.
28. M. Ingram, M. Demirkol, and D. Kim, Transmit diversity and spatial multiplexing for RF links using modulated backscatter, in *Proceedings of the International Symposium on Signals, Systems, and Electronics*, Tokyo, Japan, July 2001.
29. J. D. Griffin and G. D. Durgin, Gains for RF tags using multiple antennas, *IEEE Trans. Antennas Propag.*, Vol. 56, No. 2, pp. 563–570, 2008.
30. M. Mi, M. H. Mickle, C. Capelli, and H. Swift, RF energy harvesting with multiple antennas in the same space, *IEEE Antennas Propag. Mag.*, Vol. 47, No. 5, pp. 100–106, 2005.

31. P. Nikitin and K. V. S. Rao, Performance of RFID tags with multiple RF ports, in *Proceedings of the IEEE Antenna and Propagation Society International Symposium*, Honolulu, HI, 2007, pp. 5459–5462.

32. A. Rohatgi and G. D. Durgin, Implementation of an anti-collision differential-offset spread spectrum RFID system, in *Proceedings of the IEEE Antennas and Propagation Society International Symposium*, 2006, pp. 3501–3504.

33. M. S. Trotter, J. D. Griffin, and G. D. Durgin, Power optimized waveforms for improving the range and reliability of RFID systems, in *Proceedings of the 2009 International IEEE Conference on RFID*, Orlando, FL, April 2009, pp. 80–87.

CHAPTER 18

FADING STATISTICS FOR MULTI-ANTENNA RF TAGS

JOSHUA D. GRIFFIN
Disney Research, Pittsburgh, Pennsylvania

GREGORY D. DURGIN
School of Electrical and Computer Engineering, Georgia Institute of Technology, Atlanta, Georgia

18.1 INTRODUCTION

Backscatter radio systems—that is, systems that communicate wirelessly by modulating scattered electromagnetic fields—have found widespread use in ultra-high-frequency (UHF) and microwave radio-frequency identification (RFID) and sensor applications. In most RFID systems, backscatter communication occurs when an interrogator, or *reader*, transmits a continuous-wave (CW) signal to a small, low-power transponder, or *radio-frequency (RF) tag*. A portion of the CW signal is scattered from the RF-tag's antenna and *load-modulated* that is, the scattered signal is modulated as the RF tag changes the reflection coefficient seen by the tag antenna by changing the antenna's load impedance.

Many factors can affect the operating range and reliability of a backscatter radio system, including mulitpath fading. To understand how multipath fading occurs, consider a wave launched from a transmitting antenna. If the transmitting or receiving antenna is in the vicinity of conducting or dielectric objects, part of the transmitted signal will be scattered off the objects. These scattered waves travel different paths to the receiving antenna and, as a result, arrive with different amplitudes, phases, and angles of arrival. At some points in space, the multipath waves combine constructively, but at others they combine destructively and result in a signal *fade*. The result is that, at any given point in space, adequate power may not be available for the backscatter radio system to operate. One way to characterize these fades is with a probability density function (PDF). In general, when a line-of-sight (LOS) path

Handbook of Smart Antennas for RFID Systems, Edited by Nemai Chandra Karmakar

exists between the transmitter and receiver, the variations of the signal amplitude—or *envelope*—caused by mutlipath fading can be predicted using a Rician distribution. Otherwise, if the LOS path is obstructed, then a Rayleigh distribution can be used for the non-line-of-sight (NLOS) channel. Although Rayleigh fading is often considered to be the worst fading case, it has been shown that multipath fading in backscatter channels can have *radically* different fading statistics resulting in deeper fades [1]. Although backscatter channels often exhibit LOS propagation, heavy small-scale fading may still be present due to indoor operation, a cluttered reader environment, and the inhomogeneous nature of the tagged objects. Therefore, a thorough understanding of multipath fading statistics in the backscatter channel is imperative.

This chapter, which has been adapted from [2], [3], and [20], describes the $M \times L \times N$ dyadic backscatter channel—the most general backscatter channel—and new $M \times L \times N$ envelope fading distributions for the modulated signal received from a backscatter radio transponder. Using these new distributions, it is shown that a *pinhole diversity* gain is available to RF tags with multiple antennas to modulate backscatter. The term pinhole diversity comes from the fact that each RF-tag antenna corresponds to a pinhole in the backscatter channel that provides a set of spatially separated propagation paths, or pinhole diversity branches. While conventional techniques to mitigate multipath fading require diversity combining or coding, pinhole diversity gains only require multiple RF-tag antennas. This chapter includes two examples showing the reliability and range benefits that can be derived from pinhole diversity gains and discusses the amount of multipath fading required for them to be realized. The chapter concludes with two appendices describing commonly used multipath-fading distributions and a detailed derivation of the $M \times L \times N$ backscatter channel envelope distributions.

18.2 THE $M \times L \times N$ DYADIC BACKSCATTER CHANNEL

The most general backscatter channel is the $M \times L \times N$ *dyadic* backscatter channel*—a pinhole channel [4] that describes the propagation of signals in a backscatter radio system consisting of M transmitter, L RF tag, and N receiver antennas. In a pinhole channel, propagation paths are forced to converge at a point reducing the rank of the channel matrix and decreasing the channel capacity [4]. In RF channels, the point of convergence may be a diffracting edge [5, 6]; a hallway or tunnel [5]; a metal screen [4]; rings of scatterers separated by a long distance [7]; or the antennas of a mobile station that are found in amplify-and-forward channels [8]. In the dyadic backscatter channel, the point of convergence is the RF-tag antennas. All of these pinhole channels, except the amplify-and-forward and dyadic backscatter channels, cease their pinhole behavior under LOS conditions. The dyadic backscatter and amplify-and-forward channels remain pinholes in both LOS and NLOS conditions, making them some of the only true pinhole channels.

The hallmark of any pinhole channel is that it can be modeled as the cascade of two channels. In the $M \times L \times N$ dyadic backscatter channel, shown in Figure 18.1a,

*The term "dyadic" has a double meaning. It reflects both the two-fold nature of the channel created by the forward and backscatter links and the fact that the modulated signals are represented in matrix form.

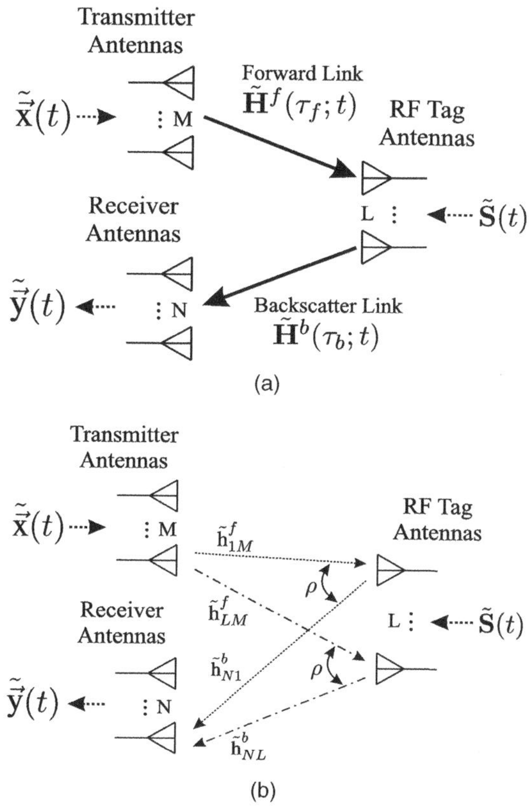

FIGURE 18.1. (**a**) The general $M \times L \times N$, dyadic backscatter channel with M transmitter antennas, L RF-tag antennas, and N receiver antennas described mathematically by Eq. (18.1). (Reprinted from Griffin and Durgin [2], copyright 2008 IEEE.) (**b**) The path from the mth transmitter antenna to the nth receiver antenna is shown where $\tilde{h}^f_{lm}$ and $\tilde{h}^b_{nl}$ are elements of the forward-link and backscatter-link matrices, $\tilde{\mathbf{H}}^f$ and $\tilde{\mathbf{H}}^b$, respectively. Elements of the forward-link and backscatter-link matrices that terminate or originate on a common RF-tag antenna have link correlation, ρ; otherwise, the matrix elements are independent (Reprinted from Griffin and Durgin [3], copyright 2007 IEEE.)

these channels are called the forward and backscatter links. The *forward link channel matrix*, $\tilde{\mathbf{H}}^f$, describes signal propagation from the M transmitter antennas to the L RF-tag antennas and the *backscatter link channel matrix*, $\tilde{\mathbf{H}}^b$, describes the propagation of signals scattered from the L RF-tag antennas to the N receiver antennas.

In mathematical terms, the received baseband signal from the $M \times L \times N$ dyadic backscatter channel is

$$\tilde{\vec{y}}(t) = \frac{1}{2} \int_{-\infty}^{+\infty} \int_{-\infty}^{+\infty} \tilde{\mathbf{H}}^b(\tau_b;t)\tilde{\mathbf{S}}(t-\tau_b)\tilde{\mathbf{H}}^f(\tau_f;t-\tau_b)\tilde{\vec{x}}(t-\tau_b-\tau_f)\,d\tau_b\,d\tau_f + \tilde{\vec{n}}(t). \tag{18.1}$$

In Eq. (18.1), $\tilde{\bar{y}}(t)$ is an $N \times 1$ vector of received baseband signals; $\tilde{\mathbf{H}}^b(\tau_b; t)$ is the $N \times L$ complex, baseband-channel impulse-response matrix of the backscatter link; and $\tilde{\mathbf{H}}^f(\tau_f; t)$ is the $L \times M$ complex, baseband-channel impulse-response matrix of the forward link. This notation is defined by Durgin [9]. Both channel matrices in Eq. (18.1) are allowed to be time-varying. $\tilde{\mathbf{S}}(t)$ is the narrowband, $L \times L$ signaling matrix, $\tilde{\bar{x}}(t)$ is an $M \times 1$ vector of signals transmitted from the transmitter antennas, and $\tilde{\bar{n}}(t)$ is an $N \times 1$ vector of noise components.

Previous work on the backscatter channel includes that by Ingram et al. in the context of RFID systems. Ingram et al. [10] were the first to propose the use of multiple antennas at both the reader and RF tag for transmit diversity and spatial multiplexing to increase the range and communication capacity, respectively. Mi et al. [11] also advocate the use of multiple RF-tag antennas for the purpose of increasing the power available to the RF tag integrated circuit (IC). Although multipath fading is a significant problem, relatively few measurements have been reported. Fading on the signal received by the RF tag have been reported by Mitsugi and Shibao [12, 13] and Polivka et al. [14]. Others have also studied fading on the signal received by the reader: Kim et al. [1] made measurements of the backscatter channel reporting envelope cumulative distribution functions (CDFs) for the modulated-backscatter signal received at the reader, and Banerjee et al. [15, 16] have presented fading measurements as well as spatial and frequency diversity gain measurements at 915 MHz.

18.2.1 The Signaling Matrix

The signaling matrix $\tilde{\mathbf{S}}(t)$ is an $L \times L$ matrix that describes the time-varying modulation that an RF tag places on radio signals absorbed and scattered by the L tag antennas. Therefore, it is natural to define the signaling matrix as the L-port scattering matrix commonly used in RF hardware design. A comprehensive discussion of the general L-port scattering matrix is given by Pozar [17]; however, two notes are worth mentioning. First, the signaling matrix of a passive or semi-passive RF tag has elements $\tilde{s}_{ij}(t)$ with magnitude less than unity, $|\tilde{s}_{ij}(t)| \leq 1$, since there is no amplification of the backscattered signal. Second, although it is not required that the signaling matrix be symmetric, most RF tags will satisfy this property.

The signaling matrix may take several different forms, depending upon the physical implementation of the modulation circuitry and RF-tag antennas.

18.2.1.1 Identity Signaling Matrix: If each RF-tag antenna is used to modulate backscatter with the same signal and no received signal power is transferred between the antennas, the signaling matrix is the normalized identity matrix,

$$\tilde{\mathbf{S}}(t) = \tilde{s}(t)\mathbf{I}_L = \tilde{\Gamma}(t)\mathbf{I}_L, \tag{18.2}$$

where $\mathbf{I}_L$ is the $L \times L$ identity matrix and $\tilde{\Gamma}(t)$ is the complex RF-tag antenna load reflection coefficient.

18.2.1.2 Diagonal Signaling Matrix: If an RF tag with L antennas modulates backscatter with *different* signals and no received signal power is transferred between the antennas, the signaling matrix is diagonal,

$$\begin{aligned}\tilde{\mathbf{S}}(t) &= \mathrm{diag}\,[\tilde{s}_{11}(t) \ldots \tilde{s}_{LL}(t)] \\ &= \mathrm{diag}\left[\tilde{\Gamma}_1(t) \ldots \tilde{\Gamma}_\mathrm{L}(t)\right].\end{aligned} \tag{18.3}$$

18.2.1.3 Full Signaling Matrix: If the RF-tag antennas modulate backscatter independently and signals are transferred between the antennas, the full signaling matrix

$$\tilde{\mathbf{S}}(t) = \begin{bmatrix} \tilde{s}_{11}(t) & \ldots & \tilde{s}_{1L}(t) \\ \vdots & \ddots & \vdots \\ \tilde{s}_{L1}(t) & \ldots & \tilde{s}_{LL}(t) \end{bmatrix}. \tag{18.4}$$

is used. Signal transfer between antennas is represented by the off-diagonal elements (i.e., s_{ij} where $i \neq j$). The fact that power can be transferred between the RF-tag antennas gives the designer an additional degree of freedom in signal scheme design; however, no present application of the full signaling matrix has been identified.

For the remainder of this chapter, it is assumed that the signaling matrix has the form of Eq. (18.2), the identity signaling matrix, with binary reflection of signals, $\tilde{\Gamma} = \pm 1$, to provide maximum signal power. Although real RF tags often switch between a matched and short-circuit load, $\tilde{\Gamma} = \pm 1$ maximizes the backscattered signal and simplifies the following analysis.

18.3 $M \times L \times N$ DYADIC BACKSCATTER CHANNEL ENVELOPE PDF

In the $M \times L \times N$ dyadic backscatter channel, each element of the forward link $\tilde{\mathbf{H}}^f$ and the backscatter link $\tilde{\mathbf{H}}^b$ is an independent, identically distributed (i.i.d.), complex-Gaussian random variable. The elements of the forward and backscatter links can be written, respectively, as $h^b = (w_b + jv_b)$ and $h^f = (w_f + jv_f)$, where $w_{f,b}$ and $v_{f,b}$ are normally distributed as $\sim \mathcal{N}(\mu_{f,b}, \sigma_{f,b}^2/2)$. In this channel model, propagation paths that originate or terminate on a common RF-tag antenna can have link correlation ρ, shown in Figure 18.1b. Link correlation simply describes the relationship between fading in the forward and backscatter links. Details on its definition may be found in Appendix A.3, Appendix B.3, and work by Griffin and Durgin [3]. Mathematically, link correlation is defined as

$$\rho = \frac{2\mathrm{Cov}(w_f, w_b)}{\sigma_b \sigma_f} = \frac{2\mathrm{Cov}(v_f, v_b)}{\sigma_b \sigma_f} \tag{18.5}$$

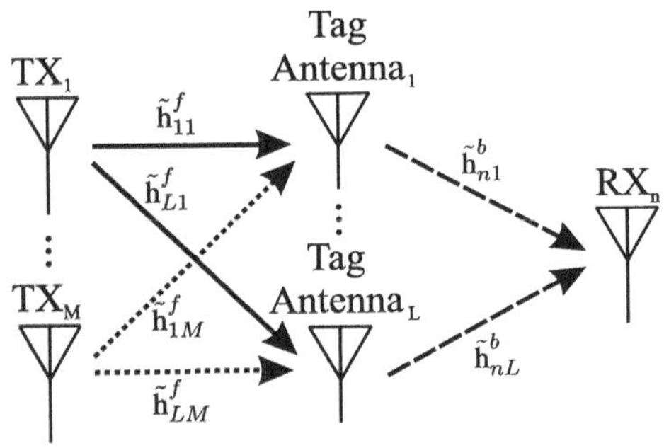

FIGURE 18.2. In a backscatter radio system, the signal received at the nth receiving antenna is the sum of ML Gaussian products. However, only L of these products are statistically independent. (Reprinted from Griffin and Durgin [2], copyright 2008 IEEE.)

where Cov(·, ·) is the covariance operator. The signal received at the nth receiver antenna is proportional to the sum of ML complex-Gaussian products, described by Eq. (18.6) and Figure 18.2,

$$\tilde{y}(t) \propto \sum_{j=1}^{L} \underbrace{\left(\tilde{h}^f_{j1} + \cdots + \tilde{h}^f_{jM}\right)}_{\tilde{g}^f_j} \tilde{h}^b_{nj}. \tag{18.6}$$

In Eq. (18.6), each element of the backscatter link matrix $\tilde{h}^b$ is multiplied by M, i.i.d. forward-link elements, $(\tilde{h}^f_{j1} + \cdots + \tilde{h}^f_{jM})$, which can be expressed as a single complex-Gaussian random variable $\tilde{g}^f_j$ with mean $M(\mu_f + j\mu_f)$ and variance $M\sigma_f^2$. Therefore, the signal received at the nth receiver antenna is proportional to the sum of L, i.i.d., complex-Gaussian products.

18.3.1 Rayleigh Fading Links

In a typical $M \times L \times N$, dyadic backscatter channel, LOS propagation dominates and the elements of the forward and backscatter matrices follow a Rician distribution ($\mu_{f,b} \neq 0$). The $M \times L \times N$ Rician envelope PDF will be discussed in Section 18.3.2, but first, the lower bound of this case, the $M \times L \times N$, dyadic backscatter channel with Rayleigh-fading forward and backscatter links, will be studied. In the Rayleigh case, the elements of the forward and backscatter links can be written, respectively, as $h^f = (w_f + jv_f)$ and $h^b = (w_b + jv_b)$ where $w_{f,b}$ and $v_{f,b}$ are $\sim \mathcal{N}(0, \sigma^2_{f,b}/2)$. The envelope PDF of the $M \times L \times N$, dyadic backscatter channel $f_{\mathbf{A}}(\alpha, \rho)$ can be derived from that of the product of two Rayleigh random variables, as shown in Appendix B.

The envelope PDF $f_{\mathbf{A}}(\alpha, \rho)$ is most easily understood by examining three special cases: the case with independent forward and backscatter links ($\rho = 0$), the case with fully correlated forward and backscatter links ($\rho = 1$), and the case with partial link correlation in the range $0 < \rho < 1$. As will be shown, exact analytical expressions can be found for the PDF with $\rho = 0$ and $\rho = 1$; however, the case of $0 < \rho < 1$ is an approximation and requires numerical techniques. Plots of the PDF for $0 < \rho < 1$ are given in Section 18.4.1.

To this point, the relationship between fading in the forward and backscatter links has been described by ρ which is defined in Eq. (18.5) as the correlation between the real and imaginary components of $\tilde{\mathbf{H}}^f$ and $\tilde{\mathbf{H}}^b$. While ρ is convenient for analysis, a more practical metric is the correlation between the envelopes of $\tilde{\mathbf{H}}^f$ and $\tilde{\mathbf{H}}^b$—*link envelope correlation*—since this correlation can be directly calculated from measured data. It can be shown that link envelope correlation ρ_e is related to link correlation by $\rho_e \approx \rho^2$ [18, 19]. In the remainder of this chapter, both ρ and ρ_e will be used when appropriate.

18.3.1.1 Zero Link Correlation $\rho = 0$: The received envelope PDF for the case of $\rho = 0$ [2] is

$$f_{\mathbf{A}}(\alpha, 0) = \alpha^L \left(\frac{2}{\sqrt{M}\sigma_b \sigma_f} \right)^{1+L} \frac{2^{1-L}}{\Gamma(L)} K_\nu \left(\frac{2\alpha}{\sqrt{M}\sigma_b \sigma_f} \right). \tag{18.7}$$

where $\mathbf{A}$ is the random channel envelope, α is the index of the PDF, $\Gamma(\cdot)$ is the gamma function, and $K_\nu(\cdot)$ is a modified bessel function of the second kind with order $\nu = 1 - L$. The mean and variance of Eq. (18.7), respectively, are

$$\mu_{\mathbf{A}} = \frac{\sqrt{M}\sigma_b \sigma_f}{2} \frac{\pi (2L)!}{4^L L!(L-1)!} \tag{18.8}$$

and

$$\sigma_{\mathbf{A}}^2 = M L \sigma_b^2 \sigma_f^2 \left(1 - \frac{L}{4} \left[\frac{\pi (2L)!}{4^L (L!)^2} \right]^2 \right). \tag{18.9}$$

Simplified envelope statistics for several practical cases are given in Table 18.1.

18.3.1.2 Unity Link Correlation $\rho = 1$: The envelope PDF for the case of $\rho = 1$ [2], which may occur only in a $1 \times L \times 1$, dyadic backscatter channel in which a single reader antenna is used to transmit and receive, is

$$f_{\mathbf{A}}(\alpha, 1) = \alpha^{L/2} \left(\frac{1}{\sigma_b \sigma_f} \right)^{1+L/2} \frac{2^{1-L/2}}{\Gamma(L/2)} K_\nu \left(\frac{\alpha}{\sigma_b \sigma_f} \right) \tag{18.10}$$

where $K_\nu(\cdot)$ is a modified Bessel function of the second kind with order $\nu = 1 - L/2$ and all other terms are as defined for Eq. (18.7). The mean and variance of Eq. (18.10), respectively, are

$$\mu_{\mathbf{A}} = \sigma_b \sigma_f \Psi \tag{18.11}$$

TABLE 18.1. Simplified Envelope Statistics for $M \times L \times N$, Dyadic Backscatter Channels with Uncorrelated ($\rho = 0$) Rayleigh-Fading Links

	$f_{\mathbf{A}}(\alpha, 0)$	$\mu_{\mathbf{A}}$	$\sigma^2_{\mathbf{A}}$
$L = 1$	$\frac{4\alpha}{\zeta^2} K_0\left(\frac{2\alpha}{\zeta}\right)$	$\frac{\pi}{4}\zeta$	$\left(1 - \frac{\pi^2}{16}\right)\zeta^2$
$L = 2$	$\frac{4\alpha^2}{\zeta^3} K_{-1}\left(\frac{2\alpha}{\zeta}\right)$	$\frac{3\pi}{8}\zeta$	$\left(2 - \frac{9\pi^2}{64}\right)\zeta^2$
$L = 3$	$\frac{2\alpha^3}{\zeta^4} K_{-2}\left(\frac{2\alpha}{\zeta}\right)$	$\frac{15\pi}{32}\zeta$	$\left(3 - \frac{225\pi^2}{1024}\right)\zeta^2$
$L = 4$	$\frac{2\alpha^4}{3\zeta^5} K_{-3}\left(\frac{2\alpha}{\zeta}\right)$	$\frac{35\pi}{64}\zeta$	$\left(4 - \frac{1225\pi^2}{4096}\right)\zeta^2$

$\zeta = \sqrt{M}\sigma_b\sigma_f$.
Source: Reprinted from Griffin and Durgin [2], copyright 2008 IEEE.

where

$$\Psi = \begin{cases} \dfrac{\pi L!}{2^L(L/2-1)!(L/2)!} & \text{for even L,} \\ \dfrac{2^L[(L/2-1/2)!]^2}{2(L-1)!} & \text{for odd L} \end{cases},$$

and

$$\sigma^2_{\mathbf{A}} = 2L\sigma_b^2\sigma_f^2\left(1 - \frac{\Lambda}{2L}\right) \tag{18.12}$$

where

$$\Lambda = \begin{cases} \left[\dfrac{\pi L!}{2^L(L/2)!(L/2-1)!}\right]^2 & \text{for even L,} \\ \left[\dfrac{2^{L-1}[(L/2-1/2)!]^2}{(L-1)!}\right]^2 & \text{for odd L} \end{cases}.$$

Simplified envelope statistics for several practical cases are given in Table 18.2.

18.3.2 Rician-Fading Links

Similar to the Rayleigh-fading case in Section 18.3.1, the PDF of the $M \times L \times N$, dyadic backscatter channel can be derived from the PDF of the product of two Rician random variables. This PDF is given in Appendix A.4 for the case of independent forward and backscatter links. However, this PDF expression does not lend itself to the derivation of an analytical expression of the $M \times L \times N$, dyadic backscatter

TABLE 18.2. Simplified Envelope Statistics for 1 × L × 1, Dyadic Backscatter Channels with Fully Correlated ($\rho = 1$), Rayleigh-Fading Links and Collocated Reader Antennas

	$f_{\mathbf{A}}(\alpha, 1)$	$\mu_{\mathbf{A}}$	$\sigma_{\mathbf{A}}^2$
$L = 1$	$\frac{1}{\delta}\exp\left(\frac{-\alpha}{\delta}\right)$	δ	δ^2
$L = 2$	$\frac{\alpha}{\delta^2}K_0\left(\frac{\alpha}{\delta}\right)$	$\frac{\pi}{2}\delta$	$4\delta^2\left(1 - \frac{\pi^2}{16}\right)$
$L = 3$	$\sqrt{\frac{2\alpha^3}{\pi\delta^5}}K_{-1/2}\left(\frac{\alpha}{\delta}\right)$	2δ	$2\delta^2$
$L = 4$	$\frac{\alpha^2}{2\delta^3}K_{-1}\left(\frac{\alpha}{\delta}\right)$	$\frac{3\pi}{4}\delta$	$8\delta^2\left(1 - \frac{18\pi^2}{256}\right)$

$\delta = \sigma_b\sigma_f$.

Source: Reprinted from Griffin and Durgin [2], copyright 2008 IEEE.

envelope PDF. Therefore, an estimate of $f_{\mathbf{A}}(\alpha, \rho)$ for different $M \times L \times N$ backscatter channels has been determined using numerical techniques. Plots and discussion of these PDFs are presented in Section 18.4.4.

18.4 DIVERSITY GAINS IN THE DYADIC BACKSCATTER CHANNEL

18.4.1 The Benefits of Multiple RF-Tag Antennas

Figure 18.3 shows plots of Eqs. (18.7) and (18.10) along with the PDF of a conventional, one-way, Rayleigh-fading channel. Each PDF has been normalized to unit power; that is, $E\{\mathbf{A}^2\} = 1$ where $E\{\cdot\}$ denotes the ensemble average.

In Figure 18.3, it can be seen that the PDF of the dyadic backscatter channel has deeper fades than that of the one-way Rayleigh channel, but improves as RF-tag antennas are added. This is evident from the fact that the area under the PDF curve for small envelope values (e.g., between zero and 0.5) decreases as RF tags are added. The area under the curve gives the probability that the received signal envelope will fall between the envelope values of interest (e.g., between 0 and 0.5). The most significant change is seen for the fully correlated channel, $\rho_e = 1$, where the PDF changes from an exponential distribution for the $1 \times 1 \times 1$ channel to a product-Rayleigh distribution for the $1 \times L \times 1$ channel ($L > 1$). The PDF changes not only as RF-tag antennas are added, but also with ρ_e, as shown in Figures 18.4a and 18.4b. These figures have the same normalization as Figure 18.3 and, as ρ_e is reduced, the PDFs improve for both the $1 \times 1 \times 1$ and $1 \times 2 \times 1$ channels. Note that the PDFs plotted in Figure 18.4a and 18.4b for $0 < \rho_e < 1$ are approximate. See Appendix B for details.

Improvements as RF-tag antennas are added can also be seen in Figure 18.5, which plots simulated, average bit-error-rate (BER) curves for various dyadic backscatter channels with independent, Rayleigh-fading forward and backscatter links. The

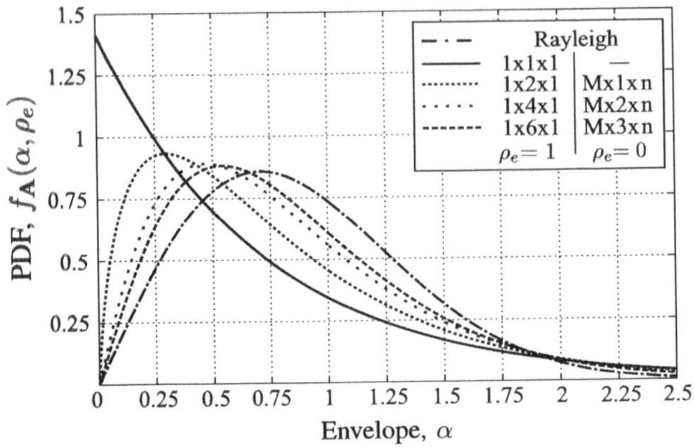

FIGURE 18.3. Plots of the envelope PDF at the nth receiver antenna, with $\rho_e = 0$ and $\rho_e = 1$, for several dyadic backscatter channels along with the PDF of a conventional, one-way Rayleigh-fading channel. Each PDF has been normalized to unit power; that is, $E\{\mathbf{A}^2\} = 1$ where $\{E\cdot\}$ denotes the ensemble average. Note that the abscissa of a PDF can be greater than one as long as the PDF integrates to one. (Reprinted from Griffin and Durgin [2], copyright 2008 IEEE.)

simulations use uncoded, coherent, binary phase shift keying (BPSK) modulation with noise and interference that is additive, white, and Gaussian. In these Monte Carlo simulations, 6.9×10^7 channel realizations were used to approximate the ensemble average of the BER, and the simulated random channel matrix was normalized by the number of transmit antennas, M, for constant transmit power.

As RF-tag antennas are added, the slopes of the BER curves increase and the curves are shifted in a manner similar to that caused by conventional diversity and coding techniques. For a BER of 10^{-4}, there is a 10-dB gain from the $1 \times 1 \times 1$ channel to the $1 \times 2 \times 1$ channel, with slightly larger gains for the $L = 3$ and $L = 4$ cases. Like the PDFs, the average BER curves are also a function of ρ_e, shown in Figure 18.6. For each dyadic backscatter channel (i.e., the $1 \times 1 \times 1$, $1 \times 2 \times 1$, and $1 \times 3 \times 1$ channels), the average BER improves as ρ_e is lowered; however, increasing the number of RF-tag antennas leads to stronger improvement.

In summary, this section has shown that the PDF and the average BER of the dyadic backscatter channel are affected by both the number of RF-tag antennas and the level of link envelope correlation. An explanation of each will be given in the following sections.

18.4.2 Pinhole Diversity Gains

The first source of PDF improvements in Figure 18.3 and the communication gains in Figures 18.5 and 18.6 is a *pinhole diversity* gain caused by the use of multiple RF-tag antennas. The term pinhole diversity is derived from the fact that each RF-tag antenna corresponds to a pinhole in the channel that provides a set of

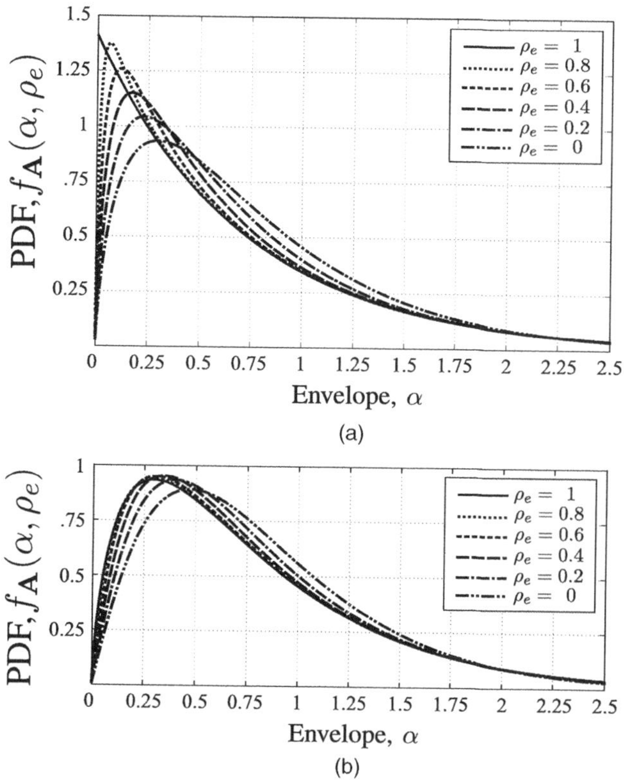

FIGURE 18.4. The PDF of the signal received at the nth reader receiver antenna for (**a**) the $1 \times 1 \times 1$ and (**b**) $1 \times 2 \times 1$ channels with different ρ_e values. The plotted PDFs are exact for $\rho = 0$ and $\rho = 1$, but only approximate for $0 < \rho < 1$. The random variables that correspond with these PDFs have been normalized to unit power (i.e., $E\{\mathbf{A}^2\} = 1$ where $E\{\cdot\}$ denotes the ensemble average). (Reprinted from Griffin and Durgin [3], copyright 2007 IEEE.)

spatially-separated propagation paths, or pinhole diversity branches. Equation (18.6) shows that, for an RF tag with L antennas, L diversity branches are formed by the independent, Gaussian product terms of the pinhole dyadic backscatter channel. As L increases, the probability that the received signal envelope will fade is reduced and, as L becomes very large, the derived channel distribution becomes Gaussian with a Rayleigh distributed envelope, as expected.

Since pinhole diversity does not require diversity branch combining at the reader, it is often assumed that it is equivalent to conventional diversity combining without channel knowledge in a one-way channel—that is, the noncoherent addition of diversity branches received through a one-way channel. However, pinhole diversity differs from both noncoherent diversity combining and conventional coherent diversity combining in two important aspects: Unlike noncoherent diversity combining, pinhole diversity causes the shape of the PDF to change and, unlike conventional

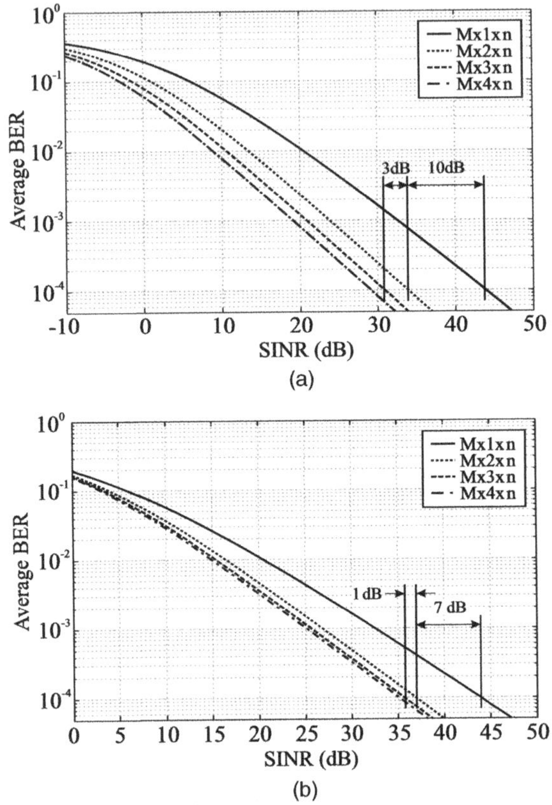

FIGURE 18.5. Average BER plots for backscatter radio systems operating in various dyadic backscatter channels with independent, Rayleigh-fading forward and backscatter links, uncoded BPSK modulation, and noise and interference that is additive, white, and Gaussian. Each curve represents the average BER of the signal received at the nth reader receiver antenna with no diversity combining. Each BER curve is plotted against the signal-to-noise plus interference ratio (SINR) at the nth reader receiver antenna in the $1 \times 1 \times 1$ channel. In (**a**), the simulated random channel matrix is normalized by M for constant transmit power, and the evident BER improvements are caused by both pinhole diversity gains and a larger tag scattering aperture as RF-tag antennas are added. In (**b**), the simulated random channel matrix is normalized by ML to show the BER improvements caused solely by pinhole diversity gains. (Reprinted from Griffin and Durgin [2], copyright 2008 IEEE.)

diversity combining, pinhole diversity requires no change to the reader hardware or signaling scheme. These are discussed in the following sections.

18.4.2.1 PDF Shape Change: Noncoherent diversity branch combining in a one-way Rayleigh fading channel only increases the power of the received signal. Pinhole diversity gains, on the other hand, are caused by the summation of terms that have a product-Rayleigh distribution, resulting in a favorable change in the shape of

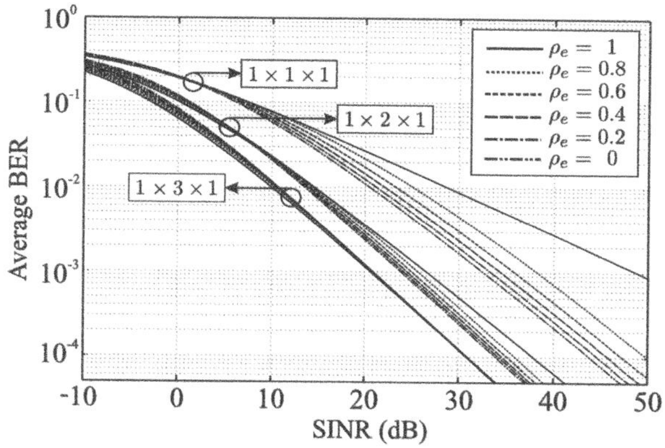

FIGURE 18.6. Average BER plots for the $1 \times 1 \times 1$, $1 \times 2 \times 1$, and $1 \times 3 \times 1$ dyadic backscatter channels for several values of link envelope correlation, ρ_e. In these Monte Carlo simulations, the following assumptions were made: Rayleigh-fading forward and backscatter links, uncoded BPSK modulation, and noise and interference that is additive, white, and Gaussian. Each curve represents the average BER of the signal received at the nth reader–receiver antenna with no diversity combining. Each BER curve is plotted against the SINR at the nth reader–receiver antenna in the $1 \times 1 \times 1$ channel. (Reprinted from Griffin and Durgin [3], copyright 2007 IEEE.)

the channel PDF. This can be seen in Figure 18.3 where each PDF has been normalized to unit power.* In fact, the number of pinholes available in the channel determines the shape of the PDF. Analysis of Eq. (18.6) shows that, when normalized to equal power, the $1 \times L \times 1$ channel with independent, uncorrelated, Rayleigh-fading links has the same PDF as the $1 \times 2L \times 1$ channel with fully correlated, Rayleigh-fading links.

18.4.2.2 Reader Design Simplification: In a one-way channel, gain combining, switch combining, or gain combining in conjunction with space–time block codes must be used to affect a favorable change in the PDF shape. Pinhole diversity gains, on the other hand, can be realized in the dyadic backscatter channel using *only* multiple RF-tag antennas to modulate backscatter. No change in the reader receiver hardware, reader transmitter hardware, or signaling scheme is required.

It should be noted that as Rayleigh products are summed, the power of the channel distribution does increase and can be attributed to an increase in the RF tag effective scattering aperture as antennas are added. This increase in scattered power can itself result in improved BER performance [10]. The BER plot shown in Figure 18.5a reflects both this increase in effective scattering aperture and improved PDF shape caused by pinhole diversity. To see the BER improvement caused solely by pinhole diversity, the simulated random channel matrix has been normalized by ML, in

*To judge whether or not the shape of the PDF has changed, comparisons must be made between distributions with equal power. Adding antennas to the tag will increase the overall power scattered back to the reader; therefore, the distributions must be normalized for comparison.

Figure 18.5b, so that the channel power is held constant with respect to both M and L. Figure 18.5b shows that pinhole diversity alone causes a 7-dB and 8-dB gain for the $1 \times 2 \times 1$ and $1 \times 3 \times 1$ channels, respectively, compared to the $1 \times 1 \times 1$ channel. Actual communication gains in a backscatter radio system are due to both pinhole diversity gains and increased effective scattering aperture; therefore, all performance comparisons should be based on Figure 18.5a.

18.4.3 Link Envelope Correlation

A second source of improvement evident in Figures 18.3, 18.5, and 18.6 is the reduction of link envelope correlation ρ_e. In a conventional, one-way channel, spatial fading correlation will hinder communication and limit available diversity gains. In a pinhole channel, such as the dyadic backscatter channel, ρ_e can have the same effect. Previous work on realistic pinhole channels has focused on situations in which the two links of the pinhole channel are likely dissimilar (e.g., outdoor propagation [7] or amplify-and-forward channels [8]), justifying the assumption of independent links. In many backscatter radio systems, however, reader transmitter and receiver antennas may be closely spaced or even collocated, causing $\rho_e \approx 1$. A high degree of link envelope correlation will occur when the dominant mechanism of wave propagation (i.e., LOS or NLOS) and the angles of arrival/departure at the reader are similar. Since a high level of link envelope correlation implies that the propagation environment of the forward and backscatter links are similar, fully correlated links can *only* occur when the reader transmitter and receiver antennas are collocated and have the same antenna patterns. If the antennas are spatially separated and/or the antenna patterns are different, ρ_e will be reduced. This fact allows the designer some control over the level of link envelope correlation. The separation distance and pattern required to reduce ρ_e to an acceptable level will vary, depending on the channel.

In the analysis presented in Section 18.3, link envelope correlation can occur only between propagation paths that terminate or originate on the same RF-tag antenna, and it is assumed that ρ_e is equal for each set of paths (see Figure 18.1b). In an actual dyadic backscatter channel, antenna coupling, close spacing of RF-tag antennas, and the scattering environment will likely cause unequal levels of correlation between *all* propagation paths.

18.4.4 Limits on Pinhole Diversity Gains

In Section 18.4.2, it was shown that a pinhole diversity gain is available in dyadic backscatter channels with Rayleigh-fading links; however, as multipath fading in the links of the channel decreases, so will pinhole diversity gains. To determine the maximum K factor at which pinhole diversity gains are possible, PDFs of the $M \times L \times N$, dyadic backscatter channel with independent, Rician-fading links are plotted in Figures 18.7 and 18.8 for $K = 0$ dB, $K = 3$ dB, and $K = 6$ dB. In Figure 18.7, it can be seen that, for K less than 3 dB, the probability of receiving a small envelope is greater in the $M \times 1 \times n$ channel than in the $M \times 2 \times n$, $M \times 3 \times n$, or $M \times 4 \times n$ channels. However, for K greater than approximately 3 dB, Figure 18.8 shows that

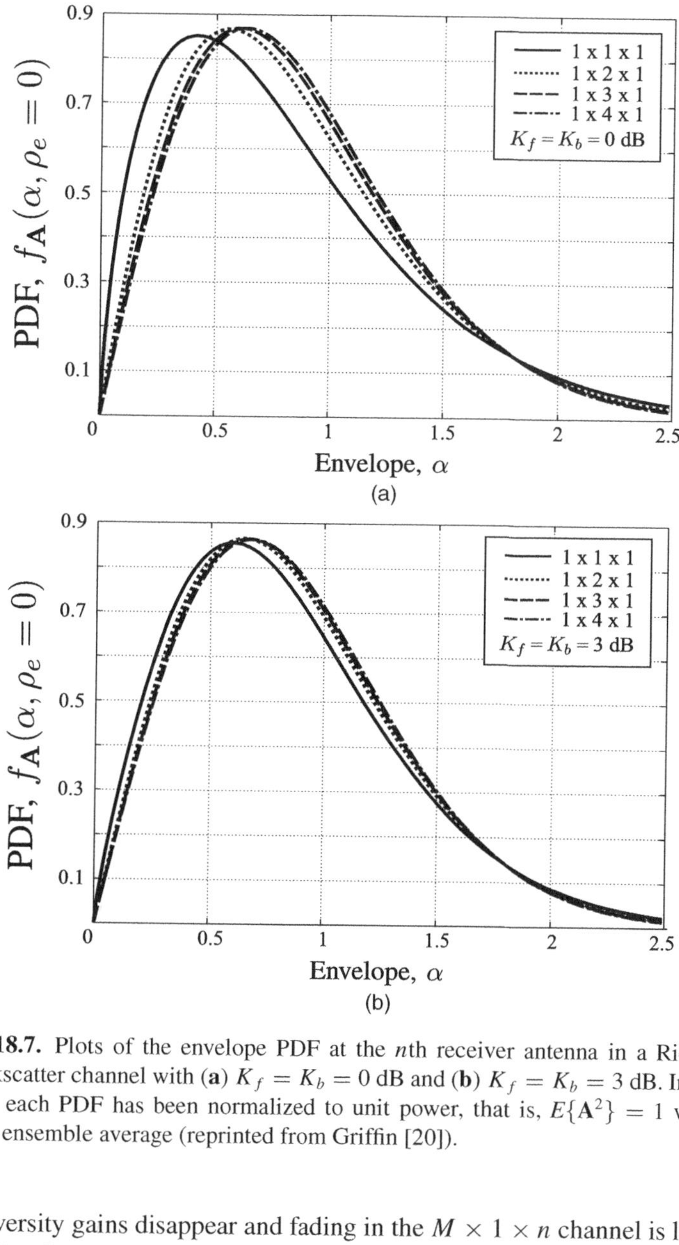

FIGURE 18.7. Plots of the envelope PDF at the nth receiver antenna in a Rician-fading dyadic backscatter channel with (**a**) $K_f = K_b = 0$ dB and (**b**) $K_f = K_b = 3$ dB. In each plot, $\rho_e = 0$ and each PDF has been normalized to unit power, that is, $E\{\mathbf{A}^2\} = 1$ where $E\{\cdot\}$ denotes the ensemble average (reprinted from Griffin [20]).

pinhole diversity gains disappear and fading in the $M \times 1 \times n$ channel is less severe than the $M \times L \times n$ channel (for $L > 1$). According to the channel model used in this chapter, the reason for this increased fading is that multiple RF-tag antennas force the strong specular signal from the reader to travel over multiple propagation paths. In other words, in a Rician backscatter channel with a large K factor, multiple RF-tag antennas create more multipath signals. Keep in mind, however, that the channel model assumes that the signals scattered from each RF-tag antenna are

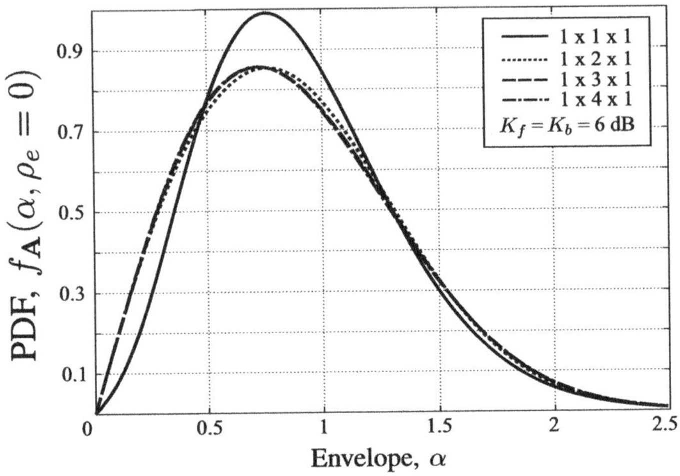

FIGURE 18.8. Plots of the envelope PDF at the nth receiver antenna in a Rician-fading dyadic backscatter channel with $K_f = K_b = 6$ dB. In this plot, $\rho_e = 0$ and each PDF has been normalized to unit power; that is, $E\{\mathbf{A}^2\} = 1$ where $E\{\cdot\}$ denotes the ensemble average (reprinted from Griffin [20]).

statistically independent. In a real backscatter channel, the signals from each RF-tag antenna will be correlated and the level of correlation will be determined by the propagation characteristics of the channel [20]. In realistic channels with high K factors, correlation may cause PDFs for $L > 1$ to approximate the PDF for $L = 1$ and reduce pinhole diversity gains. Therefore, $K = 3$ dB should be used as a rule of thumb for the upper limit to the K factor at which pinhole diversity gains can be realized and it should be noted that pinhole gains may cease for K factors less than 3 dB.

18.4.5 Conventional Diversity Gains

As discussed previously, pinhole diversity causes the shape of the channel envelope PDF to change favorably and contributes to BER improvements; however, if conventional diversity combining techniques are used at the reader, even greater gains are available. Figure 18.9 shows average BER curves for backscatter radio systems using maximal ratio combining (MRC), the optimum diversity combining technique for a fading channel, at the reader receiver. No diversity combining is performed at the RF tag. In these Monte Carlo simulations, 6.9×10^7 channel realizations were used to approximate the ensemble average of the BER, and the receiver had perfect knowledge of the independent, Rayleigh-fading forward and backscatter links.

Comparison of the BER plots for the $M \times L \times 1$ channel in Figure 18.5b and Figure 18.9 shows that MRC offers no further improvement over that caused by pinhole diversity gains (and increased scattering aperture), since the $M \times L \times 1$ channel offers only a single diversity branch to the MRC combiner at the reader receiver. However, when $N > 1$, MRC gains and pinhole diversity gains combine

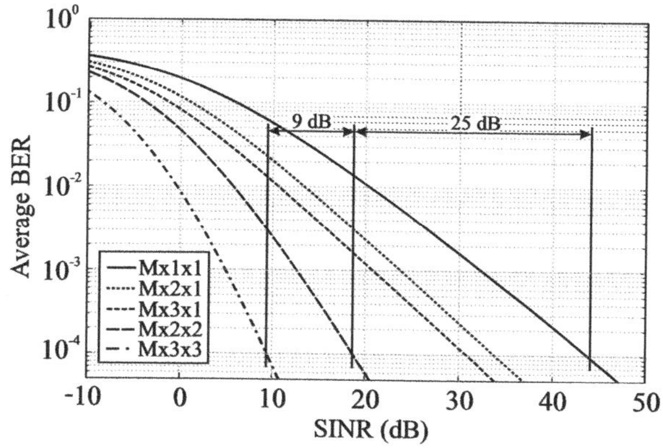

FIGURE 18.9. Average BER plots for backscatter radio systems operating in various dyadic backscatter channels with independent, Rayleigh-fading forward and backscatter links, uncoded BPSK modulation, and noise and interference that is additive, white, and Gaussian. Each curve represents the average BER of the received signal using MRC with perfect channel knowledge at the reader receiver. The random channel matrix used in this simulation has been normalized by M for constant total transmit power. Each BER curve is plotted against the SINR at the nth reader receiver antenna in the $1 \times 1 \times 1$ channel. (Reprinted from Griffin and Durgin [2], copyright 2008 IEEE.)

for significant communication performance improvement. At a BER of 10^{-4}, the $M \times 2 \times 2$ and $M \times 3 \times 3$ channels show up to a 25-dB and 34-dB gain relative to the $1 \times 1 \times 1$ channel, respectively. These gains reflect both a power gain caused by the increased RF tag scattering aperture and an improved PDF shape caused by pinhole diversity and maximal ratio combining. It should be noted that MRC diversity gains will likely decrease as multipath in the channel is decreased or, in other words, as the K factor of the channel increases.

18.4.6 Reliability and Range Improvement

In this chapter, it has been postulated that pinhole diversity gains will lead to improvements in backscatter radio range and reliability; this section will show these improvements through two examples.

18.4.6.1 Reliability Improvement: One way to show a reliability improvement is to use the BER versus SINR plots shown in Figures 18.5a and 18.9. Suppose that an RF tag is at a fixed distance from the reader such that there is adequate power to operate a passive RF tag (no such limitation is required for an semi-passive tag). If the channel worsens by the introduction of interfering signals or increased fading from additional scatterers, the SINR at the reader may drop below the threshold required for successful detection of the backscattered signal. This may happen even though the passive RF tag is still powered. As Figure 18.5a shows, an RF tag using multiple antennas will have a lower SINR threshold for a given BER. For a BER of 10^{-4}, the threshold is 10 dB lower for an RF tag using two antennas to modulate backscatter than for an RF tag using one antenna, a significant reliability increase. Figure 18.9 shows that an even greater SINR improvement – 25 dB – will be realized using MRC at the reader receiver.

18.4.6.2 Range Improvement: The following linear-scale, backscatter link budget governs the amount of modulated backscatter power P_R that is received by the reader receiver

$$P_R = \frac{P_T G_R G_T G_t^2 \lambda^4}{(4\pi)^4 r_f^2 r_b^2} \tag{18.13}$$

where

P_T is the power transmitted by the reader (W)
G_T is the load-matched, free-space gain of the TX antenna
G_R is the load-matched, free-space gain of the RX antenna
G_t is the load-matched, free-space gain of the tag antenna
λ is the carrier frequency wavelength (m)
r_f is the forward link separation distance (m)
r_b is the backscatter link separation distance (m)

This link resembles the bistatic radar equation and is simplified from that presented in Chapter 17 and [21]. In this example, it is assumed that $r = r_f = r_b$. Suppose that all parameters of the channel are held constant and the distance of the RF tag from the reader is allowed to vary. As the total transmitter-to-tag-to-receiver distance r increases, Eq. (18.13) dictates that the SINR will decrease and, at some point, will fall below the threshold required to maintain the target BER. Using multiple RF-tag antennas lowers this threshold allowing r to increase further without exceeding the desired BER. For a backscatter radio system using two RF-tag antennas and a BER of 10^{-4}, a 10-dB gain is available resulting in up to a 78% increase in the range r of the RF tag. This result was calculated using Eq. (18.13).

18.4.7 Discussion

The 10-dB SINR gain used in the previous examples may be reduced in an actual $1 \times 2 \times 1$ channel since the target BER may be greater than 10^{-4}, indoor path loss may be larger or smaller than free-space path loss [assumed in Eq. (18.13)], and the statistics of interfering signals may not be white and Gaussian (assumed in Figure 18.5a). Even so, since the source of this gain (i.e., the pinhole diversity gain coupled with an increased RF tag scattering aperture) only requires uncorrelated signal envelopes at the backscatter radio system antennas, multiple RF-tag antennas will provide gains regardless of assumptions about path loss and noise-plus-interference statistics.

It should be noted that, for passive RF tags, the range increase caused by multiple RF-tag antennas will be limited by the RF-tag chip sensitivity—that is, the amount of power required to operate the tag circuitry [22]. Such a limit does not apply to semi-passive tags, since the power to operate the tag circuitry is drawn from an onboard power source. As the power efficiency of RF tags improve, which is strongly depends on the design of the tag's RF circuitry [23], the limitation posed by the chip sensitivity will decrease.

The analysis of this chapter has assumed that the envelope of the received signal at each transmitter, receiver, and RF-tag antenna is uncorrelated; however, as noted in Section 18.4.4, envelope correlation will exist in any real antenna array as a function of the element spacing, electromagnetic coupling between array elements, and the angle spectrum of the multipath waves [9]. In a rich scattering environment, such correlation can be reduced at the reader by separating array elements by at least $\lambda/2$ [24]; however, form factor constraints on the RF tag (or reader) may require smaller antenna spacing resulting in higher envelope correlation and reduced pinhole diversity gains. At the worst, RF-tag antenna envelope correlation will reduce the $M \times L \times N$, dyadic backscatter channel to an effective $M \times 1 \times N$ channel with added power due to the increased RF-tag scattering aperture. Fortunately, research shows that diversity antennas with less than $\lambda/2$ spacing can still maintain low envelope correlation [25, 26]. Even in the case where pinhole diversity fails, MRC combining at the reader can be used to increase the RF tag range and reliability.

18.5 CONCLUSION

The $M \times L \times N$, dyadic backscatter channel is a pinhole channel with deeper small-scale fades than a conventional one-way channel. Pinhole diversity can mitigate this fading by changing the shape of the fading distribution, resulting in reduced fading and improved BER. Pinhole diversity gains are predicted for channels with Rician-fading links for K less than approximately 3 dB. Examples have demonstrated that pinhole diversity gains can lead to increased backscatter radio communication reliability and up to a 78% range increase. These gains require no channel knowledge or diversity combining at the reader, only the modulation of backscatter using multiple RF-tag antennas and separate, adequately-spaced reader transmitter and receiver antennas. Even greater range and reliability gains are available if MRC is used at the reader receiver.

APPENDIX A: IMPORTANT SMALL-SCALE FADING DISTRIBUTIONS

To benefit from the material presented in this chapter, a firm understanding of several probability density functions (PDF) is necessary. This appendix provides definitions and parameters for the Rician, Rayleigh, and product-Rayleigh PDFs based on work by Simon [27].

A.1 The Rician Distribution

The Rician distribution [28] is one of the most important fading distributions used in channel analysis. It is often used to analyze the fading that occurs on the envelope of the signal received through a transmitter-to-receiver channel whose propagation can be described by the sum of a strong, specular wave and many nonspecular, multipath waves. Such a scenario often occurs when a LOS exists between the transmitter and receiver. The nonspecular, multipath waves can be modeled as a zero-mean, complex-Gaussian random variable, and the presence of the unfading specular wave simply adds a nonzero mean. Therefore, the Rician random variable **A** is equal to the envelope of a complex-Gaussian random variable with nonzero mean,

$$\mathbf{A} = ||\mathbf{Z}|| = ||\mathbf{X} + j\mathbf{Y}|| \tag{18.14}$$

where $\mathbf{X} \sim \mathcal{N}(\mu_x, \sigma^2/2)$, $\mathbf{Y} \sim \mathcal{N}(\mu_y, \sigma^2/2)$; σ^2 is the variance of $\mathbf{A}$; $\mu_{x,y}$ are the means of and $\mathbf{X}$ and $\mathbf{Y}$; and $\mu_x = \mu_y \neq 0$. In this appendix, $\mathcal{N}(\mu, \sigma)$ denotes a normal distribution with mean μ and variance σ^2. A derivation of the distribution is given by Durgin [9, page 126], and the result is presented below. From Simon [27], the PDF of a Rician random variable is

$$f_{\mathbf{A}}(\alpha) = \frac{2\alpha}{\sigma^2} \exp\left(-\frac{\alpha^2 + a^2}{\sigma^2}\right) I_0\left(\frac{2\alpha a}{\sigma^2}\right), \qquad \alpha \geq 0 \tag{18.15}$$

where $a = \sqrt{\mu_x^2 + \mu_y^2}$ and $I_0(\cdot)$ is a modified Bessel function of the first kind. The average power of the distribution is

$$P_{\text{av}} = \sigma^2 \left(1 + \frac{a^2}{\sigma^2}\right). \tag{18.16}$$

It is common to define K, the Rician K factor, as ratio of the non-fading specular power and the nonspecular, multipath power. Using this definition, the average power can be written

$$P_{\text{av}} = \sigma^2(1 + K). \tag{18.17}$$

The cumulative density function (CDF) of a Rician random variable is

$$F_{\mathbf{A}}(\alpha) = 1 - Q\left(\frac{\sqrt{2}a}{\sigma}, \frac{\sqrt{2}\alpha}{\sigma}\right), \qquad \alpha \geq 0 \tag{18.18}$$

where $Q(\cdot, \cdot)$ is the first-order Marcum Q-function [27].

A.2 The Rayleigh Distribution

The Rayleigh distribution [29] is a special case of the Rician distribution in which no specular wave is present. In mathematical terms, a Rayleigh random variable is equal to the envelope of a zero-mean, complex-Gaussian random variable.

$$\mathbf{A} = ||\mathbf{Z}|| = ||\mathbf{X} + j\mathbf{Y}|| \tag{18.19}$$

where $\mathbf{X} \sim \mathcal{N}(0, \sigma^2/2)$ and $\mathbf{Y} \sim \mathcal{N}(0, \sigma^2/2)$. Hence, the Rician specular term $a^2 = 0$ and Eq. (18.15) reduces to the the PDF of a Rayleigh random variable

$$f_{\mathbf{A}}(\alpha) = \frac{2\alpha}{\sigma^2} \exp\left(-\frac{\alpha^2}{\sigma^2}\right), \qquad \alpha \geq 0. \tag{18.20}$$

A derivation of the Rayleigh distribution is given by Durgin [9, pages 116–118]. The average power of this distribution is

$$P_{\text{av}} = \sigma^2. \tag{18.21}$$

The CDF of a Rayleigh random variable is [27]

$$f_{\mathbf{A}}(\alpha) = 1 - \exp\left(-\frac{\alpha^2}{\sigma^2}\right), \qquad \alpha \geq 0. \tag{18.22}$$

A.3 The Product-Rayleigh Distribution

The product-Rayleigh distribution is the distribution resulting from the multiplication of two random variables with Rayleigh distributions. Therefore, for this distribution, **A** is the envelope of the product of two complex-Gaussian random variables

$$\mathbf{A} = ||\mathbf{BC}|| = ||(\mathbf{X} + j\mathbf{Y})(\mathbf{W} + j\mathbf{V})|| \tag{18.23}$$

where $\mathbf{B} = \mathbf{X} + j\mathbf{Y}$ and $\mathbf{C} = \mathbf{W} + j\mathbf{V}$; $\mathbf{X}, \mathbf{Y} \sim \mathcal{N}(0, \sigma_B^2/2)$ and $\mathbf{W}, \mathbf{V} \sim \mathcal{N}(0, \sigma_C^2/2)$; and σ_B and σ_C are the variances of **B** and **C**, respectively. The PDF of **A** is given by Simon* [27] as

$$f_{\mathbf{A}}(\alpha, \rho) = \frac{4\alpha(1 - |\rho|^2)}{\sigma_A^2\sigma_C^2(1 - \rho^2)^2} I_0\left(\frac{2\alpha\,|\rho|}{\sigma_A\sigma_C(1 - \rho^2)}\right) K_0\left(\frac{2\alpha}{\sigma_A\sigma_C(1 - \rho^2)}\right) \tag{18.24}$$

for $\alpha \geq 0$. In Eq. (18.24), I_0 is a zero-order, modified Bessel function of the first kind and K_0 is a zero-order, modified Bessel function of the second kind. The correlation between **B** and **C** is indicated by ρ, the normalized correlation coefficient, and is defined as [27]

$$\begin{aligned} \rho &= \frac{\text{Cov}(\mathbf{X}, \mathbf{W})}{\sqrt{\text{Cov}(\mathbf{X}, \mathbf{X})\text{Cov}(\mathbf{W}, \mathbf{W})}} = \frac{2\text{Cov}(\mathbf{X}, \mathbf{W})}{\sigma_A\sigma_C} \\ &= \frac{\text{Cov}(\mathbf{Y}, \mathbf{V})}{\sqrt{\text{Cov}(\mathbf{Y}, \mathbf{Y})\text{Cov}(\mathbf{V}, \mathbf{V})}} = \frac{2\text{Cov}(\mathbf{Y}, \mathbf{V})}{\sigma_A\sigma_C} \end{aligned} \tag{18.25}$$

*Equation (18.24) differs from that given by Simon [27] in that it has been normalized to satisfy $\int_0^\infty f_{\mathbf{A}}(\alpha, \rho)\, d\alpha = 1$.

where $-1 \leq \rho \leq 1$ and $\text{Cov}(\cdot, \cdot)$ is the covariance operator. It is assumed that

$$\frac{\text{Cov}(\mathbf{X}, \mathbf{Y})}{\sqrt{\text{Cov}(\mathbf{X}, \mathbf{X})\text{Cov}(\mathbf{Y}, \mathbf{Y})}} = \frac{\text{Cov}(\mathbf{W}, \mathbf{V})}{\sqrt{\text{Cov}(\mathbf{W}, \mathbf{W})\text{Cov}(\mathbf{V}, \mathbf{V})}} = 0. \tag{18.26}$$

In other words, the real parts of **B** and **C** have correlation ρ; likewise, the imaginary parts of **B** and **C** have correlation ρ. The correlation between the real and imaginary parts of **B** and **C** is zero. Since **B** and **C** are complex-Gaussian random variables, it is implied that the real and imaginary parts of **B** and **C** are independent. This correlation is defined as link correlation in Section 18.3.

A.4 The Product-Rician PDF

The PDF of the product of two independent Rician random variables is [27]

$$\begin{aligned} f_{\mathbf{A}}(\alpha) = {} & \frac{4}{\sigma_f^2 \sigma_b^2} \exp[-K_b - K_f] \\ & \times \sum_{i=0}^{\infty} \sum_{l=0}^{\infty} \frac{1}{(i!)^2 (l!)^2} \left(\frac{K_b}{\sigma_b^2}\right)^i \left(\frac{K_f}{\sigma_f^2}\right)^l \left(\frac{\sigma_b}{\sigma_f}\right)^{i-l} \\ & \times \alpha^{i+l+1} K_{(i-l)} \left(\frac{2\alpha}{\sigma_b \sigma_f}\right) \end{aligned} \tag{18.27}$$

where **A** is the random channel envelope; α is the index of the distribution; K_f and K_b are the Rician K factors of the forward and backscatter links, respectively; σ_f^2 and σ_b^2 are the variances of the forward and backscatter links, respectively; and $K_\nu(\cdot)$ is a modified Bessel function of the second kind with order ν. The power of the product of two independent Rician random variables is

$$\mathbf{E}\{\mathbf{A}^2\} = \sigma_b^2 \sigma_f^2 \exp[-(K_b + K_f)]\, {}_1F_1(2; 1; K_b)\, {}_1F_1(2; 1; K_b) \tag{18.28}$$

$$= \sigma_b^2 \sigma_f^2 (1 + K_b)(1 + K_f) \tag{18.29}$$

where ${}_1F_1(a; b; z)$ is a confluent, hypergeometric function of the first kind [30]. The average power received at the nth reader-receiver antenna through the $M \times L \times N$ backscatter channel is

$$\mathbf{E}\{\mathbf{A}^2\} = ML\sigma_b^2 \sigma_f^2 (1 + K_b)(1 + K_f). \tag{18.30}$$

APPENDIX B: THE $M \times L \times N$ BACKSCATTER CHANNEL PDF DERIVATION

This appendix derives the distribution of the signal received at the nth reader–receiver antenna through the $M \times L \times N$, dyadic backscatter channel with Rayleigh-fading links. As discussed previously, the signal received at the nth reader–receiver antenna is proportional to the sum of L, i.i.d., complex-Gaussian products. Therefore, the envelope of the signal is

$$\mathbf{A} = |h^b| \times |h^f| \tag{18.31}$$

where h^b and h^f are correlated, zero-mean, complex-Gaussian random variables with variances σ_b^2 and $M\sigma_f^2$, respectively. The PDF of the sum of L, i.i.d. random variables can be found from the product of their characteristic functions (CF). Therefore, this derivation proceeds by finding the CF of Eq. (18.24), raising it to the Lth power, and transforming the resulting CF back into a PDF.

The CF of Eq. (18.24) is found using the Hankel transform [9]

$$\Phi(\nu) = \int_0^\infty f_{\mathbf{A}}(\alpha) J_0(\nu\alpha)\, d\alpha. \tag{18.32}$$

Substituting Eq. (18.24) into Eq. (18.32), solving, and then raising the result to the Lth power yields

$$\Phi(\nu;\rho) = \left[\frac{\sigma_b^4\sigma_f^4 M^2}{16}\frac{(1-\rho^2)^4}{(1-|\rho|^2)^2}\left(\nu^2 + \frac{4(|\rho|-1)^2}{\sigma_b^2\sigma_f^2 M(1-\rho^2)^2}\right)\right.$$
$$\left.\times\left(\nu^2 + \frac{4(|\rho|+1)^2}{\sigma_b^2\sigma_f^2 M(1-\rho^2)^2}\right)\right]^{-L/2}. \tag{18.33}$$

Equation (18.33) is the exact CF of the general, $M \times L \times N$, dyadic backscatter channel for $\rho = 0$ and $\rho = 1$. The PDF of Eq. (18.33) is found using the inverse Hankel transform [9]

$$f_{\mathbf{A}}(\alpha) = \alpha \int_0^\infty \Phi(\nu) J_0(\nu\alpha)\nu\, d\nu. \tag{18.34}$$

It should be noted that, while Eq. 18.33 is exact for $\rho = 0$ and $\rho = 1$, it is only approximate for $0 < \rho < 1$. The reason is that use of the Hankel transform pair (Eqs. 18.32 and 18.34) requires that the phase of the underlying complex-Gaussian channel be uniformly distributed over 2π and statistically independent of the magnitude. This assumption is true for $\rho = 0$ and $\rho = 1$, but it has not been proven for $0 < \rho < 1$.

Therefore, in this chapter, all distributions derived from Eq. 18.33 for $0 < \rho < 1$, while useful for understanding, are considered approximate. The following sections discuss the PDF derivations for $\rho = 0$ and $\rho = 1$.

B.1 Uncorrelated Channels ($\rho = 0$)

In Eq. (18.33), if ρ approaches zero in the limit, the following CF results:

$$\Phi(\nu) = \left(\frac{4}{\nu^2 \sigma_b^2 \sigma_f^2 M + 4} \right)^L . \tag{18.35}$$

Applying the inverse Hankel transform to Eq. (18.35) yields the PDF of the $M \times L \times N$, dyadic backscatter channel with uncorrelated forward and backscatter links

$$f_{\mathbf{A}}(\alpha, 0) = \alpha^L \left(\frac{2}{\sqrt{M} \sigma_b \sigma_f} \right)^{1+L} \frac{2^{1-L}}{\Gamma(L)} K_{(1-L)} \left(\frac{2\alpha}{\sqrt{M} \sigma_b \sigma_f} \right) . \tag{18.36}$$

The mean and variance of Eq. (18.36) were found by direct application of their definitions

$$\mathbf{E}\{\mathbf{A}\} = \frac{\sqrt{M} \sigma_b \sigma_f}{\Gamma(L)} \Gamma\left(\frac{3}{2}\right) \Gamma\left(L + \frac{1}{2}\right) . \tag{18.37}$$

The variance of $\mathbf{A}$ is given by $\mathrm{Cov}(\mathbf{A}, \mathbf{A}) = \mathbf{E}\{\mathbf{A}^2\} - \mathbf{E}\{\mathbf{A}\}^2$, where $\mathrm{Cov}(\cdot, \cdot)$ is the covariance operator. Solving for the variance requires the power (or second moment) of the distribution to be known. The second moment is

$$\mathbf{E}\{\mathbf{A}^2\} = M L \sigma_b^2 \sigma_f^2 . \tag{18.38}$$

Hence, the variance of $\mathbf{A}$ is

$$\mathrm{Cov}(\mathbf{A}, \mathbf{A}) = M L \sigma_b^2 \sigma_f^2 \left(1 - \frac{\Gamma^2\left(\frac{3}{2}\right) \Gamma^2\left(L + \frac{1}{2}\right)}{L \Gamma^2(L)} \right) . \tag{18.39}$$

The mean and variance can be further reduced to Eqs. (18.8) and (18.9), respectively. It can be shown that Eq. (18.36) integrates to 1.

B.2 Fully Correlated Channels ($\rho = 1$)

In Eq. (18.33), as ρ approaches one in the limit, the following CF results:

$$\Phi(\nu) = \left(\frac{1}{\nu^2 \sigma_b^2 \sigma_f^2 M + 1} \right)^{L/2} . \tag{18.40}$$

The PDF is found using the inverse Hankel transform. Solving using the same method as for (18.36), the PDF for the $1 \times L \times 1$, dyadic backscatter channel with fully correlated forward and backscatter links is

$$f_{\mathbf{A}}(\alpha, \rho = 1) = \alpha^{L/2} \left(\frac{1}{\sigma_b \sigma_f \sqrt{M}} \right)^{1+L/2} \frac{2^{1-L/2}}{\Gamma\left(\frac{L}{2}\right)} K_{(1-L/2)} \left(\frac{\alpha}{\sigma_b \sigma_f \sqrt{M}} \right). \quad (18.41)$$

Again, the mean and variance of Eq. (18.41) can be solved by direct application of their definitions. For reference, the power of the PDF is

$$\mathbf{E}\{\mathbf{A}^2\} = 2ML\sigma_b^2\sigma_f^2. \quad (18.42)$$

The mean and variance are, respectively,

$$\mathbf{E}\{\mathbf{A}\} = \frac{2\sqrt{M}\sigma_b\sigma_f}{\Gamma\left(\frac{L}{2}\right)} \Gamma\left(\frac{3}{2}\right) \Gamma\left(\frac{L}{2} + \frac{1}{2}\right) \quad (18.43)$$

and

$$\mathrm{Cov}(\mathbf{A}, \mathbf{A}) = 2ML\sigma_b^2\sigma_f^2 \left(1 - \frac{2\Gamma^2\left(\frac{3}{2}\right)\Gamma^2\left(\frac{L}{2} + \frac{1}{2}\right)}{L\Gamma^2\left(\frac{L}{2}\right)} \right). \quad (18.44)$$

The mean and variance reduce to (18.11) and (18.12), respectively. It can be shown that Eq. (18.41) integrates to 1.

B.3 A Note on Link Envelope Correlation

The channel model for link correlation used in this chapter, which assumes that only propagation paths that originate or terminate on the same RF tag antenna can be correlated and all other propagation paths are statistically independent, is limited. For example, a $2 \times 1 \times 1$ channel cannot have $\rho = 1$ since the propagation paths of the forward link cannot be fully correlated with the backscatter link propagation path and remain statistically independent. Therefore, care must be taken in the choice of ρ so that the correlation matrix is positive semi-definite [3]. The ijth entry of the correlation matrix $\mathbf{V}$ – the correlation matrix of $\tilde{\mathbf{H}}^f$ and $\tilde{\mathbf{H}}^b$ – is

$$V_{ij} = \frac{\mathrm{Cov}(A_i, A_j)}{\sigma_i \sigma_j} \quad (18.45)$$

where A_k is the kth element of

$$\vec{A} = \left[h_{11}^f, h_{21}^f, \ldots, h_{lm}^f, h_{11}^b, h_{21}^b, \ldots, h_{nl}^b\right]^{\mathrm{T}}$$

(a column vector formed from the entries of the forward and backscatter-link matrices), σ_i is the standard deviation of the ith element of $\vec{A}$, and $\mathrm{Cov}(x, y)$ is the covariance between the scalars x and y. The positive semi-definite constraint places a limit on ρ that is a function of M; however, for the $1 \times L \times 1$ discussed in this dissertation, ρ may vary between zero and one ($0 \leq \rho \leq 1$) while satisfying this constraint.

REFERENCES

1. D. Kim, M. A. Ingram, and W. W. Smith, Jr., Measurements of small-scale fading and path loss for long range RF tags, *IEEE Trans. Antennas Propag.*, Vol. 51, No. 8, pp. 1740–1749, 2003.
2. J. D. Griffin and G. D. Durgin, Gains for RF tags using multiple antennas, *IEEE Trans. Antennas Propag.*, Vol. 56, No. 2, pp. 563–570, 2008.
3. J. D. Griffin and G. D. Durgin, Link envelope correlation in the backscatter channel, *IEEE Commun. Lett.*, Vol. 11, No. 9, pp. 735–737, 2007.
4. D. Chizhik, G. J. Foschini, and R. A. Valenzuela, Capacities of multi-element transmit and receive antennas: Correlations and keyholes, *Electron. Lett.*, Vol. 36, No. 13, pp. 1099–1100, 2000.
5. D. Chizhik, G. J. Foschini, M. J. Gans, and R. A. Valenzuela, Keyholes, correlations, and capacities of multielement transmit and receive antennas, *IEEE Trans. Wireless Commun.*, Vol. 1, No. 2, pp. 361–368, 2002.
6. V. Erceg, S. J. Fortune, J. Ling, J. Rustako, A. J., and R. A. Valenzuela, Comparisons of a computer-based propagation prediction tool with experimental data collected in urban microcellular environments, *IEEE J. Selected Areas Commun.*, Vol. 15, No. 4, pp. 677–684, 1997.
7. D. Gesbert, H. Bölcskei, D. A. Gore, and A. J. Paulraj, Outdoor MIMO wireless channels: Models and performance prediction, *IEEE Trans. Commun.*, Vol. 50, No. 12, pp. 1926–1934, 2002.
8. C. S. Patel, G. L. Stüber, and T. G. Pratt, Statistical properties of amplify and forward relay fading channels, *IEEE Trans. Veh. Technol.*, Vol. 55, No. 1, pp. 1–9, 2006.
9. G. D. Durgin, *Space–Time Wireless Channels*, Prentice-Hall, Upper Saddle River, NJ, 2003.
10. M. Ingram, M. Demirkol, and D. Kim, Transmit diversity and spatial multiplexing for RF links using modulated backscatter, in *Proceedings of the International Symposium on Signals, Systems, and Electronics*, Tokyo, Japan, July 2001.
11. M. Mi, M. H. Mickle, C. Capelli, and H. Swift, RF energy harvesting with multiple antennas in the same space, *IEEE Antennas Propag. Mag.*, Vol. 47, No. 5, pp. 100–106, 2005.
12. J. Mitsugi, UHF band RFID readability and fading measurements in practical propagation environment, in *Auto-ID Labs White Paper Series*, 2005, pp. 37–44.
13. J. Mitsugi and Y. Shibao, Multipath identification using steepest gradient method for dynamic inventory in UHF RFID, in *International Symposium on Applications and the Internet*, Hiroshima, Japan, 2007.

14. M. Polivka, M. Svanda, and P. Hudec, Analysis and measurement of the RFID system adapted for identification of moving objects, in *36th European Microwave Conference*, 2006, pp. 729–732.

15. S. R. Banerjee, R. Jesme, and R. A. Sainati, Performance analysis of short range UHF propagation as applicable to passive RFID, in *2007 IEEE International Conference on RFID*, Gaylord Texan Resort, Grapevine, TX, March 2007, pp. 30–36.

16. S. R. Banerjee, R. Jesme, and R. A. Sainati, Investigation of Spatial and Frequency Diversity for Long Range UHF RFID, in *IEEE Antennas and Propagation Society International Symposium*, San Diego, CA, July 2008, pp. 1–4.

17. D. M. Pozar, *Microwave Engineering*, 3rd edition, John Wiley & Sons, Hoboken, NJ, 2005.

18. J. N. Pierce and S. Stein, Multiple diversity with nonindependent fading, *Proc. IRE*, Vol. 48, pp. 89–104, 1960.

19. R. G. Vaughan and J. B. Andersen, Antenna diversity in mobile communications, *IEEE Trans. Veh. Technol.*, Vol. VT-36, No. 4, pp. 149–172, 1987.

20. J. D. Griffin, High-Frequency Modulated-Backscatter Communication Using Multiple Antennas, Ph.D. dissertation, The Georgia Institute of Technology, 2009 [Online]. Available at http://www.propagation.gatech.edu/Archive/PG_TR_090120_JDG/PG_TR_090120_JDG.pdf, last accessed on May 7, 2009.

21. J. D. Griffin and G. D. Durgin, Complete link budgets for backscatter radio and RFID systems, *IEEE Antennas Propag. Mag.*, Vol. 51, No. 2, pp. 11–25, April 2009.

22. P. Nikitin and K. V. S. Rao, Performance limitations of passive UHF RFID systems, in *Proceedings of IEEE Antenna and Propagation Society International Symposium*, Albuquerque, New Mexico, 2006, pp. 1011–1014.

23. G. De Vita and G. Iannaccone, Design criteria for the RF section of UHF and microwave passive RFID transponders, *IEEE Trans. Microwave Theory Tech.*, Vol. 53, No. 9, pp. 2978–2990, 2005.

24. W. C. Jakes (editor), *Microwave Mobile Communications*, IEEE Press, New York, 1974.

25. D. T. Auckland, W. Klimczak, and G. D. Durgin, Maximizing throughput with ultra-compact diversity antennas, in *Proceedings of the IEEE Vehicular Technology Conference 2003—Fall*, Vol. 1, Orlando, FL, 2003, pp. 178–182.

26. Y. Yamada, K. Kagoshima, and K. Tsunekawa, Diversity antennas for base and mobile stations in land mobile communication systems, *IEICE Trans.*, Vol. E74, No. 10, pp. 3202–3209, 1991.

27. M. K. Simon, *Probability Distributions Involving Gaussian Random Variables: A Handbook for Engineers and Scientists*, Kluwer Academic Publishers, Norwell, MA, 2002.

28. S. O. Rice, Statistical properties of a sine wave plus random noise, *Bell System Tech. J.*, Vol. 27, No. 1, pp. 109–257, January 1948.

29. S. O. Rice Mathematical analysis of random noise, *Bell System Tech. J.*, Vol. 23, pp. 282–332, July 1944.

30. E. W. Weisstein, Confluent hypergeometric function of the first kind, from MathWorld—A Wolfram Web Resource. http://mathworld.wolfram.com/ConfluentHypergeometricFunctionoftheFirstKind.html.

PART VI

MIMO ANTENNAS FOR RFID SYSTEMS

CHAPTER 19

OPTIMUM POWER ALLOCATION IN MULTIPLE-INPUT MULTIPLE-OUTPUT (MIMO) SYSTEMS UNDER INDEPENDENT RAYLEIGH FADING

JEFFREY S. FU
Department of Electronic Engineering, Chang Gung University, Taoyuan, Taiwan, Republic of China

WEIXIAN LIU
School of Electrical and Electronic Engineering, Nanyang Technological University, Singapore

NEMAI CHANDRA KARMAKAR
Department of Electrical and Computer Systems Engineering, Monash University, Clayton, Victoria, Australia

19.1 INTRODUCTION

Appearing a few years ago, a series of information theory articles were published by members of Bell Laboratories. Multiple-input multiple-output (MIMO) systems have evolved quickly to both become one of the most popular topics among wireless communication researchers and reach a spot in today's "hottest wireless technology" list. The MIMO systems, with antenna array at both transmitter and receiver sides, are able to increase spectral efficiency greatly using spatial diversity, which can be accomplished in a rich scattering channel environment [1, 2]. The radio-frequency identification (RFID) system experiences such a rich scattering channel environment where multiple tags are to be read in warehouses, retail stores, manufacturing industries, and so on. Recent research has shown that high theoretical capacity is possible since data rates as high as 40 bits/s/Hz have been demonstrated in an indoor slow-fading environment [3].

Handbook of Smart Antennas for RFID Systems, Edited by Nemai Chandra Karmakar

Optimal power allocation plays an important role for a MIMO system to achieve maximum channel capacity [4]. This problem has been solved by the well-known water-filling (WF) algorithm [5, 6], which is used to maximize theoretical channel capacity in an independent Rayleigh fading environment with given total transmission power. However, various instantaneous capacity is only suitable for some services, such as web browsing, e-mail and it is not applicable to delay sensitive services, for example, video conference. In this work, we do power allocation for a given channel capacity using WF algorithm. WF algorithm works well when power allocation is done without any constraint. However, it doesn't give optimal solution if the power allocation is under constraints of bit error rate (BER) and discrete modulation order. Then a modified water-filling algorithm called QoS-based water-filling is discussed in [7].

In this work, we are aimed to study the performance improvement in terms of capacity enhancement and power saving with optimal power allocation algorithms. We do simulations for power allocation for different total transmission power levels and find that the capacity improvement becomes less as the total transmission power gets larger with optimal power allocation comparing to equal power allocation. We also do power allocation to minimize total transmission power for various channel capacity levels using WF algorithm. We realize that the power saving becomes neglectable as the channel capacity reaches 20 bits/s/Hz. In addition, we also do simulations using a QoS-based WF algorithm with various SNRs and realize that data rates are the same as a signal-to-noise ratio (SNR) that exceeds 50 dB.

The chapter is arranged as follows. Section 19.1 has introduced the MIMO system. Section 19.2 presents the MIMO system model, capacity, power equations, and data rate analysis with QoS-based water-filling. Section 19.3 presents the simulation results and analysis of data followed by a conclusion in Section 19.4.

19.2 THEORY

19.2.1 System Model

Figure 19.1 shows a MIMO system. The multiple antenna system has m transmitting antennas and n receiving antennas. This system is referred to as an (m, n) system. In this chapter, we let $m \leq n$, meaning the number of receivers is larger than the number of transmitters. This scenario is justified in an RFID system environment where a reader (transmitter) interrogates multiple RFID tags simultaneously. Also, for the case of multiple readers, the number of reader's antennas does not exceed the number of tags. During timeslot t, for all $1 \leq k \leq m$, the transmitting antenna k transmits a signal $s_t(k)$. The receiving antenna l receives a signal $r_t(l)$, which contains a superposition of transmitted signals corrupted by added white Gaussian noise (AWGN) $v_t(l)$. The channel is assumed to be narrowband and subject to flat fading, and it is represented by an $n \times m$ matrix H_t. The (l, k)th entry of H_t, $H_t\ (l, k)$ is the (complex) gain connecting the input of the channel from the transmitting antenna k to the output of the channel to the receiving antenna l at time t. A vector representation of the

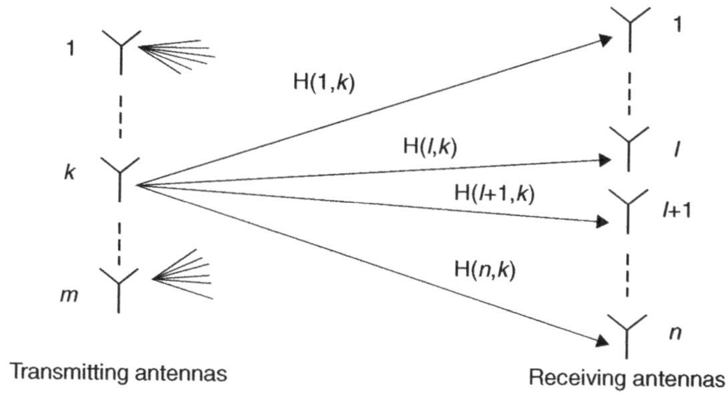

FIGURE 19.1. A MIMO system with m number of transmitting and n number of receiving antenna.

input–output relation for this discrete-time (m, n) multiple-antenna system is

$$r_t = Hs_t + v_t \tag{19.1}$$

For convenience, the time index will be dropped in the subsequent expressions.

A complex Gaussian random variable $z = x + iy$ is called circularly symmetric complex Gaussian with variance σ^2 if x and y are independent and identically distributed (i.i.d) $N(0, \sigma^2/2)$. The n components of noise v are assumed to be i.i.d. circularly symmetric complex Gaussian with variance σ_v^2. In a Rayleigh-fading environment, the individual entries of H are modeled as zero-mean, circularly symmetric complex Gaussian random variables with unit variance.

19.2.2 Capacity

We assume that communication is carried out using bursts (packets). For each burst, the channel is randomly drawn from an underlying distribution and stays fixed for the duration of the entire burst. We also assume that the burst duration is long enough that the information-theoretic idealization of infinitely long code block length is valid. In this quasi-static scenario, it is meaningful to associate a channel capacity with a given realization of channel matrix. Because the channel is random, the capacity is also considered random. Unless specified otherwise, we assume a constraint on the total transmit power, $E(s^+s) \leq P_T$. Furthermore, we are interested in the spectral utilization efficiency, the channel capacity, the transmit power, and the noise power, all of which are expressed on a per-unit-frequency basis.

The transmitted signal from the four transmitters is vector $s = [s_1 \quad s_2 \quad s_3 \quad s_4]^T$. The noise is assumed to be white Gaussian with variance σ^2. We decompose channel matrix, H, using its singular value decomposition $H = UDV^H$, where U and V are

$n \times n$ and $m \times m$ unitary matrices and Equation (19.1) becomes

$$r = UDV^H s + n \tag{19.2}$$

We may, therefore, transform the detection problem without loss of generality, as

$$r' = U^H r = U^H UDV^H s + U^H n = Ds' + n' \tag{19.3}$$

or in another form

$$r'_i = d_{ii} s'_i + n'_i \tag{19.4}$$

In the transform domain, the channel has been decomposed into min(m, n) subchannels [8]. Total capacity is the sum of the subchannels:

$$\begin{aligned} C &= \sum_{i=1}^{m} \log_2 \left[1 + \frac{p_i \lambda_i}{\sigma^2} \right] \\ P_T &= \sum_{i=1}^{m} p_i \end{aligned} \tag{19.5}$$

where p_i is the power allocated to the ith subchannel, σ^2 is the noise power, and λ_i is the ith eigenvalue of HH^{H} and it is the square of d_{ii}. P_T is the total transmission power; m represents min(m, n) in general, and it is the number of transmitter antennas in this work since we let $m \leq n$. For simplicity, we set $m = 4$ and $n = 6$. Using the well-known water-filling algorithm in [9], the power allocated to each subchannel is

$$p_i = \left(k - \frac{\sigma^2}{|d_i|^2} \right)^+ \tag{19.6}$$

where k is chosen such that $\sum_{i=1}^{m} p_i = P_T$ and $(a)^+$ represents max $(0, a)$. When the channel matrix H is unknown at the transmitter, it is normal to allocate equal power to each subchannel. The channel capacity is

$$C = \log_2 \det \left[I_n + \frac{P_T}{\sigma^2 m} HH^+ \right] = \sum_{i=1}^{k} \log_2 \left[1 + \frac{P_T}{\sigma^2 m} \lambda_i \right] \tag{19.7}$$

where rank(H) = k = min(n, m)

19.2.3 Power

Delay-sensitive services such as voice or video usually require a constant data rate. In this work, we consider a fixed channel capacity and try to minimize total transmission power using the water-filling (WF) algorithm. It is different from maximizing channel capacity with a given total transmission power. The power allocated to each subchannel is

$$p_i = \left(k - \frac{\sigma^2}{|d_i|^2}\right)^+ \tag{19.8}$$

where k is chosen such that $\sum_{i=1}^{m} \log_2(1 + p_i) = C$. It is simple for equal power allocation. It is just to find the total transmission power P_T such that

$$\sum_{i=1}^{m} \log_2 \left[I + \frac{P_T}{\sigma^2 n_t} \lambda_i \right] = C \tag{19.9}$$

19.2.4 Data Rate Analysis with QoS-Based Water-Filling

The basic idea of QoS-based WF is to use power efficiently, considering that a fixed discrete modulation order is used over a certain SNR region if there is some QoS constraint. If the modulation and the BER threshold are fixed, there is no need to increase unnecessary power in each subchannel, because extra power can only improve the SNR but not the modulation order and the total bit rate. We can keep the power to the lowest level that just meets the QoS need and therefore obtain some residual power. The residual power can then be re-allocated to other subchannels to improve their possible modulation orders. The round-off water-filling is under the constraints of BER threshold and discrete modulation order. But it just let residual power waste. The details can be found in [7].

The channel models of transmitter, and receiver are shown in Figure 19.2.

19.3 SIMULATION RESULTS AND ANALYSIS

19.3.1 Optimal Power Allocation Versus Equal Power Allocation

The total transmission power is assumed to be 6 and the noise power is assumed to be 1. The channel matrix is assumed to be independent Rayleigh fading. The simulation result is shown below. There are totally 10 samples, and each sample represents an average capacity value over 50,000 channel realizations.

According to Figure 19.3, the capacity values with optimal power allocation are between 4.6 and 4.8. However, the capacity values with equal power allocation are between 3.6 and 3.8. The difference between them is around 1. In terms of

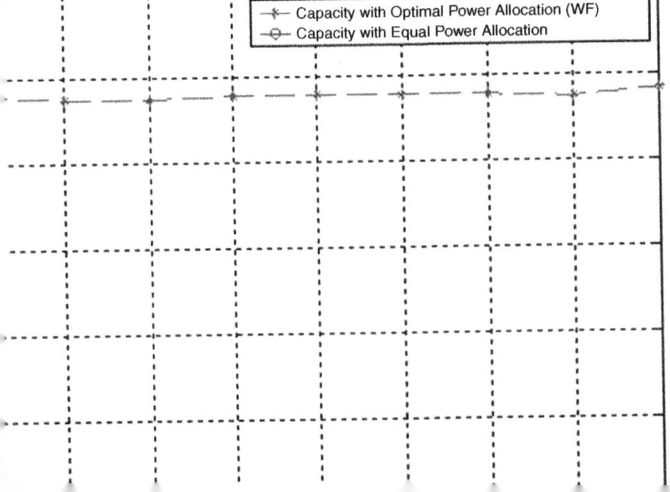
Capacity with Optimal Power Allocation (WF)
Capacity with Equal Power Allocation

dB value, it is

$$\frac{\text{Capacity with optimal power allocation}}{\text{Capacity with equal power allocation}} \approx \frac{4.75}{3.75} = 1.03 \text{ dB}$$

Comparing to equal power allocation, the reason for the capacity improvement with optimal power allocation can be explained from Eq. (19.5). The total capacity value is the sum of each subchannel's capacity. In this work, there are four subchannels.

Therefore, we have

$$C = \log_2\left(1 + \frac{\lambda_1 p_1}{\sigma^2}\right) + \log_2\left(1 + \frac{\lambda_2 p_2}{\sigma^2}\right) + \log_2\left(1 + \frac{\lambda_3 p_3}{\sigma^2}\right) + \log_2\left(1 + \frac{\lambda_4 p_4}{\sigma^2}\right) \tag{19.10}$$

As we all know, $\frac{a_1+a_2+\cdots+a_n}{n} \geq \sqrt[n]{a_1 a_2 \ldots a_n}$ if all the elements, a_i, are nonnegative and the equality is valid only

$$\left(\frac{\sigma^2}{\lambda_1} + p_1\right)\left(\frac{\sigma^2}{\lambda_2} + p_2\right)\left(\frac{\sigma^2}{\lambda_3} + p_3\right)\left(\frac{\sigma^2}{\lambda_4} + p_4\right) = \frac{2^C}{\lambda_1\lambda_2\lambda_3\lambda_4} \tag{19.11}$$

when $a_1 = a_2 \ldots = a_n$. Therefore, there is such a relationship for the left-hand side of Eq. (19.11).

$$\frac{\left(\frac{\sigma^2}{\lambda_1} + p_1\right) + \left(\frac{\sigma^2}{\lambda_2} + p_2\right) + \left(\frac{\sigma^2}{\lambda_3} + p_3\right) + \left(\frac{\sigma^2}{\lambda_4} + p_4\right)}{4} \geq \sqrt[4]{\left(\frac{\sigma^2}{\lambda_1} + p_1\right)\left(\frac{\sigma^2}{\lambda_2} + p_2\right)\left(\frac{\sigma^2}{\lambda_3} + p_3\right)\left(\frac{\sigma^2}{\lambda_4} + p_4\right)} = \sqrt[4]{\frac{2^C}{\lambda_1\lambda_2\lambda_3\lambda_4}} \tag{19.12}$$

From Eq. (19.12), we can conclude the following two results:

$$C \leq \log_2\left[\left|\sum_{i=1}^{4}\left(\frac{\sigma^2}{\lambda_i} + p_i\right)\right|^4 \lambda_1\lambda_2\lambda_3\lambda_4\right] \tag{19.13}$$

$$P_T = p_1 + p_2 + p_3 + p_4 \geq 4\sqrt[4]{\left(\frac{\sigma^2}{\lambda_1} + p_1\right)\left(\frac{\sigma^2}{\lambda_2} + p_2\right)\left(\frac{\sigma^2}{\lambda_3} + p_3\right)\left(\frac{\sigma^2}{\lambda_4} + p_4\right)} - \sum_{i=1}^{4}\left(\frac{\sigma^2}{\lambda_i}\right) \tag{19.14}$$

The equality is valid only when

$$\left(\frac{\sigma^2}{\lambda_1}+p_1\right)=\left(\frac{\sigma^2}{\lambda_2}+p_2\right)=\left(\frac{\sigma^2}{\lambda_3}+p_3\right)=\left(\frac{\sigma^2}{\lambda_4}+p_4\right)$$
$$=v=\sqrt[4]{\frac{2^C}{\lambda_1\lambda_2\lambda_3\lambda_4}} \tag{19.15}$$

Therefore the maximum capacity value and the minimum total transmission power can be obtained easily as long as either of them is given from Eqs. (19.13)–(19.15). This is an appropriate approach to obtain the optimal values. However, we use the WF algorithm in this work. The reason is that we have one more constraint on the power allocation; that is, power allocated to each subchannel must be nonnegative. Therefore we cannot use Eq. (19.15) to get answers directly. However, it provides some information for the WF algorithm $p_i=(v-\frac{\sigma^2}{\lambda_i})^+$. As the channel capacity becomes large or the total transmission power reaches a big enough value, Eq. (19.6) gets the same answer as a WF algorithm.

Figure 19.3 shows the advantage of optimal power allocation when total transmission power is given. The WF algorithm is also used to do power allocation when channel capacity is fixed. In the simulation, the channel capacity is assumed to be 6 and the channel matrix is independent Rayleigh fading. Figure 19.4 shows the simulation results. There are 10 samples, and each sample represents average total transmission power over 50,000 channel realizations.

Figure 19.4 shows that power consumed with equal power allocation is around 17.35 and power consumed with optimal power allocation is around 12.3. The difference is about 5 and it proves that the power saving is substantial, with optimal power allocation comparing to equal power allocation when the total channel capacity is 6.

19.3.2 Channel Capacity Change with Various Total Transmission Power

In Figure 19.5, total transmission power with values 10, 50, 100, and 500 have been used to do simulations with optimal power allocation and equal power allocation. There are four lines representing the simulation results of capacity improvement. The y axis is $\frac{\text{Capacity with equal power allocation}}{\text{Capacity with optimal power allocation}}$. Obviously, the y axis value is getting close to 1 as the total transmission power increases from 10 to 500, which means that the capacity values for equal power allocation and optimal power allocation are getting closer as the total transmission power increases. It can be explained by Eq. (19.15). When the total transmission power gets a large value, the WF algorithm gets the same answer as Eq. (19.15). That means the maximum capacity occurs when

$$\left(\frac{\sigma^2}{\lambda_1}+p_1\right)=\left(\frac{\sigma^2}{\lambda_2}+p_2\right)=\left(\frac{\sigma^2}{\lambda_3}+p_3\right)=\left(\frac{\sigma^2}{\lambda_4}+p_4\right) \tag{19.16}$$

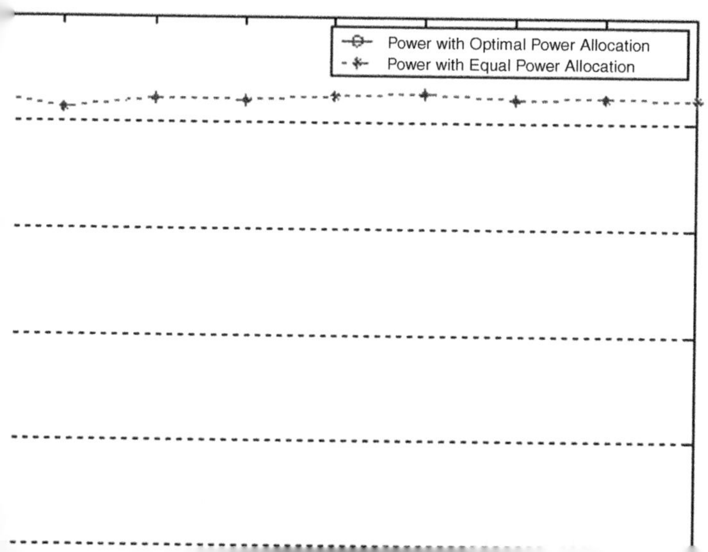
Power with Optimal Power Allocation
Power with Equal Power Allocation

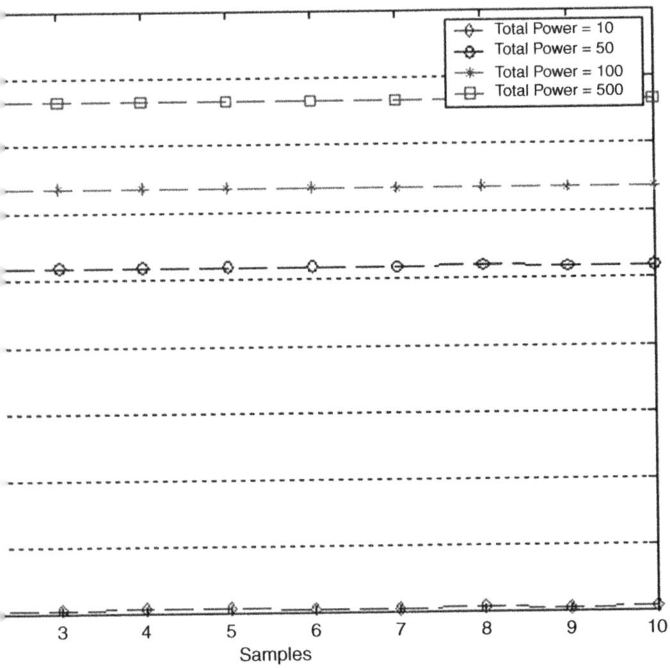

ɔarison between Optimal Power Allocation and Equal Power Allocation (Pt = 500)

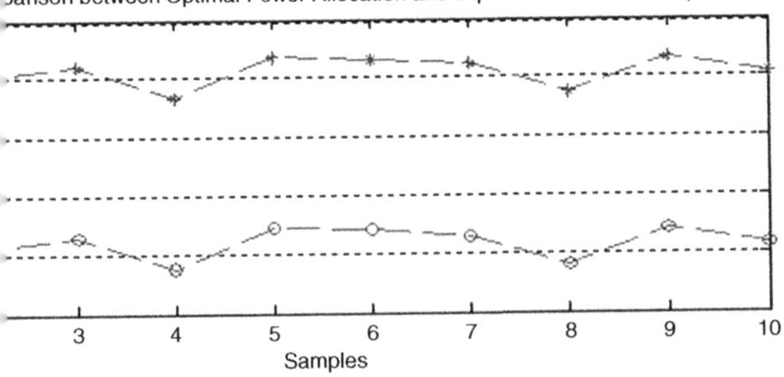

city with Equal Power Allocation/Capacity with Optimal Power Allocation

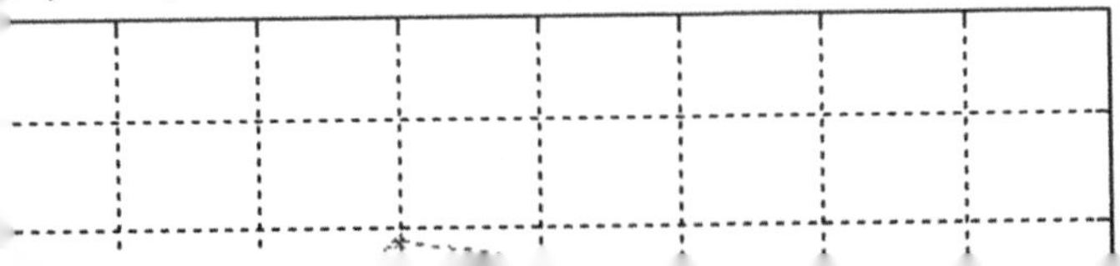

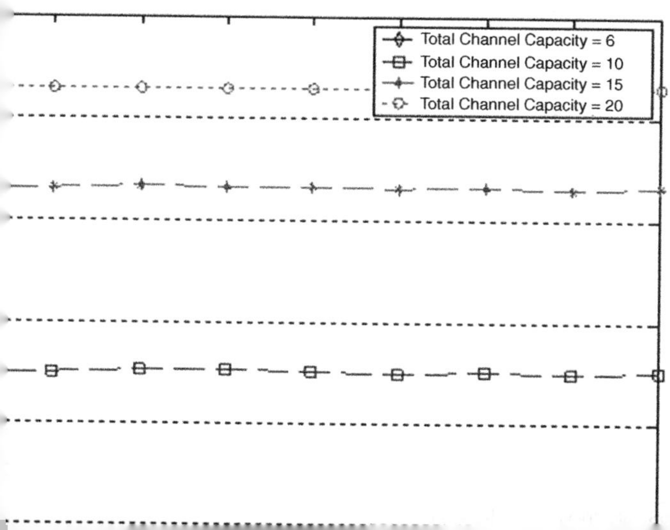

Total Channel Capacity = 6
Total Channel Capacity = 10
Total Channel Capacity = 15
Total Channel Capacity = 20

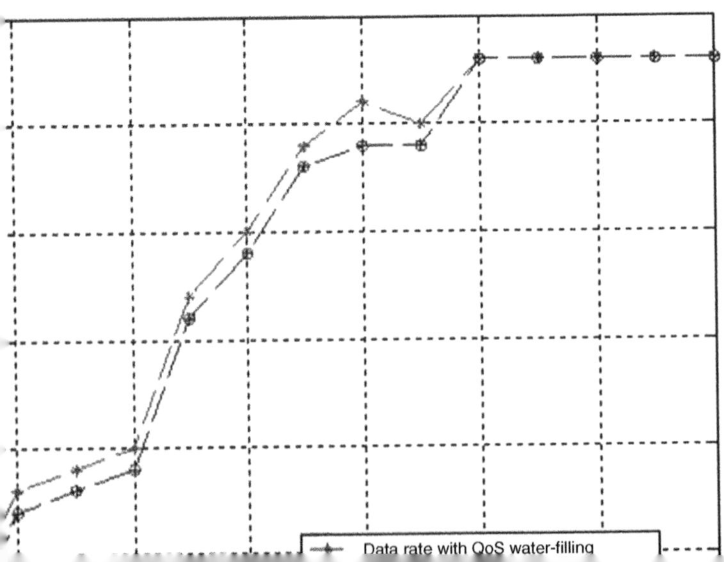

Data rate with QoS water-filling

REFERENCES

1. T. L. Marzetta and B. M. Hochwald, Capacity of a mobile multiple-antenna communication link in Rayleigh flat fading, *IEEE Trans. Inf. Theory*, Vol. 45, No. 1, pp. 139–157, January 1999.
2. R. Stridh, B. Ottersten, and P. Karlsson, MIMO channel capacity of a measured indoor radio channel at 5.8 GHz, in *Thirty-Fourth Asilomar Conference on Signals, Systems and Computers*, Vol. 1, 2000, pp. 733–737.
3. G. D. Golden, G. J. Foschini, R. A. Valenzuela, and P. W. Wolniansky, Detection algorithm and initial laboratory results using the V-BLAST space-time communication architecture, *Electron. Lett.*, Vol. 35, No. 1, pp. 14–15, January 7, 1999.
4. G. Caire, G. Taricco, and E. Biglieri, Optimum power control over fading channels, *IEEE Trans. Inf. Theory*, Vol. 45, No. 5, July 1999.
5. R. S. Cheng and S. Verdú, Gaussian multiaccess channels with ISI: Capacity region and multiuser water-filling, *IEEE Trans. Inf. Theory*, Vol. 39, No. 3, pp. 773–785, May 1993.
6. Y. Wei and J. M. Cioffi, On constant power water-filling, *IEEE Int. Conf. Commun.*, Vol. 6, pp. 1665–1669, 2001.
7. X. Zhang and B. Ottersten, Power allocation and bit loading for spatial multiplexing in MIMO systems, in *IEEE International Conference on Acoustics, Speech, and Signal Processing*, April 2003.
8. R. Knopp and G. Caire, Power control and beamforming for systems with multiple transmit and receive antennas, *IEEE Trans. Wireless Commun.*, Vol. 1, No. 4, pp. 638–648, October 2002.
9. A. Goldsmith and P. Varaiya, Capacity of fading channels with channel side information, *IEEE Trans. Inf. Theory*, Vol. 43, pp. 1968–1992, November 1997.

CHAPTER 20

LOW-COST AND COMPACT RF-MIMO TRANSCEIVERS

IGNACIO SANTAMARÍA, JAVIER VÍA, VICTOR ELVIRA, JESÚS IBÁÑEZ and JESÚS PÉREZ
Department of Communications Engineering, University of Cantabria, Santander, Spain

RALF EICKHOFF and UWE MAYER
Chair for Circuit Design and Network Theory, Technische Universitaet Dresden, Dresden, Germany

20.1 INTRODUCTION

In the last decade, multiple-input multiple-output (MIMO) wireless technology has gained considerable attention due to its potential to significantly increase spectral efficiency and/or reliability compared to traditional single-input single-output (SISO) systems [1–3]. A full-rank $n \times n$ MIMO channel is equivalent to n orthogonal SISO channels [1]; therefore, to exploit all the benefits of the MIMO channel (diversity or multiplexing gain), n parallel antenna paths must be independently acquired and processed at baseband. Consequently, the hardware costs, size, and power consumption are multiplied by a factor of n as well. Despite the numerous advantages of MIMO systems, these higher costs have delayed the wide-scale commercial deployment of multiple-antenna wireless transceivers mainly in handsets or low-cost terminals. The demand of low-cost, low-power-consumpting and compact wireless transceivers is even more important for radio-frequency identification (RFID) applications, which explains why conventional MIMO technologies have not found widespread application in RFID systems [4].

In this chapter, we describe a novel MIMO architecture that solves some of these problems by shifting spatial signal processing from the baseband to the radio-frequency (RF) front-end. The basic idea of the so-called RF-MIMO transceiver consists of performing adaptive signal combining in the RF domain by using some

Handbook of Smart Antennas for RFID Systems, Edited by Nemai Chandra Karmakar

innovative concepts that have made possible to develop a full 360° control range of the phase shifter together with an amplitude control of more than 20 dB [5, 6]. After combining, a single stream of data must be acquired and processed, and, thus, the hardware costs and the power consumption are significantly reduced. As we show in this chapter, although the multiplexing gain is limited to one (since we are transmitting a single data stream), other benefits of the MIMO channel such as full spatial diversity or full array gain are retained by the proposed RF-MIMO architecture with analog combining.

The RF-MIMO approach has some similarities to electronically steerable parasitic array radiator antennas (ESPAR) [7]. In contrast to forming the beam by signal processing in the digital baseband, both concepts use additional hardware in the analog RF domain to impact on the antenna beam by means of additional weighting circuitry or parasitic antenna elements. Both techniques adjust the beam of the antenna array by changing the amplitude and phase of the antenna currents [6, 8]. Therefore, the magnitude and direction of the main and side lobes can be controlled simultaneously in both approaches.

However, the main difference between ESPAR and RF-MIMO consists of how the antenna beam pattern is controlled and, hence, physically different system architectures result from both approaches. ESPAR uses additional parasitic elements in the antenna array for adjusting its characteristics that change the load of the central and primary antenna element. Consequently, more space is required for the array. On the contrary, RF-MIMO directly affects the voltage signals, which are steered to the antennas. Therefore, the physical configuration and size of the antenna array remain the same; only additional hardware is needed in the analog transceiver. Because silicon technologies can be used for RF-MIMO, that increase in silicon area has only limited impact on the overall system size and costs. The size of the additional weighting circuitry is usually less than $\lambda/4$ [9], which is required for the ESPAR approach. As a result, RF-MIMO enables more compact system sizes because the antenna array physically remains the same.

The proposed RF-MIMO architecture is currently being studied to increase the rate, coverage, and reliability of WLAN 802.11a systems [10], but the same concept can be applied to RFID applications. The system uses an orthogonal frequency division multiplexing (OFDM) modulation scheme due to its spectral efficiency as well as to its ability to cope with multipath fading channels typically encountered between RFID tags and readers. Also, we assume that perfect channel state information (CSI) is available at both, the transmitter (e.g., an active RFID tag) and the receiver (e.g., an RFID reader). This can be achieved by sending training sequences in a time division duplex mode.

This novel MIMO concept faces two main challenges. One is the design of RF circuits allowing precise amplitude and phase control in each branch; the other aspect is how to choose the optimal weights when OFDM transmissions are used. For changing the RF complex weights we use a vector modulator topology designed in SiGe-BiCMOS technology [5, 6]. On the other hand, we study three different criteria for optimal RF beamforming and propose an efficient iterative algorithm to find the

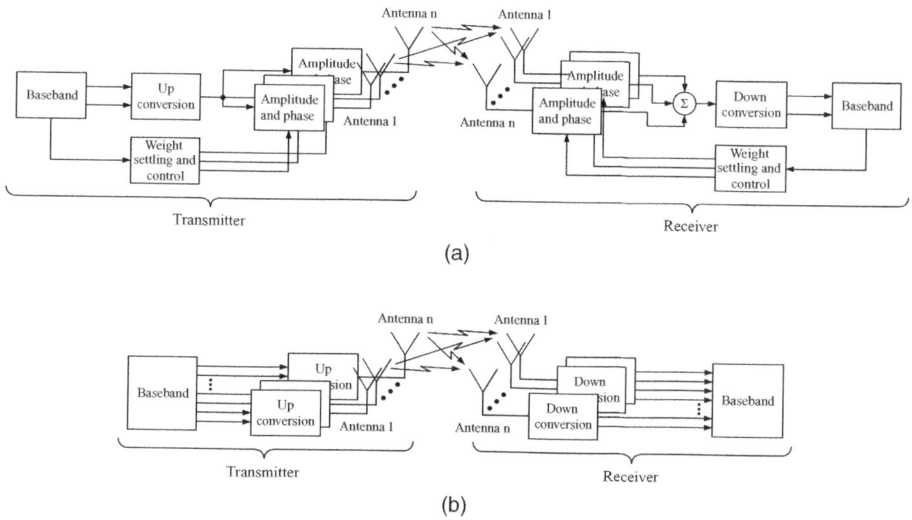

FIGURE 20.1. MIMO architectures (simplified) for transmitter and receiver: (**a**) Novel RF-MIMO architecture. (**b**) Conventional full-MIMO architecture.

weights. Specifically, we consider the maximum signal-to-noise ratio, the minimum mean-square error and the maximum capacity criteria.

This chapter is organized as follows. In Section 20.2 we present the system architecture of the novel RF-MIMO transceiver and discuss the adaptive antenna combining circuits. Section 20.3 discusses the three criteria for finding the optimal RF weights under multicarrier transmissions and frequency-selective channels. The performance of these criteria is compared in Section 20.4, followed by the concluding remarks in Section 20.5.

20.2 RF-MIMO TRANSCEIVER ARCHITECTURE

The proposed RF-MIMO architecture with adaptive antenna combining in the RF domain is shown in Figure 20.1a for the transmitter and the receiver of a MIMO system. In contrast to the conventional MIMO architecture, which performs spatial processing in the digital baseband (Figure 20.2b), the baseband processor determines a set of complex weights, which are directly applied in the RF domain by analog spatial signal processing circuits. Those complex weighted RF signals are then coherently combined and finally only one signal path is required for down-conversion, acquisition, and baseband processing in the receiver. For the transmitter the principle performs similarly.

Because the down-conversion circuitry consumes significant power and system size [11], the system costs and the power consumption of the RF-MIMO transmitter

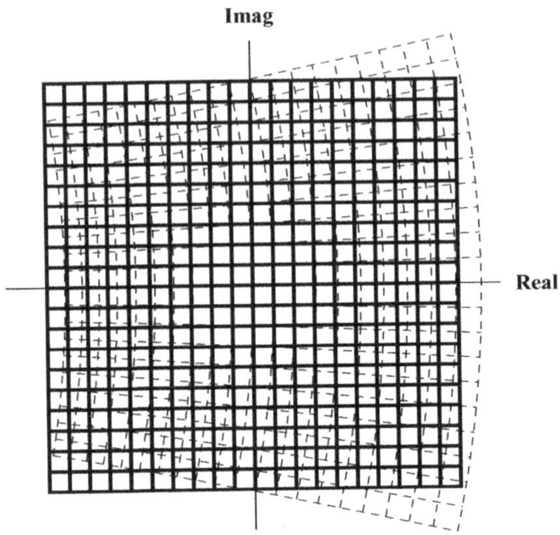

FIGURE 20.2. Effects of amplitude and phase correlation on weight settling in the complex plane (bold, no correlation; dashed, amplitude and phase correlation).

and receiver can be reduced compared to the full MIMO architecture in Figure 20.1b. The additional overhead of the spatial signal processing circuits, the weight settling, and the control unit is marginal compared to parallel operating down- or up-conversion subsystems needed for conventional MIMO approaches. Therefore, using synergies between the parallel antenna paths allows a reduction in power consumption and system complexity, but simultaneously exploiting the spatial benefits offered by MIMO systems.

Albeit providing several advantages, the architecture in Figure 20.1a is not commercially available yet, because it also faces some challenges in the circuit design. Basically, a major scientific challenge for this architecture is to achieve a precise phase control with 360° control range and an amplitude control of more than 20 dB. Also, shifting spatial signal processing from the digital baseband to the analog RF front-end requires compensating and considering impairments and process variations. Analog signal processing usually does not provide such a high reliable operation compared to digital circuit implementations because of temperature, aging or process variations effects, and, in addition, the resolution of the analog signals is limited due to thermal, flicker or shot noise processes of the used semiconductor devices [6].

Furthermore, in analog signal processing, complex weight multiplication can be achieved by adjusting the amplitudes and phases of the signals. Using analog circuitry, amplitude and phase changes are coupled and correlated; this means that when changing the amplitude, also the phase is changed, and when shifting the phase the amplitude of the signal is affected additionally [6]. As a result, the phases and amplitudes of the complex weights cannot be adjusted independently from each other

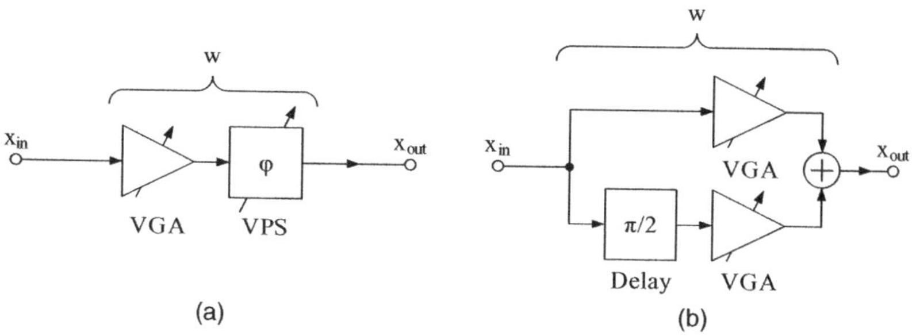

FIGURE 20.3. Complex weight multiplication for analog spatial processing: (**a**) Using a variable gain amplifier (VGA) and variable phase shifter (VPS). (**b**) Using a vector modulator approach with in-phase and quadrature signals.

and both are a function from the other variable, yielding

$$w_k = A_k(\phi_k)e^{j\phi_k(A_k)} \tag{20.1}$$

where w_k denotes the kth complex weight with magnitude A_k and phase ϕ_k. As an example, Figure 20.2 shows the effect of the coupling between amplitude and phase or real and imaginary parts, respectively. Ideally, the complex plane is uniformly covered by the weights. Analog signal processing performs a nonlinear transformation of the coordinate system due to the impairments from the physical devices. This nonlinear transformation should be properly calibrated and compensated by the control unit in Figure 20.1a in such a way that the weights calculated by the baseband processor, as will be described in Section 20.3, are those actually applied by the RF combiner.

20.2.1 RF-Weighting Topologies

Spatial signal processing by means of complex weight multiplication in the RF domain can be performed by two different concepts, either by adjusting the amplitude and phase of the RF signal [6] or by changing the amplitudes of a splitted signal with in-phase and quadrature components similarly to a vector addition [6]. These two orthogonal signals can be produced by splitting an incoming signal in two branches and applying a phase shift of $\pi/2$ to one of these two signals. In Figure 20.3, both concepts for analog spatial processing in the RF domain are shown.

Common to both approaches is the capability of adjusting the amplitude of the signals. For this purpose, a variable gain amplifier (VGA) can be used to change the amplitudes of the signal [11,12]. For the VGA, well-known architectures can be used but they have to pursue a low variation of the phase over the complete amplitude control range, decoupling both properties from each other. Basically, VGAs achieve their variable amplification either by changing the bias point of the amplification stages [e.g., 13], or by using variable attenuators [14]. Obviously, changing the bias point of an amplification stage or a component affects not only the amplitude-frequency response but also the phase characteristic of the system [11,12,15]. Therefore, this approach is not applicable for the RF-MIMO concept. Either this additional phase shift has to be compensated afterwards by an auxiliary stage [5] or the bias point of all stages has to be held constant over the complete dynamic range. The amplification range is determined by the required dynamic range of the weights that depend on the weight selecting algorithms and the environment.

Besides controlling the amplitudes, the phase of the signal has to be changed in both approaches. For the phase adjustments, either a variable phase shifter or a constant phase shifting element is needed, depending on the weight multiplication concept [6]. For the variable element in Figure 20.3a, the analog circuit has to provide a controllable phase range of $0 \leq \varphi \leq 2\pi$, without affecting the amplitude level. The vector modulator approach in Figure 20.3b requires only a constant phase shift of $\pi/2$ to produce inphase and quadrature components.

Although the requirements for both approaches seem to be very distinct from each other, the demand for a highly accurate phase shift for a mixture of sinusoidal signals typical of OFDM modulations results in a requirement of a constant group delay $(\frac{-\partial\varphi}{\partial\omega})$ for the complex envelope of the incoming signal $x_{\text{in}}(t)$. However, for the signal processing circuit in Figure 20.3a the group delay of the phase shifting element must be further controllable, whereas for Figure 20.3b only a constant shift is required.

Because the incoming signal consists of several sinusoidal signals at different frequencies, the requirement for a constant phase shift results in a constant group delay or a linear phase characteristic with respect to the frequency. Consequently, the challenge of the phase shifter development consists of achieving a constant group delay over the whole frequency band. The larger the bandwidth of the system, the larger the variations in the group delay of the phase shifter and, thus, higher deviations from a linear relationship can be observed. This nonlinear behavior arises from frequency-dependent quality factors of the silicon devices [11, 12]. Depending on the multiplication approach, the group delay has to be further controllable for the concept of Figure 20.3a. Therefore, the vector modulator concept imposes reduced challenges because the group delay does not need to be controllable over the whole frequency band.

20.2.2 RF Circuit Design

As it was described in Section 20.2.1, the major scientific challenge of controlling smart antennas in the RF domain comprises achieving precise phase control with 2π control range and a dynamic range of the weight amplitude of more than 20 dB [6]. Furthermore, both parameters have to be adjusted independently from each other to achieve the same performance as conventional baseband combining.

The spatial signal processing in the analog RF front-end is designed in a 0.25-μm SiGe-BiCMOS technology, and the development focuses on the vector modulator concept [6] shown in Figure 20.3b. However, both concepts of Figure 20.3 strongly rely on designing a VGA with low phase variations over the complete tuning range.

In the past, such circuits have been designed in III/V technology [5, 16], which has high costs, and therefore these circuits are not suitable for mass fabrication. Nowadays, these circuits are more and more integrated into silicon processes [17–19], in which stronger degradations in phase and amplitude variations are expected because of lower Q factors of the MOS varactors and the variable loads. Consequently, the challenges for circuit design are higher when using commercially attractive semiconductor processes.

20.2.2.1 Variable Gain Amplifier. It was explained in Section 20.2.1 that VGA concepts rely mainly on achieving the control range by changing the bias point of amplifier stages [12]. As a result, also the phase characteristic of the amplifier is affected. To compensate for phase deviations in the variable gain amplifier over the complete dynamic range, a constant biasing concept capable of canceling corresponding phase variations was developed and post-layout simulation results showed an amplitude control over 20 dB with phase variation below $\pm 5^\circ$ [20].

Generally, the transmission phase of an amplifier is determined by two major components: (1) its architecture (e.g., common source or common drain stage) and (2) its small signal elements. For the latter aspect, the small signal elements can be separated between constant elements and bias-point-dependent devices. Therefore, after selecting an appropriate architecture for the amplifier, the phase characteristic only depends on the small signal elements, from which only the bias-point-dependent elements cause a phase deviation when the output amplitude of the amplifier is changed.

Considering the common approach for variable gain amplifiers where the gain is set by changing the operating point, this affects the transmission phase of the amplifier and, thus, amplitude and phase cannot be controlled independently from each other if the bias point is not held constant. As a consequence, for compensated phase deviation, the amplifier has to be stabilized in its operating point independent of the chosen gain setting. Figure 20.4a shows the architecture of such an amplifier, which achieves a 20 dB amplitude range with less than 5° phase deviation.

The architecture is based on a cascode stage [6, 11, 12, 15], in which the bias point V_{o1} of the common emitter transistor is held constant; therefore, this stage cannot be used for gain control. In addition, it can be shown that the phase deviation of the cascode stage is dependent on the input impedance of the common gate stage [12], which consequently also has to be stabilized.

A constant input impedance can be achieved if parallel operating and in opposite directions steered amplifier stages are used for the common base stage (T_{G1} and T_{G2}). Regardless to the differential voltage, the sum of the current and the voltage at the source node are constant. The sum of both transconductances stays also equal resulting in a constant input impedance. The gain is controlled by applying a voltage difference (V_{G1} and V_{G2}) at the inputs of the cascode, keeping the common mode level of both inputs constant. To simplify the control circuitry, the cascode transistors can be steered with a current mirror.

The circuit was designed and fabricated in a 0.25 μm SiGe-BiCMOS with high-voltage devices (SGB25VD) from IHP [21]. Simulation results of the output voltage

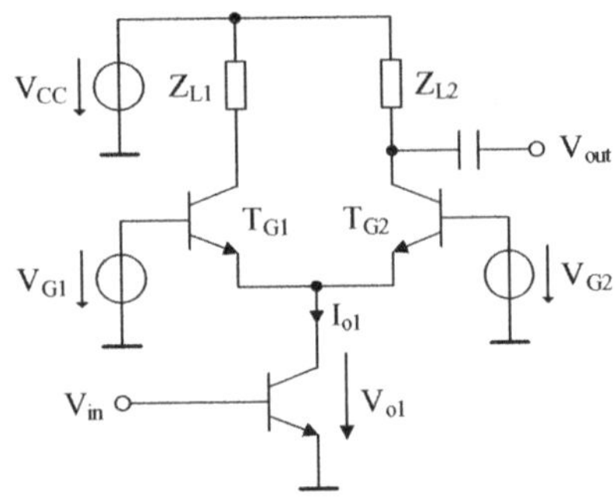

(a)

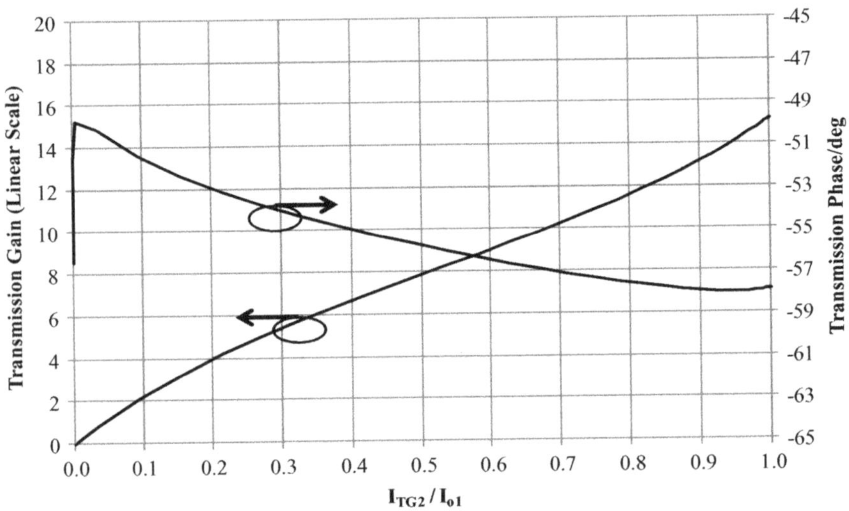

(b)

FIGURE 20.4. Constant biasing variable gain amplifier: (**a**) Simplified circuit architecture. (**b**) Simulated transmission gain and phase versus gain control range.

showed a phase variation of less than ±5° over an amplitude control range of 20 dB (see Figure 20.4b). These results perfectly match the theoretical analysis. Moreover, the VGA has a noise figure of less than 3 dB in the C-band [20].

Compared to other approaches [5, 16–19], the constant biasing concept achieves less variation in phase obtaining similar amplitude control ranges concurrently. The gain of the designed VGA can be controlled by means of an analog voltage or using a digital control word. For this purpose, a 6-bit DAC was implemented to allow direct access from the digital baseband. Because of the low phase variations of the developed VGA, this circuit is highly suited for the RF-MIMO concept, especially using the vector modulator for complex multiplication.

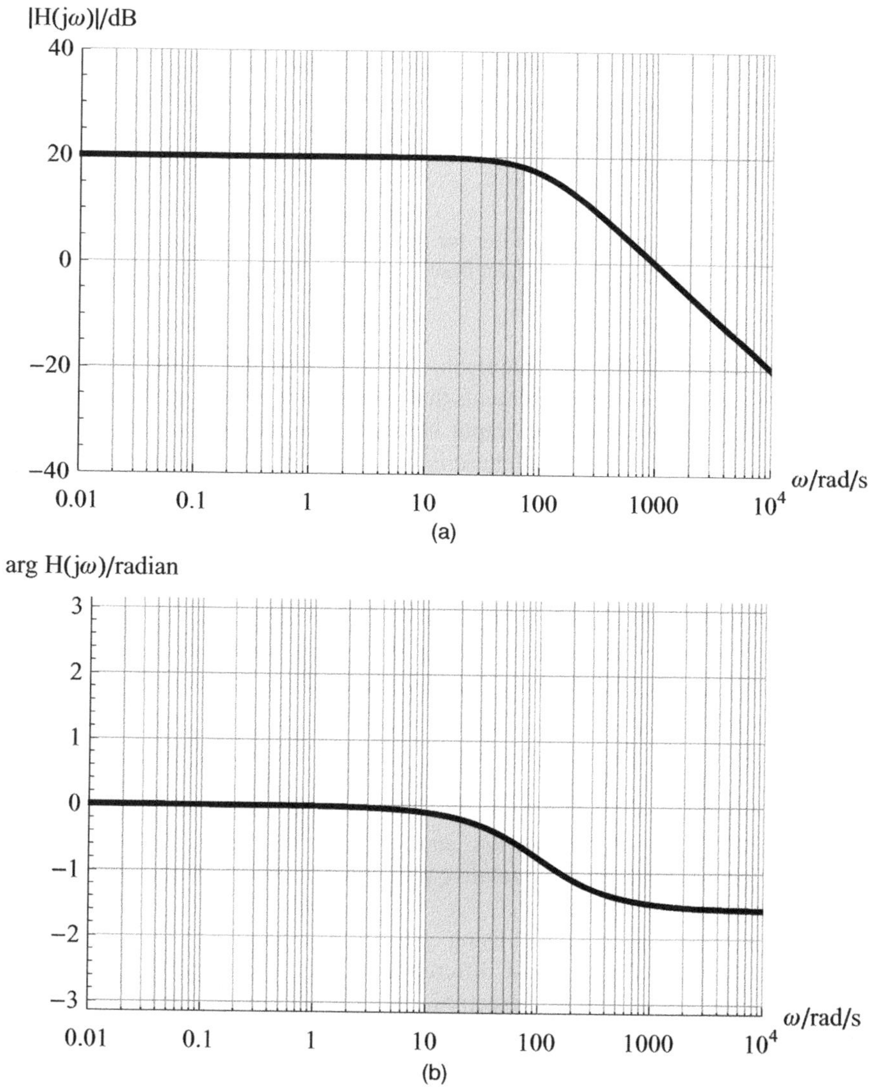

FIGURE 20.5. (**a**) Amplitude-frequency response and (**b**) phase characteristic of a low-pass filter.

20.2.2.2 Phase Shifter. According to Section 20.2.1, a constant group delay has to be provided to the incoming frequency mixture. Several concepts for circuits with constant group delays do exist, e.g. all-pass or low-pass based approaches [6, 12]. As an example, the simplest concepts are based on distributed filters [22]. Considering the transfer function of a single low-pass filter shown in Figure 20.5, the phase of the output signal depends on the characteristic length of a single filter, which

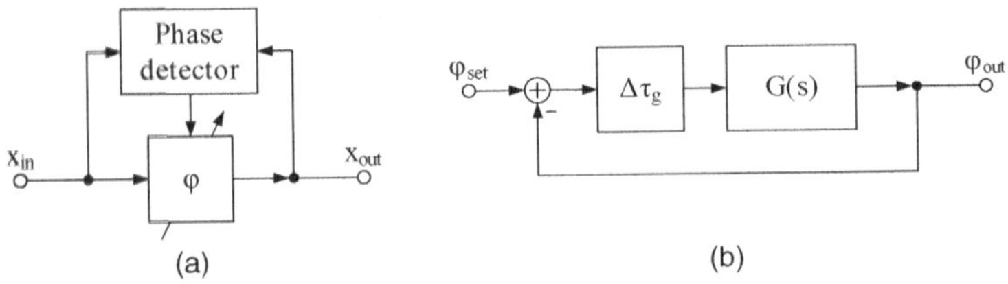

FIGURE 20.6. Control-loop phase shifter for precise group delay adjustment: (**a**) Phase-detector-based approach; (**b**) Phase detector based on control loop.

is determined by the used passive elements [6, 15]. Consequently, the phase can be changed by adjusting the passive elements that can be achieved using MOS varactors.

However, the amplitude of the output signal is also affected and decreased as can be seen in Figure 20.5a. If the phase is only controlled in a small region [e.g., 0°–30°, cf. Figure 20.5], the amplitude loss can be kept small. A distributed structure of serially connected low-pass filters allows a phase adjustment in a 360° range but keeping the amplitude losses small concurrently. Furthermore, the overall loss can be compensated by means of a larger gain of the VGA. Because only passive components are used in this approach, it is very attractive for low-power consumption applications. However, also active phase shifters, which are compensating the amplitude loss simultaneously, can be used but they will trade smaller amplitude loss off for increased power consumption.

However, besides applying concepts that still provide a nearly constant group delay, closed-loop control systems [23, 24] can be used to further linearize the phase characteristic of the phase shifter. Moreover, these control loops ensure a precise and predefined phase shift between the two antenna paths of Figure 20.1a. Here, each receive path has to ensure that the relative phase difference to its adjacent channels has to be as precise as possible.

Figure 20.6 shows the concept of using closed-loop systems [23, 24] for adjusting the phase. In Figure 20.6a, the phase difference between the incoming and outgoing signals is measured by a phase detector—for example, using a XOR gate or an analog multiplier [11, 12]. This phase difference can be used for adjusting the group delay of the phase shifter $\Delta\tau_g$ using a controller based approach. Depending on the transfer function $G(s)$ of the controller, this phase difference can be compensated with respect to the desired phase shift (Figure 20.6b). Here, $\Delta\tau_g$ denotes the adjustable delay and $G(s)$ is defined as the transfer function of the controller. $G(s)$ has to be designed to compensate the resulting phase difference. For this purpose, standard design methodologies can be used [23, 24].

However, the requirements for the control-loop systems consist of a sufficient control speed of the loop within the weight settling time, which is determined by the transmission scheme, e.g., less than 400 ns for OFDM-based transmission in 802.11a [25]. Moreover, a linear tuning range, which is independent of process, voltage, and temperature tolerances, is desired for the group delay. Requirements concerning the phase settling accuracy comprise 2% between two antenna paths. Implementations

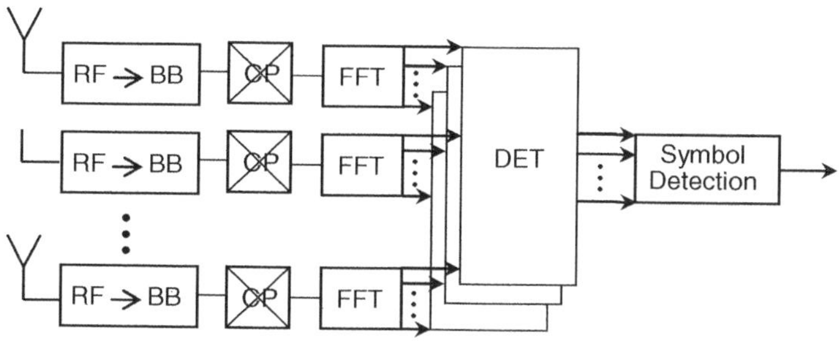

FIGURE 20.7. Schematic diagram of a conventional MIMO receiver under OFDM transmissions.

of such control-loop-based systems are currently developed based on delay-locked loop designs [26].

20.3 SELECTION CRITERIA FOR THE RF WEIGHTS

In this section, we treat the second challenge of the RF-MIMO transceiver, specifically trying to understand what optimization criteria should be used by the baseband processor (BB) to find the optimal RF or beamformer weights at the transmitter and receiver sides.

20.3.1 Problem Statement

Let us consider a MIMO point-to-point link equipped with n_T transmit and n_R receive antennas (see Figure 20.1). The MIMO channel is frequency-selective with impulse response given by [1]

$$\boldsymbol{H}(z) = \sum_{n=0}^{L-1} \boldsymbol{H}[n] z^{-n} \tag{20.2}$$

which is assumed to be known at both sides of the point-to-point link. To mitigate the adverse effects of multipath, we use an OFDM modulation [27] with N_c subcarriers and with a cyclic prefix (CP) larger than the MIMO channel order L. This allows us to decompose the original frequency-selective channel (20.2) into a set of N_c orthogonal flat fading $n_R \times n_T$ MIMO channels $\boldsymbol{H}_k$ $(k = 1, \ldots, N_c)$.

A conventional OFDM-MIMO receiver [28] computes the Fast Fourier transform (FFT) of each baseband signal and, hence, it can apply the optimal processing independently for each subcarrier (see Figure 20.7). To be more specific, the scheme shown in Figure 20.7 requires for each receiving antenna a RF to BB down-conversion chain; in the baseband domain, after frame and symbol synchronization, the CP is eliminated and an FFT per branch is computed. With this scheme, different

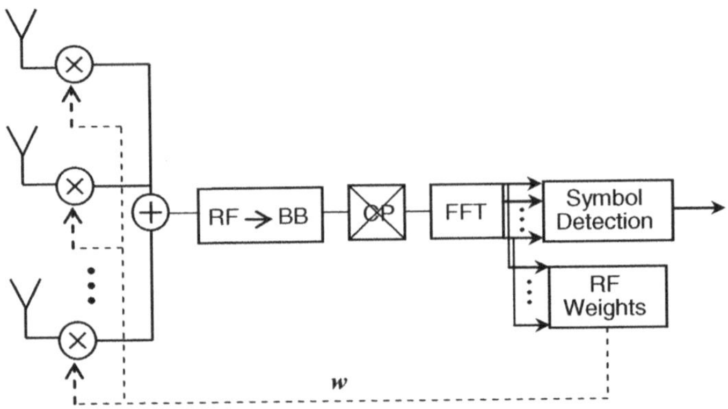

FIGURE 20.8. Schematic diagram of the proposed RF-MIMO receiver under OFDM transmissions.

subcarrier-based beamformers (in this case baseband beamformers) can be applied, thus obtaining the following set of flat fading SISO channels:

$$h_k = \boldsymbol{w}_{R,k}^H \boldsymbol{H}_k \boldsymbol{w}_{T,k}, \qquad k = 1, \ldots, N_c \tag{20.3}$$

where $\boldsymbol{w}_{T,k} \in C^{n_T \times 1}$ and $\boldsymbol{w}_{R,k} \in C^{n_R \times 1}$ represent the transmit and receive baseband unit-norm beamformers for the kth subcarrier. This scheme is sometimes called post-FFT processing [29] and we will also refer to it as full MIMO processing. Notice also that the conventional MIMO-OFDM transceiver has to compute as many FFTs as diversity branches, which can be a severe limitation in hardware complexity terms.

When perfect CSI is available at both the transmitter and the receiver sides, the optimal Tx and Rx beamformers for each subcarrier are the main right and left singular vectors of the MIMO channel $\boldsymbol{H}_k$. This transmission technique is the well-known dominant eigenmode transmission (DET) technique [30, 31]. With DET we are transmitting over the strongest spatial mode of each subcarrier MIMO matrix and hence the SNR in each equivalent SISO channel (20.3) is maximized.

However, in the proposed RF-MIMO transceiver a common beamformer must be used for all the subcarriers (see Figure 20.8). The equivalent SISO channels can be written as [30]

$$h_k = \boldsymbol{w}_R^H \boldsymbol{H}_k \boldsymbol{w}_T, \qquad k = 1, \ldots, N_c \tag{20.4}$$

where the subindex k in the Tx and Rx beamformers is not needed now since a common pair of Tx-Rx beamformers is used for all subcarriers.

A similar scheme has been proposed in the literature under the name of pre-FFT processing [29, 32–34]. The main motivation of pre-FFT processing is to reduce the computational cost due to FFT calculations. Unlike our proposed approach, all these pre-FFT techniques apply the beamforming weights in baseband, and, therefore, all antenna branches must be down-converted and acquired prior to FFT calculations.

Clearly, with these digital pre-FFT schemes we still need to down-convert and acquire independently all the RF branches; consequently, the hardware costs and size are not reduced in comparison to the full MIMO approach.

Nevertheless, we must notice that the problem of finding the optimal beamformers is the same for pre-FFT and RF-MIMO schemes. In comparison to post-FFT schemes, the design problem is much more complicated because any change in the beamformers will now affect all the subcarriers, so the N_c problems in Eq. (20.4) are coupled. In the following we find the optimal Tx and Rx beamformers under three different criteria: (i) maximum SNR (Max-SNR), (ii) minimum MSE (Min-MSE), and (iii) maximum capacity (Max-Cap).

20.3.2 Max-SNR Criterion

Previous research efforts considering pre-FFT processing, as imposed by the RF analog combining architecture, have been focused on the maximization of the received SNR. Under this criterion (denoted as Max-SNR), the optimal solution maximizes

$$\arg\max_{\boldsymbol{w}_T,\boldsymbol{w}_R} \sum_{k=1}^{N_c} |h_k|^2 = \sum_{k=1}^{N_c} \left|\boldsymbol{w}_R^H \boldsymbol{H}_k \boldsymbol{w}_T\right|^2, \quad \text{subject to (s.t.)} \quad \|\boldsymbol{w}_T\|^2 = \|\boldsymbol{w}_R\|^2 = 1 \tag{20.5}$$

Unfortunately, no closed-form solution exists for the Max-SNR criterion (20.5) under MIMO frequency-selective channels.* To find a solution in this case we have to resort to an iterative algorithm similar to the one proposed in references [32] and [33]. Here we proceed in a different way that will allow us to treat all the criteria proposed in this chapter in a unified manner. In particular, by applying the Lagrange multipliers method it is easy to see that the solution of Eq. (20.5) must fulfill the following pair of eigenvalue problems:

$$\mathbf{R}_{\mathrm{SIMO}}\mathbf{w}_R = \lambda \mathbf{w}_R \tag{20.6a}$$

$$\mathbf{R}_{\mathrm{MISO}}\mathbf{w}_T = \lambda \mathbf{w}_T \tag{20.6b}$$

where

$$\mathbf{R}_{\mathrm{SIMO}} = \sum_{k=1}^{N_c} \mathbf{h}_{\mathrm{SIMO}_k}\mathbf{h}_{\mathrm{SIMO}_k}^H \tag{20.7a}$$

$$\mathbf{R}_{\mathrm{MISO}} = \sum_{k=1}^{N_c} \mathbf{h}_{\mathrm{MISO}_k}^H\mathbf{h}_{\mathrm{MISO}_k} \tag{20.7b}$$

*Closed-form solutions exist, however, for MIMO flat fading channels and also for frequency-selective channels when a single antenna is used at either the receiver or the transmitter [29, 32, 35].

can be interpreted as correlation matrices, and

$$\mathbf{h}_{\mathrm{SIMO}_k} = \mathbf{H}_k \mathbf{w}_T, \qquad \mathbf{h}_{\mathrm{MISO}_k} = \mathbf{w}_R^H \mathbf{H}_k, \qquad k = 1, \ldots, N_c$$

are the equivalent single-input multiple-output (SIMO) and multiple-input single-output (MISO) channels obtained after fixing either the transmit or receive beamformers. In Section 20.3.5 we describe a gradient descent algorithm to find a solution for the coupled eigenvalue problems (20.6a) and (20.6b).

20.3.3 Min-MSE Criterion

Although the Max-SNR is a sensible criterion for the selection of the beamformers, a more relevant performance measure is given by the symbol or bit error rate (SER or BER, respectively) after decoding. To simplify the problem we have considered only the case in which a linear minimum mean-square error (MMSE) detector is applied. For this particular case, it is possible to obtain the following lower bound for the SER [1]

$$\mathrm{SER} \geq \alpha \operatorname{ercf}\left(\sqrt{\beta\left(\frac{\sigma^2}{\frac{1}{N_c}\sum_{k=1}^{N_c}\mathrm{MSE}_k} - 1\right)}\right) \tag{20.8}$$

where σ^2 represents the signal energy, α and β are parameters specific of the signal constellation, and MSE_k is the mean-square error in the kth subcarrier, which is given by

$$\mathrm{MSE}_k = \frac{\sigma^2}{1+\gamma|h_k|^2} = \frac{\sigma^2}{1+\gamma\left|\boldsymbol{w}_R^H \boldsymbol{H}_k \boldsymbol{w}_T\right|^2} \tag{20.9}$$

where γ is the SNR and $h_k$is the equivalent SISO channel at the kth subcarrier.

The lower bound in Eq. (20.9) is attained if and only if the symbols have been linearly precoded with an orthogonal matrix with all constant modulus entries, such as the FFT matrix [1]. Under these assumptions to minimize the SER is equivalent to minimize the average MSE across subcarriers; therefore we propose the following Min-MSE criterion [36]

$$\underset{\boldsymbol{w}_T,\boldsymbol{w}_R}{\arg\min} \sum_{k=1}^{N_c} \mathrm{MSE}_k = \sum_{k=1}^{N_c} \frac{\sigma^2}{1+\gamma\left|\boldsymbol{w}_R^H \boldsymbol{H}_k \boldsymbol{w}_T\right|^2}, \qquad \text{s. t.} \qquad \|\boldsymbol{w}_T\|^2 = \|\boldsymbol{w}_R\|^2 = 1 \tag{20.10}$$

Comparing the Max-SNR criterion given by Eq. (20.5) with Eq. (20.10), we see that the Min-MSE criterion will try to improve the performance of the worst subcarriers (those for which $\gamma \left|\boldsymbol{w}_R^H \boldsymbol{H}_k \boldsymbol{w}_T\right|^2 \ll 1$), whereas in Eq. (20.5) all the subcarriers are treated equally.

Unfortunately, a closed-form solution for the Tx and Rx beamformers that minimize Eq. (20.10) cannot be found again due to the coupling among the equivalent SISO channels. If we proceed like in the Max-SNR criterion, interestingly we arrive to the same pair of eigenvalue problems

$$\mathbf{R}_{\text{SIMO}}\mathbf{w}_R = \lambda \mathbf{w}_R, \tag{20.11a}$$

$$\mathbf{R}_{\text{MISO}}\mathbf{w}_T = \lambda \mathbf{w}_T, \tag{20.11b}$$

but now the correlation matrices for the equivalent SIMO and MISO channels are given by

$$\mathbf{R}_{\text{SIMO}} = \sum_{k=1}^{N_c} \text{MSE}_k^2 \mathbf{h}_{\text{SIMO}_k} \mathbf{h}_{\text{SIMO}_k}^H \tag{20.12a}$$

$$\mathbf{R}_{\text{MISO}} = \sum_{k=1}^{N_c} \text{MSE}_k^2 \mathbf{h}_{\text{MISO}_k}^H \mathbf{h}_{\text{MISO}_k} \tag{20.12b}$$

Comparing (20.12a) and (20.12b) with (20.7a) and (20.7b) we see that the only difference with the Max-SNR criterion is that now the correlation matrices at each subcarrier are weighted by the squared MSE.

20.3.4 Max-Cap Criterion

Although the Min-MSE criterion is optimal in terms of BER under linearly precoded symbols and linear MMSE detection, lower BERs might be attained by means of channel coding and optimal (nonlinear) receivers. Thus, we consider in this chapter a final beamforming criterion which consists of maximizing the capacity of the equivalent channel [37]

$$C \propto \sum_{k=1}^{N_c} \log\left(1 + \gamma \left|\boldsymbol{w}_R^H \boldsymbol{H}_k \boldsymbol{w}_T\right|^2\right)$$

More specifically, the Max-Cap criterion can be formulated as

$$\underset{\boldsymbol{w}_T, \boldsymbol{w}_R}{\arg\max} \sum_{k=1}^{N_c} \log\left(1 + \gamma \left|\boldsymbol{w}_R^H \boldsymbol{H}_k \boldsymbol{w}_T\right|^2\right) \quad \text{s. t.} \quad \|\boldsymbol{w}_T\|^2 = \|\boldsymbol{w}_R\|^2 = 1. \tag{20.13}$$

Again, by applying the method of Lagrange multipliers we get the same pair of coupled eigenvalue problems, but now the weights of the SIMO and MISO correlation matrices are the MSE values at each subcarrier, that is,

$$\mathbf{R}_{\text{SIMO}} = \sum_{k=1}^{N_c} MSE_k \mathbf{h}_{\text{SIMO}_k} \mathbf{h}_{\text{SIMO}_k}^H \tag{20.14a}$$

$$\mathbf{R}_{\text{MISO}} = \sum_{k=1}^{N_c} MSE_k \mathbf{h}_{\text{MISO}_k}^H \mathbf{h}_{\text{MISO}_k} \tag{20.14b}$$

20.3.5 A General Criterion for RF-MIMO Beamforming

It is interesting to point out that the three proposed criteria are just different versions of a common optimization (maximization or minimization) problem, which can be written as

$$\underset{\boldsymbol{w}_T, \boldsymbol{w}_R}{\text{optimize}} \sum_{k=1}^{N_c} f\left(\boldsymbol{w}_R^H \boldsymbol{H}_k \boldsymbol{w}_T\right) \qquad \text{s. t.} \qquad \|\boldsymbol{w}_T\|^2 = \|\boldsymbol{w}_R\|^2 = 1 \tag{20.15}$$

where the nonlinear function is $f(x) = |x|^2$, $f(x) = 1/(\gamma|x|^2 + 1)$, or $f(x) = \log(1 + \gamma|x|^2)$, for the Max-SNR, Min-MSE, and Max-Cap criteria, respectively. An even more general optimization problem can be written in terms of the MSE per subcarrier given by Eq. (20.9). Specifically, we propose the following minimization problem:

$$\underset{\boldsymbol{w}_T, \boldsymbol{w}_R}{\arg\min} \left(\frac{1}{N_c} \sum_{k=1}^{N_c} \text{MSE}_k^{\alpha-1}\right)^{\frac{1}{\alpha-1}} \qquad \text{s. t.} \qquad \|\boldsymbol{w}_T\|^2 = \|\boldsymbol{w}_R\|^2 = 1 \tag{20.16}$$

where α can be seen as a parameter controlling the trade-off between the overall channel energy and its spectral flatness. Analogously to the previous cases, the solution of Eq. (20.16) satisfies the same pair of coupled eigenvalue problems

$$\mathbf{R}_{\text{SIMO}} \mathbf{w}_R = \lambda \mathbf{w}_R \tag{20.17a}$$

$$\mathbf{R}_{\text{MISO}} \mathbf{w}_T = \lambda \mathbf{w}_T \tag{20.17b}$$

but now

$$\mathbf{R}_{\text{SIMO}} = \sum_{k=1}^{N_c} \text{MSE}_k^{\alpha} \mathbf{h}_{\text{SIMO}_k} \mathbf{h}_{\text{SIMO}_k}^H \tag{20.18a}$$

$$\mathbf{R}_{\text{MISO}} = \sum_{k=1}^{N_c} \text{MSE}_k^{\alpha} \mathbf{h}_{\text{MISO}_k}^H \mathbf{h}_{\text{MISO}_k} \tag{20.18b}$$

Obviously, for $\alpha = 0$, $\alpha = 1$, and $\alpha = 2$, this general formulation reduces, respectively, to the Max-SNR, Max-Cap, and Min-MSE criteria.

To gain some insight about the expected performance of this general criterion note that in the low SNR regime, the mean-square error associated to each subcarrier is $\text{MSE}_k = \sigma^2$; that is, all the weights are equal and hence the generalized criterion (20.16) reduces to the Max-SNR beamforming. On the other hand, in the high SNR regime the weights applied to the correlation matrices $\mathbf{R}_{\text{SIMO}}$ and $\mathbf{R}_{\text{MISO}}$ play a key role since they tend to improve the performance of the worst subcarriers at the expense of a slight degradation of the overall SNR. This expected behavior will be corroborated in the simulations section.

20.3.6 A Gradient-Descent Optimization Algorithm

To find a solution for this general optimization criterion, we propose a gradient search algorithm through the following update rules for $i = 0, 1, \ldots$:

$$\boldsymbol{w}_T^{i+1} = \boldsymbol{w}_T^i + \mu \boldsymbol{R}_{\text{MISO}}^i \boldsymbol{w}_T^i$$
$$\boldsymbol{w}_R^{i+1} = \boldsymbol{w}_R^i + \mu \boldsymbol{R}_{\text{SIMO}}^i \boldsymbol{w}_R^i$$

where μ is a constant learning rate and

$$\mathbf{R}_{\text{MISO}}^i = \sum_{k=1}^{N_c} \text{MSE}_k^{\alpha} \boldsymbol{H}_k^H \boldsymbol{w}_R^i (\boldsymbol{w}_R^i)^H \boldsymbol{H}_k$$

$$\mathbf{R}_{\text{SIMO}}^i = \sum_{k=1}^{N_c} \text{MSE}_k^{\alpha} \boldsymbol{H}_k \boldsymbol{w}_T^i (\boldsymbol{w}_T^i)^H \boldsymbol{H}_k^H$$

are weighted versions of the correlation matrices, which are obtained from the equivalent MISO or SIMO channels and hence depend on the beamformers at the *ith* iteration. We have included the superindex i to emphasize the dependency of these matrices with the iteration. Note that both updating rules are coupled.

Although the convergence analysis of the algorithm is a difficult task beyond the scope of this chapter, we have verified by means of extensive simulation examples that a proper initialization of the algorithm ensures convergence to the global optimum within a few iterations. To understand the rationale behind the proposed initialization procedure, let us first introduce the following definitions:

$$\boldsymbol{G} = \sum_{k=1}^{N_c} \text{vec}(\boldsymbol{H}_k)\text{vec}(\boldsymbol{H}_k)^H$$

$$\boldsymbol{w} = \boldsymbol{w}_R \otimes \boldsymbol{w}_T^H$$

$$\boldsymbol{W} = \boldsymbol{w}_R \boldsymbol{w}_T^H = \text{unvec}(\boldsymbol{w})$$

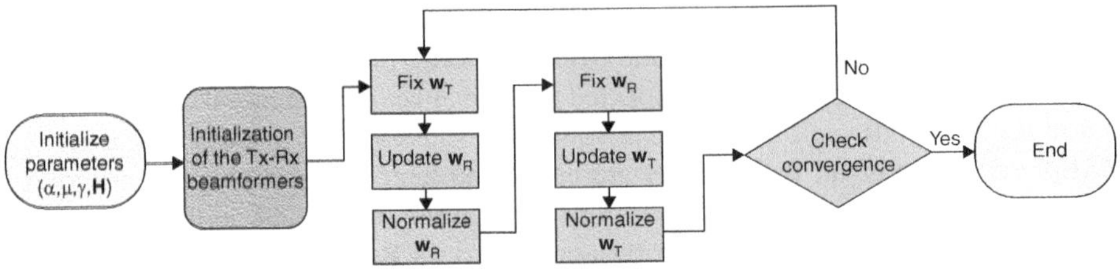

FIGURE 20.9. Flow chart for the proposed algorithm.

where vec(·) denotes the column-wise vectorized version of a matrix, unvec(·) denotes the inverse of the vec(·) operation, and ⊗ denotes Kronecker product. With the above definitions, the Max-SNR problem (20.5) can be written alternatively as

$$\arg\max_{w} \boldsymbol{w}^H \boldsymbol{G}\boldsymbol{w}$$

subject to

$$\text{rank}(\boldsymbol{W}) = \boldsymbol{w}_R \boldsymbol{w}_T^H = 1$$
$$\|\boldsymbol{w}_T\|^2 = \|\boldsymbol{w}_R\|^2 = 1 \tag{20.19}$$

This optimization problem is nonconvex due to the rank-one constraint on $\boldsymbol{W}$ [38]. However, it can be relaxed to a convex optimization problem via semi-definite relaxation (SDR), which amounts to omitting the rank-one constraint. This is the key idea that we use to efficiently initialize the proposed iterative algorithm. In particular, a good starting point can be obtained by solving the convex SDR problem from the following two-step procedure:

Step 1. Extract the eigenvector $\boldsymbol{w}$ associated with the largest eigenvalue of the $n_R n_T \times n_R n_T$ matrix $\boldsymbol{G}$.

Step 2. The initial Tx and Rx beamformers are given by the right and left singular vectors associated to the largest singular value of the $n_R \times n_T$ matrix unvec($\boldsymbol{w}$).

Finally, starting from estimates of the channel $\boldsymbol{H}_k (k = 1, \ldots, N_c)$ and the SNR γ, the proposed algorithm can be summarized as follows (a flow chart is shown in Figure 20.9):

1. Select μ (learning rate), and $\alpha = 0, 1, 2$ for the Max-SNR, Max-Cap or Min-MSE criteria, respectively.
2. Initialize the Tx and Rx beamformers, $\boldsymbol{w}_T^1$ and $\boldsymbol{w}_R^1$, according to Steps 1 and 2 described above.
3. For $i = 1, \ldots, i_{\max}$ (or until convergence)

Update $\mathbf{w}_R$

3.1. Obtain the equivalent SIMO channels $\boldsymbol{h}^i_{\text{SIMO}_k} = \boldsymbol{H}_k \boldsymbol{w}^i_T$

3.2. Update h_k and MSE_k for $k = 1, \ldots, N_c$

3.3. Obtain the weighted correlation matrix $\mathbf{R}^i_{\text{SIMO}}$

3.4. Update the beamformer $\boldsymbol{w}^{i+1}_R = \boldsymbol{w}^i_R + \mu \boldsymbol{R}^i_{\text{SIMO}} \boldsymbol{w}^i_R$

3.5. Normalize the solution: $\boldsymbol{w}^i_R = \frac{\boldsymbol{w}^i_R}{\|\boldsymbol{w}^i_R\|}$

Update w_T

3.6. Obtain the equivalent MISO channels $\boldsymbol{h}^i_{\text{MISO}_k} = (\boldsymbol{w}^i_R)^H \boldsymbol{H}_k$

3.7. Update h_k and MSE_k for $k = 1, \ldots, N_c$

3.8. Obtain the weighted correlation matrix $\boldsymbol{R}^i_{\text{MISO}}$

3.9. Update the beamformer $\boldsymbol{w}^{i+1}_T = \boldsymbol{w}^i_T + \mu \mathbf{R}^i_{\text{MISO}} \boldsymbol{w}^i_T$

3.10. Normalize the solution: $\boldsymbol{w}^i_T = \frac{\boldsymbol{w}^i_T}{\|\boldsymbol{w}^i_T\|}$

20.4 RESULTS

The performance of the proposed technique is illustrated in this section by means of some Monte Carlo simulations. In all the experiments, a 4×4 Rayleigh MIMO channel model with exponential power delay profile has been assumed. In particular, the total power associated to the nth tap is [39]

$$E[\|\mathbf{H}[n]\|^2] = \rho^n (1 - \rho) n_R n_T \tag{20.20}$$

where $n = 1, \ldots, L$ is the number of coefficients of the channel and the variable ρ controls the decay rate of the power delay profile. For the simulations we have selected $\rho = 0.4$ and $L = 16$.

In all the experiments an OFDM signal with $N_c = 64$ subcarriers using quadrature phase shift keying (QPSK) constellations has been considered. The QPSK symbols can be either uncoded or linearly precoded with the DFT matrix. The proposed iterative gradient-descent algorithm never exceeded 50 iterations, and it was compared with a SISO and a full MIMO scheme with DET in each subcarrier (denoted as full-MIMO), which can be seen as an upper bound for the performance of any system with antenna combining in the RF path.

20.4.1 Performance of the Equivalent SISO Channel

In the first set of examples, we evaluated the equivalent channel after beamforming for the Max-SNR ($\alpha = 0$), Min-MSE ($\alpha = 2$) and Max-Cap ($\alpha = 1$) criteria. Figure 20.10 shows the absolute value of the frequency response of the equivalent SISO channel at each subcarrier under the different criteria for $\gamma = 10$ dB. As can be seen, the Max-SNR criterion provides the channel with the overall largest energy

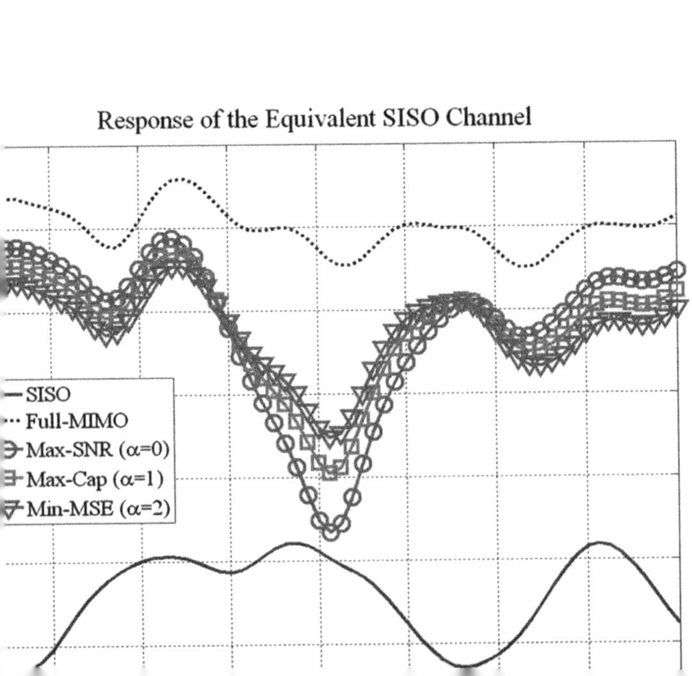
Response of the Equivalent SISO Channel
SISO
Full-MIMO
Max-SNR (α=0)
Max-Cap (α=1)
Min-MSE (α=2)

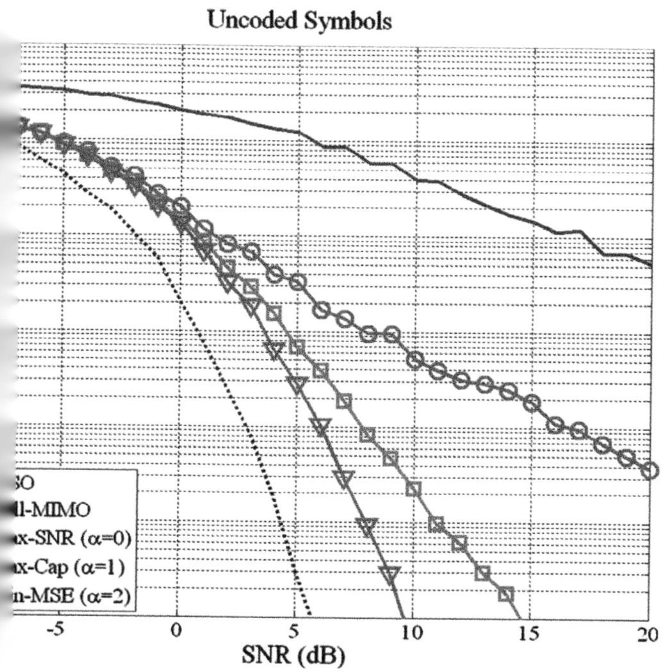

FIGURE 20.11. BER for uncoded OFDM symbols.

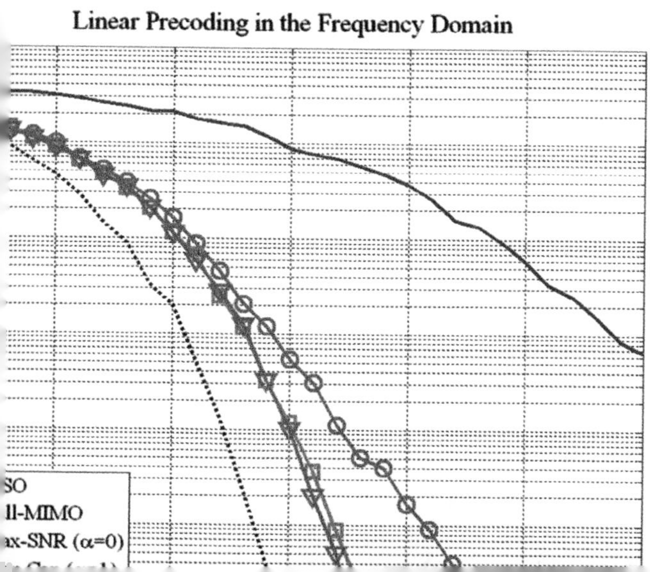

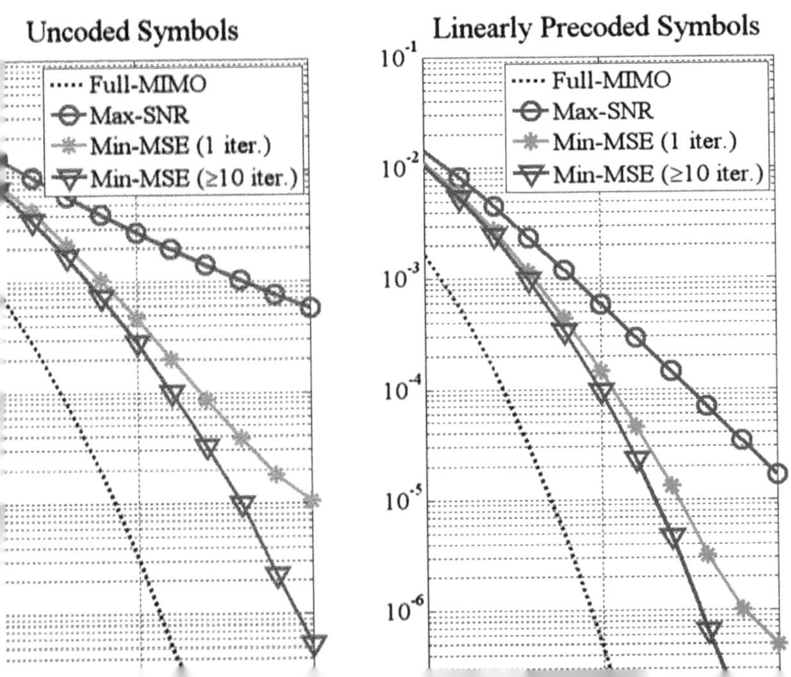

Uncoded Symbols
Full-MIMO
Max-SNR
Min-MSE (1 iter.)
Min-MSE (≥10 iter.)
Linearly Precoded Symbols
10^-1
10^-2
10^-3
10^-4
10^-5
10^-6
Full-MIMO
Max-SNR
Min-MSE (1 iter.)
Min-MSE (≥10 iter.)

this novel RF-MIMO transceiver can be significantly reduced in comparison to a conventional MIMO approach, which requires parallel processing for each antenna branch. Therefore, this novel architecture is a suitable candidate for developing enhanced RFID tags and readers with supported MIMO capabilities.

One of the main challenges for analog antenna combining consists of designing low-cost RF circuitry achieving precise phase and amplitude control. The selected topology follows a vector modulator concept, in which the incoming RF signal is split into two branches by applying a phase shift of $\pi/2$, and then a variable gain amplifier (VGA) is used to change the amplitudes of both signals. The VGA was designed in a 0.25 μm SiGe-BiCMOS technology, and post-layout simulation results showed an amplitude control over 20 dB with phase variation below $\pm 5°$. Besides, a phase shifter with a closed-loop control system that provides a nearly constant group delay was described.

A second challenge of the RF-MIMO transceiver is how to find the best RF beamforming weights, mainly when OFDM modulations are used. Under the assumption of perfect CSI at the transmitter and receiver sides, we discussed beamforming criteria that maximize the SNR, minimize the MSE, or maximize the capacity. Interestingly, the three criteria admit a common cost function that depends on a single parameter α. By changing this parameter, we obtain the Min-SNR ($\alpha = 0$), the Max-Cap ($\alpha = 1$), and the Min-MSE ($\alpha = 2$) solutions, respectively. A fast gradient-descent algorithm was proposed to minimize this common cost function. In terms of BER, several simulation examples allowed us to conclude that the min-MSE is the best performing criterion.

Since the multiplexing gain of the analog combining architecture is limited to one, the proposed RF-MIMO transceiver does not allow to read or recognize simultaneously more than one tag. However, it allows an improved reception (e.g., a lower BER for a given noise plus interference level), which translates into an increase of the reading range, or a reduction in the number of interrogations before a successful reading.

20.6 ACKNOWLEDGMENTS

The research leading to these results has received funding from the European Community's Seventh Framework Programme (FP7/2007-2013) under the MIMAX project (MIMO Systems for Maximum Reliability and Performance), Contract agreement no. 213952, as well as from the Spanish Government (MICINN) under the MULTIMIMO project (TEC2007-68020-C04-02).

REFERENCES

1. S. Barbarossa, *Multiantenna Wireless Communication Systems*, Artech House, Norwood, MA, 2005.
2. D. Tse and P. Viswanath, *Fundamentals of Wireless Communications*, Cambridge University Press, Cambridge, UK, 2005.

3. C. Oestges and B. Clerckx, *MIMO Wireless Communications: From Real World Propagation to Space–Time Code Design*, Academic Press, New York, 2007.
4. N. C. Karmakar, Smart antennas for automatic radio frequency identification readers, in *Handbook on Advancements in Smart Antenna Technologies for Wireless Networks*, C. Sun, J. Cheng, and T. Ohira, editors, Information Science Reference, London, UK, 2008.
5. F. Ellinger, U. Lott, and W. Bätchtold, An antenna diversity MMIC vector modulator for HIPERLAN with low power consumption and calibration capability, *IEEE Trans. Microwave Theory Tech.*, Vol. 49, No. 5, pp. 964–969, 2001.
6. F. Ellinger, *Radio frequency integrated circuits and technologies*, 2 ed., Berlin; Heidelberg: Springer, 2008.
7. R. F. Harrington, Reactively controlled directive arrays, *IEEE Trans. Antennas Propag.*, Vol. 26, No. 3, pp. 390–395, 1978.
8. C. Sun, A. Hirata, T. Ohira, and N. C. Karmakar, Fast beamforming of electronically steerable parasitic array radiator antennas: Theory and experiment, *IEEE Trans. Antennas Propag.*, Vol. 32, No. 7, pp. 1819–1831, 2004.
9. C. Sun, T. Ohira, M. Taromaru, N. C. Karmakar, and A. Hirata, Fast beamforming of compact array antennas, in *Handbook on Advancements in Smart Antenna Technologies for Wireless Networks*, C. Sun, J. Cheng, and T. Ohira, editors, Information Science Reference, London, UK, 2008
10. R. Eickhoff, R. Kraemer, I. Santamaria *et al.*, "Developing energy-efficient MIMO radios," *IEEE Vehicular Technology Magazine*, vol. 4, no. 1, pp. 34–41, March, 2009.
11. B. Razavi, *RF Microelectronics*, Prentice-Hall, Upper Saddle River, NJ, 1998.
12. P. R. Gray, R. G. Meyer, P. J. Hurst, and S. H. Lewis, *Analysis and Design of Analog Integrated Circuits*, John Wiley & Sons, New York, 2001.
13. J. Paramesh, R. Bishop, K. Soumyanath, and D. Allstot, A four-antenna receiver in 90-nm CMOS for beamforming and spatial diversity, *IEEE J. Solid-State Circuits*, Vol. 40, No. 12, pp. 2515–2524, 2005.
14. M. Koutani, H. Kawamura, S. Toyoyama, and K. Iizuka, A digitally controlled variable-gain low-noise amplifier with strong immunity to interferers, *IEEE J. Solid-State Circuits*, Vol. 42, No. 11, pp. 2395–2403, 2007.
15. G. D. Vendelin, A. M. Pavio, and U. L. Rohde, *Microwave Circuit Design Using Linear and Nonlinear Techniques*, Wiley-IEEE Press, New York, 2005.
16. F. Ellinger and W. Bächtold, Adaptive antenna receiver module for WLAN at C-band with low power consumption, *IEEE Microwave Wireless Compon. Lett.*, Vol. 12, No. 9, pp. 348–350, 2002.
17. F. Ellinger and H. Jackel, Low-cost BiCMOS variable gain LNA at Ku-band with ultra-low power consumption, *IEEE Trans. Microwave Theory and Tech.*, Vol. 52, No. 2, pp. 702–708, 2004.
18. H. Hayashi and M. Muraguchi, An MMIC Variable-gain amplifier using a cascode-connected FET with constant phase deviation, *IEICE Trans. Electron.*, Vol. 81, No. 1, pp. 70–77, 1998.
19. B.-W. Min and G. M. Rebeiz, A 10–50-GHz CMOS distributed step attenuator with low loss and low phase imbalance, *IEEE J. Solid-State Circuits*, Vol. 42, No.11, pp. 2547–2554, 2007.
20. U. Mayer, F. Ellinger, and R. Eickhoff, Analysis and Reduction of Phase Variations of Variable Gain Amplifiers verified by CMOS Implementation at C-Band, *IET Circuits, Devices & Systs.*, 2010.

21. IHP Microelectronics, SGB25VD 0.25 μm SiGe-BiCMOS technology, December 2007, revision 1.2 (70817), 2007.
22. F. Ellinger, J. Wagner, U. Mayer, and R. Eickhoff, Passive varactor tuned equivalent transmission line phase shifter at C-band in 0.25 μm BiCMOS, *Circuits, Devices & Systs., IET*, Vol. 2, No. 4, pp. 355–360, 2008.
23. G. F. Franklin, D. J. Powell, and A. Emami-Naeini, *Feedback Control of Dynamic Systems*, 4th edition, Prentice-Hall PTR, Upper Saddle River, NJ, 2001.
24. T. Kailath, *Linear Systems*, Prentice-Hall Information and System Science Series, Prentice-Hall, Englewood Cliffs, NJ, 1979.
25. IEEE Std. 802.11a, IEEE Standard for Local an Metropolitan Area Networks. Part 11: Wireless LAN Medium Access Control (MAC) and Physical Layer (PHY) Specifications, High-Speed Physical Layer in the 5 GHz Band, 1999.
26. J.-C. Scheytt, Phase and amplitude modulator, US Patent 7319352, 2008.
27. S. Hara and R. Prasad, *Multicarrier Techniques for 4G Mobile Communications*, Artech House, Norwood, MA, 2003.
28. A. Paulraj, D. A. Gore, R. U. Nabar, and H. Bolcskei, An overview of MIMO communications—A key to gigabit wireless, *Proc. IEEE*, Vol. 92, No. 2, pp. 198–218, 2004.
29. M. I. Rahman, S. S. Das, F. H. P. Fitzek, and R. Prasad, Pre- and post-DFT combining space diversity receiver for wideband multi-carrier systems, in *Proceedings of the 8th Conference on Wireless Personal Multimedia Communications*, Aalborg, Denmark, 2005.
30. J. B. Andersen, Array gain and capacity for known random channels with multiple element arrays at both ends, *IEEE J. Selected Areas Commun.*, Vol. 18, No. 11, pp. 2172–2178, 2000.
31. T. Dahl, N. Christophersen, and D. Gesbert, Blind MIMO eigenmode transmission based on the algebraic power method, *IEEE Trans. Signal Processing*, Vol. 52, No. 9, pp. 2424–2431, 2004.
32. M. Okada and S. Komaki, Pre-DFT combining space diversity assisted COFDM, *IEEE Trans. Veh. Technology*, Vol. 50, No. 2, pp. 487–496, 2001.
33. D. Huang and K. B. Leitaief, Symbol-based space diversity for coded OFDM systems, *IEEE Trans. Wireless Commun.*, Vol. 3, No. 1, pp. 117–127, 2004.
34. D. Huang and K. B. Leitaief, Pre-DFT processing using eigenanalysis for coded OFDM with multiple receive antennas, *IEEE Trans. Commun.*, Vol. 52, No. 11, pp. 2019–2027, 2004.
35. S. Sandhu and M. Ho, Analog combining of multiple receive antennas with OFDM, in *Proceedings of the IEEE Internationl Conference on Communications (ICC 2003)*, Anchorage, Alaska, Vol. 5, pp. 3428–3432, 2003.
36. J. Via, V. Elvira, I. Santamaria, and R. Eickhoff, Minimum BER beamforming in the RF domain for OFDM transmissions and linear receivers, in *IEEE International Conference on Acoustics, Speech and Signal Processing (ICASSP 2009)*, Taipei, Taiwan, 2009.
37. J. Via, V. Elvira, I. Santamaria, and R. Eickhoff, Analog antenna combining for maximum capacity under OFDM transmissions, in *IEEE International Conference on Communications (ICC 2009)*, Dresden, Germany, 2009.
38. S. Boyd and L. Vandenberghe, *Convex Optimization*, Cambridge University Press, Cambridge, UK, 2004.
39. J. G. Proakis, *Digital Communications*, 4th edition, McGraw-Hill, New York, 2000.

CHAPTER 21

BLIND CHANNEL ESTIMATION IN MIMO FOR MC-CDMA

ABDUR RAHIM and NEMAI CHANDRA KARMAKAR
Department of Electrial and Computer Systems Engineering, Monash University, Clayton, Victoria, Australia

KAZI M. AHMED
Telecommunications Program, School of Engineering and Technology, Asian Institute of Technology, Thailand

21.1 INTRODUCTION

21.1.1 Background

Wireless communication systems are witnessing a rapid growth in the number of subscribers and in the range of services. However, the radio spectrum available for wireless services is a limited resource. A challenge for wireless technology is to meet the increase in demand for wireless services within the available spectrum. Therefore, techniques that increase spectrum efficiency are of great commercial interest. The concept of Multiple-Input Multiple-Output (MIMO) was first introduced by Jack Winters in 1987 [1] and has received great attention by researchers. MIMO architectures facilitate powerful techniques for improving the capacity of wireless communication systems, especially in rich multipath environments. Space–time coding, such as Space–Time Block Coding (STBC), relying on multiple antenna transmissions and appropriate signal processing in the receiver was proposed by Alamouti [2]. It provides diversity and interesting coding gains compared to uncoded single antenna transmissions. Thus, this new transmit diversity scheme can improve the capacity of wireless communication systems. Multicarrier transmission methods for CDMA communication systems have been proposed by Hara and Prasad [3] as an efficient technique to combat multipath propagation. Multicarrier (MC) systems split the transmitted digital information in multiple carriers and achieve high transmission speeds using large symbol periods. As a consequence, MC systems can simply

Handbook of Smart Antennas for RFID Systems, Edited by Nemai Chandra Karmakar

remove the intersymbol interference (ISI), thereby introducing a short guard time between symbols and avoiding the utilization of a channel equalizer.

Channel estimation can be carried out by training sequence-aided as well as blind techniques. For a given bandwidth, use of training sequences decreases the effective information rate. Moreover, the training signals consume a part of the bandwidth, thereby reducing data throughput to some extent. On the other hand, no training sequences are used in blind techniques. So, there has always been an improved interest and attention for the blind channel estimation techniques.

Recently, radio-frequency identification (RFID) has attracted much interest as the means of short-range wireless communication for automatic identification and tracking. RFID in mobile and crowded environments is vivid in toll collection of vehicles in central business districts (CBDs). Multiple access schemes with MIMO antennas can be implemented easily in gantry-based readers and vehicle-items-based RFID tags, respectively. Therefore, the application areas of blind channel estimation in MIMO-based MC-CDMA will have great impact on capacity improvement and quality of services for RFID systems.

21.1.2 Statement of Problem

Due to bandwidth limitations and multipath propagation, the transmission channel distorts the signal being transmitted, leading to intersymbol interference (ISI). The receiver needs to identify this channel distortion and equalize it [4]. Classical system identification techniques require the use of both system input and output, which leads to the transmission of a training sequence—that is, a set of fixed data that are known to both transmitter and receiver. As mentioned earlier, for a given bandwidth, use of training sequences decreases the effective information rate. Moreover, training-sequence-based methods are expensive in terms of implementation complexity and reduced data rate. Also the accuracy of training-based estimation largely depends on frequency of the training signal—that is, the rate at which the estimation is carried out. Higher estimation frequency gives a better channel estimate with the penalty of capacity reduction due to the training overhead. In blind channel estimation and equalization, no training sequences are used. These methods are also more robust if the signal undergoes severe fading during training. Furthermore, the blind technique can be used to reduce the interferences [5], which could be present during training-sequence-based estimation.

Duong [6] studied channel estimation in MIMO MC-CDMA systems employing training-sequence-based channel estimation. Blind estimation of channel and modulation schemes in adaptive-modulation-based OFDM-CDMA systems was carried out by Chatterjee [7], where the MIMO technique was not considered. In this chapter, the channel has been estimated blindly using the subspace-based estimation algorithm for the space–time-coded MIMO MC-CDMA systems.

21.1.3 Objectives

The main objective of this chapter is to estimate the channel blindly in a two-transmit-antenna, two-receive-antenna system in MC-CDMA environment. The specific objectives of this chapter are outlined as follows:

1. To implement subspace-based blind estimation algorithm for the space–time-coded MIMO MC-CDMA system.
2. To estimate the channel coefficients using subspace-based blind estimation technique and to study the mean square error (MSE) performance of the proposed scheme.
3. To investigate the bit error rate (BER) performance of subspace-based algorithm for MC-CDMA in the proposed system.
4. To compare the system performance with those of the some existing systems under different scenarios.

This chapter has been organized as follows. Section 21.2 represents a detailed survey of the various topics concerned with this research. Section 21.3 deals with the system model and related methodologies. It also presents various system specifications used for the simulations. The results of the computer simulations are presented and discussed in Section 21.4, followed by the conclusion of the chapter in Section 21.5.

21.2 MULTIPLE-INPUT MULTIPLE-OUTPUT (MIMO) SYSTEM

Jack Winters introduced MIMO in 1987 for two basic communication systems. The first one was for communication between multiple mobiles and a base station with multiple antennas; the second one was for communication between two mobiles, each with multiple antennas [1]. MIMO products are now under research; it is expected that by putting multiple antennas at the transmitter and receiver, the capacity will remarkably increase when compared with the single antenna system. The MIMO technique, which provides a high spectral efficiency, is also known as the space–time processing method. This system is inherently proper for high-speed data communications and particularly for the frequency-selective fading channels since the MIMO can gain from spectral diversity of the frequency-selective fading channel.

21.2.1 Capacity of MIMO Channel Systems

A MIMO channel with antenna arrays is shown in Figure 21.1. It has recently been reported that when these dual arrays are deployed in a suitably rich scattering environment, there is considerable potential for obtaining high spectral efficiencies, provided that a suitable space–time code is used [8]. Hence we have used a space–time block code in our system. If n antenna elements exist at each end of the link, then it is possible to create n parallel channels between the transmitter and receiver elements with a corresponding increase in spectral efficiency.

The full channel response of a narrowband system, consisting of n_T transmit elements and n_R receive elements, can therefore be described by an n_R-by-n_T matrix, $\mathbf{G}$ [9]. The elements of the matrix $\mathbf{G}_{ij}$ represent the narrowband amplitude and phase response between each receive element i and each transmit element j. The maximum

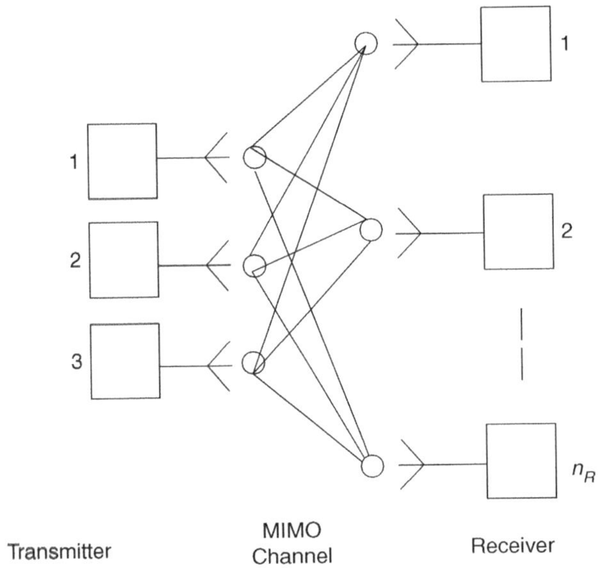

FIGURE 21.1. MIMO Channel and antenna arrays.

possible capacity of this channel [9] is then given by

$$C = \log_2 \left[\det \left(\mathbf{I}_{n_R} + \frac{r}{n_T} \mathbf{H}\mathbf{H}^* \right) \right] \qquad \text{bits/s/Hz} \tag{21.1}$$

where **H** is the n_R-by-n_T complex channel matrix ($\mathbf{H}_{ij}$ is the normalized transfer function from transmit element j to receive element i), n_T and n_R are the number of transmit and receive elements, respectively, r is the average signal-to-noise ratio at each receiver branch, **I** is the identity matrix, det is the determinant, and * is the complex conjugate transpose. The channel response matrix, **H**, is normalized to remove the path loss component and only show the relative variation in the path responses between all n_R-by-n_T elements [8]. MIMO systems employ multiple antennas at both the transmitter and receiver as shown in Figure 21.2. They transmit independent data

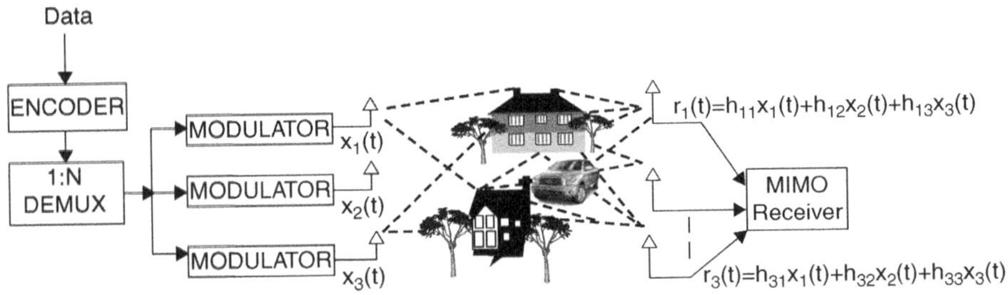

FIGURE 21.2. A generic MIMO system [10].

(say $x_1, x_2, \ldots, x_N$) on different transmit antennas simultaneously and in the same frequency band. At the receiver, a MIMO decoder uses $M \geq N$ antennas. Assuming N receive antennas and representing the signal received by each antenna as r_j, the received signals can be expressed as [10]

$$\begin{aligned} r_1 &= h_{11}x_1 + h_{12}x_2 + \cdots + h_{1N}x_N \\ r_2 &= h_{21}x_1 + h_{22}x_2 + \cdots + h_{2N}x_N \\ &\vdots \\ r_N &= h_{N1}x_1 + h_{N2}x_2 + \cdots + h_{NN}x_N \end{aligned} \tag{21.2}$$

As can be seen from the above set of equations, in making their way from the transmitter to the receiver, the independent signals $\{x_1, x_2, \ldots, x_N\}$ are all combined. Traditionally, this combination has been treated as interference. However, by treating the channel as a matrix, we can in fact recover the independent transmitted streams $\{x_i\}$. To recover the transmitted data stream $\{x_i\}$ from the $\{r_i\}$, we must estimate the individual channel weights h_{ij} and construct the channel matrix $\mathbf{H}$. Having estimated $\mathbf{H}$, multiplication of the vector $\mathbf{r}$ with the inverse of $\mathbf{H}$ produces the estimate of the transmitted vector $\mathbf{x}$. This is equivalent to solving a set of N linear equations in N unknowns.

21.2.2 Space–Time Coding

A space–time-coded system utilizes multiple transmit and receive antennas to increase the information capacity of a wireless communications system [11]. This is achieved by a coding scheme that introduces temporal and spatial correlations into signals transmitted from different antennas, in order to provide diversity gain at the receiver without sacrificing bandwidth. Unlike single antenna systems where scattering is a problem, it becomes a bonus and indeed is necessary for the system to work.

Space–time block codes (STBCs) can be constructed for any number of transmit antennas. An STBC provides diversity gain, with very low decoding complexity. It was originally proposed by Alamouti [2] as a full rate code for two transmit antennas. Consider a wireless communication system with n antennas at the base station and m antennas at the remote. The path gain from transmit antenna i to receive antenna j is defined to be $\alpha_{i,j}$. The received signal at time t at antenna j is given by Tarokh et al. [11]

$$r_t^j = \sum_{i=1}^{n} \alpha_{i,j} c_t^i + \eta_t^j \tag{21.3}$$

where the noise samples η_t^j are independent samples of a zero-mean complex Gaussian random variable. The encoding and decoding algorithm for an STBC are described as follows.

21.2.2.1 Encoding Algorithm. A space–time block code is defined by a $p \times n$ transmission matrix $\mathbf{G}$. The entries of the matrix $\mathbf{G}$ are linear combinations of the

Tx-Ant=1
Rx-Ant=1

over all possible values of s_1 and s_2. It is noted that due to the quasi-static nature of the channel, the path gains are constant over two transmissions. The minimizing values are the receiver estimates of s_1 and s_2, respectively. We expand the above metric and delete the terms that are independent of the codewords and observe that the above minimization is equivalent to minimizing [11].

$$-\sum_{j=1}^{m}\left[r_1^j\alpha_{1,j}^*s_1^* + (r_1^j)^*\alpha_{1,j}s_1 + r_1^j\alpha_{2,j}^*s_2^* + (r_1^j)^*\alpha_{2,j}s_2 - r_2^j\alpha_{1,j}^*s_2\right.$$
$$\left.-(r_2^j)^*\alpha_{1,j}s_2^* + r_2^j\alpha_{2,j}^*s_1 + (r_2^j)^*\alpha_{2,j}s_1^*\right] + (|s_1|^2+|s_2|^2)\sum_{j=1}^{m}\sum_{j=1}^{2}|\alpha_{i,j}|^2 \quad (21.6)$$

The above metric decomposes into two parts, one of which

$$-\sum_{j=1}^{m}\left[r_1^j\alpha_{1,j}^*s_1^* + (r_1^j)^*\alpha_{1,j}s_1 + r_2^j\alpha_{2,j}^*s_1 + (r_2^j)^*\alpha_{2,j}s_1^*\right]$$
$$+ |s_1|^2\sum_{j=1}^{m}\sum_{j=1}^{2}|\alpha_{i,j}|^2 \quad (21.7)$$

is only a function of s_1, and the other one

$$-\sum_{j=1}^{m}\left[r_1^j\alpha_{2,j}^*s_2^* + (r_1^j)^*\alpha_{2,j}s_2 - r_2^j\alpha_{1,j}^*s_2 - (r_2^j)^*\alpha_{1,j}s_2^*\right] + |s_2|^2\sum_{j=1}^{m}\sum_{j=1}^{2}|\alpha_{i,j}|^2 \quad (21.8)$$

is only a function of s_2. Thus the minimization of (21.6) is equivalent to minimizing these two parts separately. This in turn is equivalent to minimizing the decision metric [11]

$$\left|\left[\sum_{j=1}^{m}\left(r_1^j\alpha_{1,j}^* + (r_2^j)^*\alpha_{2,j}\right)\right] - s_1\right|^2 + \left(-1+\sum_{j=1}^{m}\sum_{i=1}^{2}|\alpha_{i,j}|^2\right)|s_1|^2 \quad (21.9)$$

for detecting s_1 and the decision metric [11]

$$\left|\left[\sum_{j=1}^{m}\left(r_1^j\alpha_{2,j}^* - (r_2^j)^*\alpha_{1,j}\right)\right] - s_2\right|^2 + \left(-1+\sum_{j=1}^{m}\sum_{i=1}^{2}|\alpha_{i,j}|^2\right)|s_2|^2 \quad (21.10)$$

for decoding s_2. This is the simple decoding scheme described in Alamouti [2].

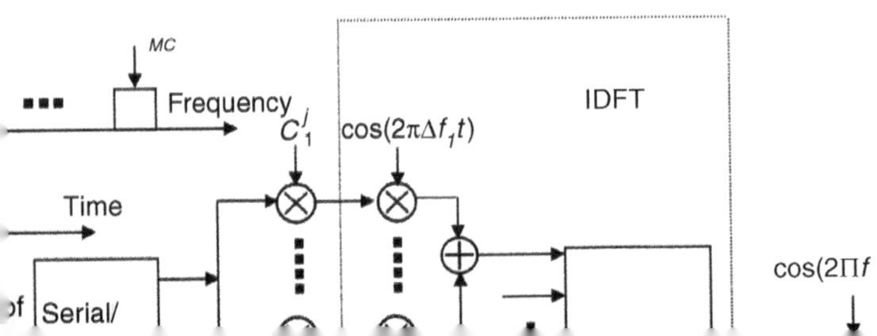
MC
Frequency
Time
C_1^j
$\cos(2\pi\Delta f_1 t)$
IDFT
$\cos(2\Pi f$
Serial/

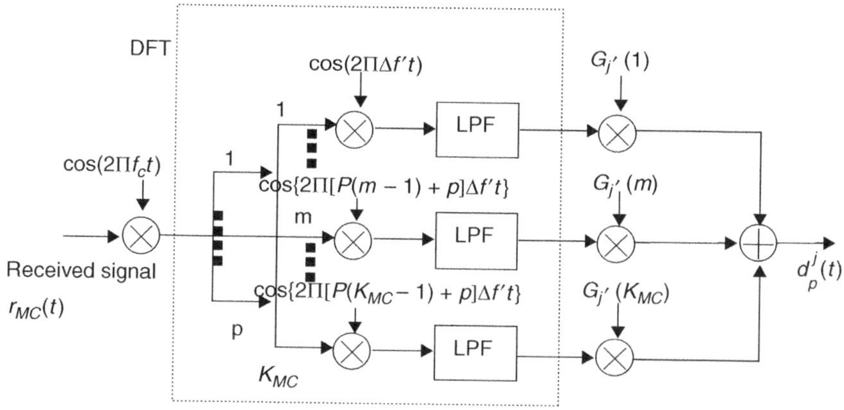

FIGURE 21.5. MC-CDMA receiver [13].

caused by multipath fading, and finally the signal is transmitted after radio-frequency (RF) up-conversion. The transmitted signals for multiusers (with the number of users, j) at time index $t = 0$ will be [13]

$$S_{MC} = \sum_{p=1}^{P} \sum_{m=1}^{K_{MC}} \sum_{j=1}^{j} d_p^j C^j(m) \tag{21.11}$$

where $[C^j(1), C^j(2), \ldots, C^j(k_{MC})]$ is the spreading code with length K_{MC}. If the original symbol duration is T_s, the symbol duration after serial-to-parallel conversion will be PT_s. Let the symbol duration at the subcarrier level be $T_s' = PT_s + \Delta$, where Δ is the guard interval and we take its value to be 10% of the original symbol duration, PT_s. So minimum subcarrier frequency separations for our system will be $\Delta f' = 1/T_s' - \Delta$ and the bandwidth of the transmitted signal spectrum can be written as [13]

$$B_{MC} = \frac{P \cdot K_{MC} - 1}{T_s' - \Delta} + \frac{2}{T_s'} \tag{21.12}$$

The last term in Eq. (21.12) is involved because we are adding the every guard interval after the transmission of P symbols. The symbol duration at the subcarrier level is P times as long as the original symbol duration due to serial/parallel conversion, and the total number subcarriers will be P times K_{MC}.

MC-CDMA Receiver Structure. A basic MC-CDMA receiver is illustrated in Figure 21.5. The receiver reverses the operation of the transmitter. First, the received signal is demodulated by multiplying this signal with the orthogonal carrier frequencies and then low-pass filtering the resulting signals. It requires coherent detection for successful despreading operation, and this causes the structure of MC-CDMA receiver to be very complicated. After down-conversion, the m-subcarrier components ($m = 1, 2, \ldots, K_{MC}$) corresponding to the received data $d_p^j(i)$ is the first coherently detected with discrete Fourier transform (DFT) and then multiplied with the gain $G^j(m)$ to combine the energy of the received signal scattered in the frequency domain.

The received signal after passing through channel will be [13]

$$r_{MC} = \sum_{p=1}^{P} \sum_{m=1}^{K_{MC}} \sum_{j=1}^{J} z_{m,p}^j d_p^j C^j(m) + n \qquad (21.13)$$

where n is zero mean AWGN noise and $z_{m,p}^j$ is the received complex envelope (fading coefficient) at the $(mP + p)$th subcarrier of the jth user. To get this fading coefficient in frequency domain, we need to do FFT for each pair of the time-domain fading coefficients with K points DFT.

$$z_{m,p}^j(f) = \text{FFT}\{z_l^j(t)\} \qquad (21.14)$$

If we want the received complex envelope for all users to be the same ($z_m = z_m^1 = z_m^2 = \cdots = z_m^j$) and omit the subscription p, we can rewrite Eq. (21.13) as follows [13]:

$$y(m) = \sum_{j=1}^{J} z_m d^j C^j(m) + n \qquad (21.15)$$

Let the desired user be j'. To combine the energy of the received signal scattered in the frequency domain, we need to multiply with the gain $G^{j'}(m)$. The estimated coded symbol of the desired user is given by

$$\hat{d}^{j'}(m) = G^{j'}(m) y(m) \qquad (21.16)$$

The gain can be calculated by using any of the combining methods described in Hara and Prasad [13].

21.2.4 Channel Estimation

There are two aspects of channel estimation. The channel impulse response needs to be estimated in order to employ coherent demodulation of the received CDMA symbols and compensate for the distortions introduced into the signal in the channel. An accurate channel response estimate allows the use of multiuser detection and power loading to improve the system performance [14]. The other aspect of channel estimation is its importance in adaptive modulation-based systems as the modulation format is changed in the transmitter according to the condition of the channel.

The channel impulse response $h(t)$ can be expressed in terms of the transmitted signal $s(t)$, the received signal $r(t)$, and the additive noise $w(t)$ as [15]

$$r(t) = s(t)^{*}h(t) + w(t) \tag{21.17}$$

If $S(k)$ and $Y(k)$ are, respectively, the transmitted and received signals in the discrete frequency domain, $H(k)$ is the channel transfer function, and $N(k)$ is the AWGN noise sample, then we can write the receive signal as

$$Y(k) = H(k)S(k) + N(k) \tag{21.18}$$

Then, the estimated channel response in frequency domain can be given in terms of the transmitted and received signal as

$$\hat{S}(k) \gg \frac{Y(k)}{\hat{H}(k)} \tag{21.19}$$

where $\hat{S}(k)$ and $\hat{H}(k)$ are the estimated signal and channel samples, respectively. $\hat{H}(k)$ (or $\hat{h}(t)$) can be estimated in the receiver in two ways. The first way is to transmit some known signals (pilot signals) at regular intervals and then estimate the channel by analyzing these signals. This is commonly known as pilot-based channel estimation. The other way is to estimate the channel blindly. In other words, the channel is estimated by the receiver completely on its own, without taking any help from the transmitter.

We use the blind estimation technique in this chapter. Numerous researchers have actively studied blind techniques over the past few decades. There are several reasons for studying blind algorithms, which are as follows:

- Bandwidth is conserved by eliminating or reducing training sets.
- In rapidly time-varying channels, training is not efficient.
- Severe multipath fading during the training period can lead to poor channel estimates.
- Training requires synchronization, which may not be feasible in multiuser scenarios.
- In distributed networks, sending training signals each time a new communication link is to be set up may not be feasible.

In GSM, 20% of the bandwidth is devoted to training, and blind algorithms can be used to reduce the length of the training sequence [5]. Blind methods are also more robust if the signal undergoes severe fading during training. The scenario is common for many practical applications where RFID has been brought forward as a solution; such scenarios are airport baggage tagging and tracking, toll collection for vehicles using gantry-based readers, busy warehouses, assembly lines in manufacturing environment, and so on.

21.2.4.1 Blind Channel Response Estimation in MC-CDMA. Subspace-based channel estimation has been proposed for CDMA and subsequently for MC-CDMA systems by number of researchers. In Bensley and Aazhang [16], by exploiting the eigenstructure of the received signal's sample correlation matrix, it has been shown that the observation space can be partitioned into a signal subspace and a noise subspace without prior knowledge of the unknown parameters. The channel estimate is formed by projecting a given user's spreading waveform into the estimated noise subspace and then either maximizing the likelihood or minimizing the Euclidean norm of this projection. Liu and Xu [17] also proposed a similar approach for a synchronous CDMA system, but instead of applying eigenvalue decomposition, it uses singular value decomposition (SVD) to the received data matrix to obtain the orthogonal subspace. The channel vector is then estimated by solving a linear equation set obtained as a result of the orthogonality. Escudero et al. [18] presented a blind channel identification method for downlink MC-CDMA based on the subspace decomposition, which exploits the orthogonality property between the noise subspace and the received user codes to obtain a channel identification algorithm. The channel coefficients of the desired user are estimated by projecting the desired user vector onto the noise subspace. The same algorithm has been used in this chapter to estimate the channel blindly for the proposed system.

21.3 METHODOLOGY

In this section, the blind estimation of channel in MIMO for MC-CDMA system is studied. The block diagram of low-pass equivalent representation of the overall simulation model is shown in Figure 21.6. A general description of the transmitter and receiver MC-CDMA is discussed in Section 21.2. More detailed description of each block of the system model is presented here.

21.3.1 The System Model

Figure 21.6 shows the MC-CDMA transmitter and receiver of the nth user with the space–time code, two transmit antennas, and two receive antennas. Table 21.1 shows the summary of simulation parameter of the system.

21.3.2 Space–Time Encoding and Decoding

The details of the space–time block encoding and decoding algorithm are described in Section 21.2.2. In this study, the space–time block code is adapted to two transmit antennas as shown in Figure 21.7. The two adjacent modulated data symbols x_1 and x_2 presented at the input of the space–time encoder for nth user are sent to antenna one and antenna two, respectively. During the next period, $-x_2^*$ is sent to antenna one, and signal x_1^* is sent to antenna two. Each symbol at the output of the space–time encoder is multiplied with its user-specific Walsh–Hadamard (WH) spreading

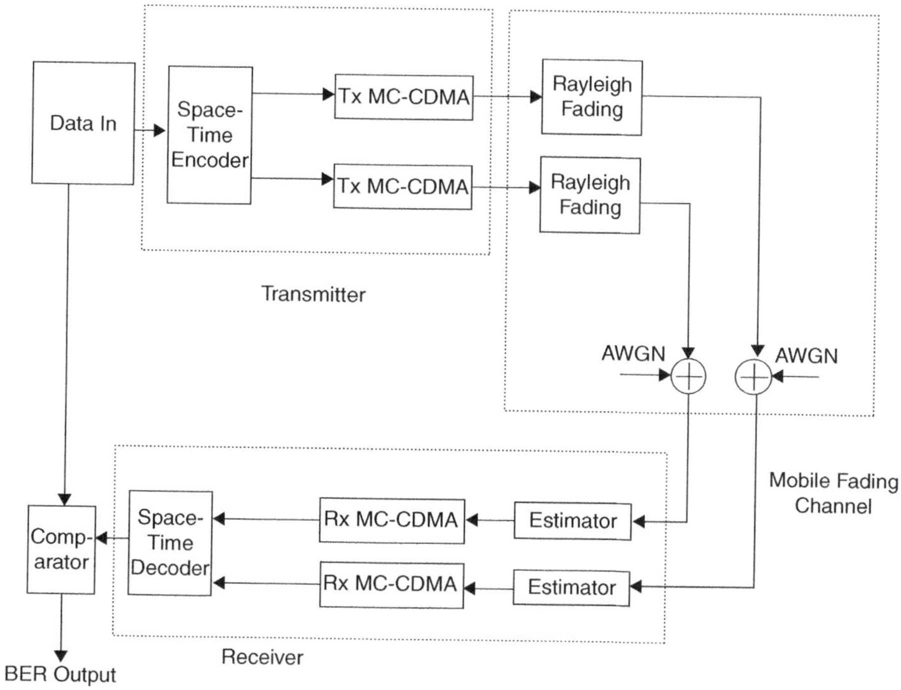

FIGURE 21.6. Block diagram of the system model.

TABLE 21.1. Summary of Simulation Parameters

MC-CDMA System Design (based on Hara and Prasad [13])	
Serial-to-parallel conversion factor	$P = 4$
Number of subcarriers	128
Processing gain	32
FFT/IFFT Points	128
Subcarrier bandwidth	<128 kHz
Total MC-CDMA bandwidth	16.5 MHz for S/P factor of 4 with above configuration
Other Parameters	
Source symbol rate	640 ksymbols/s
Chip rate	4.096 Mchips/s
Carrier frequency	2 GHz
Spreading sequences	Walsh–Hadamard orthogonal sequence of length 32
Channel type	Rayleigh fading channel; delay spread 1 μs; 6 taps exponential delay profile
Mobile velocity	100 km/h, 50 km/h, and 5 km/h
BER Simulation	Monte Carlo method
Total symbol duration	50 μs
Circular prefix	6–10% of total frame length

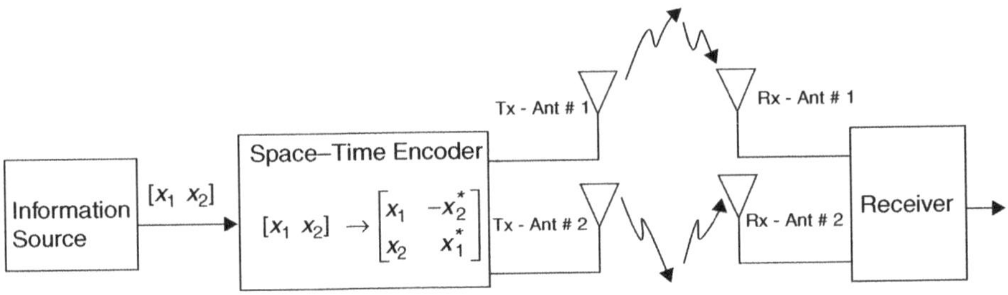

FIGURE 21.7. Space–time encoding system.

code assigned to user n with kth chip [19]. The decoding algorithm is described in Section 21.2.2.

21.3.3 Transmitter

The transmitter section of each user generates the samples at the chip rate. A brief description of each component is given with respect to the block diagram of Figure 21.8. The user data with symbol rate of 512 ksymbols/s is serial to parallel converted with serial to parallel conversion factor of 4. This results in a rate of 128 ksymbols/s in each of the parallel converted branch. Each of the branches is then subjected to parallel spreading using Walsh–Hadamard (WH) codes of length 32. Resulted streams are subjected to interleaving by a factor of 4 and fed to IFFT block to ensure that the subcarrier frequency separation for the same data bit is increased by a factor of 4. The interleaving here works similar to the normal interleaving phenomena such that because of the frequency diversity, the chance of simultaneous data loss in each of the subcarrier is minimized. The IFFT output is parallel to serial converted and cyclic prefix is inserted in between the OFDM frames. The cyclic prefix is determined such that it is greater than or equal to the channel impulse response length whereby ISI is minimized. For our case, it is taken as 6% to 10% of the total frame length.

21.3.4 Channel Model

Since it is not possible to always have a direct line-of-sight (LOS) between mobile station (MS) and base station (BS), the most appropriate channel model for this

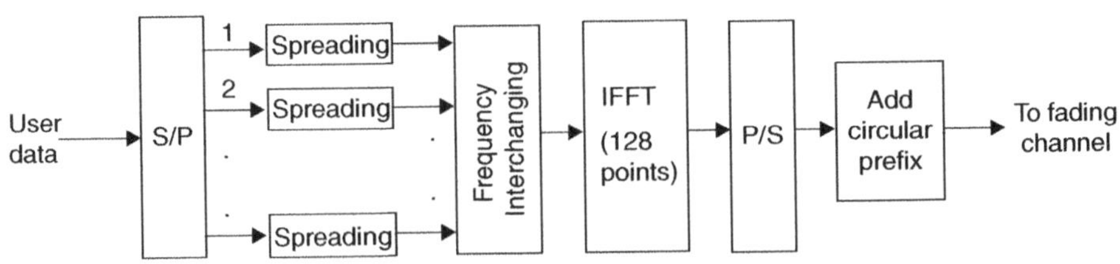

FIGURE 21.8. Transmitter model.

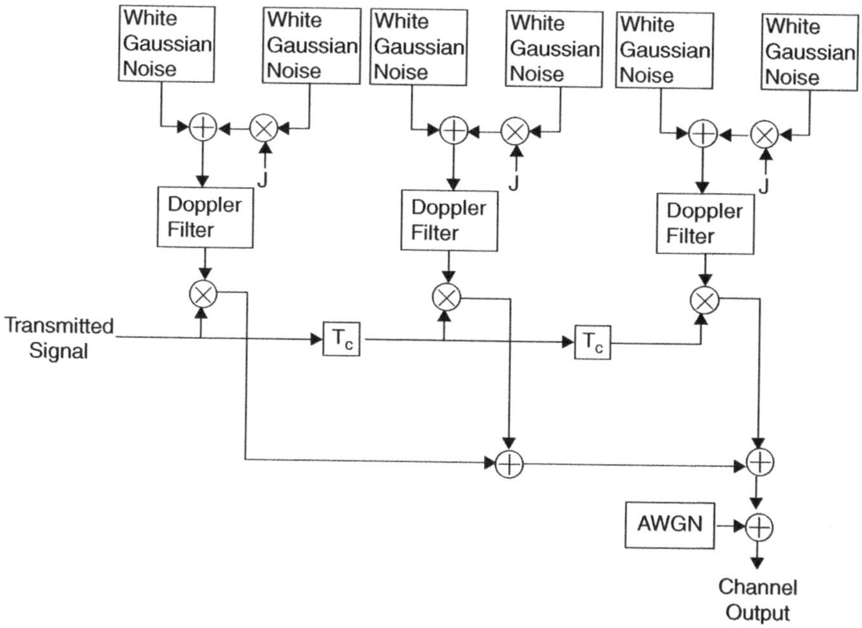

FIGURE 21.9. Rayleigh channel model [20].

case is the Rayleigh fading model consisting of multiple narrowband channel paths separated by delay elements as illustrated in Figure 21.9 [20].

The discrete-time frequency-selective Rayleigh fading channel model is used in the simulation. Typical multipath environment with delay spread of 1 μs is considered [21]. The impulse response of such a time-variant frequency-selective channel can be expressed as

$$h(kT_c) = \sum_{l=0}^{L-1} \beta_l \partial(kT_c - lT_c) \tag{21.20}$$

where β_l are channel tap coefficients and are complex random process having Rayleigh distributed amplitudes and uniformly distributed phases. The number of taps L is determined using the following equation for CDMA systems [15]:

$$L = \left\lfloor \frac{T_{\max}}{T_c} \right\rfloor + 1 \tag{21.21}$$

where $\lfloor x \rfloor$ is largest integer contained in x and T_c is the chip duration. The delays can also be approximated to be of the integer multiple of the chip duration. In our case, for maximum delay spread of 1 μs and $T_c = 1/(4.096 \times 10^6)s$, we can model the channel with $L = 6$ taps. Here we have considered only the paths with delays less than $6T_c$. The mobility of the users is modeled by filtering the time-variant channel tap coefficients through Doppler filters. The channel model is illustrated in Figure 21.9. This channel model represents a complete channel for a single-user scenario. For a multiuser scenario, the AWGN should be added at the received front end of the array as shown in the simulation model of Figure 21.6.

The power delay profile for the system under consideration can be approximated by a one-sided exponentially decaying function of the form $S(\tau) = \kappa \exp(-\tau/\sigma_{rms})$. To realize this power delay profile, different variances to noise source is assigned for each tap. Total averaged power of the channel is normalized to unity. It is seen from Figure 21.9 that two Gaussian noise sources are used to generate Rayleigh fading envelopes. Let the variances of each noise source be σ_l^2. The average power of the lth path Rayleigh fading envelope is given by its mean square value,

$$S(l) = E\left(\alpha_l^2\right) = 2\sigma_l^2 \tag{21.22}$$

$$S(l) = \exp(-lT_c/T_d)/\left(\sum_{l=0}^{6} \exp(-lT_c/T_d)\right) \tag{21.23}$$

Here we have assumed the decaying factor to be 1 μs to make most of the power concentrated within the maximum delay spread, even though the total is normalized to unity. The standard deviation of noise sources for each tap is given by

$$\sigma_1 = \sqrt{\frac{S(l)}{2}} \tag{21.24}$$

The Doppler filters used to model the movement of the users are the same for all taps, because it is determined by the carrier frequency and mobile speed. The Doppler power spectral density function characterizes the time-variant nature of the channel and is given by

$$S(v) = \frac{1}{\pi f_D \sqrt{1 - \left(\frac{v}{f_D}\right)^2}} \tag{21.25}$$

where v denotes the frequency shift and f_D is the maximum Doppler frequency given by $f_D = v/\lambda$, where v denotes the relative mobile velocity and λ is wavelength. The frequency response of the Doppler filter with the above power spectral density function is expressed as

$$H(f) = \sqrt{\frac{1}{\sqrt{1 - \left(\frac{f}{f_D}\right)^2}}} \tag{21.26}$$

In our simulation model, f_D is 185.2 Hz for 100 km/h, 92.6 Hz for 50 km/h, and 9.3 Hz for 5 km/h mobile velocities. Implementation of a Doppler filter for a slow varying channel necessitates the high sampling frequency, because the maximum Doppler frequency is very low compared to the chip rate. Designing the filter with such a high frequency would require a very-high-order FIR filter. However, in our case we have assumed that the channel is invariant over an OFDM frame. This makes the sampling rate to be 32 kHz. For 100 km/h, we design the filter with initial sampling rate of 800 Hz, which is well above the Nyquist criterion. The filter is designed as a 127th-order FIR filter with 128 taps. Hence we can interpolate it by a factor of 40 in a single interpolating stage. Similarly for 50 km/h and for 5 km/h, we adjust the initial sampling rate and interpolation factor accordingly. The filter coefficients are then normalized to unity power gain to maintain the average power of all paths as

$$h_D[n] = \frac{h_D[n]}{\sqrt{\sum_n |h_D[n]|^2}} \tag{21.27}$$

21.3.5 Blind Channel Impulse Response Estimation

The estimator block of the receiver side is the main concern of this chapter. The received MC-CDMA signal from the Raleigh fading channel is fed to the channel response estimator, which blindly estimates the channel impulse response, $\widehat{h}(t)$, by a subspace-based estimation algorithm. This channel impulse response estimate is converted to its frequency-domain counterpart, called the channel transfer function, using the DFT matrix. There are various methods that are commonly used to estimate the channel impulse response blindly. A subspace-based channel estimation algorithm is proposed and derived by Escudero et al. [18]. In this research, this channel estimation algorithm has been used in the case of MIMO for MC-CDMA, and its performance in terms of MSE and BER has been analyzed and compared with those of the some existing systems. The basic steps of the algorithm are summarized below:

- Find the autocorrelation matrix, $\mathbf{R}$, of the DFT output in the MC-CDMA receiver.
- To separate the signal and noise subspaces, perform eigenvalue decomposition (EVD) of $\mathbf{R}$.
- A set of linear equations are obtained to exploit the orthogonality between the signal and noise subspaces.
- From the solutions of these equations the channel response vector $\widehat{h}$ is derived.
- By multiplying the estimated channel impulse response with the DFT matrix, the channel transfer function $\widehat{H}$ can be obtained.

In our system, we have calculated the normalized mean square error of each subchannel, which is defined as $\text{MSE} = \frac{1}{N}\sum_{i=1}^{N} \|\hat{\mathbf{H}}(i) - \mathbf{H}\|^2/\|\mathbf{H}\|^2$, where N is the number

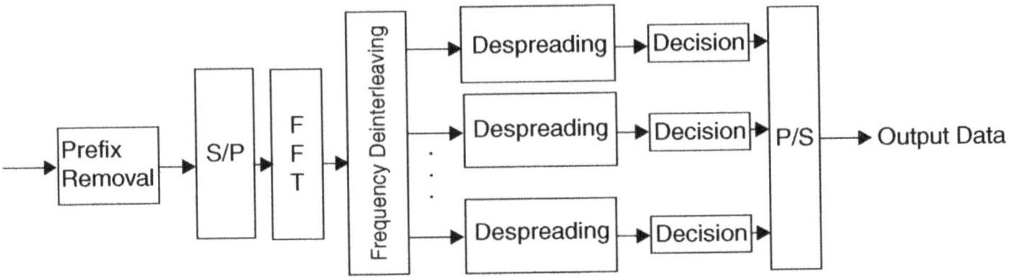

FIGURE 21.10. MC-CDMA receiver block.

of Monte Carlo trials in the simulation. The matrix $\hat{H}(i)$ is the estimate of **H** from the ith trial. The overall normalized mean square error is obtained by averaging overall subchannels. The estimated channel coefficients are fed to the decoder to estimate the transmitted bits.

21.3.6 Receiver

The basic MC-CDMA receiver block is shown in Figure 21.10. The received signals are sampled and serial to parallel converted to 128 streams and fed to FFT block to demodulate the data information in each subcarrier. These parallel streams are then deinterleaved exactly opposite to the process of interleaving to reassemble the appropriate data information in a group. Each group is then despreaded using the Walsh–Hadamard (WH) codes of length 32.

21.4 RESULTS AND DISCUSSIONS

In this section, the performance of blind estimation of channel in MIMO technique is studied through computer simulations for MC-CDMA systems. The performance of investigated estimation techniques in Rayleigh fading channel is presented and discussed in this section. Apart from the main simulation work with blind estimation techniques of channel, comparisons of mean square error (MSE) and bit error rate (BER) performances have been made also with those of the some existing systems.

21.4.1 Blind Channel Response Estimation

A subspace-based algorithm, as discussed in Section 21.3.5, is used to obtain an estimate of the channel impulse response for the MIMO MC-CDMA systems. Results are shown for systems using a 16-QAM-modulation scheme for different channel conditions. Table 21.2 shows the accurate and estimated values of the channel impulse response coefficient of one subchannel for the considered MC-CDMA systems in a Rayleigh fading channel with six-path exponential power distributions for channel conditions of 10 dB and 20 dB. We have considered four users in the simulation for this section. The table indicates that the estimated channel impulse response is differing from the actual for lower value of channel conditions. However, for higher

TABLE 21.2. Magnitude of the Estimated and Actual Values of the Channel Impulse Response for the MC-CDMA System in Rayleigh Fading Channel with 16-QAM Modulation Scheme

Channel Impulse Responses	SNR = 10 dB		SNR = 20 dB	
	Actual Values	Estimated Values	Actual Values	Estimated Values
$h(1)$	0.5285	0.3071	0.5285	0.5098
$h(2)$	0.0372	0.0766	0.0372	0.0421
$h(3)$	0.2926	0.4588	0.2926	0.3116
$h(4)$	0.1264	0.2216	0.1264	0.1392
$h(5)$	0.2122	0.3062	0.2122	0.2367
$h(6)$	0.0245	0.0788	0.0245	0.0389

value of SNR with the same simulation environment, the estimated and actual values of the channel coefficients are found to be comparable.

The results obtained in this section indicate that the estimated channel impulse response is comparable to that of the actual values, thus validating the correct implementation of the blind estimation algorithm for the proposed system.

21.4.2 Performance of Subspace Algorithm for MIMO MC-CDMA System

In this section, we use simulation to study the performance of subspace algorithm in terms of mean square error (MSE) for the proposed system. For each experiment, we have used several modulation schemes. After estimating the MIMO channel impulse responses, we have calculated the normalized mean square error of each subchannel. The overall normalized mean square error is obtained by averaging overall subchannels. All results concerning MSE are ensemble averages of 100 independent Monte Carlo runs in the simulation.

The mean square error (MSE) performance of channel for the proposed system under different number of symbols for various modulation schemes is shown in Figure 21.11. The number of subcarriers is chosen to 128 to satisfy the required symbol rate and combat the multipath delay spread. Walsh–Hadamard orthogonal sequences of length 32 are used as a spreading factor, and the SNR is fixed to 20 dB. The simulation setting with 14 active users is considered. It is observed that the MSE of the channel is almost the same for all modulation formats, and MSE is decreasing significantly with an increasing number of symbols.

Figure 21.12 shows the simulated MSE encountered in the channel impulse response estimation for various modulation schemes and under different SNRs of the received users. The simulation environment is the same as before. It is also observed that the MSE is decreasing with increasing SNR. Comparing the simulated results of Figures 21.11 and 21.12 to the results obtained by Escudero et al. [18], the simulated performance of subspace algorithm in terms of MSE for the proposed systems is found to be a bit improved in our systems.

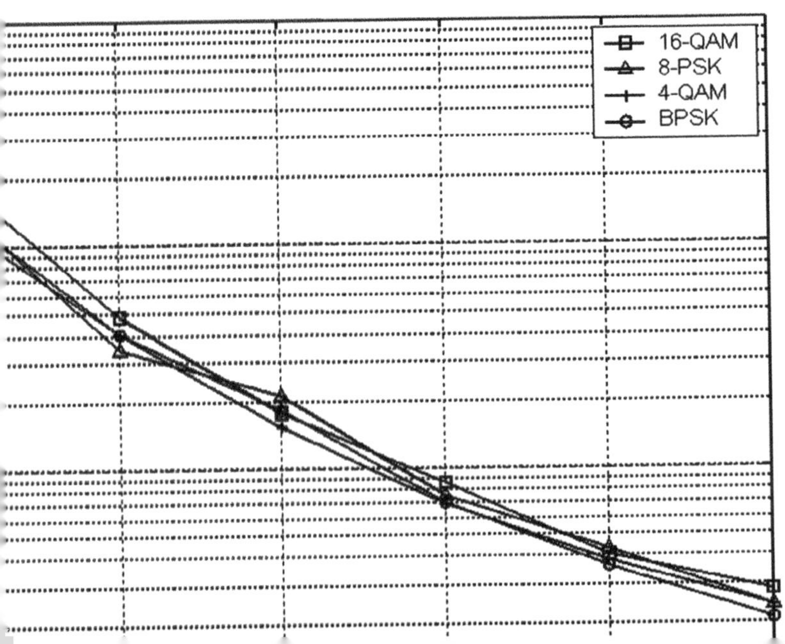
16-QAM
8-PSK
4-QAM
BPSK

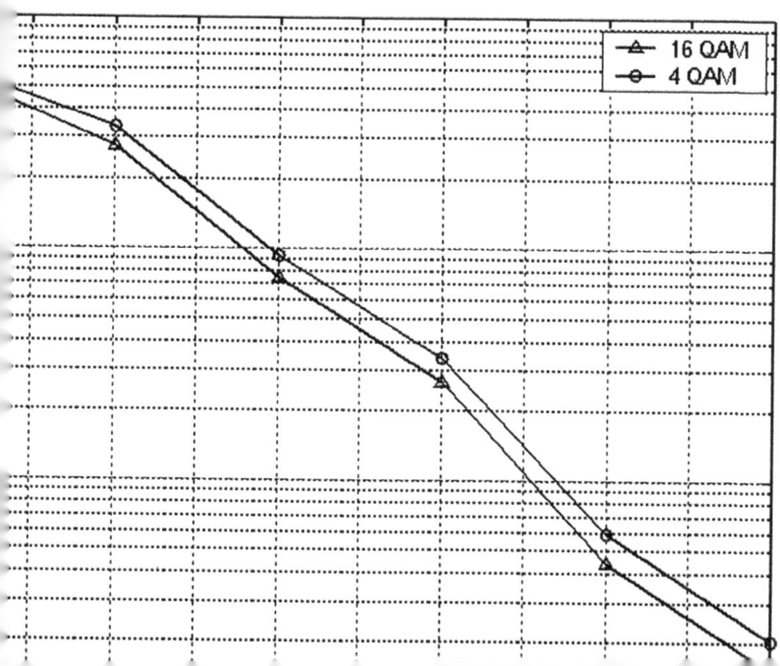
16 QAM
4 QAM

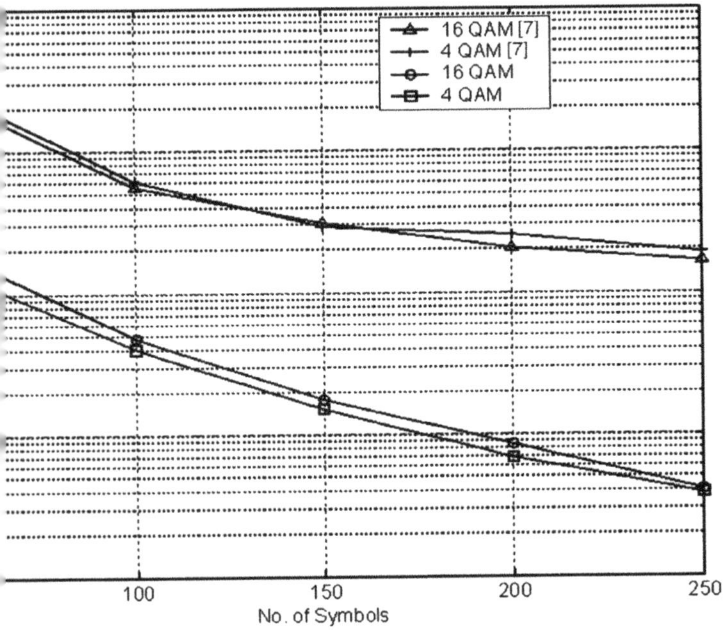

3. MSE performance of subspace algorithm for an MC-CDMA system in two under different symbols, SNR = 20 dB.

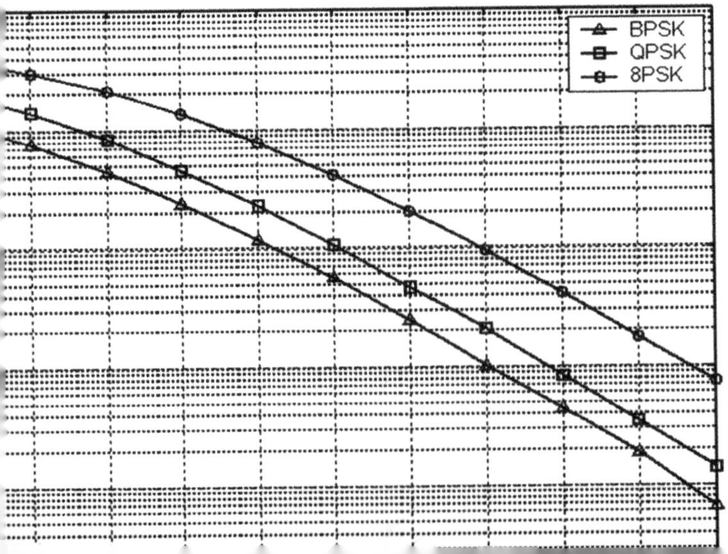

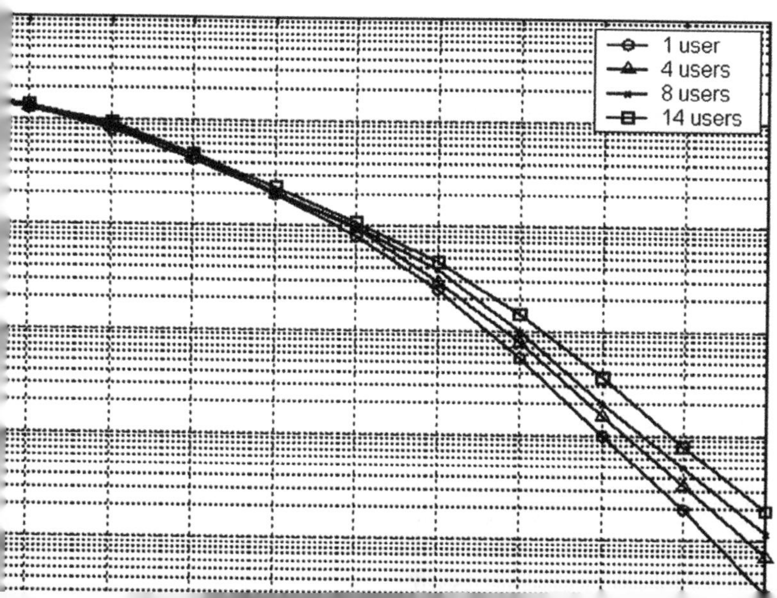
1 user
4 users
8 users
14 users

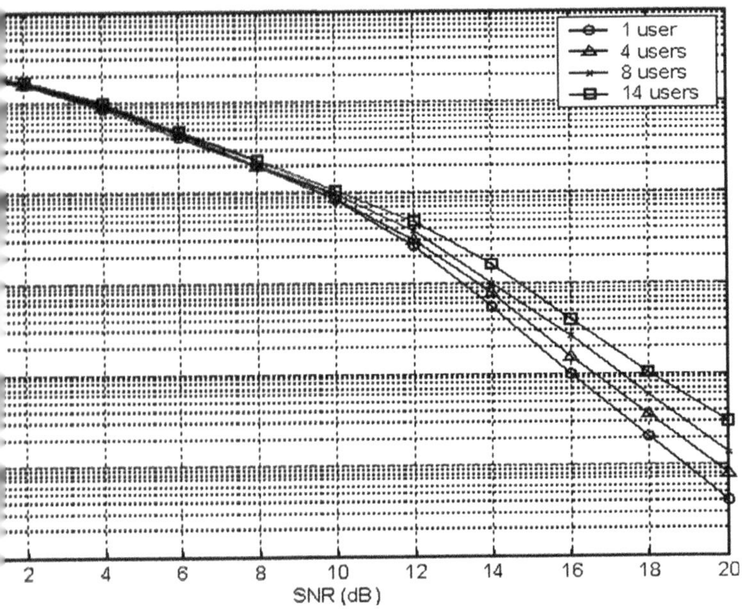

16. The space–time coding MC-CDMA performance under various numbers of
ing a 16-QAM modulation scheme for 50-km/h mobile velocity.

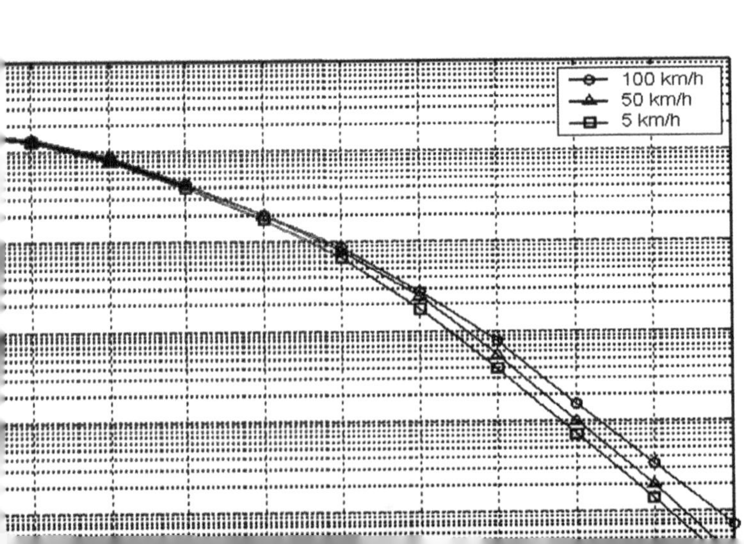

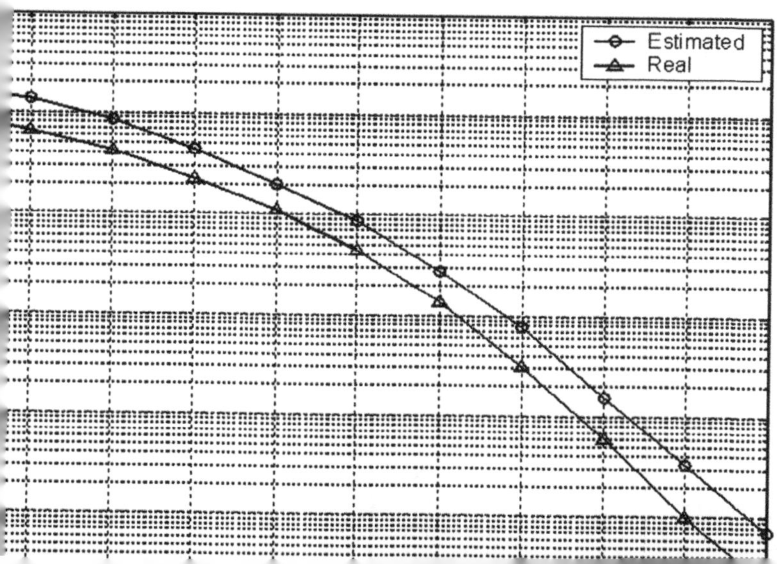

Estimated
Real

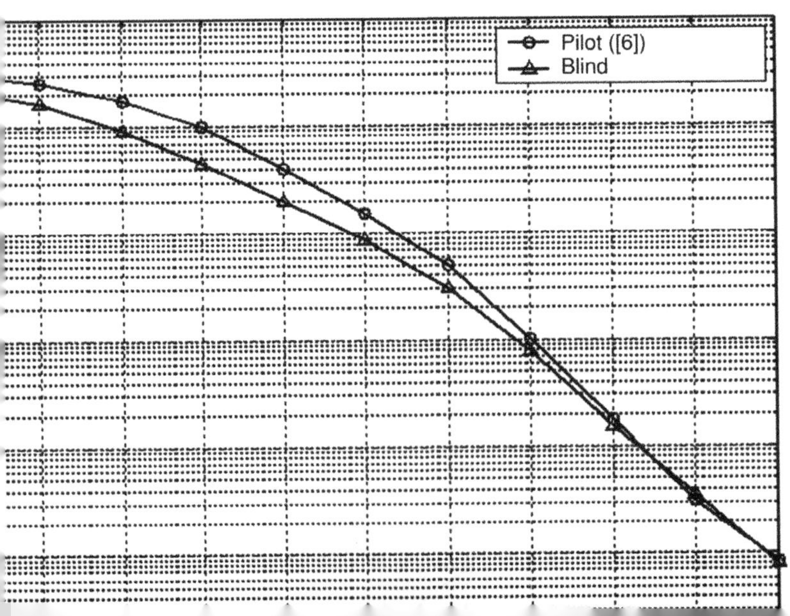

Pilot ([6])
Blind

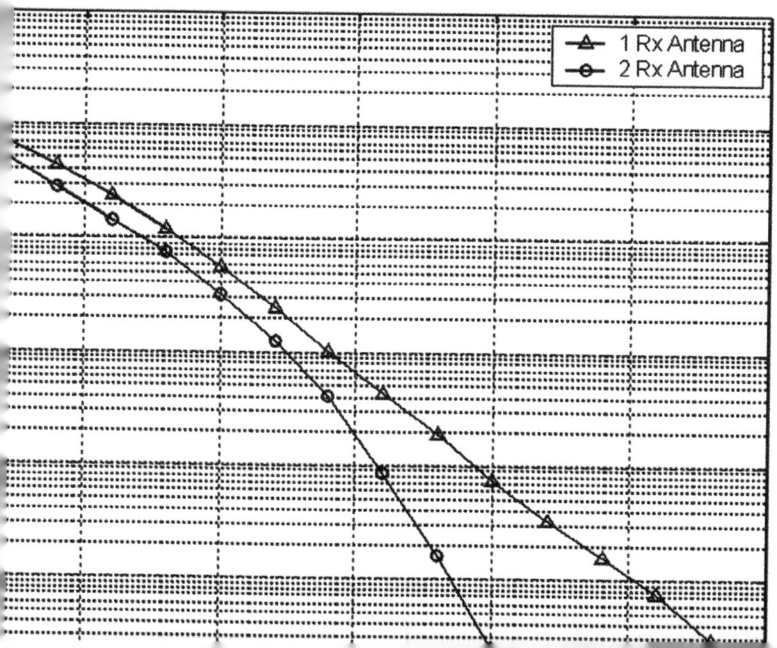

1 Rx Antenna
2 Rx Antenna

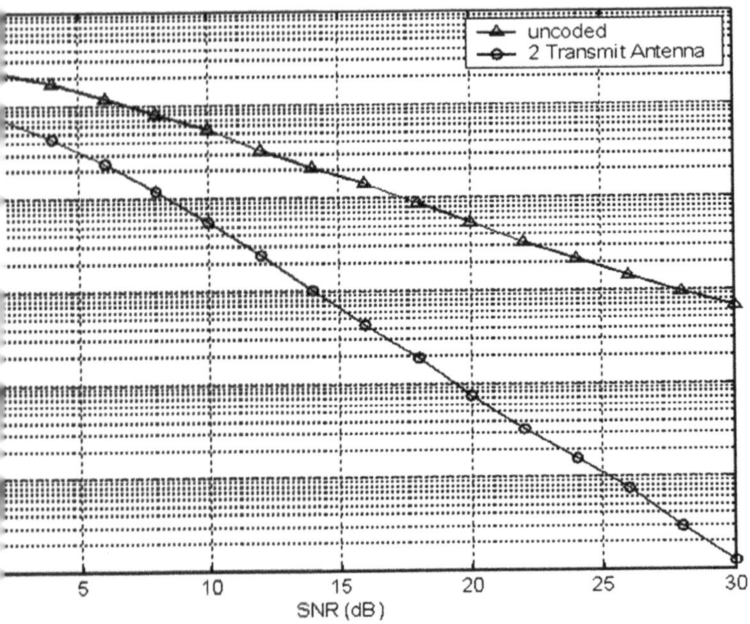

21. The performance of STBC with rate 1 bit/s/Hz for uncoded and two transmit r one receive antenna for a BPSK modulation scheme.

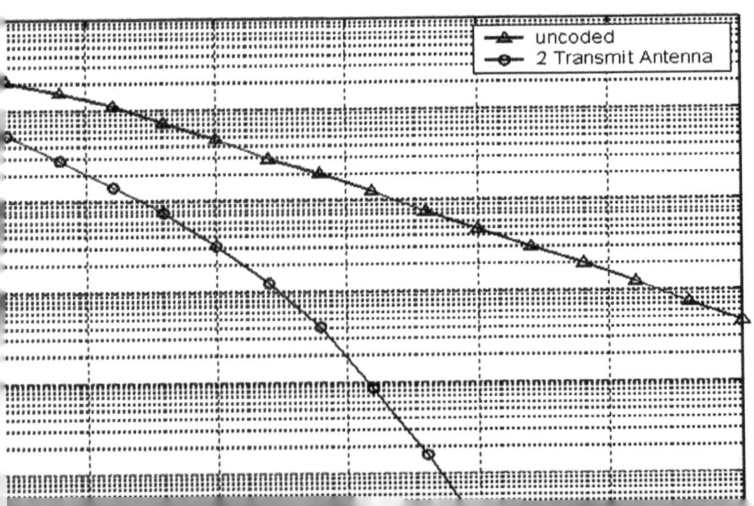

MIMO systems. The digital modulation technique, called MC-CDMA, is introduced and implemented as a multiplexing scheme applied to the system in the downlink transmission. The simulations of the two transmit and one and two receive antennas system are carried out to verify the performance of the proposed estimation method for the multipath channel.

The channel impulse response has been estimated blindly using subspace-based algorithm in the case of MIMO MC-CDMA systems. In the implementation of subspace-based algorithm in the simulation, eigenvalue decomposition of an autocorrelation matrix has been performed to separate the signal and noise subspaces. After performing this operation, a set of linear equations has been obtained. From the solutions of these equations, the channel impulse response has been estimated. In the estimation of channel coefficient values, it has been observed that there is a degradation of channel coefficient for smaller values of SNR. However, for higher values of SNR, there is no significant difference between the estimated and the actual value.

The mean square error (MSE) of the channel in terms of SNR and symbols were found to be better than the existing systems where MIMO were not considered. The bit error rate (BER) performance of the space–time-coded MIMO MC-CDMA systems using blind channel estimation is compared with that of the pilot-symbol-based estimation. It is observed from simulation results that the BER performance of the systems based on blind channel estimation is almost similar to those based on pilot-based channel estimation. The BER performance of the STBC over Rayleigh fading channel shows that the BPSK and QPSK modulation schemes give a better performance than the 8-PSK modulation scheme. The STBC for transmission efficiency of 1 bit/s/Hz with two transmits antennas and one and two receive antennas has been implemented. From the simulated results, it is observed that the system with two receive antennas has a better performance than the system with one receive antenna. The BER performance over multipath channel using the proposed channel estimation and ideal channel for one user under the 4-QAM modulation scheme has been observed in the simulation. It is shown that the proposed estimation scheme requires about 2 dB more than the ideal scheme. Based on the observation above, the following recommendations are made:

1. Frequency offset and peak-to-average power ratio, which are main serious problems of multicarrier modulation scheme, are not considered. An investigation into these problems is recommended for the proposed systems
2. Subspace-based channel estimation algorithm is applied to the space–time-coded MIMO MC-CDMA systems. The effect of other channel estimation algorithms may be tested on these systems for any improvement in the BER performance.
3. We have used STBC for the MIMO MC-CDMA system. STBC provides diversity gain, with very low decoding complexity, whereas space–time trellis codes (STTCs) provide both diversity and coding gain at the cost of higher decoding complexity. The STTCs may be used to improve the BER performance of the proposed systems.

REFERENCES

1. J. Winters, On the capacity of radio communication systems with diversity in a Rayleigh fading environments, *IEEE J. Selected Areas Commun.*, Vol. SAC-5, pp. 871–878, 1987.
2. S. M. Alamouti, A simple transmitter diversity scheme for wireless communications, *IEEE J. Selected Areas Commun.*, Vol. 16, No. 8, pp. 1451–1458, October 1998.
3. S. Hara and R. Prasad, Overview of multicarrier CDMA, *IEEE Commun. Mag.*, December 1997.
4. C. Bonnet, Mobile Communications Department Activity Report, Institute Eurecom, France. Available at http://www.eurecom.fr/Mobile/Research/RapportRech-erche2002.pdf. Accessed October 2004.
5. S. Talwar, Blind Space–Time Algorithms for Wireless Communication Systems, Ph.D dissertation, Scientific Computing and Computational Mathematics, Stanford University, January 1996.
6. T. A. T. Duong, Channel Estimation in a MIMO multi carrier CDMA system, Master's Thesis, Asian Institute of Technology, Bangkok, Thailand. Thesis No. TC-01-7, 2001.
7. S. Chatterjee, Blind Estimation of Channel and Modulation scheme in Adaptive Modulation based OFDM-CDMA Systems, Master's Thesis, Asian Institute of Technology, Bangkok, Thailand. Thesis No. TC-03-7, 2003.
8. G. J. Foschini, and M. J. Gans, On limits of wireless communications in a fading environment when using multiple antennas, *Wireless Personal Communications*, Vol. 6, No. 3, pp. 311–335, March 1998.
9. K. Z. Castro, W. G. Scanlon, and F. Tofoni, Dynamic capacity estimation for the indoor wireless channel with MIMO arrays and pedestrian traffic, in *Proceedings of the 1st Joint IEI/IEE Symposium on Telecommunications Systems Research*, November 2001.
10. B. Daneshrad, Introduction to MIMO and MIMO-OFDM, available at http://www.mimo.ucla.edu/summaries/INTRO_MIMO&OFDM.pdf.Accessed November 5, 2004.
11. V. Tarokh, H. Jafarkhani, and A. R. Calderbank, Space–time block coding for wireless communications: Performance results, *IEEE J. Selected Areas Commun.*, Vol. 17, No. 3, pp. 451–460, March 1999.
12. N. Yee, and J. P. Linnartz, Multi-carrier CDMA in an indoor wireless radio channel, in *Proceedings of IEEE PIMRC'93*, Yokohama, Japan, pp. 109–113, September 1993.
13. S. Hara, and R. Prasad, (1999), Design and performance of multicarrier CDMA system in frequency-selective Rayleigh fading channels, *IEEE Trans. Veh. Technol.*, Vol. 48, No. 5, pp. 1584–1595, September 1999.
14. U. Tureli, D. Kivance, and H. Liu, Channel estimation for multicarrier CDMA, in *Proceedings of IEEE International Conference in Acoustics, Speech, and Signal Processing (ICASSP)*, Istanbul, Turkey, Vol. 5, 5–9 June, pp. 2909–2912, 2000.
15. J. G. Proakis, *Digital Communications*, 4th edition, McGraw-Hill, New York, 2001.
16. S. E. Bensley and B. Aazhang, Subspace-based channel estimation for code division multiple access communication systems, *IEEE Trans. Commun.*, Vol. 44, No. 8, pp. 1009–1020, August 1996.
17. H. Liu, and G. Xu, A subspace method for signature waveform estimation in synchronous CDMA systems, *IEEE Trans. Commun.*, Vol. 44, No. 10, pp. 1346–1354, October 1996.

18. C. J. Escudero, D. I. Iglesia, and L. Castedo, A novel channel identification method for downlink multicarrier CDMA systems, in *Proceedings of the 11th IEEE International Symposium on Personnel, Indoor and Mobile Radio Communications*, PIMRC-2000, London, UK, pp. 103–107, September 18–21, 2000.

19. B. Vucetic, and J. Yuan, *Space–Time Coding*, John Wiley & Sons, Hoboken, NJ, 2003.

20. T. S. Rappaport, *Wireless Communications—Principles and Practice*, Prentice-Hall, Englewood Cliffs, NJ, 1996.

21. S. Sigdel, Performance Evaluation of MC-CDMA Uplink System with Antenna Arrays and Multiuser Detection, Master's Thesis, Asian Institute of Technology, Bangkok, Thailand. Thesis No. TC-01-6, 2001.

PART VII

ANTI-COLLISION ALGORITHM AND SMART ANTENNAS FOR RFID SYSTEMS

CHAPTER 22

RFID ANTI-COLLISION ALGORITHMS WITH MULTI-PACKET RECEPTION

JEONGKEUN LEE
Hewlett-Packard Laboratories, Palo Alto, California

TAEKYOUNG KWON
Seoul National University, Seoul, Korea

22.1 INTRODUCTION

The radio-frequency identification (RFID) system is to identify and track objects by attaching a small RFID tag to the objects. Each RFID tag stores information about the object, such as its unique identification number. When these tags reside within a reader's radio field, they transmit this information to the reader and the objects eventually become identified. An important question is, How are the RFID tags powered? An *active tag* uses a battery to power the circuitry and transmits/receives a signal. On the other hand, a *passive tag* draws power entirely from the electromagnetic waves sent by an RFID reader; hence it only responds after receiving a radio-frequency (RF) signal from the reader.

Due to the order of magnitude difference in cost, passive tags are expected to be used much more widely. However, they can respond only once for each RFID reader's signal, which limits the design of a wireless communication protocol between readers and tags. This limitation has become a crucial issue when a large number of tags need to be identified simultaneously, thus leading to what is called an *anti-collision* problem. The RFID anti-collision problem is usually classified into reader collision and tag collision. Reader collision occurs when the coverage areas of multiple RFID reader overlap [1]. The readers may interfere with each other directly or may interfere with the transmission and/or reception of tags in the overlapped coverage area. The reader collision problem has been addressed in the literature [1–3] that propose to use nonoverlapping time or frequency for readers to avoid collision.

Handbook of Smart Antennas for RFID Systems, Edited by Nemai Chandra Karmakar

Tag collision happens when multiple tags simultaneously respond to a reader's request signal. To tackle this problem, many tag anti-collision algorithms have been proposed in the literature and they can be grouped into two broad protocol families: Aloha-based protocols [4–6] and tree-based protocols [5, 7, 8].

The goal of this chapter is to show the impact of smart antenna technologies that enable a reader to recognize multiple tags simultaneously without collision on the tag-reading performance of RFID systems. We select two representative tag anti-collision algorithms—the slotted Aloha (S-Aloha) algorithm of EPCglobal [4] and the binary tree splitting algorithm of Law et al. [7]—and show their performance improvements with various smart antenna capabilities via performance modeling and simulation. Because most Aloha and tree-based protocols run orthogonally on top of underline antenna and physical layer technologies, the results from this chapter can be projected to many other tag anti-collision algorithms that seek innovations in the protocol layer.

22.1.1 Multi-Packet Reception

In this chapter, we discuss the idea of incorporating multi-packet reception (MPR) capability into an RFID reader in order to enhance the RFID tag reading rate. We present a probabilistic model and a Markov chain model that analyze the binary tree splitting algorithm and the S-Aloha algorithm with MPR capability. The preliminary ideas of using multi-input and multi-output (MIMO) and adaptive array antennas to access RFID tags have been discussed in reference 9, and this chapter extends the discussion by using a more generalized antenna system model of MPR that embraces the MIMO and adaptive array antenna models.

Traditionally, the MAC layer has been using simple collision models: packets transmitted at the same time are destroyed, and retransmissions are required. Recently, the advent of sophisticated signal processing has changed many of the underlying assumptions made by conventional MAC techniques. For example, the use of an array antenna enables multi-packet receptions, which allows the reception of multiple responses transmitted by tags at the same time [10]. We propose to equip an RFID reader with the array antenna and leverage its MPR capability. Ward and Compton [11, 12] analyzed the throughput improvement of S-Aloha packet radio networks with adaptive array antenna models. Their Markov chain analysis requires the Markov chain states to reach an equilibrium by assuming infinite tag traffic input of general packet radio networks. However, an RFID system deals with a finite number of tags at a certain moment: The objective is to recognize all the tags in a given area as quickly as possible. Hence, a packet arrival rate of an RFID system depends on the number of remaining, unrecognized tags in contrast to the infinite traffic assumption of Ward and Compton [11, 12]. In that sense, we cannot use the same Markov chain model of Ward and Compton [11, 12] to compute the performance of the RFID system with a finite number of tags.

With these in mind, we analyze the throughput improvement of binary-tree-splitting and slotted Aloha of an RFID system with MPR capability, and we validate the analysis by simulation. The analysis and simulation results show that the tag-reading throughput of RFID systems can increase by the factor of three when a four-antenna MPR reader is employed and by the factor of six when an eight-antenna reader is used. Moreover, we show that RFID reader antenna design and signal separation techniques play an important role in improving RFID system performance with MPR capability.

Multi-packet reception can also be achieved by code division multiple access (CDMA) scheme [13]. However, the collision probability due to a random selection of CDMA codes still bounds the number of tags successfully decoded at a reader. In addition, the MPR capability with array antenna systems is different from that of CDMA systems: the presented throughput analysis model in this chapter considers the generalized MPR capability of array antenna systems.

22.2 SYSTEM MODEL

We consider a simple RFID communication system in which an RFID reader reads multiple tags in its vicinity. An RFID reader broadcasts request messages to the tags. Since only the reader transmits on the downlink, there is no contention on the downlink request message. Upon receiving the request, each tag "selectively" sends (or holds) a response message, according to the anti-collision algorithm. For simplicity, we will describe the two anti-collision algorithms on a slot basis assuming equal length for collided, idle, and successful slots. The number of tags in our model is assumed to be fixed during each runtime of the algorithm since our focus here is to reduce the number of slots required to identify all the tags in the target volume by using MPR.

We assume the reader is equipped with an array of multiple antenna elements. A general model of M responding tags (out of N tags) and a reader with an array of K antenna elements corresponds to an $M \times K$ multi-input and multi-output (MIMO) channel as shown in Figure 22.1. Although tags are physically separated and they are not antenna elements which belong to one transmitter, M responding tags are synchronized in packet transmission because they transmit in response to the request message broadcast from the reader. This MIMO channel model of distributed

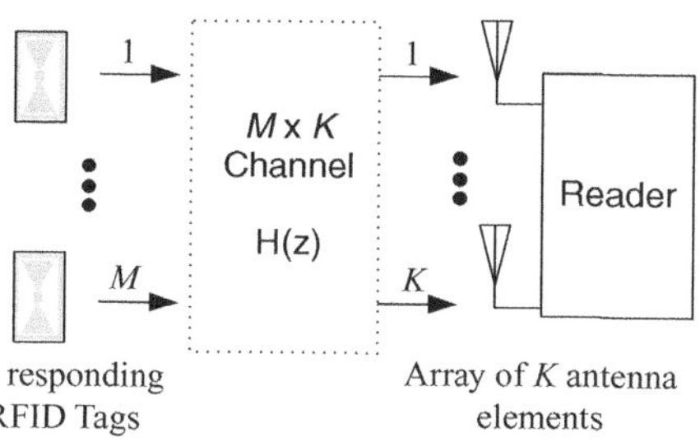

FIGURE 22.1. RFID tag-to-reader MPR model.

transmitters is similar to the cooperative diversity channel model of [14, 15] except that RFID tags do not relay messages of other tags in our model.

We use a simplified MPR model from [16]. The MPR capability of a reader is characterized by a pair of positive integers $\{F, K\}$, where $F \leq K$: F and K indicate *reception capability* and *collision-free capability*, respectively. Let i be the number of tags responding in a given slot. Besides the case of the idle slot ($i = 0$), three more cases are possible with an MPR model of $\{F, K\}$: (1) if $1 < i \leq F$, all i tags will be correctly recognized, (2) if $F < i \leq K$, only F tags among i tags will be successfully recognized, (3) when $i > K$, collision occurs and no signal can be decoded. If $i \leq K$, the reader is assumed to know the number of responding tags i possibly due to channel estimation and signal separation techniques (see the survey of papers in reference 17). Ward and Compton [11, 12] also provide exemplary channel estimation and signal separation techniques.

In general, a receiver MPR system cannot separate signals whose number is more than the number of antenna elements. Thus, K is assumed to be equal to the number of antenna elements, and F is bounded by K. The weakest MPR is the conventional collision channel $\{1, 1\}$, and the strongest MPR is $\{K, K\}$ ($K > 1$) that models perfect multi-packet reception. In between, F can take various numbers between 1 and K as a function of the channel conditions and signal separation algorithms. In the example of a beamforming array antenna system, even though we have multiple antennas (say, K antennas), at most one signal can be received while the other simultaneously transmitted signals are nullified. This is the case where the *reception capability* F is one and its MPR model is characterized as $\{1, K\}$.

In this chapter, we assume that F is a constant while all the tags are read. This generalized model of various F makes the MPR model different from the adaptive array antenna models of Ward and Compton [11, 12] and enable us to investigate on the performance improvement of the RFID system with various MPR capabilities.

22.3 BINARY TREE SPLITTING

22.3.1 Algorithm Description

A communication procedure between an RFID reader and RFID tags consists of a series of message triples (request, response, ACK), where each triple is completed in one slot. Each tag has a globally unique identifier (ID) represented by a string of bits. The reader specifies the range of tag IDs in the request message to which the tags falling under that range must respond. Upon correctly receiving responses, the reader acknowledges the recognized tags through the ACK message. The acknowledged tags hold their transmission of response message even though they fall in the ID range specified in the request message.* In the first slot, the reader requests all relevant tags in the reading volume to respond.

*This feature may be realized by the use of an "inventoried" flag such as the one employed in reference 4.

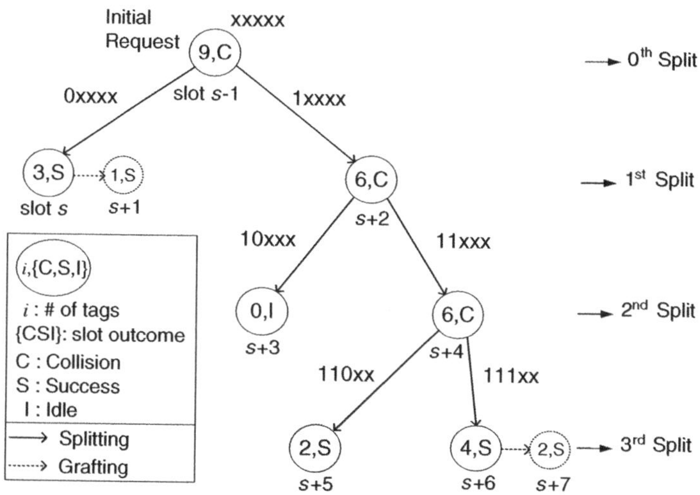

FIGURE 22.2. Binary tree splitting example ($F = 2$, $K = 4$).

When a collision occurs (the number of responding tags is greater than K), say in the $(s - 1)$th slot (or slot $s - 1$), all tags involved in the collision are split into two subsets as illustrated in Figure 22.2. The reader uses the successive bits of the original ID field to make a narrowed-down choice of the ID range.

In Figure 22.2, for example, the range [00000, 11111] (or xxxxx) will be split into two parts, 0xxxx and 1xxxx where x can be either 0 or 1. The reader requests the first subset to respond in slot s. Let i be the number of tags allotted to the first subset. If slot s is idle ($i = 0$), the second subset is requested to respond in slot $s + 1$; if slot s is successful ($0 < i \leq K$), the reader reads all i tags using $\lceil i/F \rceil$ slots and the second subset is requested to respond in slot $s + \lceil i/F \rceil$. Particularly, if $F < i \leq K$, in addition to the first slot, $\lceil i/F \rceil - 1$ more slots are required to recognize all tags in a subset: we call these additional slots as "graft" slots. In the case of $F = 2$ and $i = 3$, for example, $\lceil 3/2 \rceil = 2$ slots are consumed with one graft slot and the second subset is examined starting from slot $s + 2$ as shown in Figure 22.2. If $F = 1$, the reader reads all i tags in the first subset one by one by using i slots, and then, the second subset is asked to respond in slot $s + i$. On the other hand, if another collision occurs in slot s (i.e., $i > K$), the first subset splits again, while the second subset waits for the resolution of the first subset. This splitting mechanism is recursively iterated until no further collision occurs.

22.3.2 Throughput Evaluation

We evaluate the throughput (measured by the number of successfully recognized tags per slot) of the binary tree splitting algorithm. Let R be the number of bits indicating the initial ID range of interest, and ρ be the tag density in the ID range of interest ($\rho = N/2^R$). Assume k be the number of bits mapped to K, ($K = 2^k$). The probability that i tags fall in the range of the jth split, $P_i(j)$, and the probability of collision in that split, $P_{col}(j)$, are given by

$$P_i(j) = \binom{2^{R-j}}{i}(1-\rho)^{2^{(R-j)}-i}\rho^i \qquad (0 \leq i \leq 2^{R-j}) \tag{22.1}$$

$$P_{col}(j) = 1 - \sum_{i=0}^{K} P_i(j). \tag{22.2}$$

Then, the expected number of slots in the jth split, $s(j)$ for $0 \leq j < R - k$, is expressed recursively as

$$\begin{aligned} s(j) = 1 + &\sum_{i=F+1}^{K} \left\lceil \frac{i}{F} - 1 \right\rceil P_i(j) \\ &+ 2 \cdot s(j+1) \cdot P_{col}(j) \end{aligned} \tag{22.3}$$

with a boundary condition

$$s(R-k) = 1 + \sum_{i=F+1}^{K} \left\lceil \frac{i}{F} - 1 \right\rceil P_i(R-k)$$

One slot is required by default, regardless of the possible outcomes: *Idle, Success, or Collision*.

Then, the expected number of slots to read all tags is $s(0)$ and the expected number of recognized tags per slot is given as $N/s(0)$.

22.4 SLOTTED ALOHA (S-ALOHA)

22.4.1 Algorithm Description

A communication procedure between an RFID reader and RFID tags in S-Aloha algorithm is somewhat different from that of binary tree splitting algorithm. The reader initiates the communication by sending a request message and then a series of message pairs follows. Each pair consists of simultaneous responses from tags and an ACK from the reader. We assume that each pair is completed in a slot. When the reader requests tags to respond, each tag holds the transmission of its data (ID) until expiration of a counter whose value is generated randomly and independently among tags. The reader announces the beginning of each slot by presenting a gap pulse (e.g., no RF field for some designated time) at which the random number counter of each tag is decremented. When a collision occurs, each tag discovers the collision in the absence of an ACK message from the reader and becomes backlogged. Each backlogged tag again waits for a random number of slots before retransmitting.

22.4.2 Throughput Evaluation

Let $P_{n,m}$ be the probability of having n tags ($0 \leq n \leq N$) read successfully until the mth slot, where N is the total number of tags to read. To analyze the throughput of S-Aloha, we employ a two-dimensional Markov chain (n, m) where each of n and m corresponds to a dimension. $P_{n,m}$ is the probability that the system may stay at a state (n, m). The Markov chain always starts from the state (0, 0), and its state transition stops when n becomes equal to N—that is, when all the tags in the system are recognized. At each state transition, m is always incremented by one as the slot time progresses one by one, while the increment of n at each state transition varies from zero to F as the increment denotes the number of successfully recognized tags in the mth slot. Each slot and the number of recognized tags in the slot determine a state transition and a state (n, m) is a result of all the possible transitions from states $(n - F, m - 1) \ldots (n, m - 1)$. A state (n, m) is reached when the mth slot has just finished. The goal of this Markov chain analysis is to compute the expected number of slots to read all N tags and it will be given as

$$\sum_{m=\lceil N/F \rceil}^{\infty} m \cdot P_{N,m} \tag{22.4}$$

at the end of this analysis.

If the initial transmissions (responses) from the tags and the retransmissions from backlogged tags are sufficiently randomized, it is plausible to approximate the total number of retransmissions and initial transmissions in a given slot as a Poisson random variable [18] with a total tag-responding rate $(N - n)\lambda$. λ is a response rate of each tag and is a configuration parameter of slotted Aloha system. Since the number of tags are finite, in other words, the tag-responding rate in a slot is determined by the number of unrecognized tags, $N - n$, multiplied by λ. The reader announces the value of λ in the request message.

Let $p(n, i)$ be the probability of i responding tags when n tags have already been successfully recognized ($i \leq N - n$):

$$p(n, i) = \frac{((N - n)\lambda)^i}{i} e^{-(N-n)\lambda} \tag{22.5}$$

Please recall that the condition to recognize the entire or some part of responding tags in a slot without a collision is $i \leq K$, where i is the number of responding tags in the slot and K is the *collision-free capability*. If $F < i \leq K$, only F out of i tags are recognized but still there is no collision. Hence, having n tags recognized, the probability of *successful* recognition in a slot, $P_s(n)$, is the summation of the probabilities that the number of responding tags out of the remaining $N - n$ tags is one to K. $P_s(n)$ is independent of the slot number m. If the number of remaining tags $N - n$ are less than K, the summation range will be one to $N - n$. Thus, $P_s(n)$ is

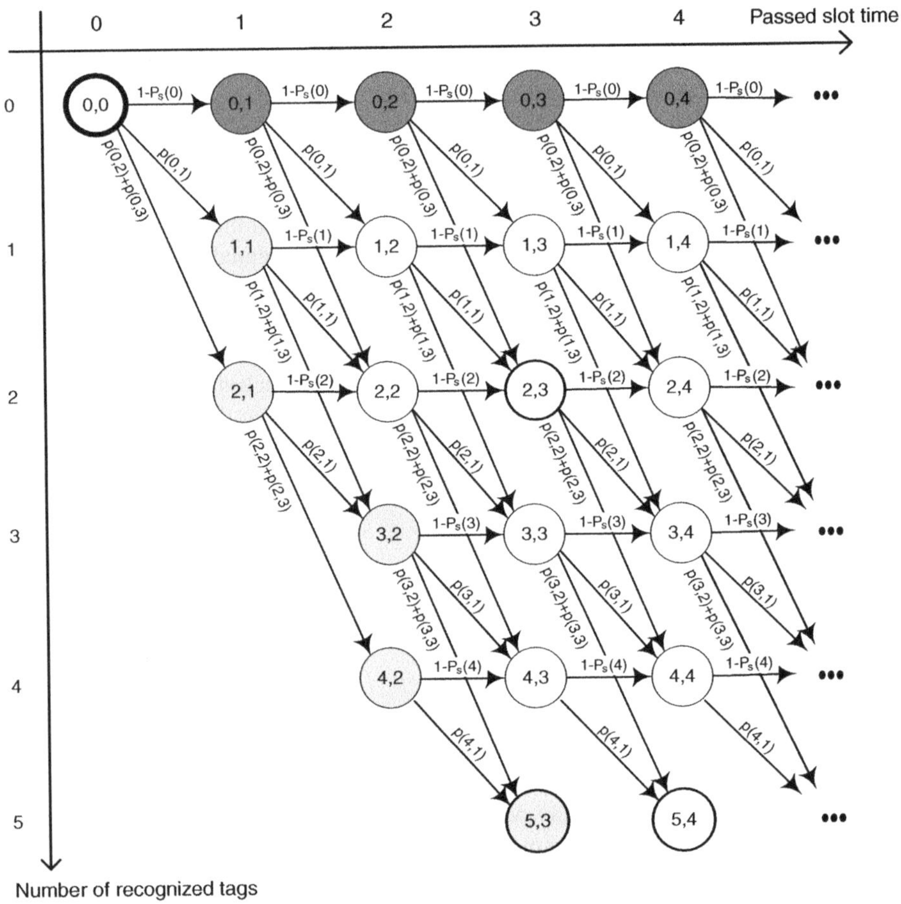

FIGURE 22.3. Markov chain for slotted Aloha ($N = 5$, $F = 2$, $K = 3$).

given by

$$P_s(n) = \sum_{i=1}^{\min(K, N-n)} p(n, i) \tag{22.6}$$

Figure 22.3 shows an example state transition diagram when $N = 7$, $K = 3$ and $F = 2$. If the slot m is successful ($1 \leq i \leq K$), the state (n, m) transits to $(n + 1, m + 1)$ when $i = 1$ or to $(n + 2, m + 1)$ when $F \leq i \leq K$ (diagonal transitions); otherwise it transits to $(n, m + 1)$ (horizontal transition) with a probability $1 - P_s(n)$. In S-Aloha, two slot outcomes, idle and collision, result in the same horizontal state transition: The slot time increases by one with zero tag recognition. The state transition continues until all the five tags are read—that is, until one of the states $(5, m')$ is reached, where

$m' \geq \lceil N/F \rceil = 3$. As shown in Figure 22.3, at least three slots are needed to read all the five tags because up to two tags can be read in one slot ($F = 2$).

We define $P_{n,m}$ iteratively from the initial state (0,0) by enumerating the possible ways that each transition may occur.

- $n = 0, m = 0$: The initial state

$$P_{0,0} = 1 \tag{22.7}$$

- $n > mF$ or $n > N$: Not possible, since at most $m \cdot F$ tags can be read until the mth slot and n is limited by the total number of tags N.

$$P_{n,m} = 0 \tag{22.8}$$

- $n = 0, m \geq 1$: No tags have been read. The upper-most slots in Figure 22.3 fall in this case. Only horizontal transitions have occurred from the initial state.

$$P_{n,m} = P_{n,m-1} \cdot (1 - P_s(n)) \tag{22.9}$$

- $(m - 1)F < n \leq min(mF, N)$: Have no horizontal transition from idle or collided previous slot but diagonal transitions from successful previous slots. In the Markov chain of Figure 22.3, the leftmost states of each row (the gray circles) correspond to this case.

$$\begin{aligned}
P_{n,m} &= \begin{cases} A & \text{if } n < F \\ A' + B & \text{if } n \geq F \end{cases} \\
A &= \sum_{i=1}^{n} P_{n-i,m-1} \cdot p(n-i, i) \\
A' &= \sum_{i=1}^{F} P_{n-i,m-1} \cdot p(n-i, i) \\
B &= \sum_{i=F+1}^{min(K,N-(n-F))} P_{n-F,m-1} \cdot p(n-F, i)
\end{aligned} \tag{22.10}$$

Both A and A' represent the transitions that all responding tags are successfully recognized because the number of responding tags, i, is F or less. B represents the transition from state $(n - F, m - 1)$ to (n, m) when only F tags are recognized among i transmitted tags ($F + 1 \leq i \leq K$). As for B, if the number of remaining tags in the previous slot, $N - (n - F)$, is less than K, the summation range will be $F + 1$ to $N - (n - F)$.

From the example of Figure 22.3, the state (3, 2) can be stemmed from the two states (1, 1) and (2, 1). The transition (1, 1) $\rightarrow$ (3, 2) happens when

$i = 2$ (A') or when $i = 3(B)$ in the second slot. The transition (2, 1) → (3, 2) happens when $i = 1$ (A) in the second slot.

- $0 < n < min((m-1)F+1, N)$: Have both diagonal (A, B) and horizontal transitions (C).

$$P_{n,m} = \begin{cases} A + C & \text{if } n < F \\ A' + B + C & \text{if } n \geq F \end{cases} \tag{22.11}$$

$$C = P_{n,m-1} \cdot (1 - P_s(n)) \tag{22.12}$$

This case addresses the states of white circles in Figure 22.3 which have horizontal transitions from their previous states on the left-hand side. For example, the state (3, 3) can be stemmed from the three states (1, 2), (2, 2), and (3, 2). The transition (1, 2) → (3, 3) happens when $i = 2$ (A') or when $i = 3$ (B) in the third slot. The transition (2, 2) → (3, 3) happens when $i = 1$ (A) in the third slot. The transition (3, 2) → (3, 3) happens when $i = 0$ or $i > 3$ (C) in the third slot.

Then, the expected number of slots to read all N tags is calculated as

$$\sum_{m=\lceil N/F \rceil}^{\infty} m \cdot P_{N,m} \tag{22.13}$$

22.5 NUMERICAL RESULTS

Figures 22.4 and 22.5 show the throughput of binary-tree-splitting algorithm and S-Aloha with various $\{F, K\}$ combination. The throughput of S-Aloha is maximized over the tag response rate λ.*

Throughput is plotted in terms of F. Each plot has a legend: the value of K and a character indicating analysis (A) or simulation (S). Note that $K = 1$ indicates the baseline case without MPR capability. In both graphs, the throughput of the proposed analytical model is almost equal to that of simulation results. The throughput of both algorithms can increase by the factor of three when a four-antenna MPR reader is employed $\{4, 4\}$ and by the factor of six when an eight-antenna reader is used $\{8, 8\}$. Increasing K alone can enhance the throughput even with small F; however, F plays a critical role to improve the throughput as K increases. In other words, both *reception capability* F and *collision-free capability* K must be large in order to achieve high throughput. The throughput with $\{2, 4\}$ MPR capability is higher than that with $\{1, 8\}$ MPR capability. Therefore, channel estimation and signal separation techniques (which determine the F) play an important role in improving RFID system performance as much as the number of receiving antennas (K) does.

*The optimal λ depends on the current number of remaining tags to read. Estimating the exact number of tags is important in Aloha-based protocols and there are a number of papers addressed this problem [6].

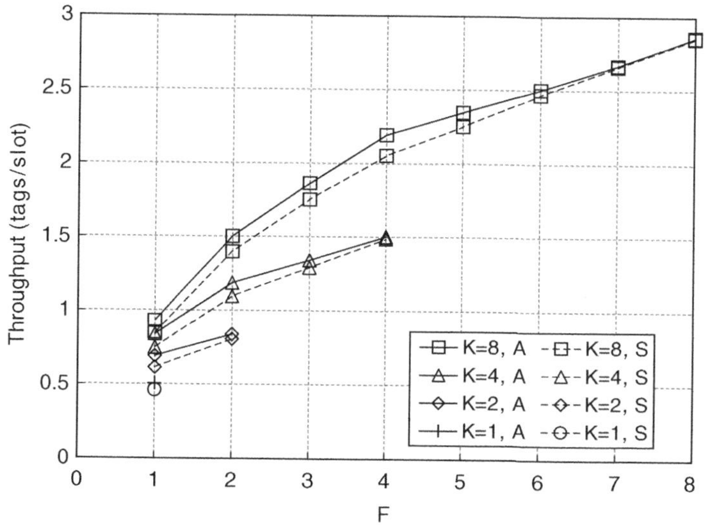

FIGURE 22.4. Throughput of binary tree ($R = 12$, $\rho = 0.5$).

When the throughput of binary-tree-splitting is analyzed and simulated in Figure 22.4, R is set to 12 and the tag density ρ is set to 0.5, in order to consider large RFID systems. If we change the parameters R and ρ, the shape of plots may slightly change, but we have observed that the overall tendencies mentioned in the previous paragraph do not change. In Figure 22.5, the throughput of S-Aloha is maximized

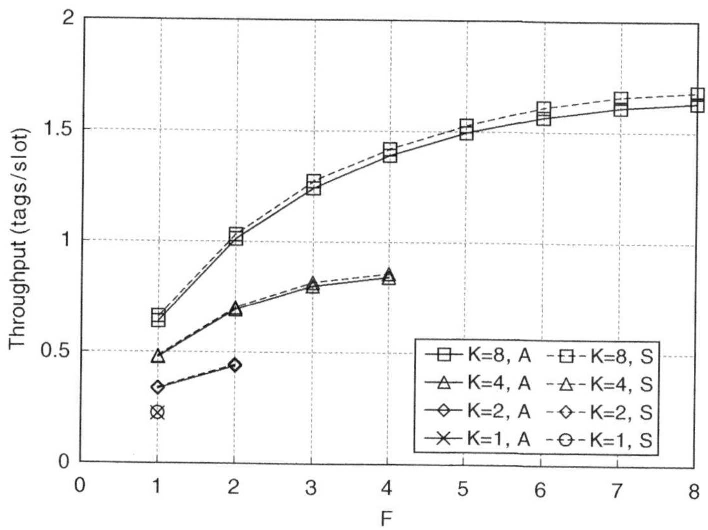

FIGURE 22.5. Throughput of S-Aloha (optimal λ and $N = 150$).

over the tag response rate λ. Although the overall tendencies of S-Aloha throughput still remain with different λ values, λ is controllable parameter and must be controlled on-line during a tag-reading process. In our experiments, for simplicity, λ is set at the beginning and fixed as a constant while all tag are read.

22.6 SUMMARY

This chapter shows how the multi-packet reception (MPR) capability from the recent advance on array antenna systems can improve the tag-reading throughput of RFID systems. By analysis and simulation, we show that both *reception capability* F and *collision-free capability* K must be large in order to achieve high throughput. Thus, the importance of channel estimation and signal separation techniques is emphasized as well as the importance of antenna design.

Although we have focused on the impact of MPR capability on the tag collision problem in this chapter, the MPR capability can also be exploited to address the reader-collision problem [4]. The use of multiple readers with partially overlapping coverage areas increases the RFID system performance and the MPR capability can further improve the performance by forming a cooperative MIMO channel between multiple readers. The work of Vaidya and Das [20] discusses the issues of reader diversity as well as antenna diversity with the focus on beamforming antennas.

REFERENCES

1. D. W. Engels, The Reader Collision Problem, Technical report, Auto-ID Center, Boston, MA, November 2001.
2. J. Waldrop, D. W. Engels, and S. E. Sarma, Colorwave: An anticollision algorithm for the reader collision problem, in *IEEE International Conference on Communications (ICC)*, 2003, pp. 1206–1210.
3. K. S. Leong, M. L. Ng, A. R. Grasso, and P. H. Cole, Synchronization of RFID readers for dense RFID reader environments, in *International Symposium on Applications on Internet Workshops*, 2006.
4. EPCglobal, *EPC™ Radio-Frequency Identity Protocols Class-1 Generation-2 UHF RFID Protocol for Communications at 860 Mhz–960 Mhz, Version 1.0.9*, January 2005.
5. H. Vogt, Efficient object identification with passive RFID tags, *Lecture Notes Comput. Sci.*, Vol. 2414, pp. 98–113, 2002.
6. J. Zhai and G. Wang, An anti-collision algorithm using two-functioned estimation for RFID tags, *Lecture Notes in Computer Science*, Vol. 3483, pp. 702–711, 2005.
7. C. Law, K. Lee, and K.-Y. Siu, Efficient memoryless protocol for tag identification, in *Proceedings, 4th ACM International Workshop on Discrete Algorithms and Methods for Mobile Computing and Communications*, August 2000.
8. J. Myung, W. Lee, J. Srivastava, and T. K. Shih, Tag-splitting: Adaptive collision arbitration protocols for RFID tag identification, *IEEE Trans. Parallel Distrib. Syst.*, Vol. 18, pp. 763–775, 2007.

9. J. Lee, T. Kwon, Y. Choi, S. K. Das, and K.-a Kim, analysis of RFID anti-collision algorithms using smart antennas, in *Proceedings, ACM SenSys (Refereed Poster)*, Baltimore, MD, November 2004.

10. L. Tong, Q. Zhao, and G. Mergen, Multipacket reception in Random Access Wireless Networks: From Signal Processing to Optimal Medium Access Control. November 2001.

11. J. Ward and R. T. Compton, Jr., Improving the performance of a slotted ALOHA packet radio network with an adaptive array, *IEEE Trans. on Communications*, Vol. 40, No. 2, February 1992.

12. J. Ward and R. T. Compton, Jr., High throughput slotted ALOHA packet radio networks with adaptive arrays, Vol. 41, No. 3, March 1993.

13. Y. Fukumizu, S. Ohno, M. Nagata, and K. Taki, Communication scheme for a highly collision-resistive RFID system, *IEICE Trans. Fundamentals*, Vol. E89-A(2), No. 2, pp. 408–415, February 2006.

14. J. Lee, S. Kim, H. Suman, T. Kwon, Y. Choi, J. Shin, and A. Park, Downlink node cooperation with node selection diversity, in *Proceedings, IEEE VTC 2005-Spring*, Stockholm, Sweden, May 2005.

15. M. Dohler, E. Lefranc, and H. Aghvami, Virtual antenna arrays for future wireless mobile communication systems, in *Proceedings, IEEE ICT*, Beijing, China, June 2002.

16. S. Ghez, S. Verdu, and S. Schwartz, Optimal decentralized control in the random access multipacket channel, Vol. 3, No. 11, pp. 1153–1163, November 1989.

17. R. Liu and L. T. Ed, Special issue on blind system identification and estimation, *Proceedings of the IEEE*, Vol. 86, Issue 10, October 1998.

18. D. Bertsekas and R.Gallager, *Data Networks. Section 4.2.* Prentice-Hall, Englewood Cliffs, NJ, 1992.

19. Chen Qian, Hoilun Ngan, and Y. Liu, Cardinality estimation for large-scale RFID systems, in *IEEE International Conference on Pervasive Computing and Communication (PerCom)*, 2008, pp. 30–39.

20. N. Vaidya and S. R. Das, RFID-based networks: Exploiting diversity and redundancy, *SIGMOBILE Mob. Comput. Commun. Rev.*, Vol. 12, No. 1, pp. 2–14, 2008.

CHAPTER 23

ANTI-COLLISION ALGORITHM AND SMART ANTENNAS FOR RFID SYSTEMS

QI JING TEOH and NEMAI CHANDRA KARMAKAR
Department of Electrical and Computer Systems Engineering, Monash University, Clayton, Victoria, Australia 3800

23.1 INTRODUCTION

The growing popularity in the radio-frequency identification (RFID) system is pushing for more efficient and reliable multiple access methods to automatically identify multiple tagged objects at once. Optical barcodes have dominated the identification and tracking market from their inception in 1970s. The limitations of optical barcodes and several available technologies for identifications are well known and have been discussed in the preceding chapters. RFID has the capability to identify objects uniquely and has the feature of automated non-line-of-sight reading. However, reliability becomes an issue when the reader attempts to read multiple tags in close proximity. This is due to the fact that data collision occurs during simultaneous transmission of multiple tags. To mitigate the collision problem of reading multiple tags at once by a reader or a set of readers connected to a data processing unit, various anti-collision algorithms have been proposed. A comprehensive account of RFID anti-collision protocols can be found in [1]. In this chapter a comprehensive overview of anti-collision protocols is presented first. The influence of smart antennas, especially MIMO antennas, to improve the anti-collision features of RFID tags is presented next. The frame-slotted Aloha protocol and smart antennas to improve the throughput of the RFID system is presented, followed by conclusion.

23.2 ANTI-COLLISION ALGORITHM

In many existing applications, reading of a single tag at a time is acceptable. Examples include sorting of items in a sequential order. However, in many automated

Handbook of Smart Antennas for RFID Systems, Edited by Nemai Chandra Karmakar

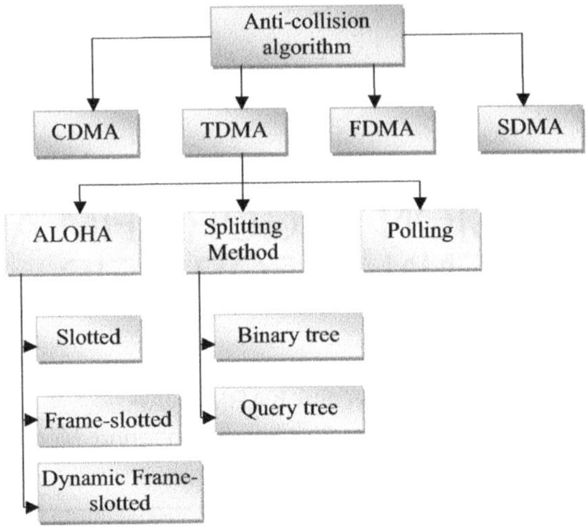

FIGURE 23.1. Classification of anti-collision protocols.

systems, reading of multiple active tags in an RFID reader's reading zone is essential. Examples include library database management systems, airline baggage handling and dispatch, asset management in a warehouse, apparel hanging tags, and so on. In these applications, anti-collision protocols help elevate reading of multiple active tags at once. The anti-collision protocol is also known as the multiple access schemes. It is a link-layer software that is installed in the reader to separate individual participants' signals to identify multiple objects accurately and efficiently. To effectively read all tags within the reader's reading zone without significant delay, the reader requires a desirable anti-collision algorithm. There are many different anti-collision protocols, each of which has its pros and cons. Figure 23.1 illustrates the classification of the anti-collision algorithm [1]. Following are the detailed descriptions of the RFID anti-collision algorithm used in an RFID system.

SDMA. The space division multiple access (SDMA) scheme uses the technique of dividing the reader's interrogating zone in spatially separated areas. Multiple readers and antennas are combined together to form an array that can sense the presence of tags in different locations [1]. This can also be done by using the digital multi-beamforming based on the smart antenna [2]. Basically, the reader consists of directional antennas that can differentiate each tag in the reading zone by the tags' angular positions. This multiple access scheme is able to increase the efficiency of the tag reading process. However, it requires a complex reader and antenna system that implies higher implementation cost.

FDMA. The frequency division multiple access (FDMA) scheme relates to the technique of the interrogator using multiple transmission channels to communicate with the transponders. Each transmission channel operates on different carrier frequencies. This allows the tag to transmit data to the reader on one frequency and receive data from another frequency channel. Also, several tags can communicate simultaneously on different carrier frequencies. The disadvantage of this type of multiple access scheme is high system cost to install receivers for different frequency channels [1].

CDMA. The code division multiple access (CDMA) scheme relates to the technique using the spread spectrum and unique coding method [3]. This allows the interrogator to multiplex all the incoming tags signals over the same channel. Then, the tags are recognized based on their unique codes. CDMA brings significant benefit in anti-jamming for security purposes. However, it requires a very complicated system and coding scheme which could be too complex to the RFID tags [1].

TDMA. The time division multiple access (TDMA) scheme is most commonly used in designing the anti-collision algorithm. It is a technique that divides the time into multiple timeslots to be accommodated by tags. The types of TDMA multiple access schemes are as follows.

ALOHA. The ALOHA protocol was created in University of Hawaii in 1970 for the data communication with bursty traffic [4–9]. It is a simple tag-driven protocol where the tags randomly transmit their ID to the reader once inside the reading range. On the reader side, it does not require any complex mathematical calculations for the algorithm but only to monitor the tag transmission process for successful and collided slots. Many previous researches have proven theoretically that the maximum throughput for a pure ALOHA-based protocol is 18.4% [5]. If the time period of a data is T, then the time required for a tag to guarantee a successful transmission is 2T (T period of time transmission until the end of the packet). It is as illustrated in Figure 23.2. There are three popular Aloha protocols that are described below.

(a) Slotted ALOHA. Slotted ALOHA protocol was created for throughput improvement from the pure ALOHA protocol. It is theoretically proven that the maximum

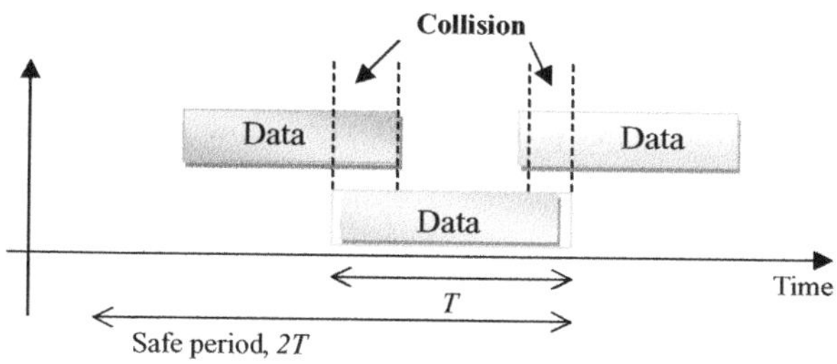

FIGURE 23.2. ALOHA protocol.

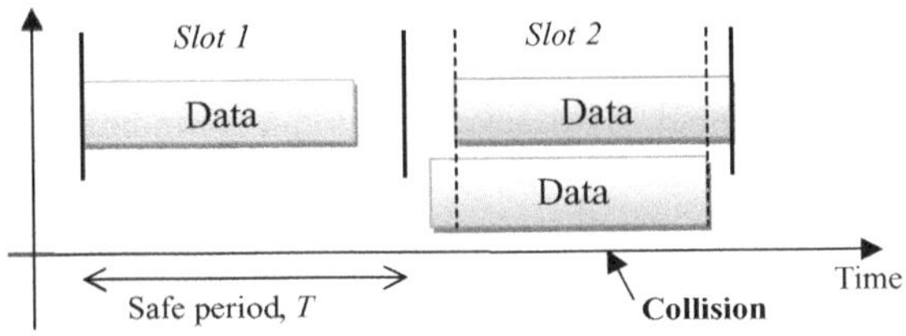

FIGURE 23.3. Slotted ALOHA protocol.

throughput is 36.8% [4]. The only difference between slotted ALOHA and pure ALOHA protocol is that there is a restricted timeslot for each data transmission. If a collision happened, the tag will have to randomly transmit at the next timeslot. In this case, if the timeslot has T time period, then a tag is successfully read as long as no collision occurs within the T period. It is as illustrated in Figure 23.3.

(b) Frame-Slotted ALOHA. Frame-slotted ALOHA protocol uses the same concept as slotted ALOHA, and the only major difference is that it imposes a limit for the number of timeslots in one frame (read cycle). With the number of tag transmission opportunities being restricted, it also imposes a limit on the number of retransmission attempts. That is, the tag is only allowed to transmit only once in a time frame. In slotted ALOHA, the reader sends acknowledgment after each successful transmitted slot. However, in frame-slotted ALOHA, the reader notifies all successful tags at the end of each frame [6].

(c) Dynamic Frame-Slotted ALOHA. An improvement of the frame-slotted ALOHA protocol is the dynamic frame-slotted ALOHA algorithm. In this algorithm, the length of the frame is adjusted according to the number of tags. This ensures minimal time delays and a higher system throughput. There were many protocols proposed using this type of transmission control strategies such as the Bayesian transmission strategy [4] and the Q algorithm [5, 6].

Polling. With the polling anti-collision algorithm, the reader is required to have a pre-list of all tags to be read before the tag identification process begins. This is because the tags are being identified bit by bit. The reader will first call out the first bit of binary data. If there is a collision, it will move on to another bit until only a single tag is recognized. This procedure is very slow and inflexible, but it can guarantee that all tags are identified [1].

Splitting Method. The splitting method is an anti-collision protocol that is introduced by Capetanakis [7] to use the "Tree-search" method to resolve collision. In this protocol, the tag's ID will be split into subsets during a collision. All tags in the first subset will transmit their ID to the reader first. When the reading of the first subset completes, the reader will then continue with the rest of the subsets. However, if the subsets encounter another collision, it will continue to split into more subsets [1]. All of the tags are successfully identified when all subsets are cleared.

(a) Binary Tree. Binary tree protocol works by splitting tags into two subsets according to the first collided data bit. Thus, in this algorithm, the reader requires a system that can track the bit position of the collided binary strings. When there is a collision, the reader will silence the tags that have a collided bit "1". Then, those tags that have collided bit "0" will continue to send their ID to the reader on the next read request [8]. This will continue until only one unique tag is recognized. After that, the reader will start to identify the rest of tags by using the First-in first-out (FIFO) method [1].

(b) Query Tree. The query tree protocol resolves collision by sending out string prefix to the tags that begins with a "1" or "0". The tag will reply if its ID contains the prefix. If there is more than one tag response, the reader will append a bit "1" or "0" to the original prefix [8]. One bit will be added each time there is a collision, and the prefix will become longer each time. This identification process will continue until there are only one tag replies. With this technique, the protocol can recognize all tags present in the reading range.

23.3 SMART ANTENNAS FOR ANTI-COLLISION PROTOCOL

One of the most popular multiple access protocols as discussed above is the ALOHA-based anti-collision algorithm such as slotted ALOHA protocol. Liu and Lai [9] investigated several different tag estimation methods in the slotted ALOHA protocol to show that the system achieves maximum throughput when the number of tags equals the timeslots. Gao [10] showed that the theoretical maximum throughput for slotted ALOHA is 36.8%. Both works were based on the conventional single-packet reception system, which consists of a single-input single-output (SISO) antenna model.

However, in the wireless network environment, the SISO system is vulnerable to the noise and interference caused by the multipath wave propagation. This can be reduced by using the smart antenna technology. A smart antenna system can adjust its directionality and optimize its radiation pattern toward desired users and null toward interferers. This will dramatically increase the system performance and eliminate the interference from unintended users [11].

In the following sections, different smart antenna systems that augment the anti-collision protocol and improve throughput are presented. In this chapter, the frame-slotted Aloha protocol is presented followed by throughput calculations of the system. The slotted ALOHA algorithm and the system design are presented in Chapter 22.

The communication between tag and reader can be classified according to the types of smart antenna: single-input multiple-output (SIMO), multiple-input single-output (MISO), and multiple-input multiple-output (MIMO).

As shown in Figure 23.4, SIMO is a smart antenna technology that employs multiple antennas at the receiver side and a single antenna at the transmitter side

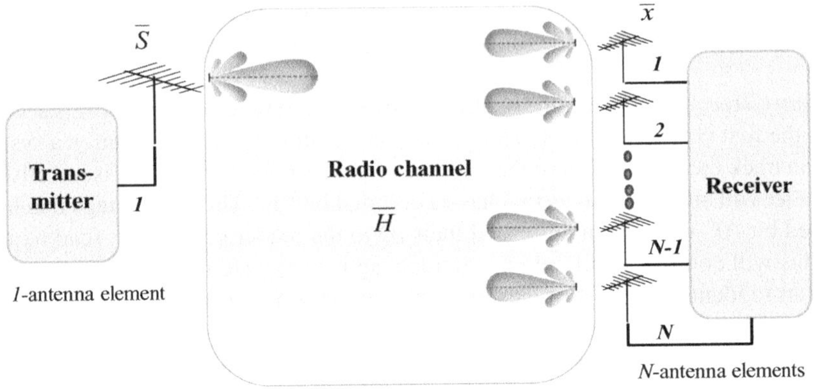

FIGURE 23.4. SIMO antenna system.

[12]. Here, the transmitter has only one antenna, where $\overline{S}$ is the complex transmit vector, $\overline{H}$ is the complex channel matrix relating to one input and N outputs, and $\overline{x}$ is the complex receive vector. The advantage of such multiple antenna systems is to exploit space and time diversity at the same time [13, 14]. While multipath fading in the wireless channel is a problem for SISO system, the MIMO antenna system exploits the multipath fading as an advantage. This main goal of the MIMO system is to combine signals in both the transmitter and receiver ends so that the system capacity improves; bit error rates (BER) reduce and intersymbol interference (ISI) in the case where the signal bandwidth is larger than the channel bandwidth [15]. For the MIMO antenna system applied to the RFID system the multiple antennas, which decode the same tag, a higher-quality signal-to-noise ratio (SNR) with a lower bit error rate (BER) can be achieved. This could happen when a particular antenna is located in a certain location or orientation that enables it to receive a better signal. Besides that, when the first antenna is transmitting and receiving the data while the second antenna is only receiving the data, and there happened to be a jammer, the SNR of the first antenna will be degraded by the jammer but the second antenna could receive a better signal [16].

As shown in Figure 23.5, MISO is a smart antenna technology which uses multiple antennas to transmit data while a single antenna receives the data. The probability of losing signals due to the channel fading environment decreases proportionally with the number of antenna elements located at the source. This will substantially improve the reliability of the system. Some of the applications that use the MISO technology are: the digital television and wireless local area networks [17].

As shown in Figure 23.6, MIMO is a smart antenna technology system that has multiple antennas to decode the same tag. This will minimize the error and ensure that a high-quality signal-to-noise-ratio (SNR) can be achieved. Besides that, MIMO antenna utilizes spatial multiplexing that allows data receiving and transmitting simultaneously. Thus, MIMO is able to produce significant data rates and increase the channel capacity without additional signal bandwidth [18–20].

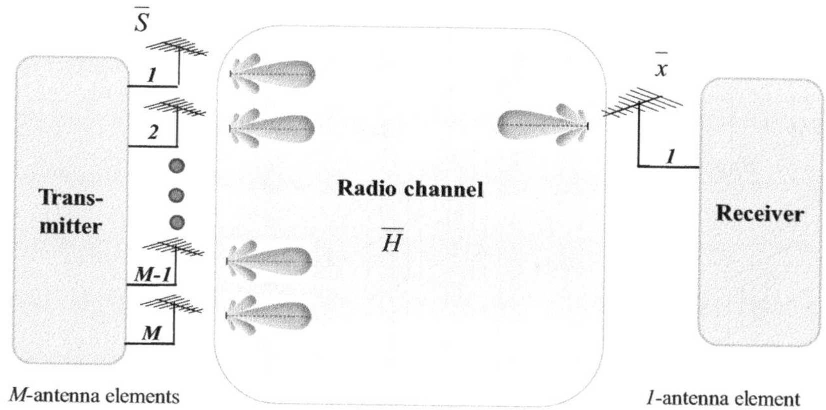

FIGURE 23.5. MISO antenna system.

Figure 23.7 illustrates the sequence of data transmission and reception between the tag and the reader of SISO and MIMO antenna. As can be seen, the SISO antenna system only exploits the time diversity. Contrary to the SISO model, the MIMO system exploits both time and spatial diversities. Therefore, while comparing the two systems, the MIMO system is able to produce significant gain and larger channel capacity, proportional to the number of antennas [20]. With K number of antennas, the MIMO system has K degrees of freedom (DoFs) and it receives a superposition of the simultaneous incoming signals.

These multiple signals will then be separated with different signal signatures using signal processing techniques [21]. The primary aim of the MIMO is to reduce the signal correlation at the receiver [22, 23]. As long as the number of simultaneous transmission from the M responding tags (out of N total tags) does not exceed K

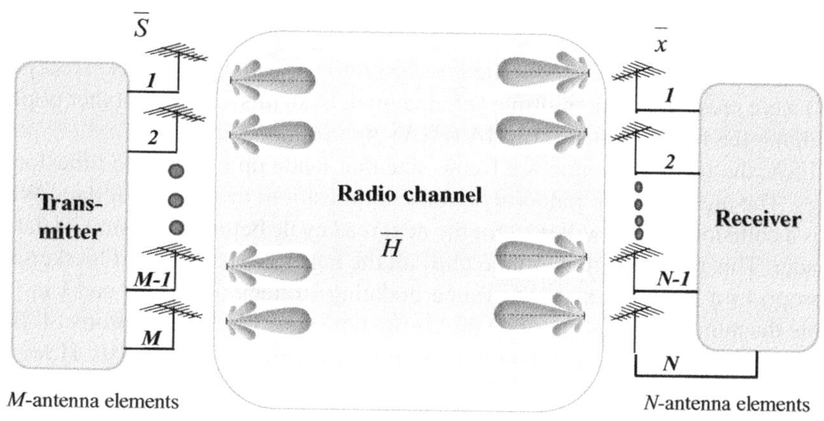

FIGURE 23.6. MIMO antenna system.

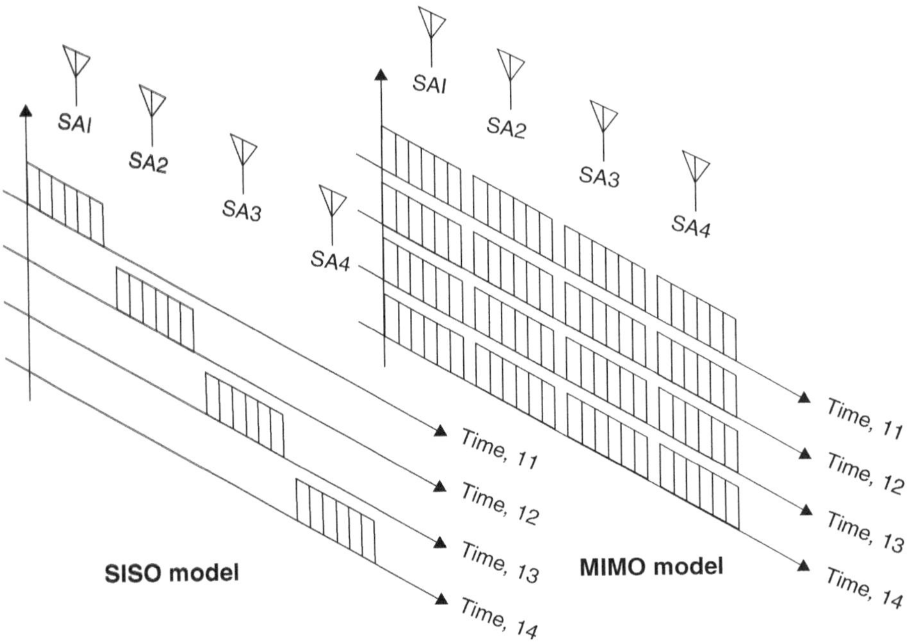

FIGURE 23.7. Data transmission in SISO and MIMO antenna systems.

(K DoFs), the packets could be correctly received. Otherwise, there will be collision and the reader receives zero packets when the number of transmission is more than K. The MIMO system in this chapter is assumed to be implemented only at the receiver (reader) side.

23.4 FRAME-SLOTTED ALOHA ALGORITHM

In Chapter 22, the binary tree and slotted Aloha protocols using multi-packet reception (MPR) were presented with multiple antenna models. In this section, another popular algorithm—the frame-slotted ALOHA (FSA) algorithm—is discussed.

In FSA, the reader announces a frame size that made up of multiple timeslots to the tags. The tags will then randomly choose a timeslot to transmit their data. When there is a collision, the tag will wait for the next read cycle before retransmit its data to the reader. This process will continue until all the tags have been read. Floerkemeier [24] proposes a Bayesian frame-to-frame updating strategy and Chen and Lin [25] estimate the number of tags based on the empty time slots for the FSA protocol. Both works have proven that the maximum throughput achievable is 36.8%. However, they were based on the conventional single-packet reception system, which consists of a single-input single-output (SISO) antenna model. With the recent advancement in smart antenna technologies and sophisticated signal processing techniques technology, the capability to deal with multiple packets simultaneously in the physical (PHY) layer is made possible. This multiple packet reception (MPR) capability at the PHY layer can enhance the traditional single-packet reception channel in resolving packet collision. The system with MPR can accommodate a defined number of M simultaneous transmissions. All transmitted packets can be received correctly when they are equal or less than M. On the other hand, no data or collision occurs when the number of transmitted packet is more than the M [26].

In this section, the MPR capability is studied and incorporated into the FSA anti-collision algorithm to evaluate the throughput of the system. Then, an estimation model based on the collided timeslot in a SISO model proposed by Cha and Kim [27] is studied and showed that it can be employed in the MIMO system to improve the tag recognition process.

23.4.1 System and Mathematical Model

An RFID system in which the reader is capable of reading multiple tags within its read range simultaneously is considered. Let L be the number of total timeslots in a read cycle. The way that the tags are allocated into the timeslots is described as a binomial distribution. Let N be the total number of tags to be read and let k be the number of transmission attempts in a timeslot. Each of these tags has a $1/L$ transmission probability since they randomly choose a timeslot to transmit their data. Thus, the probability of obtaining k tags in a timeslot can be given by

$$P_{\text{coll}} = \begin{bmatrix} N \\ k \end{bmatrix} \left[\frac{1}{L}\right]^k \left[1 - \frac{1}{L}\right]^{N-k} \tag{23.1}$$

Let M be the MPR capability of the reader. Since the reader can recognize up to M tags simultaneously, the successful transmission probability is the summation of one tag to M tags responding probabilities. It can be given by [28]

$$P_{\text{coll}} = P_t\{[k \leq M]\} = \sum_{k=1}^{M} \begin{bmatrix} N \\ k \end{bmatrix} (P_t)^k (1 - P_t)^{N-k} \tag{23.2}$$

Similarly, the probability of an idle timeslot can be expressed by [28]

$$P_{\text{idle}} = P_t\{k = 0\} = (1 - P_t)^N \tag{23.3}$$

When the numbers of k attempts exceed M, all the transmitted data packets are lost as a result of collision. The probability of a collision can be given by [28]

$$\begin{aligned} P_{\text{coll}} &= P_t\{k > M\} = 1 - P_{\text{succ}} - P_{\text{idle}} \\ &= \sum_{k=M+1}^{N} \begin{bmatrix} N \\ k \end{bmatrix} (P_t)^k (1 - P_t)^{N-k} \end{aligned} \tag{23.4}$$

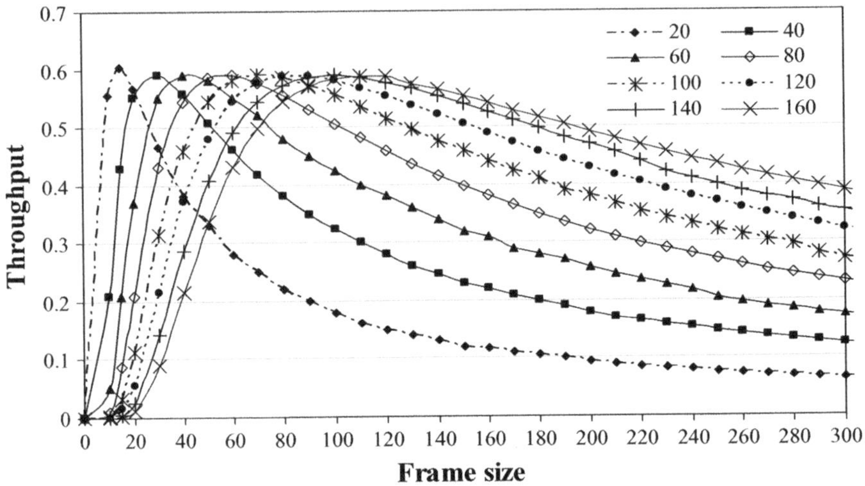

FIGURE 23.8. Throughput versus frame size for MPR capability = 2. The total number of tags is varied from 20 to 160.

The throughput, T, of the system is measured by the probability of up to M successfully received tags, and it can be expressed as [27]

$$T = \frac{P_{\text{succ}}}{P_{\text{succ}} + P_{\text{coll}} + P_{\text{idle}}} \tag{23.5}$$

The throughput versus different frame size and total number for system with MPR capabilities 2 and 3 is shown in Figures 23.8 and 23.9. It can be seen that the optimal frame length is different for the total number of tags and MPR. The maximum throughput is close to 0.6 and 0.73 at the optimum frame size for system with MPR capabilities 2 and 3.

23.4.2 Estimation Model

In this section, a system model that consists of five stages as shown in Figure 23.10 is proposed. This system model consists of the throughput analysis discussed in the previous section and an estimation model to estimate the number of unread tags. This estimation process makes use of the single reception model from the Cha and Kim's [27] previous works to estimate up to M number of tags, depending on the reader's MPR capability. First of all, the reader will check the number of collided timeslots after completing the first read cycle.

Next, based on the number of collided timeslots, the reader calculates the collision ratio (CR) from the ratio of total number of collision slots to the frame size. The CR

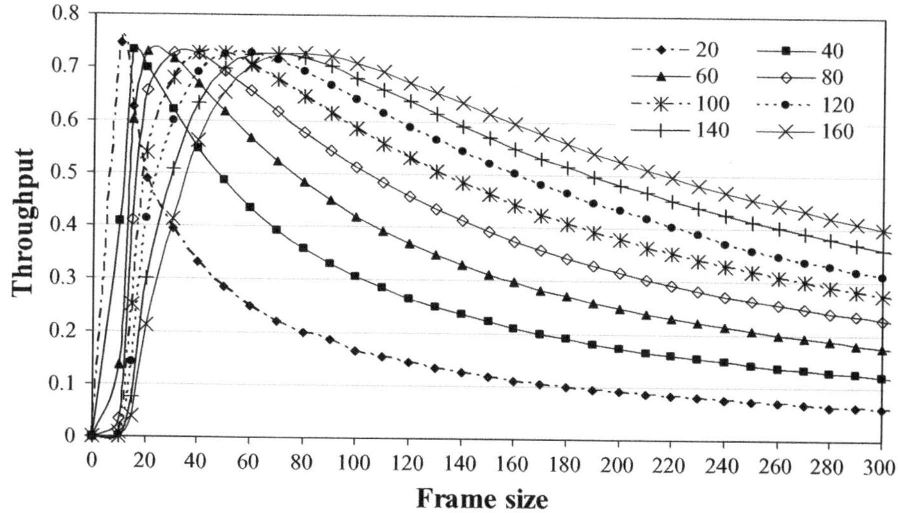

FIGURE 23.9. Throughput versus frame size for MPR capability = 3. The total number of tags is varied from 20 to 160.

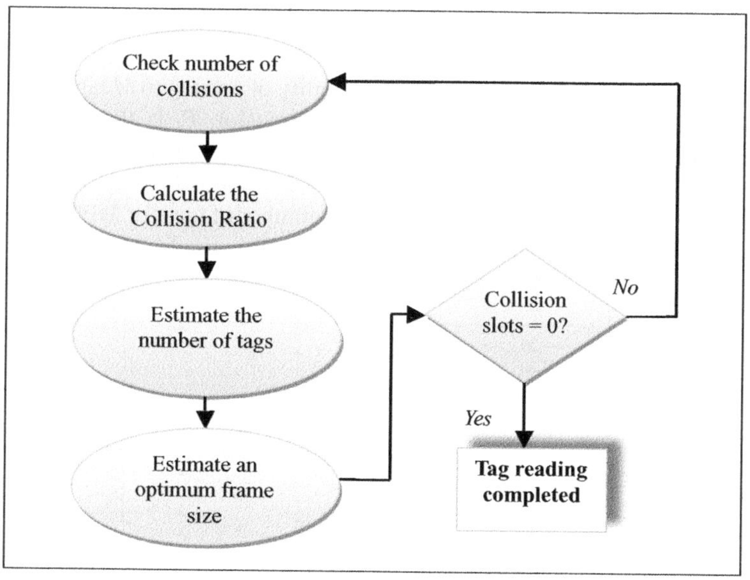

FIGURE 23.10. System flow chart for the estimation model.

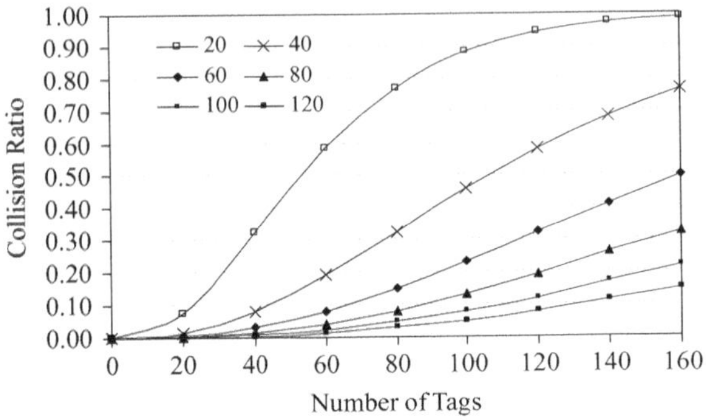

FIGURE 23.11. Collision ratio versus number of tags for MPR capability = 2. The total number of tags (left-hand corner) and frame length are varied from 20 to 120.

can be given by [27]

$$\begin{aligned}\text{Collision Ratio} &= \frac{\text{Slots with collision}}{\text{Frame length}} \\ &= 1 - P_{\text{succ}} - P_{\text{idle}} \\ &= 1 - \sum_{k=1}^{M} \begin{bmatrix} N \\ k \end{bmatrix} (P_t)^k (1 - P_t)^{N-k} - (1 - P_t)^N \end{aligned} \tag{23.6}$$

where P_{succ} is the successful transmission probability of one tag to M tags responding probabilities, P_{idle} is the probability of an idle timeslot, P_t is the probability of obtaining k tags in a timeslot, N is the total number of tags, and K is the number of transmission attempts in a timeslot.

The simulated graphs of collision ratio versus number of tags for MPR capabilities 2 and 3 are as shown in Figures 23.11 and 23.12. The graph was generated from

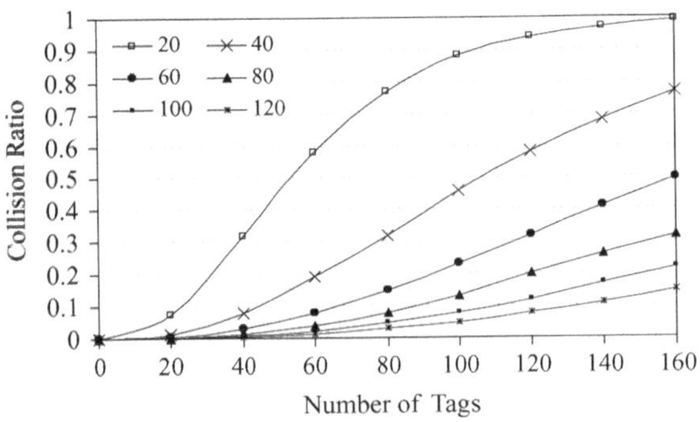

FIGURE 23.12. Collision ratio versus number of tags for MPR capability = 3. The total number of tags (left-hand corner) and frame length are varied from 20 to 120.

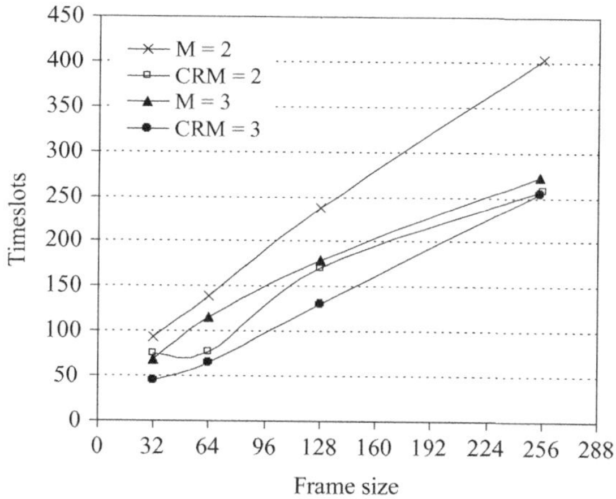

FIGURE 23.13. Timeslots versus frame size required to complete all tag identifications. ($M = 2$ and $M = 3$ depicts the conventional FSA algorithm with a fix frame size. CRM = 2 and CRM = 3 indicates the FSA algorithm with dynamically adjusted frame size.)

using Eq. (23.6), which takes into account the total number of tags to be read and the reader's MPR characteristic. Thus from Figure 23.11, when MPR = 2, the collision ratio detected by the reader is 0.2 with a frame size of 80. This gives the estimated number of tags to be around 120.

Then, based on the collision ratio, the reader estimates the total number of tags. The number of unread tags is obtained by deducting the tags that are successfully recognized from the estimated total number of tags. Next, based on the number of unread tags, the reader will estimate the optimum frame length from the throughput analysis above. Finally, the reader will check if there is still any collision slot in the next read cycle. The reading is complete when no tags collided. This system model has demonstrated a flexible process of estimating the number of unread tags and hence determined the most suitable frame size for each read cycle. In order to examine the throughput analysis and the estimation model, the total number of timeslots required is tested with C++ language. The simulation was carried out with 100 tags to be read, and the frame size varies: 32, 64, 128 to 256 slots. In Figure 23.13, $M = 2$ and $M = 3$ results depict the FSA algorithm, which uses a fixed frame size for the entire tag reading process. The CRM = 2 and CRM = 3 results were simulated using the collision ratio that dynamically adjusted the frame size for each read cycle as discussed.

23.5 CONCLUSION

This chapter introduces the anti-collision protocol and the smart antenna technologies which include the MIMO antennas. A Markov chain model was presented in terms of the smart antenna's reception capability and degree of freedom. The mathematical model is explained in details with the aid of diagrams and flow charts. From the results, it shows that both the reception capability and number of antennas play an important factor in achieving high system throughput. This verifies that the MIMO antenna, which uses the spatial multiplexing technique, has a larger channel capacity than the MISO, SIMO, and SISO systems.

The MPR is studied and the probabilities for successful, empty, and collided timeslots are analyzed and integrated with the frame-slotted ALOHA anti-collision protocol. The results show that a maximum throughput as high as 0.6 and 0.73 can be achieved with MPR capabilities of 2 and 3 by employing the MIMO antenna. At last, the throughput analysis is validated together with a simulation to estimate the number of tags. The estimation model calculates the number of collided timeslots for every read cycle, and the optimum timeslot is generated from the throughput analysis. The results show that a smaller number of timeslots is required to read all tags by dynamically adjusting the frame length for each read cycle.

REFERENCES

1. D. H. Shih, P. L. Sun, D. C. Yen, and S. M. Huang, Taxonomy and survey of RFID anti-collision protocols, *Comput. Commun.*, Vol. 31, No. 10, pp. 100–103, 2006.
2. J. X. Yu, K. H. Liu, X. Huang, and G. Yan, An Anti-collision algorithm based on smart antenna in RFID system, *ICMMT Microwave Millimeter Wave Technol.*, Vol. 3, No. 21–24, pp. 1149–1152, 2008.
3. P. Wang, A. Hu, and W. Pei, The design of anti-collision mechanism of UHF RFID system based on CDMA, *IEEE Asia Pacific Conf.*, No. 4–7, pp. 1703–1708, 2006.
4. C. Floerkemeier, Bayesian transmission strategy for framed ALOHA based RFID protocols, *IEEE Int. Conf. RFID*, No. 26–28, pp. 228–23, 2007.
5. Laynetworks,Pure Aloha Protocol [Online]. Available at http://www.laynetworks.com/Pure%20Aloha%20Protocol.htm.
6. J. E. Wieselthier, A. Ephremides, and L. A. Michaels, An exact analysis and performance evaluation of framed ALOHA with capture, *IEEE Trans. Commun.*, Vol. 37, No. 2, pp. 125–137, February 1989.
7. J. I. Capetanakis, Tree algorithms for packet broadcast channels, *IEEE Trans. Inf. Theory*, Vol. 25, No. 5, pp. 505–515, 1979.
8. B. Feng, J. T. Li, J. B. Guo, and Z. H. Ding, ID-Binary Tree Stack Anticollision Algorithm for RFID, in *Proceedings of the 11th IEEE Symposium on Computers and Communications*, 26–29, pp. 207–212, 2006.
9. L. Lui and S. Lai, ALOHA based anti-collision algorithms used in RFID system, *ICMMT Microwave Millimeter Wave Technol.*, Vol. 3, pp. 1149–1152, 2008.
10. J. Gao, Analysis of ALOHA and slotted ALOHA, Department of Computer Science, Stony Brook University, Stony Brook, NY, 2007.
11. The International Engineering Consortium, Smart Antenna System [Online]. Available at http://searchmobilecomputing.techtarget.com/sDefinition/0,,sid40_gci1026138,00.html.

12. Whatis.com, Smart Antenna, [Online]. Available at http://searchmobilecomputing.techtarget.com/sDefinition/0,,sid40_gci1026138,00.html.

13. M. Shafi, D. Gesbert, Da-shan Shiu, P. J. Smith, and W. H. Tranter, MIMO systems and applications: Part I, *IEEE J. Selected Areas Commun.*, Vol. 21, No. 3, pp. 277–280, April 2003.

14. M. Shafi, D. Gesbert, Da-shan Shiu, P. J. Smith, and W. H. Tranter, MIMO systems and applications: Part II, *IEEE J. Selected Areas Commun.*, Vol. 21, No. 5, pp. 681–683, June 2003.

15. F. Ross, *Smart Antennas for Wireless Communications with MATLAB*, McGraw-Hiil, New York, 2005.

16. Y. Gregory, Multiple reader coordination in RFID system. [Online], available at www.freepatentsonline.com/y2007/0001813.html.

17. B. Bahlmann, MISO—Multiple input single output [Online], available at http://www.birds-eye.net/definition/acronym/?id=1158767683.

18. P. Suvikunnas, J. Salo, L. Vuokko, J. Kivinen, K. Sulonen, and P. Vainikainen, Comparison of MIMO antenna configurations: Methods and experimental results, *IEEE Trans. Veh. Technol.*, Vol. 57, No. 2, pp. 1021–1031, March 2008.

19. R. W. Heath, Jr., Antenna design and analysis for MIMO communication systems, [Online], available at http://users.ece.utexas.edu/~rheath/research/mimo/antenna/.

20. Smart Antenna [Online], available at http://searchmobilecomputing.techtarget.com/sDefinition/0,,sid40_gci1026138,00.html.

21. S. M. Ross, *Introduction to Probability Models*, 6th edition, Academic Press, New York, 2007.

22. J. Lee, T. Kwon, Y. Choi, S. K. Das, and K. Kim, Poster abstract: Analysis of RFID anti-collision algorithms using smart antennas, *SenSys '04 ACM*, November 3–5, 2004.

23. J. Lee, Wireless Network Characterizations: Interference, Collision and Localization, Ph.D. Thesis, August 2007.

24. C. Floerkemeier, Bayesian transmission strategy for framed ALOHA based RFID protocols, in *IEEE International Conference on RFID*, pp. 228–235, 2007.

25. W. T. Chen and G. H. Lin, An efficient anti-collision method for tag identification in a RFID system, *IEICE Trans. Commun.*, Vol. E89-B, No. 12, pp. 3386–3392, December 2006.

26. Y. J. Zhang, P. X. Zheng, and S. C. Liew, How does multiple-packet reception capability scale the performance of wireless local area networks?, *IEEE Trans. Mobile Comput.*, Vol. 8, No. 7, pp. 923–935, 2008.

27. J. R. Cha and J. H. Kim, Novel anti-collision algorithms for fast object identification in RFID system, in *IEEE 11th International Conference on Parallel and Distributed Systems*, Vol. 2, pp. 63–67, 2005.

28. Y. J. Zhang, S. C. Liew, and D. R. Chen, Delay analysis for wireless local area networks with multipacket reception under finite load, in *IEEE Global Telecommunications Conference*, pp. 1–6, February 2008.

CHAPTER 24

ANTI-COLLISION OF RFID TAGS USING CAPTURE EFFECT

QI JING TEOH and NEMAI CHANDRA KARMAKAR
Department of Electrical and Computer Systems Engineering, Monash University, Clayton, Victoria, Australia

24.1 INTRODUCTION

RFID systems operates in a complex environment where many tags reside in one or multiple readers' reading zones. The transponders within the reading zone do not respond with the same signal strength to the reader. The signal strength varies from transponder to transponder due to their different physical location, architecture, data packet structures, and temporal variation in responding to the interrogation signal from the reader. When more than one transponder is transmitting simultaneously to an interrogator within its reading zone, it leads to a collision at the interrogator. In a real communication environment, the reader is able to successfully identify the strongest data packet in a collision. This phenomenon is called the *capture effect* [1–3]. However, most works on anti-collision protocols have made a simple assumption that when a collision occurred between tags, all data packets are considered lost. This assumption is ignored in capture effect calculation.

As discussed in the previous chapter, in the frame-slotted Aloha (FSA) algorithm, there are a fixed number of timeslots for each read cycle. The tags will randomly choose a timeslot from the time frame to transmit their data. Figure 24.1 illustrates the process of tag recognition in FSA for Reader#1 (without capture effect) and Reader#2 (with capture effect). The picture shows that, when Tags 1, 2, 3, and 4 communicated with the first reader, each of them collided in Read cycle 1. However, because of the capturing effect, Reader#2 is able to recognize the stronger Tag 1 and Tag 4, while Reader#1 assumes that all data packets are lost due to collision. Thus, Reader#2 completes reading all tags in two read cycles, whereas Reader#1 requires more timeslots.

Handbook of Smart Antennas for RFID Systems, Edited by Nemai Chandra Karmakar

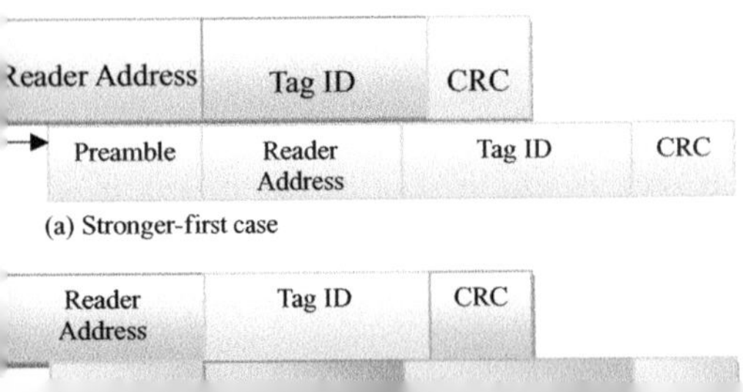
Reader Address
Tag ID
CRC
Preamble
Reader Address
Tag ID
CRC
(a) Stronger-first case
Reader Address
Tag ID
CRC

The chapter is organized as follows: Section 24.1 presented the advantages of the capture effect in multiple tag reading scenario. Section 24.2 presents the theoretical model of slotted Aloha protocols in three different distributions: binomial distribution—an ideal case of allocating a fixed number of tag responses in appropriate timeslots; Rician fading environment—in which direct line-of-sight (LOS) paths exist between multiple tags, the interference, and the reader; and finally, Rayleigh fading model—in which the signal propagation occurs under non-LOS (NLOS) conditions—as, for example, in an urban environment where the signal propagates between buildings and obstacles. Subsequently, the numerical results of the models are presented. Section 24.3 presents the Agilent ADS simulation setup of the capturing effect for multiple tag reading, followed by a conclusion in Section 24.4.

24.2 THEORY

24.2.1 Throughput Evaluation

In FSA, there is a fixed time frame for each read cycle. The tags randomly choose a timeslot from the time frame to transmit their data. Previous works [1–5] have proven that the maximum throughput of FSA is 36.8%. We shall show that with the capturing effect, the throughput can be improved significantly. For a given L number of timeslots, the way the number of tags allocated into a timeslot is described as the binomial distribution with N number of total tags. The tag randomly chooses a number from the total L number of timeslots to transmit its data. Thus, each k number of tag(s) in any one timeslot has $1/L$ transmission probability. The transmission probability can be expressed by

$$P_t = \binom{N}{K}\left(\frac{1}{L}\right)^k\left(1-\frac{1}{L}\right)^{N-k} \tag{24.1}$$

In the Rician fading environment, the signal arrives at the receiver by two or more different paths. Rician fading occurs when there is a direct LOS path in between the tags, reader, and interferences. The capture probability in a Rician environment is given by [3]

$$P_{\mathrm{ric}} = e^{-kR}e^{-Az}\sum_{i=1}^{\infty}\frac{(kR)^i}{i!}\sum_{i=0}^{k-1+i}\frac{(Az)^k}{i!} \tag{24.2}$$

where R is the Rician factor: $R = (P_d \ / \ P_s)$ and $P = (P_d + P_s)$, where P is the average received power, P_d is the direct power, P_s is the scattered power, and $A = (1 + R) \ / \ P$.

To evaluate the capture performance in an NLOS environment, a Rayleigh fading model is considered. Rayleigh fading often occurs in the urban environment where the signal propagates across buildings and trees. The capture probability in a Rayleigh

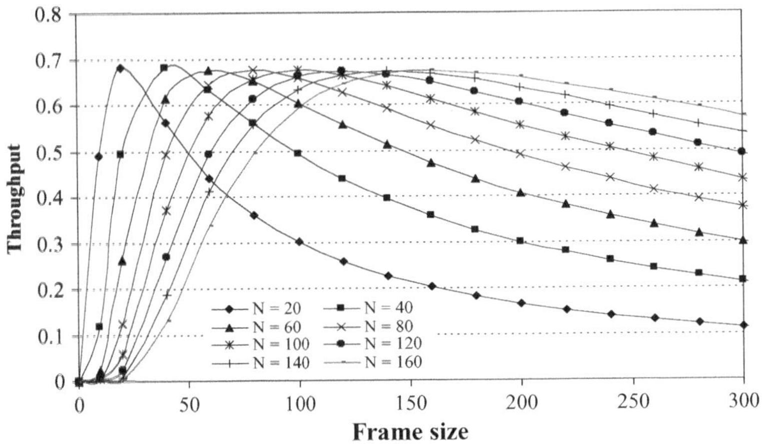

FIGURE 24.3. Throughput versus frame size for a Rician fading environment with total tags ranging from 20 to 160.

environment is given by [3, 4]

$$P_{\text{ray}} = (kz + 1)^{-1} \tag{24.3}$$

The throughput in the presence of capture effect is defined as the average number of successful transmission in a timeslot and is given by [6]

$$T = \sum_{k=1}^{\infty} P_t(k)\, P_c(k) \tag{24.4}$$

where P_c is the capture probability. The capture probability of Rician and Rayleigh fading channel, P_{ric} and P_{ray}, is then applied in Eq. (24.4), and the values ($R = 1$, $A = 1$, $z = 1$) were used to generate the optimum throughput.

24.2.2 Numerical Results

The simulated results of throughput versus frame size for the Rician and Rayleigh fading channels is shown in Figures 24.3 and 24.4, respectively. It is evident from the result that the throughput depends on the number of appropriate frame size and the total number of tags, N. The results show that the maximum throughput close to 0.7 (Rician) and 0.6 (Rayleigh) can be achieved with capture effect in the Rician and Rayleigh environments, respectively. Also, it can be seen that the highest throughput occurs when the time frame equals the number of total tags ranging from 20 to 160 tags.

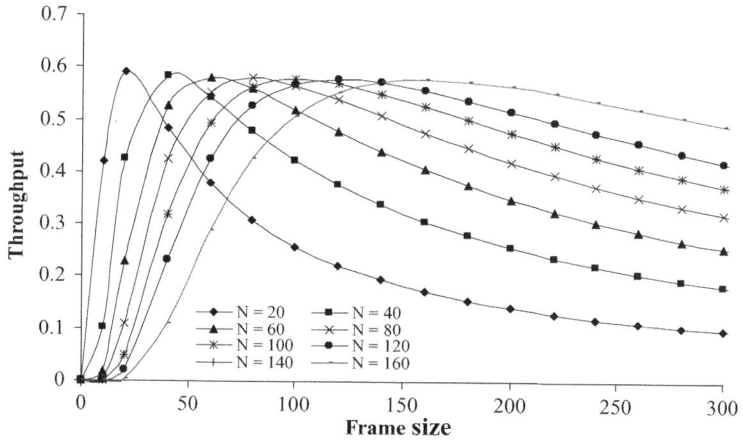

FIGURE 24.4. Throughput versus frame size for a Rayleigh fading environment with total tags ranging from 20 to 160.

In Figure 24.5, the simulated results of the throughput versus the frame size without and with capture effect in the Rician and Rayleigh fading channel are compared. The highest throughput is obtained from the Rician model as the reader and tag communicate in an LOS environment. From the results it can be inferred that the receiver's capturing ability and a suitable time frame are important in the tags' reading process. In the following section a virtual experimental model of an RFID system with multiple tags and a reader with the capture effect is developed using Agilent ADS.

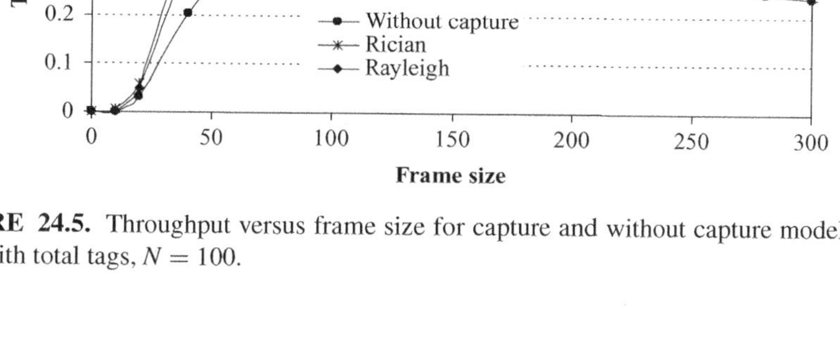

FIGURE 24.5. Throughput versus frame size for capture and without capture model simulated with total tags, $N = 100$.

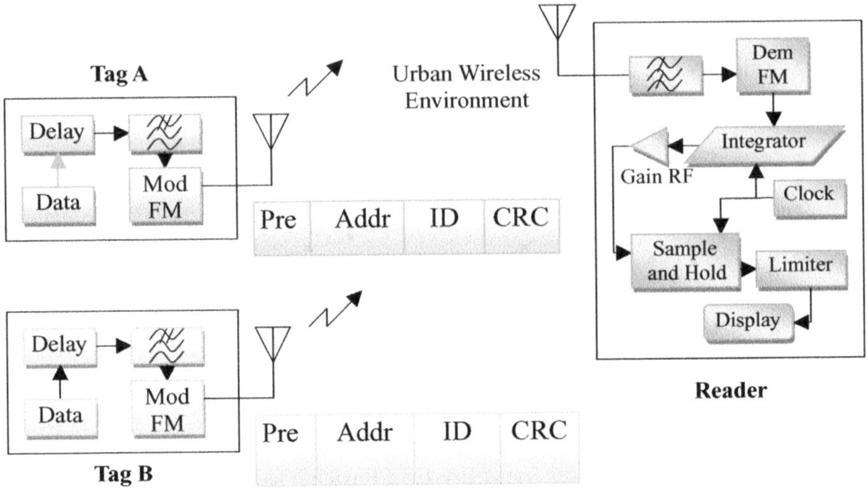

FIGURE 24.6. Simulation setup.

24.3 EXPERIMENTAL METHODOLOGY

24.3.1 Simulation Setup

A virtual test bench is set up and simulated in the ADS environment [7]. All of the components were designed according to the RFID UHF frequency band at 2.45 GHz. This includes two sets of tags (Tag A and Tag B) and a reader. RF front ends such as the transceivers, filter, amplifier, and the antennas that are connected to the air interface channels are also shown in the figure. The simulation is carried out in a real-time wireless urban environment as shown in Figure 24.6.

The length of the data packet used in this experiment is 28 μs and 14 bits long. This consists of 2 bits of preamble, 4 bits of reader address, 6 bits of payload with the first 4 bits as the tags address, and then CRC for the last two bits. A short data length is used in this experiment for the purpose of investigation on the capture effect observed at the reader's side.

All of the components are designed in blocks in ADS as shown in Figure 24.7.

Tag Components. The ADS design blocks of the tag as shown in Figure 24.8 include [8]: Data, DelayRF, LPF_GaussianTimed, FM_Mod, and AntMobile. In this simulation, the ID of Tag A and Tag B is set to be 10100000 and 10010000. The delay block is used to introduce some time delays into the signal. It is used to create an arrival time difference between two tags in the *Stronger-First* and *Stronger-Last* case evaluation.

Reader Components. The ADS design blocks of the reader as shown in Figure 24.9 include [8] AntBase, FM_Demod, IntDumpTimed, GainRF, SampleAndHold,

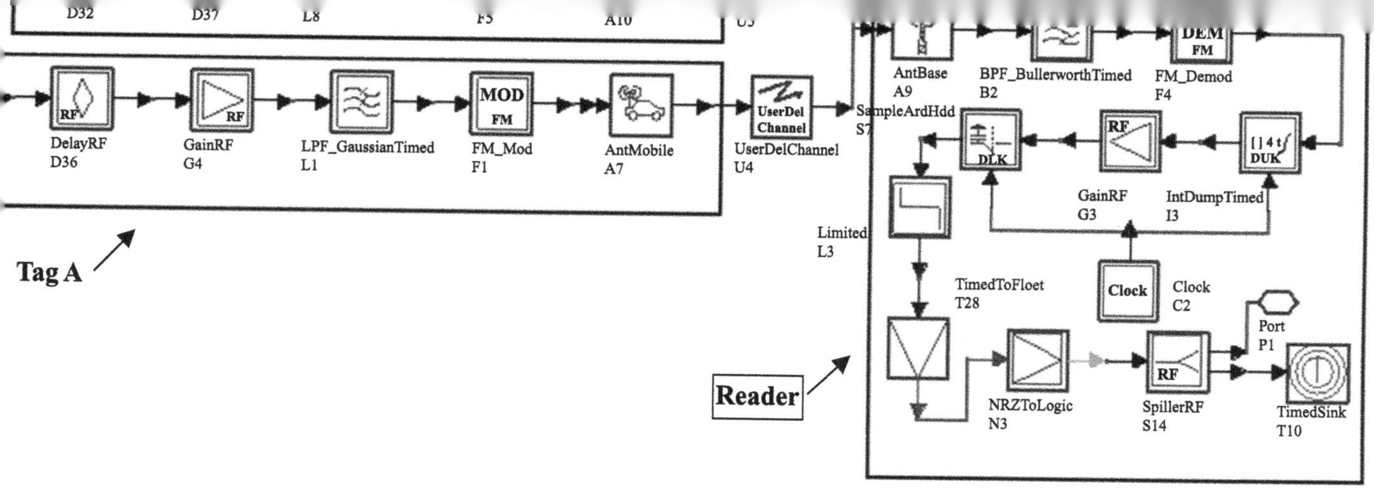
D32
D37
L8
F5
A10
U5
RF
DelayRF
D36
RF
GainRF
G4
LPF_GaussianTimed
L1
MOD
FM
FM_Mod
F1
AntMobile
A7
UserDel
Channel
UserDelChannel
U4
Tag A
AntBase
A9
BPF_BullerworthTimed
B2
DEM
FM
FM_Demod
F4
SampleArdHdd
S7
DLK
RF
GainRF
G3
DUK
IntDumpTimed
I3
Limited
L3
TimedToFloet
T28
Clock
Clock
C2
Port
P1
Reader
NRZToLogic
N3
RF
SpillerRF
S14
TimedSink
T10

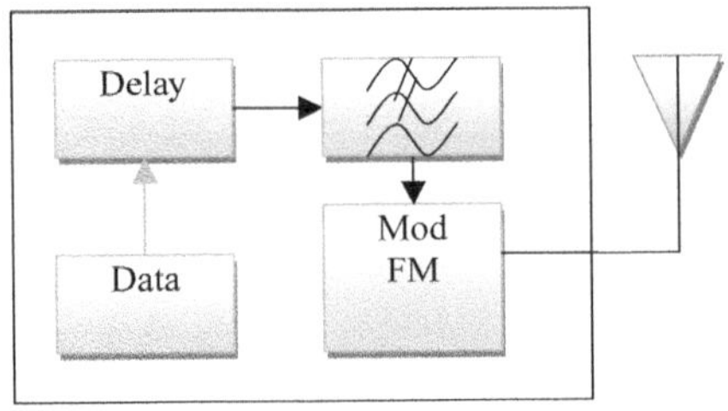

FIGURE 24.8. Tag components.

Clock, Limiter, TimedToFloat, NRZToLogic, SplitterRF, and UserDefChannel. The component, TimedToFlat, converts a timed signal into a real floating point signal. It is used to connect the demodulated signal in timed format to the NRZToLogic component in logic format. NRZToLogic converts an NRZ level signal to a logic format signal. If the signal amplitude is larger than 0, the output logic equals 1 while any signal lesser than 0 will result in a logic 0.

24.3.2 Stronger-First Case (SFC) and Stronger-Last Case (SLC) Evaluations

When a collision between two or more tags occurs, the stronger signals among all can be read by the reader. As shown in Figure 24.10. This is called the capture capability of the reader. It can be applied into the anti-collision algorithm to improve the efficiency of the tag recognition process. In this experiment, Tag B is designed to have a higher output power at −5 dBm than Tag A. Both tags transmit from the same

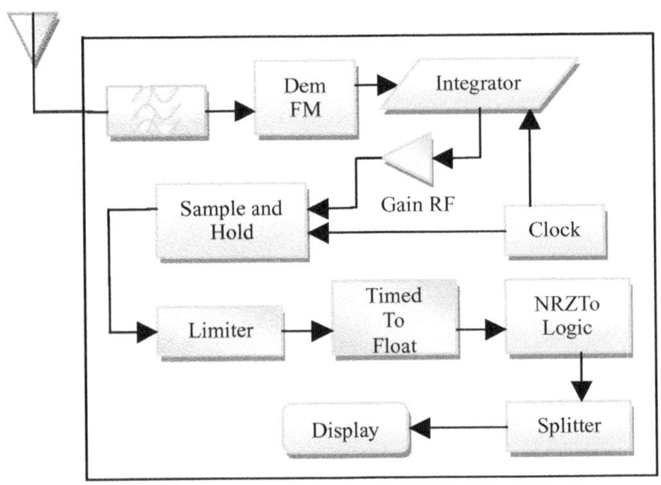

FIGURE 24.9. Reader components.

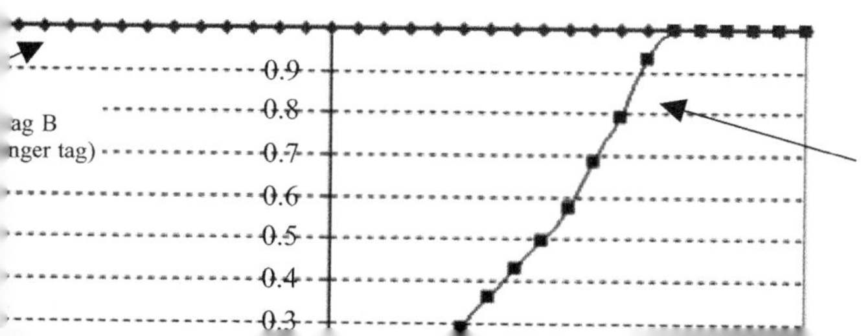

ag B
nger tag)
0.9
0.8
0.7
0.6
0.5
0.4
0.3

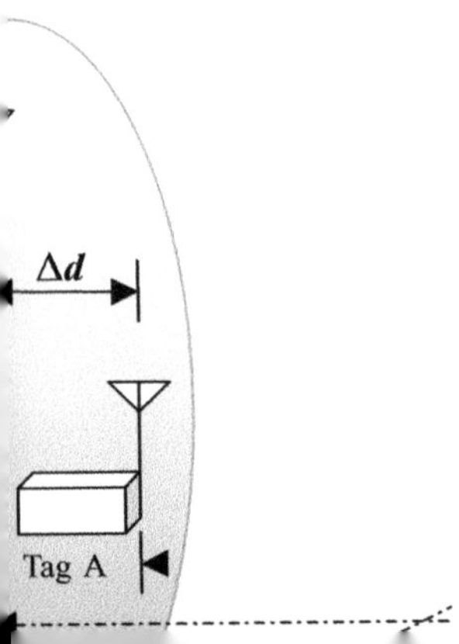

Δd
Tag A

TABLE 24.1. Theoretical and Simulation Results

		SIR = 1 dB		SIR = 2 dB		SIR = 3 dB	
Δd	$z^{1/\alpha}$	T	P	T	P	T	P
0 m	1	X	X	X	X	X	X
1 m	1.11	√	√	X	X	X	X
2 m	1.25	√	√	√	√	√	√
3 m	1.43	√	√	√	√	√	√
4 m	1.67	√	√	√	√	√	√

X = no data detected from Tag B, √ = data detected from Tag B.

system, z. From [3], the receiver can successfully capture the signal when,

$$P_b/P_a = z, \quad \text{where} \quad z \geq 1 \tag{24.5}$$

Let d_A and d_B be the distances of Tag A and Tag B from the reader. According to Goodman and Saleh [10], for the receiver to successfully capture the stronger signal over a weaker signal, it must satisfy

$$d_A/d_B > z^{1/\alpha} \tag{24.6}$$

where $2 < \alpha < 5$ in the UHF operation. $\alpha = 4$ is used in this experiment because the simulation is carried out in an urban wireless environment.

The simulated results are shown in Table 24.1. If SIR = 3 dB, then SIR in Eq. (24.4) is $z = 10^{0.3} = 2$. T represents the theoretical data from Eq. (24.5), while P indicates the ADS simulation data. At SIR = 1 dB and capture index = 1.11, Tag B cannot be received when Tag A is at the same distance with Tag B. At SIR = 2 dB and capture index = 1.25, Tag B cannot be received when Tag A is 1 m further away from Tag B and closer to the reader.

24.4 CONCLUSION

In this chapter, the capturing effect occurred at the RFID reader is analyzed in Rician and Rayleigh fading environments. The capture probability in both fading environment are first discussed and integrated into the RFID frame-slotted ALOHA algorithm. The results show a maximum throughput of 0.6 and 0.7 achieved at the optimum frame length in both Rayleigh and Rician environments, respectively. This indicates that a higher throughput than the conventional 36.8% can be achieved with capturing effect considered in the anti-collision algorithm. Then, a different method of analyzing the capturing effect using the Agilent ADS simulation tool is presented. Two sets of RFID tags and a reader were tested at different scenarios such as the near/far effect and stronger first (SFC) and stronger last (SLC) cases. From the near/far effect evaluation, the results has indicated the correlation between signal-to-interference ratio (SIR) and distances by presenting the levels of SIR required at different distances to enable capture by the reader. Then, from the SFC and SLC evaluations, the results have shown how the capturing effect can increase the efficiency of RFID tag recognition process. Therefore, this chapter has shown that the capture effect can be used as the design considerations for the anti-collision protocol to improve the throughput of the RFID system [11–12].

ACKNOWLEDGMENT

Agilent ADS academic license to Monash University is acknowledged.

REFERENCES

1. K. Finkenzeller, *RFID Handbook*, 2nd edition, John Wiley & Sons, Hoboken, NJ, 2003.
2. K. Whitehouse, A. Woo, F. Jiang, J. Polastre, and D. Culler, Exploiting the capture effect for collision detection and recovery, *The Second IEEE Workshop on Embedded Networked Sensors*, pp. 45–52, May 2005.
3. J. S. Garcia and D. R. Smith, Capture probability in Rician fading channels with power control in the transmitters, *IEEE Trans. Commun.*, Vol. 50, No. 12, pp. 1889–1891, December 2002.
4. W. T. Chen and G. H. Lin, An efficient anti-collision method for tag identification in a RFID system, *IEICE Trans. Commun.*, Vol. E89-B, No. 12, pp. 3386–3392, December 2006.
5. J. R. Cha and J. H. Kim, Novel anti-collision algorithms for fast object identification in RFID system, in *IEEE 11th International Conference on Parallel and Distributed Systems*, Vol. 2, No. 22–22, 2005, pp. 63–67.
6. H. Zhou and R. H. Deng, Capture model for mobile radio slotted ALOHA systems, *IEE Proc. Commun.*, Vol. 145, No. 2, pp. 91–97, April 1998.
7. Agilent Advanced Design Systems 2005A Documentation.
8. G. Nerlekar, Development of 2.4 GHz active radio frequency identification system, M.S. thesis, Monash University, Australia.
9. J. Lee, W. Kim, S. J. Lee, D. Jo, J. Ryu, T. Kwon, and Y. Choi, An experimental study on the capture effect in 802.11a networks, *WINTECH*, September 10, 2007.
10. D. J. Goodman and A. A. M. Saleh, The near/far effect in local ALOHA radio communications, *IEEE Trans Veh. Tech.*, Vol. VT-36, No. 1, pp. 9–27, February 1987.
11. B. Zhen, M. Kobayashi, and M. Shimizu, Framed ALOHA for multiple RFID objects identification, *IEICE Trans. Commun.*, Vol. E88-B, No. 3, March 2005.
12. Y. J. Zhang, P. X. Zheng, and S. C. Liew, How does multiple-packet reception capability scale the performance of wireless local area networks? *IEEE Trans. Mobile Comput.*, pp. 923–935, 2008.
13. R. A. Santos, V. Rangel Licea, L. Villaseñor and A. Edwards, Wireless propagation characteristics for vehicular ad-hoc networks in motorway environments, in *Ingenieria Investigation y Tecnologia, Facultad-de Ingenieria Unam*, 2009, pp. 295–302.

INDEX

Handbook of Smart Antennas for RFID Systems, Edited by Nemai Chandra Karmakar

Printed and bound by CPI Group (UK) Ltd, Croydon, CR0 4YY
18/09/2026
14965004-0001